Recreation ecology

Recreation ecology

The ecological impact of outdoor recreation and ecotourism

Michael Liddle

Senior lecturer, Environmental Sciences,
Griffith University, Queensland, 4111 Australia

CHAPMAN & HALL

London · Weinheim · New York · Tokyo · Melbourne · Madras

Published by Chapman & Hall, 2–6 Boundary Row, London SE1 8HN

Chapman & Hall, 2–6 Boundary Row, London SE1 8HN, UK

Chapman & Hall GmbH, Pappelallee 3, 69469 Weinheim, Germany

Chapman & Hall USA, 115 Fifth Avenue, New York, NY 10003, USA

Chapman & Hall Japan, ITP-Japan, Kyowa Building, 3F, 2-2–1 Hirakawacho, Chiyoda-ku, Tokyo 102, Japan

Chapman & Hall Australia, 102 Dodds Street, South Melbourne, Victoria 3205, Australia

Chapman & Hall India, R. Seshadri, 32 Second Main Road, CIT East, Madras 600 035, India

First edition 1997

Typeset in 10/12pt Sabon by Florencetype Ltd, Stoodleigh, Devon
Printed in Great Britain at the University Press, Cambridge

ISBN 0 412 26630 X

A catalogue record for this book is available from the British Library

Library of Congress Catalog Card Number: 97–66867

♾ Printed on permanent acid-free text paper, manufactured in accordance with ANSI/NISO Z39.48-1992 and ANSI/NISO Z39.48-1984 (Permanence of Paper).

This book is dedicated to Gaia and all those who believe in her survival

Cover design and artwork: Jonathan Manopavane, from the Liveworm Studio, Queensland College of Art, Griffith University. Creative Director: Sebastian Di Mauro.

Cover image: Yellow poplar or tuliptree (*Liriodendron tulipfera*) at the Hambidge Center in the Appalachian Mountains, North Georgia. Photography by Dr Michael Liddle.

Contents

Managment Strategy

Management
stratagy

Foreword

The science of recreational ecology has developed only during the past 30 years or so as a response to the increasing public use of our declining resource of unspoilt wilderness and countryside. Natural areas no longer exist in many countries, apart from some coastal formations, due to the long process of 'taming' wild nature as part of the development of civilization and exploitation of the environment to serve the needs of man.

With industrialization, the expansion of urban areas, population growth and modern agricultural technology, the rate of change in the undeveloped areas is remaining accelerated and a few enlightened minds realized that the unspoilt environment was fast disappearing and that what remained should be protected. Even in the vast area of North America, the same events were taking place. John Muir, the Scot who was one of the first dedicated conservationists in the USA wrote in 1901: 'Only 30 years ago, the great central valley of California, 500 miles long and 50 miles wide, was one bed of purple and gold flowers. Now it is ploughed and pastured out of existence, gone for ever – scarcely a memory of it left in fence corners and along the bluffs of streams. The gardens of the Sierra also and the noble forests in both the reserve and unreserved portions are sadly hacked and trampled notwithstanding the ruggedness of the topography. In the noblest forests of the world, the ground, once divinely beautiful, is desolate and repulsive, like a face ravaged by disease. This is true also of many other Pacific coast and Rocky Mountain valleys and forests. The same fate, sooner or later, is awaiting them all, unless awakening public opinion comes forward to stop it.'

Public opinion did awaken, though much later than John Muir hoped. Today most unexploited nature is confined to national parks, nature reserves, forest and marine parks, together with numerous wildlife areas specifically designed to create favourable habitats for animals and plants. However, these areas attract increasing numbers of the public and one of the management problems now is how to safeguard the habitats and quality of the wilderness environment from the very people who enjoy them but who may not be aware of the effect of their trampling and other activities. Michael Liddle was one of the first ecologists in the UK to recognize the need for research on recreational ecology; he also helped to found the Recreational Ecology Research Group in Britain in the early 1970s. Since then he has worked on problems of the human impact on terrestrial, aquatic and marine wildlife habitats in Britain and Australia. Allied to this research is the need to understand how to recreate habitats once they have been destroyed.

This type of research is developing in several countries with particular emphasis in Britain on recreating heathland and herb-rich grassland on former agricultural land and various habitats in the USA and other countries. Much more work is needed because urban and industrial expansion continues, often destroying nature reserves and sites scheduled for their interest. This occurs especially where new roads are built, where land is reclaimed for agriculture, or wetlands drained by water abstraction from nearby boreholes, which lower the ground water-table and also dry out rivers and streams.

Modern economies are still driven by the need to expand and priority is usually given to developments thought likely to enrich the country, improve standards of living and provide more jobs. These are laudable objectives, but a balance must be struck so that remote wilderness and quiet rural areas are maintained where public use is in harmony with wildlife conservation. H.G. Wells in his *The Outline of History* (1920) wrote that 'human history becomes more and more a race between education and catastrophe'. This is even truer today than 76 years ago. Not only must we, the public, be kept informed of the impact of new developments but we must also ensure that our political decision-makers take into account our views, convictions and values before they introduce policies likely to change the remaining natural environment and which may add to the long list of past catastrophes. Michael Liddle's book will provide a firm scientific base of information to enable us to do this.

Eric Duffey
August 1996

Preface

My earliest memories are of hiding from my parents under a gorse bush when we were camping on the downs in what is now the Cuckmere Country Park in Sussex, England. Since then my strongest memories and much personal recreation have come from experiences in nature. As a schoolboy I first experienced the thrill of a sail filling with wind and the gentle lapping of water as my small boat moved through the current of the River Thames. About this time I also experienced the wonder of the never-ending purple of the heather in bloom on the Yorkshire moors. As a young man I tested myself on Tryfan mountain in Snowdonia and gained new confidence when I found myself at the summit in slanting rain and a thick mist that required a compass to direct us back to the valley.

When, with a bit more maturity, I turned to ecology and found the pleasure that the scientific knowledge of nature can bring, I also discovered salt marshes and experienced the thrill of hearing the silent swish of wings as a flock of geese landed near us in the dusk as we returned from a full day of field work. Who can forget the quiet movement of the high tides through small cracks and pools and the small noises made by water and land in their slow relentless struggle, or the reflected light of dawn or dusk on wet mud or sand?

On nearing my first half century I was able to travel and experience the thunder of the Pacific surf after a storm, the tall, misty bird calls of a tropical rainforest, the smell of eucalyptus after rain and the rainbow hues of the small fish on a coral reef. My debt to nature is deeper than this book can repay, but it is with this attitude and the joy of knowledge that it is written and I hope it may help to preserve some small part of our natural environment.

Michael Liddle

Acknowledgements

When I started to work on my Ph.D Professor, Peter Grieg-Smith, suggested that human trampling would be an interesting area of study. My thanks go to him for his unstinting advice and encouragement; he provided a thorough and precise basis for my research which later became a career in recreation ecology. I am also grateful for my continuing interactions with Dr Neil Bayfield, Dr David Cole and Professor Fred Kuss who have stimulated and sustained this research. I thank all of them for comments on various parts of the manuscript. Dr Eric Duffy OBE was involved from the early stages of the work and I am deeply grateful for his constant support and comments on the entire work.

My gratitude also goes to Dr S. Niven, Professor Geoffrey Wall, Dr Darryl Jones, Dr Barrie Goldsmith and Professor Georg Grabherr who have all read and commented on parts of the manuscript.

My thanks also go to Professor Kees Blom of Nijmegen University and Dr Michael Hutchings of Sussex University who both provided me with accommodation and support in their institutions during the early stages of the writing. Fellowships at The Sitka Centre for Art and Ecology, Origon, The Hambidge Center, Georgia, and the hospitality of Helen and Mort Meddors of Rabun Gap, Georgia, were instrumental in the final stages of manuscript preparation. Without the continuing support of my own institution, Griffith University, and the various deans of Environmental Sciences, this project would not have been possible.

Discussions with Professor Valerius Guise, Dr Howard Wilshire, Dr Neil West and many other ecologists, rangers and land managers in many countries, have also contributed greatly to the final book.

Mrs Maggie Barber, who typed the whole manuscript, showed remarkable dedication, patience and tolerance; her faith is much appreciated. Ms Maureen Evans prepared most the figures with care and cheerfulness that made our cooperation a pleasure. I also thank my first ecology teacher, Dr John R. Packham, whose enthusiasm and dedication to ecology gave me a remarkable introduction to the subject and the impetus to complete this project.

My deepest thanks go to my wife, the artist Dawne Douglas, whose commitment and tolerance of the travails of the years of accumulating, sorting and organizing such a book as this, and the constant intrusion of piles of paper into the domestic and studio environment made this book possible.

I would also like to acknowledge the many authors whose work is quoted in the book and whose figures and tables appear in the text. While I have not been able to contact all of them, and many do not own the copyright for their work, I am grateful for their generosity and support.

Ultimately, this book depends on the environment and the inspiration it has given me and my thanks go to all those living organisms who have inspired me throughout my life.

My sincere thanks also go to the copyright holders listed below who have given me permission to reproduce their drawings and tables.

(a) Figs 2.3, 2.4, 2.5, 4.9, 5.11, 5.12, 5.13, 5.14, 5.15, 5.16, 7.4, 7.5, 7.6, 7.7, 11.12, 15.14, 16.4, 18.1, 18.8, 18.9, 28.5, Tables 5.1, 5.2, 5.3, 5.4, 5.5, 5.6, 7.1, 18.1, 27.2, 27.3, Academic Press; Figs 6.8, 14.2, 15.1, Table 26.6, Addison-Wesley Longman Group UK Ltd; Figs 8.3, 17.4, Australian Environmental Sciences, Griffith University, Fig. 14.7, Agriculture and Food Research Council; Fig. 4.10, American Association for the Advancement of Science; Fig. 27.2, American Fisheries Society; Fig. 21.7, American Institute of Biological Sciences (fee required); Fig. 2.6, *American Journal of Veterinary Research*; Figs 2.8, 11.3, 11.6, 11.7, 11.8, 11.10, 11.11, 12.1, 12.2, 12.3, 12.9, 12.18, American Society for Agricultural Engineering; Tables 6.11, 13.1, American Society of Agronomy; Table 26.2, American Society of Civil Engineers (fee required); Fig. 27.1, Table 27.1, Asian Fisheries Society; Fig. 29.1, Asian Institute of Technology.

(b) Figs 1.1, 3.6, 3.18, 3.19, 3.20, 3.21, 4.1, 4.3, 4.5, 4.6, 4.11, 5.8, 5.9, 6.1, 6.33, 6.34, 6.35, 8.12, 9.2, 9.5, 11.1, 11.2, 11.13, 11.18, 11.19, 12.4, 12.11, 12.12, 12.13, 12.14, 12.15, 12.16, 12.17, 14.6, 14.9, 14.11, 15.11, 15.12, 18.2, 21.5, 21.6, 24.2, 24.3, Tables 4.8, 6.3, 6.12, 9.3, 12.3, 24.2, Blackwell Scientific Publications, London; Fig. 8.1, Blackwell Scientific Publications, Boston; Figs 3.1, 3.4, 3.12, Dr D. Bludhorn; Fig. 6.26, The Botanical Society of America; Fig. 23.7, E.J Brill; Table 4.9, The Botanical Society of the British Isles; Fig. 8.13, British Ecological Society; Fig. 26.2, K.J. Brown; Table 6.8, Business Publications Ltd.

(c) Fig. 12.10, California Turf Grass Culture; Figs 6.5, 6.6, California State University, Berkeley; Fig. 7.9, Cambridge Philosophical Society; Figs 4.16, 13.4, Tables 6.2, 6.10, 9.1, Cambridge University Press; Fig. 14.5, *Canadian Journal of Plant Science*; Fig. 4.8, Carlton University, Ottawa; Fig. 6.27, Chapman & Hall; Figs 20.4, 20.9,

20.10, Cooper Ornithological Society, Berkeley; **Figs** 3.8, 15.8, Countryside Commission; **Table** 4.11, Verlag vin J. Cramer; **Figs** 6.19, 13.1, 25.4, **Table** 27.4, CSIRO.

(**d**) **Fig**. 2.8, Defence Research Board of Canada; **Fig**. 13.2, Dawne Douglas.

(**e**) **Figs** 5.4, 5.5, 6.13, Editions Gunther-Villars; **Figs** 6.31, 6.32, Edward Arnold; **Figs** 2.9, 2.10, 3.3, 3.7, 5.7, 6.4, 6.14, 6.21, 8.6, 8.7, 8.8, 8.15, 10.3, 10.5, 13.3, 13.8, 16.1, 16.2, 17.6, 17.7, 18.3, 18.4, 18.6, 18.7, 19.1, 19.3, 20.1, 20.2, 20.3, 20.5, 20.6, 20.16, 20.17, 20.18, 20.19, 20.20, 21.4, 24.1, 25.1, 26.1, 26.3, 26.4, 26.5, 26.9, 28.2, 28.3, 28.6, **Tables** 6.1, 6.4, 8.2, 10.1, 13.5, 19.1, 20.1, 20.5, 23.1, 26.1, 26.3, 26.4, Elsevier Applied Science; **Figs** 6.9, 6.16, 8.2. Dr W.M.H.G. Engelaar; **Figs** 10.4, 10.6, 20.11, 20.12, **Tables** 11.2, 11.4, Environmental Conservation Foundation.

(**f**) **Fig**. 21.3, Finnish Zoological and Botanical Publications Board, **Fig**. 26.7, Fisheries Council of Canada; **Fig**. 9.1, Dr E.A. Fitzpatrick; **Fig**. 24.5, Florida Academy of Sciences; **Table** 4.3, Floristisch-Soziologische Arbeitgemeinschaft EV.

(**g**) **Figs** 18.10, 18.12, 18.13, 18.14, 18.15, 18.16, Great Barrier Reef Marine Park Authority; **Figs** 13.6, 26.10, **Table** 26.8, Gordon & Breach Publishing.

(**i**) **Table** 26.7, Institute of Biology; **Fig**. 3.13, International Association for Vegetation Science; **Figs** 4.7, Iowa State University Press; **Fig**. 18.5, **Table** 18.3, Interperiodica, Moscow.

(**k**) **Figs** 3.16, 5.6, Mr P.A. Kendal; **Figs** 14.13, 27.3, **Table** 27.5, Kulwer Academic Publishing.

(**l**) **Figs** 10.7, 12.19, Dr F. Leney; **Fig**. 13.5, **Table** 13.3, Lesovdenie; **Figs** 14.3, 28.4, **Table** 3.2, Macmillan; **Fig**. 21.2, Missoula Press & Publication Co.; **Fig**. 15.18, **Table** 15.9, Dr R.P.C. Morgan; **Fig**. 21.2, Mountain Press Publishing Co.

(**n**) **Fig**. 22.1, **Table** 20.6, National Audubon Society; **Fig**. 6.22, Nature Conservancy; **Fig**. 4.15, Natuur-en Landschapbescherning; **Fig**. 22.5, New York Zoological Society.

(**o**) **Table** 4.10, Oregon State University; **Fig**. 23.5, **Table** 25.1, Ottawa Field Naturalists' Club; **Fig**. 7.3, Oxford University Press.

(p) **Fig.** 17.5, Plenum Press; **Table** 20.7, Pennsylvania Academy of Science.

(r) **Fig.** 6.18, Recreation Ecology Group; **Figs** 15.4, 15.10, The Research Council of Alberta; **Fig.** 26.8, Johns Hopkins University.

(s) **Fig.** 6.2, Sedgwick & Jackson; **Figs** 6.23, 6.24, 11.15, **Tables** 6.9, 15.3, Society for Range Management; **Table** 19.2, Society for the Study of Amphibians and Reptiles; **Fig.** 4.13, Societé Scientifique de Torun; **Fig.** 3.23, **Table** 12.1, Society of American Foresters; **Figs** 3.14, 22.2, 22.3, 22.4, **Table** 3.3, Society of Forestry in Finland; **Figs** 12.5, 15.16, **Table** 15.7, Soil and Water Conservation Society; **Fig.** 14.12, **Table** 16.1, Soil Science Society America; **Fig.** 4.4, SPB Academic Publishing; **Figs** 2.1, 7.8, 11.16, 12.7, 15.5, 15.13, Springer Verlag, Berlin; **Figs** 3.10, 4.2, 8.5, 15.17, 16.3, 18.11, 19.2, **Tables** 3.4, 4.4, 4.5, 6.5, 6.7, 11.3, 15.1, 15.2, 15.8, 15.10, Springer Verlag, New York; **Fig.** 15.3a, Dr D. Streeter; **Fig.** 13.7, **Tables** 4.1, 13.4, Swedish University of Agricultural Science; **Fig.** 20.6, Swiss Academy of Science.

(t) **Fig.** 7.2, Tohoku University; **Fig.** 4.12, **Tables** 4.6, 4.7, The Wisconsin Academy of Science, Arts and Letters.

(u) **Fig.** 25.3, Unione Zoological Italiana; **Figs** 8.4, 8.9, 8.10, 10.8, 11.9, 15.3b, c, 15.7, **Table** 9.4, United States Department of Agriculture; **Tables** 20.3, 20.4, University of California; **Figs** 5.10, 8.11, University of Chicago Press (fee requested); **Fig.** 6.20, University of Queensland Press; **Fig.** 22.10, University of Wyoming; **Table** 6.6, **Figs** 6.11, 6.12, Ustav Experimental Biologie A Ekologie.

(v) **Fig.** 9.6, Verlag Eugen Ulmer; **Fig.** 6.25, Dr A.E. Vines.

(w) **Fig.** 1.2, Professor G. Wall; **Fig.** 7.10, Walter de Gruyter, Berlin; **Table** 26.5, Water Environment Federation; **Table** 13.2, West Virginia University; **Fig.** 17.5, Dr F.A. White; **Figs** 20.13, 20.14, 20.15, 22.6, 22.7, 22.8, 22.11, 23.2, 23.3, 23.6, 25.2, **Tables** 20.2, 20.9, 25.2, 28.1, The Wildlife Society, Washington; **Fig.** 24.4, Wildlife Research Unit, Iowa State University Press; **Figs** 20.7, 20.8, Wildfowl and Wetlands Trust; **Fig.** 9.3, 14.1, 15.2, 17.2, **Tables** 4.2, 12.2, 15.6, Wiley, New York; **Fig.** 14.10, Williams & Wilkins.

(y) **Table** 14.8, Yokohama National University.

Introduction 1

The impacts of outdoor recreation, including ecotourism, are extensive and increasing, focusing more and more on the world's remaining natural areas. This presentation is aimed at helping to reduce these impacts through informed research and management. It is, I believe, the first book to present this level of data on the ecological effects of recreation.

/ update

1.1 The concept of outdoor recreation

Outdoor recreation has only arisen as a definable activity since houses and other buildings were created in which it was possible to spend most of the working day, and leisure out of doors became a luxury. This development could presumably be traced back several millennia to the cities of the Middle East and, more recently, to the castles and palaces of Europe when, for example, hunting was the relaxation of the élite.

Outdoor recreation as an activity for the populace became a feature of industrial towns in the eighteenth and nineteenth centuries, where parks were provided for perambulations in the summer evenings and on Sundays. As attitudes to nature in its 'untamed' state changed from fear to admiration, so more people ventured further and more frequently to explore natural areas, climb mountains and generally to create recreation impacts. With the advent of the motor car, between the First and Second World Wars, there was an increasing use of the 'countryside' and appreciation of natural history. The 1960s saw the beginning of the major increase in car ownership in the UK and conservation became an issue. This brought perhaps the greatest single influx of people to the countryside and their associated impact. This influx caused the increase in footpaths at Aberffraw dunes on Anglesey seen during that decade (Fig. 1.1).

Since then there has been a steady increase in number of people using our natural resources for pleasure, as the world human population approaches six billion people. Coincident has been the rise in international travel and the development of the ecotourism industry. This is perhaps exemplified by the rapid increase in the number of ecotourists

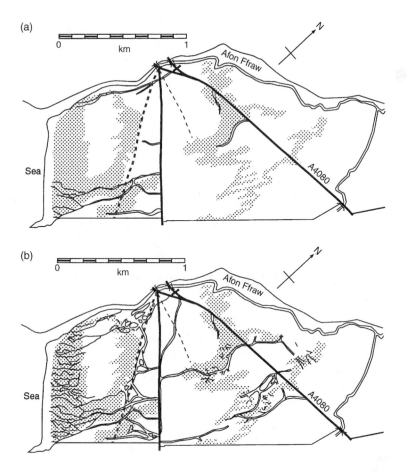

Fig. 1.1   Paths and tracks in the Aberffraw sand dune system in (a) 1960 and (b) 1970. In 1960 there were 3.2 km of track and 2.2 km of footpaths. These had increased to 11.7 km of track and 16.5 km of footpaths by 1970. ▬▬▬, Surfaced roads; ═══, tracks; ──, footpaths; ▬▬ disused tracks; -----, disused footpaths; stipple, sand dunes. (From Liddle and Greig-Smith, 1975a.)

visiting the Antarctic, where the only activity is environmentally based (Fig. 28.6); but, also, increasing numbers of organized groups are visiting all kinds of national parks, from desert to rainforest, and from sea-shore to high mountains. They pursue all kinds of environmentally based activities, from mountaineering and rafting to bird watching and photographing big game. It is this recent change in the pattern of impacts, to repeated short-term visits with high concentrations of visitors, that provides new problems for recreation ecologists.

Recreation ecology is in a sense a redefining of an age-old activity. Doubtless, humans moving through the wilderness or bush have always observed the impacts of others who may have gone before them. So, as with many areas of enquiry, it is hard to give a date when recreation ecology began. The earliest record that could be classed as recreation ecology, that I have come across, is by Stillingfleet (1759), who noted the presence of annual meadow grass (*Poa annua*) on footpaths in the Malvern Hills, England. Since the surrounding areas had fairly acid soils, this showed him that trampling had created more mesic conditions where this grass could survive.

Between the two World Wars Braun Blanquet developed his community classification schemes, which included a trampled 'association' (Braun-Blanquet, 1928, 1932). In England Bates (1935, 1938) published two papers on the 'vegetation of footpaths, sidewalks, cart-tracks and gateways' and in the USA Meinecke (1928) observed 'The effect of excessive tourist travel on the California Redwood Parks'.

The number of ecologists interested in this study increased exponentially from the 1960s until the early 1980s. Bayfield (1971b), who worked on what have become recognized as the central problems of this field of study, was probably the first of this period, followed by myself (Liddle, 1973b) and Cole (1978). The earlier paper by Goldsmith, Munton and Warren (1970) provided an exemplary study of the effects of visitor use and impacts on the Scilly Isles, UK, and the papers of Chappell *et al.* (1971) and Frenkel (1972) on chalk grasslands and 'floristic convergence in the tropics under trampling effects', respectively, should not be overlooked. Blom's population studies (1977, 1979) and the work by Kuss on soils (1983, 1986) provided alternative views on the central problems of the processes of impacts and how to manage them.

The focus has been on the effects of walking (or trampling), camping, horse riding, off-road vehicle and trail-bikes and the effects of development and clearance to facilitate skiing. These were all plant and/or soil studies, but at the same time concern was growing about the effects of recreation on animals. Geist's paper entitled 'Is big game harassment harmful?' (1971a) and the work of Craighead and Craighead (1971) on bears in Yellowstone National Park are indicators of the parallel zoological studies.

Global environmental problems have become increasingly evident from the start of the 1980s and younger ecologists have turned to widespread habitat approaches, such as integrated catchment studies, although the recent body of work by Dan Sun (Sun and Liddle, 1993a,b,c,d) marks a valuable contribution.

The contents list of this book provides a guide to what is known about recreation ecology, but there are many problems still to be solved

1.2 Recreation ecology

(see Chapter 29) and I am optimistic that these will be tackled by ecologists of the future.

1.3 Subdivisions in recreation ecology

The summary of recreation impacts by Wall and Wright (1977) divided the areas of investigation and knowledge into four groups, namely plants, soils, wildlife and aquatic situations (Fig. 1.2). The first three are clearly discipline-based and each area has its own slightly different paradigm. These divisions are generally reflected in the academic disciplines and their associated journals, such as *The Journal of Vegetation Science*, *Plant and Soil* and the *Journal of Wildlife Management*. Aquatic studies are also a specialist area of knowledge (see, for example, the journals *Freshwater Biology*, *Water Research* or *Regulated Rivers*). Each discipline has arisen naturally in response to the acquisition of facts that are associated more with each other, and a particular view of the natural world, than with other constellations of knowledge.

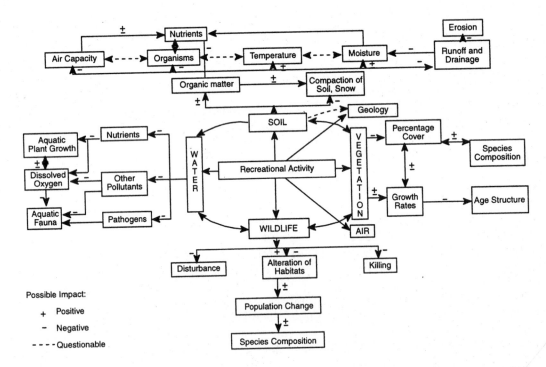

Fig. 1.2 Interrelationships between major recreation impacts (from Wall and Wright, 1977).

Conventional disciplines are therefore historical and reactive in nature. At times when new problems arise it is necessary to ~~cross~~ boundaries to bring together information from a number of different disciplines and to approach the definitions of research areas in a creative way. This has resulted in the discipline of recreation ecology, although a true blending of disciplines to form a new 'science', perhaps problem- or habitat-centred, has yet to be achieved. With the present state of our knowledge this approach to describing recreation ecology would result in a series of very incomplete accounts which would be of relatively little use as a reference. At this stage it is likely that knowledge will be sought on an organism basis, thus a description of ungulates' responses to recreation disturbance would be more useful than a series of partial accounts from a range of problems or habitats.

[handwritten margin note: ASSESS management strategies from alternate view points so they be assed and reformulated in order to function]

For these reasons I have adopted the discipline-based approach for ease of understanding and reference. But it must be borne in mind that in taking a reductionist approach to recreation impacts there is a danger that the holistic effects on all parts of the environment at the same time might be overlooked. Where possible, cross-references are provided.

1.4 Sources of information

There is a considerable body of writing on recreation impacts in the international journals such as *Biological Conservation*, *Environmental Management* and the *Journal of Wildlife Management*, but there is also a large amount of information in not so easily obtained reports and various government publications. The work in both of these sources provides the examples used in this book.

I have chosen the examples from as wide an international base as possible. By far the greater number of papers comes from the USA, UK and northern Europe and this has biased the content towards those countries. However, the general principles may often be applied to similar situations in other parts of the world, as long as the local conditions are taken into account.

Similarly, constraints on space in one book do not allow all the possible examples of any particular impact to be discussed. It would, for example, be possible to write several books just on the disturbance to birds; so, again, the examples have been selected that point to general principles rather than attempting a comprehensive treatment of those groups.

The opposite problem exists with respect to the physiological responses to recreation impacts, especially by plants. In this case I have assembled data from various sources in order to sketch an outline of some likely effects. Hopefully, this will stimulate further research in this area.

Recreation impacts to plants and soils have been fairly well documented with respect to temperate, mesic and montane situations. Tropical and subtropical data are hard to find and, apart from North American oil exploration areas, Arctic and Antarctic situations have been little studied in the recreation context. However, there is a reasonable amount of information from montane situations in North America, Scotland and Austria which has some bearing on colder habitats. Sessile plants and animals in marine situations have only been investigated in selected areas such as the Great Barrier Reef and the Florida Keys. There is information on only a few marine animals, such as turtles, dolphins, some other cetaceans and commercial fish populations.

In defining the information that should go into this book, the primary criteria, apart from the quality of the material, was to fit the ideas and examples into as coherent a story as possible. However, with such a wide-ranging topic as recreation impacts it was also necessary to draw boundary lines and exclude material that, ideally, would have been included. For example, the body of research on sports turf and plant breeding to produce resistant grasses had to be largely omitted. This decision was made partly because there are already several books on the subject (Beard, 1982; Emmons, 1984; Turgeon, 1991). Similarly, the whole subject of hunting various mammals and of fishing is only touched on where it directly affects the animals' behaviour. The problems relating to visitor behaviour and their reactions to the environment deserve a separate book, as does the subject of environmental management at all levels.

Wherever possible I have used the common names of plants and animals, as in my experience these are generally used by managers and field staff concerned with land management. The binomial is given where the particular plant or animal is first mentioned in the text.

Mechanical forces exerted by various recreation activities and the concept of protection 2

This section deals with the mechanical forces created by the various recreation activities which have an immediate impact upon the environment in which they take place. The most obvious of these are the forces created by the human foot, either standing still and only subject to the vertical force of gravity, or the greatly increased horizontal and vertical forces created by the action of walking and running, not to mention the lesser ones resulting from the whole body sitting or rolling in the grass. But there are also much larger forces exerted by the combined effects of a rider and his horse, and much larger forces again are created by the use of trail-bikes, four-wheel-drive vehicles and other machines in which several people and their equipment may be transported over a rough and sometimes fragile terrain. Finally, I consider the forces created by boats propelled by various means, especially in lakes and rivers where the natural waves are normally quite small.

2.1 Static forces

The simplest first approach to understanding the different impacts of various recreation activities and of other animals and machines is to compare their different ground pressures. These may be derived by dividing the weight by the area in contact with the ground and expressing it as $g\,cm^{-2}$. Walking is one of the most commonest recreation activities and walkers come in many sizes and with a wide variety of footwear. In one impact study of back-country hikers in the Yosemite National Park California, most were wearing deeply cut rubber 'Vebram lug-sole' boots (Holmes and Dobson, 1976). Nowadays, I suspect that a high proportion might be wearing running or jogging shoes, however. It was found that the raised pattern area comprised 45% of the total area of the Vibram sole, and this amounted to 88 cm² for males (size 9½) and 78 cm² for females (size 7½). These authors then derived a formula which gave a weighted average ground pressure as follows:

$$\frac{\text{(\% of males in group) (average male weight)}+\text{(\% females in group) (average female weight)}+\text{(weight of boots and clothes)}}{\text{(\% males in group)(male boot area, cm}^2)+ \text{(\% females in group)(female boot area, cm}^2)}$$

$$= \frac{\text{the average weight}}{\text{average boot area}} = \text{weighted average pressure}$$

therefore (for this example):

$$\frac{(0.664)\ (70) + (0.336)\ (58) + (4.5)}{(0.664)\ (88) + (0.336)\ (78)} = \frac{70.468}{84.64} = 0.833 \text{ kg cm}^{-2}$$

(or 833 g cm⁻²).

This pressure is the average static pressure exerted on hard ground by a stationary person wearing Vibram-soled boots. This simple approach is supported by Webb's (1983) recalculated data of Liddle and Greig-Smith (1975a). He concluded that the bulk density increase caused by both a saloon car and walkers was directly related to the applied pressure. He multiplied the applied pressure by the number of passes independently of the type of impact, and found that both forms of impact produced similar changes in bulk density, dependent only upon the applied pressure (Fig. 2.1).

Examination of the various ground pressures exerted by different sorts of recreation shows ranges from 7 g cm⁻² (human on a snow-mobile) to 4380 g cm⁻² (human on a horse walking on hard ground) (Table 2.1). However, in general the use of an animal or vehicle for transport increases the ground pressure by 5–10 times in comparison to that exerted by walkers, and the amount of change in the soil is multiplied accordingly. Note that the x axis of Fig. 2.1 is on a log scale, so the measured increase in bulk density will not have a linear relationship with applied pressure. The particular pressures that stand out are the high pressures exerted by the shoes of horses, although they are quickly reduced once the shoe has cut into the soil and the whole foot is in contact. Mechanical transport generally has a high ground pressure, although the snowmobiles and hovercraft are obvious exceptions. At the other end of the scale, the low pressures exerted by equipment used to negotiate snow or soft ground, including the special Alaskan transport machinery, skis and snowshoes, are also noticeably low.

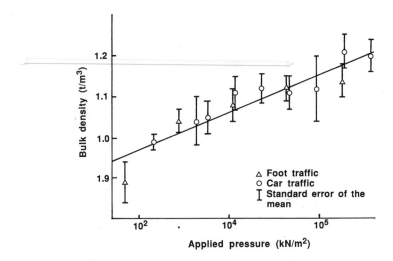

Fig. 2.1 Variation in the bulk density of a sandy soil with applied pressure, or the pressure per pass times the number of passes (adapted from Liddle and Greig-Smith, 1975a, by Webb, 1983).

The general principle is that for a given load the ground pressure is inversely related to the area in contact with the ground. Hence the use of wide, low-pressure tyres or other load-spreading devices such as snowshoes, to prevent humans or machines from sinking into the soft sand or snow. The same load-spreading principle is applied when taking advantage of many stationary elements of the environment. Hill walkers often tread on grass or rush tussocks to cross boggy areas, and in so doing they spread their weight over a greater area by means of the stems and root mass of the plant. Neolithic humans also exploited this principle 4870 years ago, in the construction of trackways across bogs. These were made of bundles of twigs and small branches which were laid at right angles to the path and then sunk with layers of stones and soil to raise the surface of the track above the water level (Coles and Hibbert, 1968) (Fig. 2.2).

Modern pavements or wooden-surfaced and stone-surfaced pathways also spread the walkers' load on a soft substrate, except where piles are used to transmit the forces to lower, more compacted levels of soil. The way in which particulate matter such as wood chips (Slaughter *et al.*, 1990) protect the surface is less clear, but is probably also the result of spreading the load distribution.

While ground pressure gives a good indication of likely changes in bulk density, other actions, such as the cutting edge of skis or the side draft from hovercraft, should not be overlooked.

Table 2.1   Total weights, ground contact area pressure and calculated ground pressure associated with various outdoor recreation activities, vehicles and animals

Static ground pressure	Average of total weight (g)	Ground contact area (g cm²)	Pressure (g cm²)	Source of data
Human				
Bare-footed on hard ground	73 000	262	297	Liddle (unpublished)
Shoes	73 000	406	180	Liddle and Greig-Smith (1975a)
Football boots (with studs)	75 000	75	1 000	Canaway (1976)
Vibram-soled boots on hard ground	70 500	166	416	Holmes and Dobson (1976)
Boots with whole sole in contact with the ground				
Man	80 000	388	206	Liddle (1973b)
Woman	57 000	356	160	Liddle (1973b)
On snow skis	75 000	2 660	28	Liddle (unpublished)
On snowshoes	75 000	2 310	33	G. Wall (personal communication) (area); Liddle (unpublished)
Animals				
Sheep	80 000	85	941	Liddle (1973b)
Sheep	43 200	63	690	Ssemakula (1983)
Cow	440 000	300	1 467	Liddle (1973b)
Cattle	306 300	314	980	Ssemakula (1983)
Goat	39 900	55	730	Ssemakula (1983)
Oryx	126 900	149	860	Ssemakula (1983)
Eland	225 600	235	1 090	Ssemakula (1983)
Horse and rider (whole foot)	613 000	478	1 282	Liddle (unpublished)
Horse and rider (shoes only)	613 000	140	4 380	Liddle (unpublished)
Vehicles				
Trail-bike	229 000	114	2 008	Eckert et al. (1979) (weight) Liddle (area)
Small, personal-use, three-wheeler, all-terrain cycle	105 000	1 050	100	Slaughter et al. (1990)
Four-wheeler, all-terrain cycle	140 000	1 400	100	Slaughter et al. (1990)
Snowmobile	75 000	10 800	7	Greller, Goldstein and Marcus (1974) (weight); G. Wall (personal communication) (area)
Saloon car and driver on hard ground	1 282 000	855	1 500	Liddle and Greig-Smith (1975a)
Four-wheel-drive Toyota, empty on hard ground	2 100 000	1 355	1 550	Liddle (unpublished)

Table 2.1 *continued*

Static ground pressure	Average of total weight (g)	Ground contact area (g cm^2)	Pressure (g cm^2)	Source of data
Four-wheel-drive Toyota, loaded 4 people and gear on hard ground	2 500 000	1 483	1 686	Liddle (unpublished)
Four-wheel-drive Toyota, empty on 'supa digger' tyres on hard ground	2 100 000	2 106	997	Liddle (unpublished)
Jeep	1 180 000	526	2 240	Slaughter *et al.* (1990)
Weasel	1 200 000	17 143	70	Slaughter *et al.* (1990)
Caterpillar D-7	15 800 000	22 571	700	Slaughter *et al.* (1990)
Roligon Alaskan transport machine, large low-pressure pneumatic tyres	2 631 000	12 469	211	Burt (1970)
Yukon flat truck caterpillar tractor	8 074 000	38 265	211	Burt (1970)
Hovercraft SK5 air-cushion vehicle	7 264 000	484 266	15	Rickard and Brown (1974)

2.2 Active forces in terrestrial situations on smooth ground

The static ground pressures are only the beginning of the story. Outdoor recreation involves activities of many kinds, but they all obey the laws of physics and any force exerted is countered by an equal force in the opposite direction. Thus all methods of propulsion apply an active force to the earth's surface, whether it is just walking, skiing or travelling on some complex machine. Any force applied at an angle to the earth's surface may be analysed in terms of its horizontal and vertical components, or vectors. Thus human walking and vehicle movements involve horizontal forces to propel the person or vehicle along, to slow down or turn a corner. The vertical forces of gravity are modified by other elements of movement, for example as a walker raises his or her foot off the ground and places it down again. The forces exerted by walkers have been measured by setting up a 'force platform' that is equipped with sensors calibrated to give electrical signals in three orthogonal directions or vectors (Fig. 2.3) (Harper, Warlow and Clarke, 1961; Quinn, Morgan and Smith, 1980). When walking straight forward on level ground an average vertical force of 1100 g cm^{-2} (maximum: 3100 g cm^{-2}) was recorded for the whole foot and 12 000 g cm^{-2} for the heel at the time of impact (Harper, Warlow and Clarke, 1961). These forces vary throughout the time of a single step, with an initial peak as the heel makes contact with the ground and the second peak as the front of the foot raises the body for the next step (Fig. 2.4). Quinn, Morgan and Smith (1980) recorded vertical forces of 8000–9000 newtons at these two peaks, and corresponding horizontal forces of up to 150 newtons in both the initial decelerating

(a)

(b)

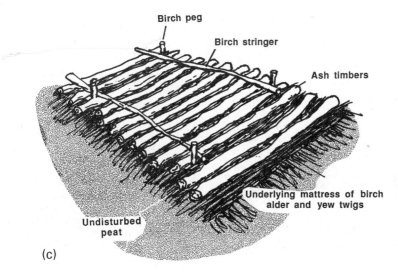

Fig. 2.2 Construction of old trackways. (a) and (b) Track through a reed swamp in the reconstructed neolithic village at Lelys, Denmark. The underlying mattress of small branches can be seen amongst the water plants, together with the vertical pegs and longitudinal branches; this example has been covered with soil and rocks (photographs by M.J. Liddle). (c) The construction used on the Somerset Levels 3000 years BC. (Based on information in Coles and Hibbert, 1968.)

phase and the subsequent accelerating phase of the step. As the ground becomes steeper, the peak force of the component at right angles to the ground's surface decreases, and the peak force of the shearing action parallel to the surface increases (Fig. 2.5). These measurements are of the total force exerted by a single step, and the area of contact over which the force is distributed is not considered.

The measurements of force recorded from a horse's front hoof using a special shoe equipped with force transducers show surprisingly low forces (Frederick and Henderson, 1970). They were lowest when the animal was walking on grass, where there would have been a considerable yielding of the surface and the contact time is also longer (Fig. 2.6). The registered forces rose on harder substrates and as the horse moved into a trot and then a gallop. The contact time also decreases in these modes. This force would have been distributed over a relatively small area ($c.$ 140 cm^2) and therefore produced very high ground pressures.

Tangential forces exerted by vehicles are clearly much more than 'one horsepower' discussed above, so although the ground pressures

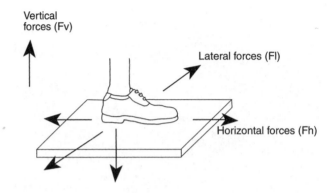

Fig. 2.3 The direction of forces created by the foot in walking (measured by Quinn, Morgan and Smith, 1980). (See Figs 2.4 and 2.5.)

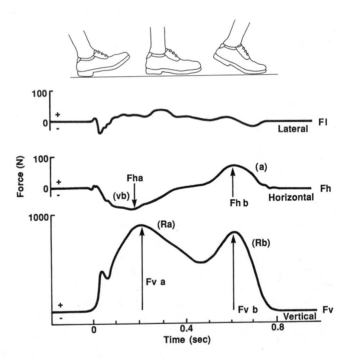

Fig. 2.4 The lateral, horizontal and vertical forces exerted by a single footstep on a slope of 10.5° (Quinn, Morgan and Smith, 1980). (See also Figs 2.3 and 2.5.)

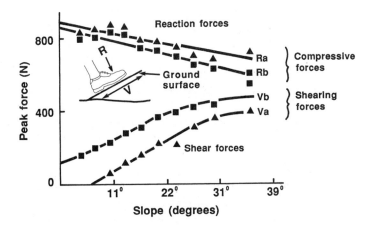

Fig. 2.5 Relationship between forces and slope angle. Ra = Fha sinΘ + Fva cosΘ = the first peak at right angles to the ground surface. Rb = Fvb cosΘ + Fhb sinΘ = the second peak at right angles to the ground surface. Va = Fha cosΘ + Fva sinΘ = the first peak parallel to the ground surface. Vb = Fvb sinΘ + Fhb cosΘ = the second peak parallel to the ground surface. (After Quinn, Morgan and Smith, 1980). (See also Figs 2.3 and 2.4.)

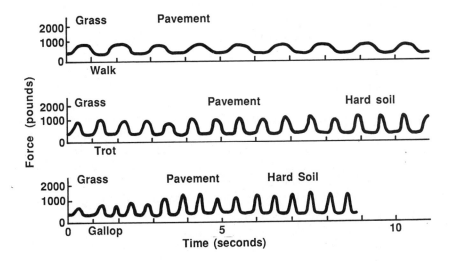

Fig. 2.6 Record of the force exerted by a front hoof of a horse for three gaits on three different surfaces (from Fredrick and Henderson, 1970).

exerted may be less than the pressure of a horse's hoof, the vehicles have the power to disrupt vegetation, especially intermediate-sized woody stems or exposed roots, to a much greater degree.

In all cases discussed above, only forward movement on a level or sloping surface has been considered. When accelerating, decelerating or turning are considered, the magnitude of the forces increases greatly. Irregular or unusual movements, such as the sliding tackle in football or a walker jumping off a style, will also greatly increase the force exerted.

2.3 Modifications to ground pressures on rough ground

In this discussion all the forces have been considered as occurring on smooth, relatively hard ground. In practice, the surface may often be uneven with tree roots crossing the path or stones protruding from the soil. A person's weight may be distributed over an area of less than 20 cm^2 giving a static pressure of over 3650 g cm^{-2} and a much higher active force. This is enough to remove the bark from a tree root after a very few steps. Coral colonies are also a special case, a person's whole weight may be applied to a horizontal branch of a coral of, say, 1 cm diameter and 10 cm length, giving a bending moment of, say, $730\,000 \text{ g}$. This is more than any coral can withstand (Liddle and Kay, 1987).

Soft ground has the reverse effect to small roots or stones, allowing a greater contact area and thus lowering the ground pressure. A human wearing boots with Vibram soles has only 45% of the sole area in contact on hard ground, giving a pressure of 833 g cm^{-2} (Holmes and Dobson, 1976). This is reduced to 375 g cm^{-2} when all the cleats sink into the mud, and the whole sole is in contact. Likewise, Liddle (1973b) showed that when a normal saloon car was driven on to soft substrate the contact area rose from 855 cm^2 to 1350 cm^2, reducing the ground pressure from 1500 g cm^{-2} to 950 g cm^{-2}. This process has been studied in detail for a tracked vehicle with cross-links across each tread. The track sank into the mud so the force exerted by the cross-links was steadily reduced as the rest of the track touched the ground, eventually reaching very low pressures (Fig. 2.7) (Kevan, 1971). In all these cases, there is, of course, a disruption of the surface layer of soil and possible damage to plants and surface roots.

Forces exerted on the soil surface are distributed down the profile of the soil, where they are dissipated as frictional heat as the particles rub together, or the forces are stored as elastic deformation and are released as the soil reforms after the walker or vehicle has passed. The distribution of forces under a tyre and a tracked vehicle was studied by Reaves and Cooper (1960). The highest pressures occurred at the surface and there was considerable transmission to 1 m under

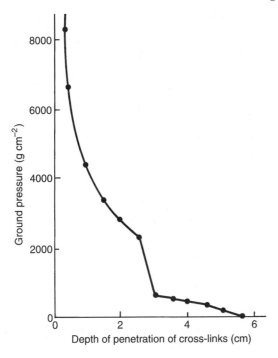

Fig. 2.7 Forces exerted on the ground by track cross-links as they become buried and the area of contact of the track increases (after Kevan, 1971).

the tyre and 55 cm under the track, which had a lower surface pressure (Fig. 2.8). There was also a small sideways transmission of forces under both treatments. This transmission of force down the soil profile brings about the changes below the surface that we recognize as soil compaction.

2.4 Aquatic situations

Much of the following discussion is based on Liddle and Scorgie (1980). Physical forces in aquatic situations mainly originate from boats, although swimming and scuba diving may be important in some situations. When a boat or swimmer is in the water they displace a mass of water equivalent to their weight, and thus their static force is distributed over the floor of the entire lake or other water body in which they are floating. Once they start to move there are displacement and reaction effects creating wash and other turbulence in the surrounding water, so they may indirectly affect living organisms in this way. Boats and swimmers may also come in direct contact with macrophytes or other living organisms or sometimes, unfortunately, with each other. Trampling effects may also occur in places where

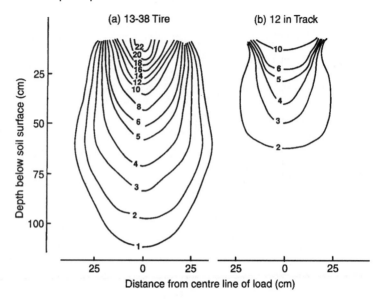

Fig. 2.8 Equal vertical stress lines, perpendicular to the direction of travel, for a similarly loaded tyre and track on Congavee silt loam (Cohron, 1971, after Reaves and Cooper, 1960).

people do not usually walk, when boats are being launched or swimmers are gaining access to the water.

WASH

Forces generated by boats may be considerable. The power required to drive a boat must be dissipated in the surrounding water, which in turn directs it on to the beds and banks of water bodies, in some cases causing severe erosion (cf. Constantine, 1961). The energy transmitted by a boat's wash depends, among other things, on the speed and power of the boat, the shape of its hull and its displacement. As a boat moves forward, water is piled up in front of it and the level at the stern falls (Fig. 2.9a). The size of the wash will also depend upon both the speed and design of the boat and the energy used to drive it along. Surface waves are generated at the bow tangentially to the direction of movement, and at right angles to the direction of movement at the stern (Fig. 2.9b). The maximum height of these waves occurs where they intersect, and the height indicates the relative amount of energy transmitted by them. The height of the waves has been shown, in experiments on the river Rhône, to increase with the speed of the boat to a nearly constant value when travelling at 13 km h^{-1} against a mean current speed of 0.67 m sec^{-1}; and to increase exponentially above 13 km h^{-1} through the water, when the direction of travel was the same as that of the current (Fig. 2.9c) (cf. Bruschin and Dysli, 1973). Transom sterns will create large transverse waves (Fig. 2.9b) when the boat is overpowered or run at high speeds, but tunnel sterns enclosing a propeller driven from an inboard engine cause less wash (British Transport Docks Board, 1972). The net effect of these factors is that the water level at the margin of a waterway at first rises slightly as a boat approaches, then drops sharply as the boat passes and finally there is a series of waves, usually less than half of the depth of the initial fall (Fig. 2.9d).

When a boat travels along a narrow canal its maximum speed is limited by water displaced at the bow having to pass to the stern of the boat. This is controlled by the ratio of its cross-sectional area below the water line to the cross-sectional area of the canal (Fig. 2.9e) (the blockage factor). For any particular design, this speed can be calculated in relation to the width and depth of the canal (Fig. 2.9f), and it can be related to the wash characteristics of that design. It is thus possible to consider a speed-limiting system based on canal (or waterway) size and the limiting speeds of each type of boat; the British Transport Docks Board (1972) consider 0.66 of the limiting speed to be reasonable.

DIRECT IMPACT

Turbulence in the water may be created by propeller action, and the extent of water movement will depend upon the size of the propeller, its design, position in relation to the hull and the power of the motor driving it. The edges of the propellers can also act as a set of rotating knives when they come in contact with submerged macrophytes or other organisms (Liddle and Scorgie, 1980). Boats propelled by oars or paddles impart relatively little energy to the water, although direct contact of the oars with lake bed or banks can have some effect. Boats may also dissipate their kinetic energy by direct collision with marginal vegetation, banks or beaches. Forces of natural origin may be transmitted to the biotic environment by anchored boats, both in the form of a downwash striking the hull (Liddle and Scorgie, 1980) and by the dragging of anchor chains over the lake or sea floor (Fig. 2.10) (Davis, 1977).

Operational characteristics such as the number of traffic lanes, depth requirements of various types of boats and the way in which they are driven (e.g. stops, turns, etc.) were taken into account in the survey system proposed by Jaakson (1988) (Chapter 26).

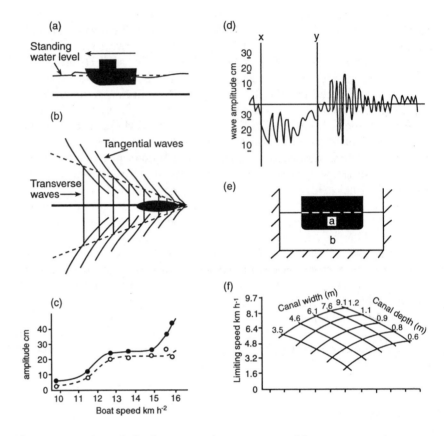

Fig. 2.9 Wash and displacement characteristics of boats. (a) Displacement of water by moving boat (after Constantine, 1961). (b) Type of waves generated by a moving boat; dotted lines indicate points of intersection (after Bruschin and Dysli, 1973). (c) Maximum amplitude of waves generated by a barge and 'pushing tug'; ○, moving against the current; ●, moving with the current (after Bruschin and Dysli, 1973). (d) Variations in water level at a stationary point 1.8 m from the side of a moving boat, x = bow passes point, y = stern passes point (J.W. Richardson, personal communication). (e) Cross-section of a boat in a canal: l/b = blockage factor, where 1= the whole cross-sectional area of the canal (a + b) (after Constantine, 1961). (f) Limiting speeds for a typical canal cruiser (after British Transport Docks Board, 1972). (From Liddle and Scorgie, 1980.)

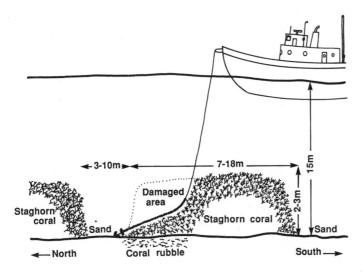

Fig. 2.10 Typical anchor deployment on the staghorn coral reef south-west of Dry Tortugas, Florida (Davis, 1977).

At first sight, air may seem an unlikely transmitter of forces. However, very high speed air movements are created by aircraft taking off, hovercraft and aerial-propeller-driven watercraft, such as the air boats used in the Florida Everglades (Schemnitz and Schortemeyer, 1974). In the first and last cases, the air is moved horizontally over the substrate and will remove at least some plant litter, sand and small soil or ice particles, with some environmental effect. The downdraft from a hovercraft must equal its weight, so the air cushion on which, for example, an SK5 air-cushion vehicle rides will have a pressure of 15 $g\,cm^{-2}$ (Table 2.1), but the main damage is apparently caused by the strong horizontal motion of the air as it escapes under the skirt of the hovercraft, disturbing any loose material on the soil surface.

2.5 Mechanical forces in air

The physical forces reaching the environment as a result of recreation activities depend upon the amount of energy utilized, the time over which the impact takes place and the area over which the force is expended. A human walking exerts about 180 $g\,cm^{-2}$ of ground pressure, and a horse and rider about 10 times that amount. Tangential forces from human and animal feet, and torque effects from vehicles, can also cause considerable damage. Wash from boats and direct impacts can also reduce protecting bank vegetation, some macrophytes are particularly vulnerable (Chapter 26). Hovercraft have their particular effect of blowing away the loose surface particles.

2.6 Summary

The biotic changes brought about by the various impacts depend upon the magnitude of these forces and the fragility of the environment. The management of biotic resources involves techniques that either reduce the level of the physical forces or reinforce the environment so that it may absorb the energy of recreation with little consequent change. Much of this book discusses the nature and extent of the changes brought about in the living environment by physical forces, but chemical changes are also discussed in Chapters 9 and 26.

Major changes in vegetation as a result of wear 3

Examination of any area of natural or semi-natural vegetation that is used for recreation will show signs of impact, usually differences in the size and quantity of plants in the used area. This effect may range from a minor pathway across a pasture, which can only be detected by a slight lowering in height of the grass and other plants, to the partial or complete clearance of trees and the rest of the ground cover in a forested area, to provide space for parking cars, camping or other activities. The amount of impact is influenced by the amount of use, vegetation fragility, vegetation density and the type and distribution of the recreation activity (cf. Cole, 1992). In all cases one simple way of measuring the degree of change is to record the weight or biomass, percentage cover or height of the undisturbed vegetation, and to compare the results with the biomass, percentage cover or height of the plants in the disturbed area.

3.1 Cover and biomass

Cover of vegetation is, at its simplest, the amount or percentage of ground, within a defined area, that is covered by plants. So the cover of grass viewed from a normal standing position, or the cover of forest viewed from an aeroplane, may both be expressed in this way. The remaining uncovered area is usually referred to as bare ground, which will, of course, have a reciprocal relationship with cover.

Plant cover may be subdivided into dead and alive material, and the dead, detached portions of plants may be recorded as litter. Cover of the various species may also be estimated. Bare ground may also be subdivided according to the type of substrate, e.g. soil and rocks.

Cover is usually estimated by eye within a defined area, or quadrat. Common sizes are 1 m × 1 m, and 4 m × 4 m; oblong quadrats of 25 cm × 1 m are often used to fit within the normal 45 cm width of a single-file path (Huxley, 1970). Estimates are often recorded as a percentage of the quadrat area and the records are then expressed as the mean of a number of replicate quadrats from a larger area. Domin or Braun-Blanquet scales (Table 3.1) may also be used. The Domin system includes further information on the abundance of plants. The quantity of plants making up the ground cover may also be estimated, by passing a vertical rod or 'pin' through the vegetation and noting how many times it makes contact with the plants. Cover is not only reduced by wear within the fixed quadrats, but paths will also widen with increasing amounts of use.

Ground cover was measured in a study of the wear taking place around picnic tables in a subtropical rainforest clearing in Brisbane Forest Park. Bluhdorn (1985) found that the mean percentage of bare ground was 53% at the edge of the concrete slabs on which the tables were mounted, and that mean percentage of ground covered with vegetation and plant litter increased to 96% at a distance of 1–1.25 m from the edge of the tables (Fig. 3.1).

The biomass of living organisms may be defined as 'a quantitative estimate of the total mass of organisms comprising all or part of a specified unit (e.g. dry weight of plants m^{-2}) at a given time, measured as volume, mass or energy' (Lincoln, Boxshall and Clark, 1982). In practice, several samples of vegetation from small areas, often only 1 m^2 or less, are cut and weighed, either fresh (fresh weight) or, more

Table 3.1   Domin and Braun-Blanquet estimates of vegetation cover

Categories	Scale
Domin scale: for estimating cover and abundance of plant species	
Single individual	+
Very few individuals	1
Sparsely distributed, less than 1% cover	2
Frequent but less than 4%	3
4–10% cover	4
11–25% cover	5
26–33% cover	6
34–50% cover	7
51–75% cover	8
76–90% cover	9
91–100% cover	10
Braun-Blanquet scale: for estimating cover of plant species	
Under 1% cover	+
1–15% cover	1
6–25% cover	2
26–50% cover	3
51–75% cover	4
76–100% cover	5

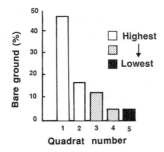

Fig. 3.1 Vegetation loss around a picnic table in Brisbane Forest Park, Australia, expressed as a mean of six transects, each starting adjacent to the concrete slab (quadrat number 1) (from Bluhdorn, 1985.)

often, after oven-drying (dry weight). The sample may be subdivided into standing dead and alive fractions and the detached dead material (litter) is usually weighed separately. The sample may also be subdivided into layers (e.g. shrubs and ground layer, and perhaps roots if they have been dug out of the soil). Plants are sometimes divided into separate species before drying. Total biomass may range from 80 kg m^{-2} in an undisturbed tropical rainforest to 1.6 kg m^{-2} in a natural temperate grassland and 0.02 kg m^{-2} in extreme desert conditions (Table 3.2).

The total biomass change is interesting when considering clearance of woodland, but where the effect of vehicles, walking or campers

Table 3.2 Plant biomass for world ecosystem types (after Whittaker, 1975)

	Biomass per unit area (dry kg m^{-2})	
Ecosystem type	Normal range	Mean
Tropical rainforest	6–80	45
Tropical seasonal forest	6–60	35
Temperate evergreen forest	6–200	35
Temperate deciduous forest	6–60	30
Boreal forest	6–40	20
Woodland and scrubland	2–20	6
Savannah	0.2–15	4
Temperate grassland	0.2–5	1.6
Tundra and alpine	0.1–3	0.6
Desert and semi-desert scrub	0.1–4	0.7
Extreme desert, rock and ice	0.1–0.02	0.02
Cultivated land	0.4–12	1
Swamp and marsh	3–50	15

is important the biomass changes of the separate layers becomes more important. Vegetation may be divided into defined horizontal layers (e.g. ground layer, 0–2 m; shrubs, 2–4 m; lower canopy, 4–13 m; upper canopy, 13–25 m; and emergents, over 25 m; Richards, 1952), but only the ground layers, shrubs and the roots of the individuals of the upper layers will be affected directly by mechanical wear.

The total undisturbed biomass recorded in recreation studies is naturally varied; two examples are 794 g m^{-2} in sand dunes at Cape Cod (Godfrey, 1975) and 50 g m^{-2} from a snow-bank community in the Alpine Olympic National Park, Washington (Bell and Bliss, 1973). These weights were reduced to 150 g m^{-2} as a result of 675 passages by a four-wheel-drive vehicle, and approximately 2 g m^{-2} after 1200 passages by walkers, respectively. The reduction that occurs in any particular site would depend on both the nature of the habitat and the intensity and duration of use.

A study of an existing track used by cars and a path used only by walkers was made at the Aberffraw dunes on the Isle of Anglesey, Wales (Fig. 3.2) (Liddle, 1973b). The biomass was collected from 25 cm × 25 cm quadrats, dried and weighed. In this case the tyre ruts were not completely bare of vegetation, although the ground surface was depressed. The final transect shown in this figure was measured on a line across a cow track leading from one grazing area to another and here the edge is clearly defined, although in this case the marginal vegetation was not sampled. An increase in biomass at the sides of the track and pathway is clearly shown. The abrupt-edged pathway seems to be common for some herbivores as the deer path recorded by Carlson and Godfry (1989) has the same characteristics (see their Fig. 3).

The reduction in plant cover that occurs as a result of wear by walkers or vehicles has, in nearly all of the cases that I have been able to examine, a curvilinear relationship with the number of walkers or vehicles walking or driving on the vegetation. These data are derived from wear experiments carried out in the field (Fig. 3.3) (see also Grabherr, 1985; Sun and Liddle, 1993a). Trailside vegetation in a rainforest in Costa Rica was also found to give a curvilinear response, suggesting that a similar response may be found in most types of habitat (Boucher et al., 1991). The exception to the curvilinear relationship between cover and wear appears to occur when the same plot is measured sequentially, rather than when a series of plots receive different levels of wear and are measured at one time (Bayfield, 1971b, 1979b; Holmes and Dobson, 1976). A further exception appears to be the response of very resistant communities, such as sand dune pasture (Liddle, 1973b) and a *Carex nigricans* snowmelt meadow at 2050 m in the Cascade Mountains of Washington, which had an

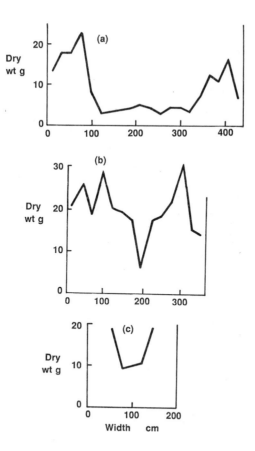

Fig. 3.2 Vegetation loss on sand dune pasture in Wales on transects across (a) a car track, (b) a footpath and (c) a cow path, showing changes in biomass that occurred as a result of mechanical wear (from Liddle, 1973b).

understorey dominated by short, wiry sedges and rushes (Cole and Bayfield, 1993) (Fig. 3.3). Cole (1992) also commented that in camp-site areas 'The curvilinear relationship between amount of use and amount of impact can be explained by the tendency for activities to become increasingly concentrated as amount of use increases'.

It would appear that the shape of the response curves is often curvilinear but not in all situations. Whether the deductions in Cole's (1992) paper are correct or not, the important point is that this is one of the few attempts to provide a rigorous theoretical basis for recreation ecology and Cole points the way for future research.

The curvilinear response is probably because there is usually an initial sharp decline in cover as the more vulnerable plants are eliminated by trampling, and then a slower attrition of those resistant

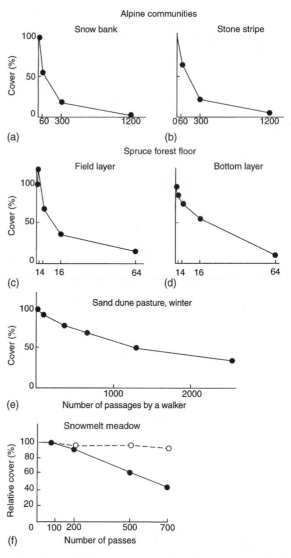

Fig. 3.3 Loss of vegetation cover as a result of experimental trampling in a variety of habitats: (a) and (b) alpine habitat, Olympic National Park, Washington, USA (Bell and Bliss, 1973); (c) and (d) spruce forest floor, near Helsinki, Finland (Kellomaki, 1973); (e) sand dune pasture in winter, Anglesey, Wales (Liddle, 1973b); (f) *Carex nigricans* snowmelt meadow at 2050 m in the Cascade Mountains, Washington; ●, immediately after trampling; ○, 1 year after trampling (Cole and Bayfield, 1993).

individuals that are left, until at some point, often beyond the range of the experimental treatments, no living vegetation remains on the path or track. Some examples are given in Fig. 3.3. The long extension of the 'resistant' part of the curve has led many authors to use logarithmic values of the number of passages on the *x* axis. Biomass

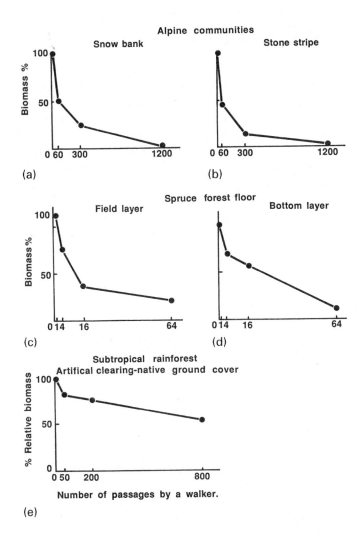

(a)

(b)

(c)

(d)

(e)

Fig. 3.4 Loss of vegetation biomass as a result of experimental trampling in a variety of habitats, expressed as percentage of the original biomass or (e only) of the adjacent biomass. (a) and (b) alpine habitat, Olympic National Park, Washington, USA (Bell and Bliss, 1973); (c) and (d) spruce forest floor near Helsinki, Finland (Kellomaki, 1973); (e) subtropical rainforest clearing, Brisbane Forest Park, Queensland, Australia (Bluhdorn, 1985).

responds to wear in a very similar way to cover, with the initial sharp decline and slower removal of the more resistant material (Fig. 3.4). It might be expected that biomass would decline before any reduction of cover was detected, but comparison of coyer and biomass measurements where both parameters have been recorded at the same time, shows that the two measurements are almost directly related, presumably as a result of a lack of homogeneity in the environment so that some areas are denuded as soon as wear commences (Fig. 3.5). This would not be expected where the wear was very slight and 100% cover remains after some reduction of biomass.

The relationship between the width of the bare ground defining the pathway and numbers of people using the path is also curvilinear, with an initial fast rise and then a slower but steady increase in width as the numbers of walkers increased (Fig. 3.6). This process may take place over a long period of time; Lance, Baugh and Love (1989) found that the index of extent (see Bayfield and Lloyd, 1971 and Chapter 15 for methods) steadily increased over a period of 11 years (Fig. 3.7). In the northern Rocky Mountain parks of Gallatin Valley, Grand Teton and Yellowstone, a steady increase of usage generally doubled

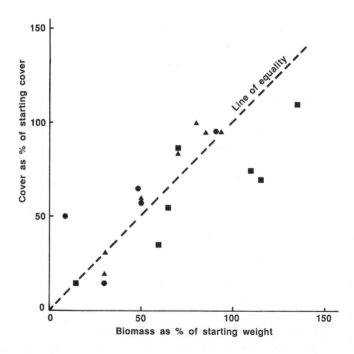

Fig. 3.5 Relationship between biomass and cover measurements recorded in wear experiments: ●, human walking (Šomšák *et al.*, 1979); ■, human walking (Kellomaki, 1973); ▲ artificial tramp (Kellomaki, 1973).

the width of the trail for every tenfold increase in the number of users (Dale and Weaver, 1974) (Fig. 3.6). However, chalk grasslands appear to be more vulnerable as the width increased at the rate of 4.4 times for every tenfold increase in use (Fig. 3.8) (Satchell and Marren, 1976). Figure 3.6 also suggests that for a given number of users, trails are wider in meadows than in forests, at least when the level of use is below 1500 visits yr^{-1}. Soil wetness and roughness may also influence path width (Bayfield, 1973) (see also Chapter 12). The creation of multiple paths side by side as walkers avoid boggy or eroded sections has also been recorded frequently (Bayfield, 1973; Lance, Baugh and Love, 1989).

Campsites offer a different spatial distribution to paths, and Cole (1992) has modelled them as a circle (see below and Fig. 3.10). He proposes a standard hypothetical model and then manipulates the 'data' to examine the influence of individual variables on the amount of impact. These models suggest that the vegetation impact 'curve' would be a function of use concentration. Cole claims that 'In the absence of spatially concentrated activities vegetation loss will increase as the square of any increase in use'. This makes the assumption that the lower remaining portions of plants are as easily damaged as the upper, in my opinion, more fragile, shoots and stems.

With many researchers approaching the problem of defining the consequences of wear from different philosophical positions, it has

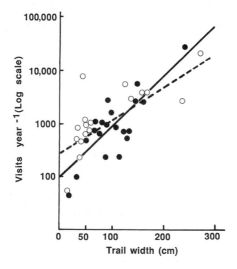

Fig. 3.6 Relationship between number of walkers and trail width (Dale and Weaver, 1974). Note that using a log scale on the *y* (vertical) axis converts the figure to linear form but not the relationship between the measurements, which remains curvilinear. ●, Forest trails; ○, meadow trails.

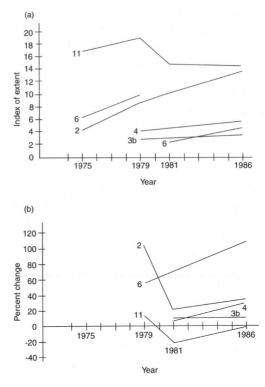

Fig. 3.7 (a) Changes in the index of extent (I), 1975–86. (b) Variations in extent expressed as a percentage of the index (I) at the preceding date. Path numbers are given for comparative purposes (Lance, Baugh and Love, 1989).

been difficult to make succinct comparisons between their results. There has been general agreement that the amount of wear imposed can be defined in terms of the number of passages of a walker or vehicle along a path or track, or more simply the number of people or vehicles passing a fixed point.

A definition of a recognizable and comparable amount of damage to vegetation has proved more difficult. One approach adopted by Kellomaki (1973), and expanded by Kellomaki and Saastamoinen (1975), was to define the mean proportion of the vegetation that remains after each passage of a walker. They defined this as P, a parameter which expresses the coverage (or biomass) after one trampling passage. In its simplest form, the equation is:

$$y = P^j . x + e$$

where y is the biomass of the vegetation after trampling; x, the original biomass of the vegetation; e, the uncontrolled variance and j is

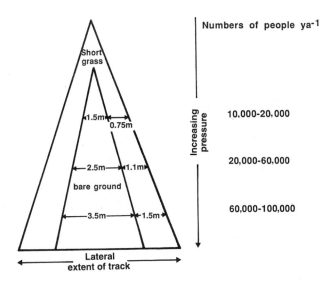

Numbers of people ya^{-1}

Increasing pressure

10,000-20,000

20,000-60,000

60,000-100,000

Fig. 3.8 Extension of the width of tracks as indicated by bare ground and short turf, under increasing trampling pressure on chalk grassland in south-east England (from Satchell and Marren, 1976, based on a figure by Goldsmith and O'Connor, 1975).

the rate of trampling. As shown in Figs 3.3 and 3.4, the relationship between wear and remaining vegetation is generally curvilinear.

These authors then presented values of P calculated in relation to unity (1), and compared these calculated values with the actual value recorded in a field experiment. In all cases there is a significant correlation between the predicted P value for the community and the actual figure obtained in the field, for both percentage cover and biomass (Table 3.3). The utilization of either cover or biomass is appropriate and, as noted, there is in practice a good correlation between the two measurements (Fig. 3.5). So it is not inappropriate to consider both types of measure in the same general formula.

In an alternative approach, Liddle (1975b) took a fixed reduction of 50% in the amount of vegetation present, expressed as cover or biomass, and from the data published by a number of authors, estimated the number of passages required to reduce the vegetation to this level. The number of passages was then compared between different vegetation types to assess their relative vulnerability (Fig. 3.9). On this basis comparisons may also be made between the effect of different types of vehicles, or the data may be utilized for management or other theoretical purposes. Within the available data there is a range of over 32 to 1 between the resistance of sand dune pasture at sea-level, at 53°N, when trampled in summer, and alpine habitats

Table 3.3 The trampling tolerance (*P*) values calculated by Kellomaki and Saastamoinen (1975), the *r* value and significance of the correlation of *P* with observed damage to vegetation in the field

Forest site type	Horizontal strata	Percentage cover			Biomass		
		P value	r	Significance	P value	r	Significance
Vaccinium	Field layer	0.974	0.87	**	0.992	0.89	**
myrtillus	Bottom layer	0.987	0.89	**	0.993	0.91	**
	Total biomass	–	–	–	0.993	0.77	*
Vaccinium	Field layer	0.982	0.98	***	0.994	0.93	***
vitis-idaea	Bottom layer	0.973	0.95	***	0.975	0.92	**
	Total biomass	–	–	–	0.983	0.97	***
Calluna	Field layer	0.835	0.94	***	0.93	0.95	***
	Bottom layer	0.818	0.74	*	0.865	0.89	**
	Total biomass	–	–	–	0.896	0.93	***

P = The proportion of the vegetation remaining after one passage of a walker.

at 2500 m, at 48°N as studied by Bell and Bliss (1973) (Fig. 3.9). However, the most vulnerable type appears to be the ground flora of a dry sclerophyll open eucalyptus forest at Brisbane, Australia which was reduced to the 50% level by 12 passages of a walker.

A very practical model, which expresses differences in amount of cover as 'area of vegetation loss', was proposed by Cole (1992). His campsite variables included the amount of use, vegetation fragility, vegetation density and the degree to which on-site traffic is concentrated. His 'standard model' campsite is circular and the levels of use grade outwards from the centre (Fig 3.10). This is used as a basis for calculation of area of vegetation loss in which, for example, an area of 188 m² of vegetation with a cover of 0.5% equals 94 m² (Table 3.4). The various input factors listed above can be altered independently. As Cole says 'this model does permit, for the first time, an analysis of how each explanatory variable affects the amount of impact'. Cole and Bayfield (1993) have also developed an experimental protocol for assessing resistance, resilience and tolerance of vegetation to trampling (Chapter 8).

3.2 Height and structure

Although differences in height are often the first visual indicator of wear taking place, height has not often been measured in wear experiments, perhaps because interest has been focused on the more dramatic and aesthetically displeasing consequences, such as bare ground and erosion. However, height is important when the processes of damage are considered, especially with respect to the response of separate species. A pioneer of trampling studies, A. Klecka (1937), published

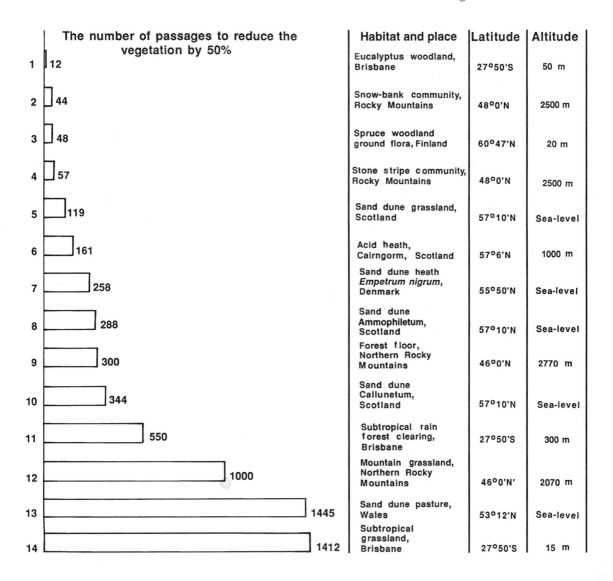

The number of passages to reduce the vegetation by 50%	Habitat and place	Latitude	Altitude
1 — 12	Eucalyptus woodland, Brisbane	27°50'S	50 m
2 — 44	Snow-bank community, Rocky Mountains	48°0'N	2500 m
3 — 48	Spruce woodland ground flora, Finland	60°47'N	20 m
4 — 57	Stone stripe community, Rocky Mountains	48°0'N	2500 m
5 — 119	Sand dune grassland, Scotland	57°10'N	Sea-level
6 — 161	Acid heath, Cairngorm, Scotland	57°6'N	1000 m
7 — 258	Sand dune heath *Empetrum nigrum*, Denmark	55°50'N	Sea-level
8 — 288	Sand dune Ammophiletum, Scotland	57°10'N	Sea-level
9 — 300	Forest floor, Northern Rocky Mountains	46°0'N	2770 m
10 — 344	Sand dune Callunetum, Scotland	57°10'N	Sea-level
11 — 550	Subtropical rain forest clearing, Brisbane	27°50'S	300 m
12 — 1000	Mountain grassland, Northern Rocky Mountains	46°0'N'	2070 m
13 — 1445	Sand dune pasture, Wales	53°12'N	Sea-level
14 — 1412	Subtropical grassland, Brisbane	27°50'S	15 m

Fig. 3.9 The resistance to walking of different habitats, expressed as the number of passages required to reduce the cover or biomass of the vegetation by 50%. The data sources are: 1, Thyer (1982); 2 and 4, Bell and Bliss (1973); 3, Kellomaki (1973); 5, 8 and 10, Leney (1974); 6, Bayfield (1971b); 7, Hylgaard and Liddle (1981); 9 and 12, Weaver and Dale (1978); 11, Bluhdorn (1985) (mean of two frequencies of wear); 13, Liddle (1973b) (mean of summer and winter wear); 14, Kendal (1982) (extrapolated from results where vegetation was reduced to 60% cover).

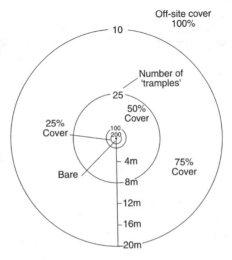

Fig. 3.10 The standard model campsite, showing isopleths of trampling intensity and zones of vegetation cover. Campsite area is 1256 m², vegetation loss is 29%, area of vegetation loss is 364 m². See Table 3.4 for calculations. (From Cole, 1992.)

Table 3.4 Calculation of area of vegetation loss for standard model campsite see Fig. 3.10 for spatial design (Cole, 1992)

Vegetation cover class	Outer boundary		
%	Distance (m) from centre	Number of tramples	Area (m²)
0	1	200	3
25	2	100	9
50	8	25	188
75	20	10	1056

Campsite area $= 20$ m radius $= 1256$ m²

Vegetation cover $= [(3\ m^2 \times 0) + (9\ m^2 \times 0.25) + (188\ m^2 \times 0.5) + (1056\ m^2 \times 0.75)] \times 1256\ m^2 = 71\%$

Vegetation loss $= 100\% - 71\% = 29\%$

Area of vegetation loss $= 1256\ m^2 \times 0.29 = 364\ m^2$

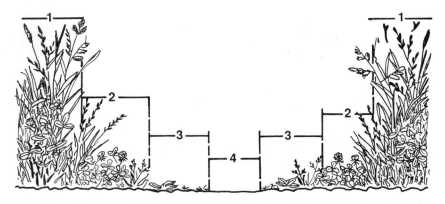

Fig. 3.11 A transect across a path in Czechoslovakia showing the height and zones of the trampled vegetation. Zone 1, untrampled, with *Trifolium repens*, *Lolium perenne*, *Poa annua*, *Polygonum aviculare*, *Plantago major*, *Achillea millefolium* and *Agrostis canina*. Zone 2, lightly trampled, with *Poa annua*, *Polygonum aviculare*, *Plantago major* and *Trifolium repens*. Zone 3, heavily trampled, with *Poa annua* and *Polygonum aviculare*. Zone 4, bare ground. (From Klecka, 1937.)

an attractive figure of a transect across a path, illustrating the effects of trampling on the height and composition of vegetation (Fig. 3.11). He clearly shows the trend in height from the bare path through a gradient of decreasing trampling intensity to the untrampled bordering vegetation. This presentation has also been adopted in some more recent publications (e.g. Roxburgh, Watkins and Bastow, 1993).

The height of vegetation is usually recorded in relation to the ground immediately beneath the plant being measured, and is expressed as the mean of a number of random measurements within a given area or quadrat or as the mean of the maximum heights, again within defined areas. For example Bluhdorn (1985) recorded a reduction from 45 cm to 12 cm after 800 passages on the grass floor of a subtropical rain-forest clearing (Fig. 3.12).

The structure of the vegetation is important for aesthetic reasons and as a matrix that determines the environment of other plants and animals. The structure or spatial arrangement of the components of the vegetation is often considered in terms of the different layers defined on the basis of height. The tallest layers composed of trees may be deliberately cleared to create campsites, picnic areas, boat-launching ramps, tracks, ski-runs or other site changes, and this will drastically change the whole ecosystem. Trees may also be damaged by people collecting wood for fires, by vandalism or collision with vehicles.

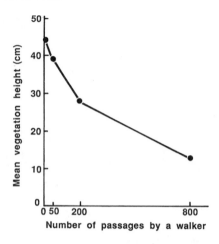

Fig. 3.12 Reduction in height of the vegetation of a subtropical rainforest clearing with increasing numbers of passages (data from Bluhdorn, 1985).

Mortensen (1989) recorded that at 70% of the campsites on the Knobstone Trail, Indiana, trees had broken branches and were scarred, or saplings had been cut down. Cole and Fichtler (1983) found that 96% of the campsites in the Eagle Cap Wilderness, Oregon, had injured trees, 25% had scarred trees, 28% had felled trees and 32% had trees with exposed roots. Marion and Merriam (1985) indicated that dying trees show a strong correlation with the tree damage and root exposure in the Boundry Waters Canoe Area, Minnesota.

Casual wear from vehicles may also markedly change the shrub and, as previously discussed, the ground layers. For example Vollmer *et al.* (1976) recorded that over 90% of desert shrubs were damaged when they were growing on a newly created track driven 42 times over a seven-month period by four-wheel-drive vehicles. Shrubs are particularly vulnerable to damage from vehicles, for two reasons. First, because of their height they are struck by the whole width of the vehicle and damage is not confined to the wheel ruts. Secondly, by definition, shrubs are woody and therefore rarely have the flexibility to bend and spring back after a vehicle has driven over them, especially when they begin to mature. Those that do bend may also have the bark scraped from the side of the stem which is in contact with the vehicle.

Recent work has shown that even the vegetation of a 3 cm-high mown lawn has a vertical stratification, with the various species each having a measurable position within the 'canopy' (Roxburgh, Watkins and Bastow, 1993) (Fig. 3.13). While the relative levels of some species

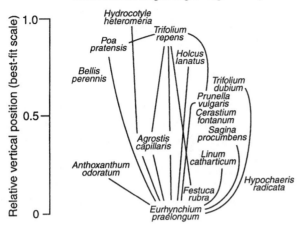

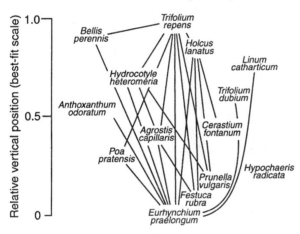

Fig. 3.13 Stratification in a mown lawn. Lines connect species pairs that are significantly different in relative vertical position. Rare species (<20 occurrences) with no significant relations are omitted. (From Roxburgh, Watkins and Bastow, 1993.)

does change immediately after mowing, there is remarkably little change, and this throws doubt on the appropriateness of the concept of succession as applied to a 16-day mowing cycle. The authors concluded that the vertical stratification was 'dependent largely on the intrinsic properties of the species, their height and recovery from defoliation'.

3.3 Productivity and prediction   The productivity of vegetation has been related to its tolerance to wear. To do this, it is necessary to define a particular consequence of wear so that comparisons may be made. A quantitative comparative system, such as the 50% level of damage (Liddle, 1975b), has the advantage that it can be used in a search for features that show correlation with the data collected in the field, or derived from experiments. The correlations may either be used to explain the results or provide a basis for prediction, for example of the vulnerability of habitats that have not been measured directly (Liddle, 1975b; Bowles and Maun, 1982). Kellomaki and Saastamoinen (1975) discussed evaluation of vulnerability and, in particular, these authors related their P value to the fertility of their experimental sites. Fertility was measured in terms of the height attained by Scots pine (*Pinus sylvestris*) over a period of 90 years. This is essentially the productivity of the site. The P values for different sites (Table 3.3) were plotted against the site quality index, and a good correlation was found between the two factors of fertility and wear resistance (Fig. 3.14). The relationship is curvilinear, with the poorest and richest sites having lower trampling tolerance.

Liddle (1975b) has also investigated the relationship between the number of passages required to reduce the cover or biomass by 50% and the productivity of the particular habitat in g m^{-2} yr^{-1} (Table 3.5).

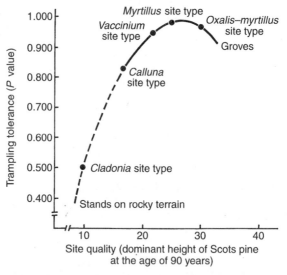

Fig. 3.14 The relationship between trampling tolerance of the ground cover and site fertility in a Scots pine (*Pinus sylvestris*) plantation, Finland. These points were based on calculations of vulnerability of species typical of these site types, measured at other locations (Kellomaki and Saastamoinen, 1975). *Myrtillus* refers to sites dominated by *Vaccinium myrtillus* and *Vaccinium* to sites dominated by *V. vitis-idaea*.

There was a close positive correlation between the two characteristics when they were expressed on a log scale (Fig. 3.15). The predictive equation based on the data is:

log number of passages to reduce cover or biomass by 50% = (1.178) × (log productivity measured in g m^{-2} yr^{-1}) − 0.496

where 1.178 defines the slope of the line and 0.496 the intercept with the y axis.

It is therefore possible to predict the number of passages a given vegetation type will withstand if the annual productivity is known. The relationship in Fig. 3.15 may be used to estimate the number of passages that vegetation types not considered in Table 3.5 will withstand.

This kind of prediction must be considered in relation to local conditions so that such features as loose soil, uneven drainage, rocky substrates, the morphology of monospecific stands of plants and other cultural activities are taken into account. Cole and Bayfield (1993) have

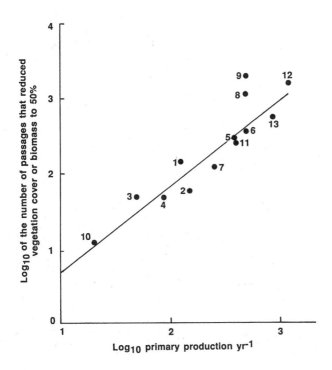

Fig. 3.15 The relationship between the number of walking passages that reduce cover or biomass by 50% and primary productivity. Note both axes are logged. See Table 3.5 for sources of data.

Table 3.5 The number of walking passages required to reduce vegetation to 50% of the original cover or biomass and the associated primary production (from Liddle, 1975b, with additions)

Author, vegetation type and location	Total number of passes to reduce vegetation by 50%	Primary productivity g m^{-2} yr^{-1}	Details of experimental method		
			Numbers of passes at each treatment (intensities)	Numbers of treatments (frequencies)	Time in weeks between first treatment and measurement of cover or biomass
Bayfield (1971b)					
1 Acid heath, Scotland	161	140[a]	40, 80	1, 3	16
Bell and Bliss (1973)					
2 Stone stripe community	57	152[b]	5, 25, 100	12	4
3 Snow-bank community, alpine flora, Washington, USA	44	49[b]			
Kellomaki (1973)					
4 Woodland ground flora (*Vaccinium myrtillus* L.) Finland	48	90[c]	1, 4, 16, 64	8	8
Leney (1974)					
5 Ammophiletum	288	400[d]	40	8	16
6 Dune grassland	344	500[a]			
7 Callunetum, Scotland	119	251[e]			
Liddle (1973b) Sand dune pasture					
8 Winter	1061	500[a]	16, 32, 64, 128	3, 5, 10, 20	20
9 Summer, North Wales	1828	500[a]			
Liddle and Thyer (1986)					
10 Eucalyptus woodland, Brisbane	12	20[f]	4, 8, 16, 32	17	17
Bowles and Maun (1982)					
11 Lake Huron sand dunes	273	400[g]	50, 300	4	12
Kendal (1982)					
12 *Imperata cylindrica* mixed forest ground cover, Brisbane	1475	1200[h]	8, 12, 24, 32, 48, 96	8, 16, 24	8, 1
Bluhdorn (1985)					
13 Subtropical rainforest ground cover, Brisbane	896	550[i]	10, 40, 160	5	5

[a] Whittaker (1975) (precise data were not available so a generalized figure for temperate grassland was utilized for sand dune pasture).
[b] Bell and Bliss (1973).
[c] Kellomaki (1973) (productivity data were not available so standing crop measured at the experimental site was utilized for woodland ground flora.
[d] Estimated by L.A. Boorman (personal communication).
[e] Gimingham (1972).
[f] Students' estimate.
[g] Bowles and Maun (1982).
[h] R.M. Jones (personal communication).
[i] Two times the total standing biomass.

proposed an alternative index based on the mean relative cover after 0–500 passages. Although it cannot be applied to much of the existing data, it does have the considerable advantage of incorporating responses at different levels of trampling (Chapter 8).

An interesting point is that several studies have recorded a slight increase in cover, or especially biomass, in areas receiving very light levels of trampling (see Liddle, 1973b (Fig. 3.2a,b); Kellomaki, 1973 (Fig. 3.3c)). The cause of this increase is uncertain but it is possible that the shift in species composition that occurs with light levels of wear may be associated with a greater utilization of available space and other resources, thus leading to a greater biomass and, where the cover was incomplete, to an increase in cover.

Many workers suggest or imply that vegetation change is directly related to the intensity of wear (Wagar, 1964; Burden and Randerson, 1972; Trew, 1973; Wynn and Loucks, 1975). However, the effectiveness of productivity as a factor in predictive calculations suggests that we also need to consider the features of both standing crop or cover at one point in time and the potential for regrowth (and death) during the period of the damage experiment. High productivity may mean that there is a larger volume or biomass of vegetation to be damaged and that the upper layers of the ground flora may protect those underneath them. It may also mean that there is a greater regrowth of plants between damage treatments before the final measurement has been made. Thus it is important to consider not only the total number of passages walked or driven in any experiment (the intensity) but also their distribution in time (their frequency) in

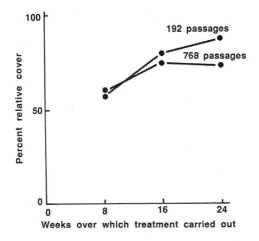

Fig. 3.16 Relationship between percentage cover of the remaining vegetation and the duration of wear. Two intensities of trampling totalling 192 and 768 passages (from Kendal, 1982).

relation to the growing season. It has been observed that at low intensities, wear carried out at one time causes a greater reduction in cover
than when the same number of passages are distributed over a longer
period (Fig. 3.16) (Liddle, 1973b; Leney, 1974; Hylgaard, 1982;
Kendal, 1982). Kendal (1982) was able to show that the growing
conditions were exceptionally good during the final 8 weeks of
his 24-week experiment. The shorter duration treatments were carried
out at the start of the 24 weeks and so missed the most productive conditions. Another, more extreme, example is where a sand dune
pasture was trampled for 20 weeks in summer and, on different plots,
for 20 weeks in winter (Liddle, 1973b). It took 1828 passages to reduce
the relative cover by 50% in summer, but only 1061 passages to cause
the same level of damage in the winter months.

The influence of soil types and conditions as well as the species
composition and management factors should all be taken into account
when applying the results discussed in this section to specific habitats
around the world.

3.4 Comparison of the effects of walking, horse riding, driving vehicles and other activities

The comparisons of the effect of wear on different habitats described
above were all based on the effects of walkers on the vegetation.
However, the forces exerted on the environment are very varied, as
shown by the author's observations of a churned-up, muddy, Sussex
bridal-way used by horse riders and an adjacent smooth, grassy
pathway used only by walkers, or similar observations of a deeply
eroded area used for car parking and adjacent vegetated footpaths in
a Welsh sand dune pasture. For direct comparisons of the effects of
the different kinds of use it is again necessary to turn to experimental
results.

The way in which people walk is affected by the degree of protection provided by the clothing they are wearing and, in particular, the
type of shoes or boots they have on, or indeed if they are barefooted.
Walkers' behaviour will also depend on how accustomed they are
to the degree of protection their clothing provides and to the habitat
in which they are walking. This was shown nicely by Nickerson
and Thibodeau (1983) when they compared the effects of people
walking, both shod and barefooted, on a monoculture of beach grass
(*Ammophila breviligulata*) in sand dunes. These authors refer to their
results as change in height but, since this was determined by placing
a 1 m^2 aluminium sheet on the vegetation and measuring the height
at which each corner was held off the ground, the results should in
my opinion be regarded as a change in bulk, or even strength, of the
vegetation. However, the results are clear enough and the beach grass
was less damaged by the barefooted walkers than by walkers wearing
shoes (Fig. 3.17). Nickerson and Thibodeau commented that it was

realized early on that if the experimenter walked through the bare-
foot test plots with any degree of self-interest, no plant would ever
have been trodden on. Walking was therefore done with the walker
not watching where he was stepping. However, 'withdrawal occurred
rapidly from foot placement which resulted in the foot being abruptly
skewered'. This not only indicates that walking barefooted is probably much less damaging to the vegetation than their results suggest,
but also that scientists concerned with these problems are very dedicated to their work!

Different types of activity may also have different effects. A comparison by Leney (1974) of the amount of cover remaining after walking
or running 320 times, and 200 sitting-down and standing-up actions,
over a period of 16 weeks in summer, showed that there was little
difference between running and walking, but that sitting on the
vegetation had less of a 'depressing' effect. However, Leney noted that
even sitting once on mature Ammophiletum damaged it so severely
that the effect was easily visible 2 years later. She added that running
removed a great deal of dead material as well as living shoots, especially on sloping ground, possibly due to the greater force exerted; but
the measured results tend to be similar as she made only three steps
on the plot when running, compared to four steps when walking.

Sloping ground increased the damaging effect of walkers on pastures,
where the vegetation was able to withstand just over 1000 passages

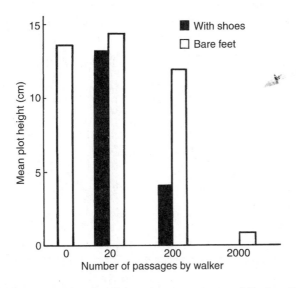

Fig. 3.17 Reduction in height of beach grass (*Ammophila breviligulata*) by
walking with (■) or without (□) shoes. (From data of Nickerson and Thibodeau,
1983.)

before being reduced to 50% cover on level ground but only 700 passages had the same effect on a 15% slope (Weaver and Dale, 1978). The same authors found that corresponding figures for a forest floor vegetation were 300 passages and 50 passages, thus this vegetation was six times more vulnerable on a 15° slope than when it grew on flat ground. The effect of sloping ground was also recorded on existing paths at Cairngorm, Scotland, where Bayfield (1973) measured the steepness of the slope and the associated path widths along two unbranched pathways which had similar levels of use. He found a positive correlation between slope and path width to an angle of 20° (Fig. 3.18). The narrower path had a shallow angle of correlation, suggesting that the true relationship may be curvilinear, with a greater increase in width on the steeper slopes.

The ground pressure of a horse's hoof when a rider is on its back may be as much as 27 times that of a walker's shoe and the horizontal forces are likely to be correspondingly greater than those generated by a person walking (Chapter 2), so it is not surprising to find that horse riders on level ground reduce the cover of a meadow about twice as fast as walkers, and three times as fast on a slope of 15° (Weaver and Dale, 1978) (Fig. 3.19). The ratio between the effect of horses and walkers on a more vulnerable shrubby forest understorey was also about three to one on level ground but on a 15° slope both horses and walkers reduced the cover to less than 50% after 100 passages. Horses had eliminated all cover after only 200 passes while walkers took 1000 passes for the cover to be totally destroyed. Results from the same experiment showed that, on average, 1000 walkers

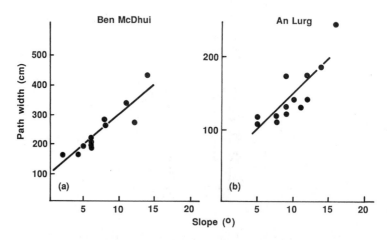

Fig. 3.18 Two different positive correlations between slope and path width at (a) Ben McDhui; and (b) An Lurg, Scotland (from Bayfield, 1973).

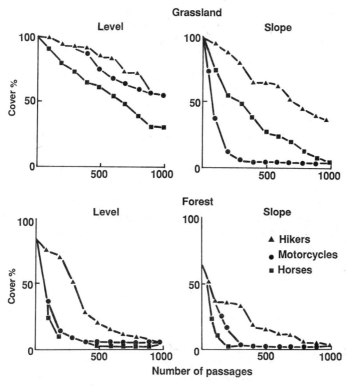

Fig. 3.19 The effects of wear by hikers (▲), motorcycles (●) and horses (■) on the vegetation cover of level and sloping, grassland and forest sites (from Weaver and Dale, 1978).

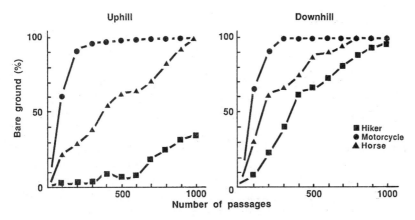

Fig. 3.20 The effects of uphill and downhill traffic on the amount of bare ground at a meadow site with a 15° slope. ■, Hiker; ●, motorcycle; ▲, horse. (From Weaver and Dale, 1978.)

created 35% bare ground when going uphill but this rose to 95% bare ground when they were walking downhill (Fig. 3.20), whereas horses had roughly the same effect in both directions. This is to be expected as more sliding occurs when walkers are going down slopes.

The increased damage caused by horses was also recorded in forested campsites in the Bob Marshall Wilderness, Montana (Cole, 1983b). In this case use by casual horse riders and outfitters was differentiated. Outfitters, that is organized tours on horseback which may be considered a part of the ecotourist industry, caused less reduction in vegetation cover (30% v. 33%), a greater increase in soil exposure (32.9% v. 9.3%) and introduced more weedy species (61 v. 43) than did casual horse campers.

The amount of impact a walker has on the vegetation depends in part on the number of steps he or she makes on a given length of path, which depends on his or her pace length. Bayfield (1973) observed the different pace lengths of walkers on three sections of an unfenced Cairngorm gravel path which varied in slope from 13° to a gentle 3°. He found that the mean pace length increased from 60 cm on the steepest slope to 72 cm on the gentle slope. He also found that walkers going uphill on a 13° slope had a mean pace length of 65 cm, which reduced to 55 cm when they were descending. When descending, 27% of walkers left the path, compared to 6% when going uphill, thus paths tended to be widened more by downhill travel than by those ascending. Weaver and Dale (1978) also measured the width of trails on which ground cover was completely destroyed. They found that horse riders made wider trails than those made by walkers, even after the first 100 passages, and after 1000 passages the horse trails were generally twice as wide as those created by walkers (Fig. 3.21).

As the reader will have noted, the effects created by riding a trail-bike over the same ground were also measured in this study by Weaver and Dale (1978). On level grassland the reduction in cover was about twice that caused by walkers in the intermediate-use level, but initially and after 1000 passages, the reduction in cover was similar after both types of use, and half that caused by the horse riders. However on the 15° grassland slope after 400 passages, the trail-bike had destroyed all cover within the measured area while there was still 35% cover on the horse trail and 65% cover on the walking trail. This gives a clear demonstration of the potential that powerful machines have for great effects on the environment and, in particular, the deleterious effect of torque when applied as a lateral force to the vegetation, especially considering that the static ground pressure of a trail-bike on soft ground is much less than the ground pressure exerted by horses' shoes, although not the whole foot (Table 2.1). The damage caused by the trail-bike in the forest was essentially the same as that caused by the horse rider. The effects of a trail-bike on trail width again depended on the habi-

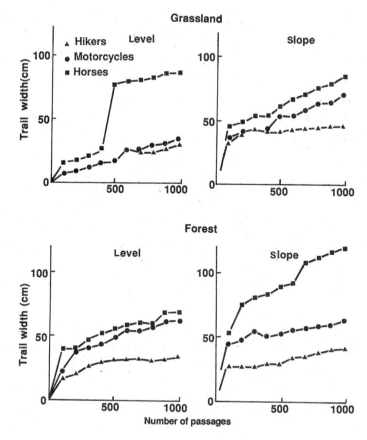

Fig. 3.21   The effects of wear by hikers (▲), motorcycles (●) and horses (■) on the trail width of level and sloping, grassland and forest sites (Weaver and Dale, 1978).

tat. The trail-bike track had a similar width to the walker's pathway on level grassland but was about twice that of the walker's path in the level woodland (Fig. 3.21). On sloping ground the bike trail was approximately one and a half times as wide as that created by walkers and three-quarters of the width of the horse trail. The differential effect of walking and riding trail-bikes and horses in different habitats can be summarized using the 50% level of vegetation cover (Fig. 3.22).

Liddle (1973b) compared the effects of a 760 kg van (including driver) and walkers (average weight 68.5 kg) on a sand dune pasture, spread over a period of 20 weeks in winter and, separately, for 20 weeks in summer. After 640 passages the vehicle had reduced cover to 10%, while cover was still at a mean of 70% on the walking tracks. The 50% level after summer wear was reached after 203 passages of

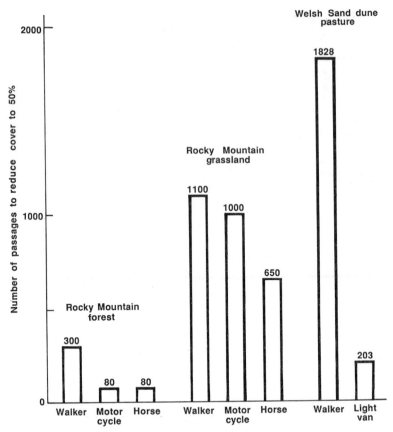

Fig. 3.22 Comparison of the number of passages required to reduce the vegetation cover to 50% by walkers, motorcycles, horses and a light van. The relative effects of different forms of wear may vary in different habitats. Note that the Welsh sand dune measurements were made in the growing season. (Data from Liddle, 1973b; Weaver and Dale, 1978.)

the van but 1828 passages of the walkers – the effect of the van being nine times that of the walkers.

In an unusual study a comparison was made of the effects of air boats and half-track vehicles on the biomass of saw grass (*Cladium jamaicensis*) in the Everglades, Florida, USA (Schemnitz and Schortemeyer, 1974). The measurements were made in June, 5 months after treatment, so there was a considerable period for regrowth. Under these conditions they found that the biomass of the plots run over five times by an air boat did not differ significantly from the controls, while plots treated by five passages of a half-track vehicle were significantly reduced to a mean of 34% of the biomass of the controls. Although the low number of passages makes the 50% level rather

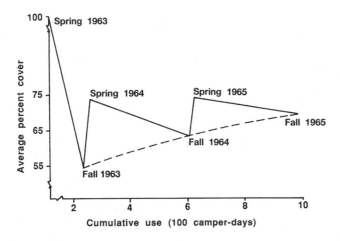

Fig. 3.23 Increasing resistance of campsite vegetation in the first 2 years of use. The increase was attributed to a growing proportion of resistant species. (From La Page, 1962.)

meaningless, it is clear that the floating air-boat hull is less damaging than the heavier half-track vehicles running on the submerged soil. Since this is not easy walking country, there is no comparison with the effects of walkers.

Caution should be used in the interpretation of the results given in this section, especially as they were derived largely from experimental impact on undisturbed vegetation. In practice, an area that has been subject to trampling forces for a longer period may have changed its species composition, with more resistant types invading and strengthening the ground layer (La Page 1962) (Fig. 3.23).

3.5 Summary

The effect of trampling and other activities is clearly to reduce the biomass, cover and height of plants, and the relationship between wear and these parameters is generally curvilinear. Vehicles and horses cause a greater reduction of vegetation per passage than walkers, and the vulnerability of habitats has been shown in Table 3.5 to range from 12 to 1828 passages to reduce vegetation cover or biomass by 50% (note that the figure of 1828 passages was the result of an experiment carried out during the growing season). Vegetation types with a ground flora of low productivity, such as eucalyptus forests, are the most easily damaged, while highly productive pastures are able to withstand the greatest wear. The impact of wear is least in the growing season and this is also the time when the greatest recovery can take place if the area is given a rest period. However, the vegetation types with low productivity may take hundreds of years to recover from wear (Chapter 8).

4 Changes in communities and plant species as a result of wear

As I walk along an unpaved footpath through natural or semi-natural vegetation the reason that I'm able to follow the path is that I can see the lower height and reduced biomass of the grasses and forbes (Chapter 3). But I will also notice that the path and its edges are covered with different plants from those dominating the adjacent vegetation. It is this change in species composition that brings into existence the 'trampling communities', the results of gradients of species quantity and dominance.

As many readers will have been conditioned by the western European type of education system, they will be aware that we handle broad spectra of information by putting the separate items into some kind of fixed order or juxtaposition. Someone with a more flexible philosophy may not feel the need to make this mental step which, in this case, probably arises from the desire to understand and possibly manipulate the plant communities in recreation areas. Those who are not used to the categorical form of thinking can move happily through an unstructured environment extracting their needs from the surprises and originality that this approach brings. However, the aspect of recreation ecology that I am concerned with in this book is based on ordering of our knowledge of the environment, and in this chapter the ordering of the species composition of areas used for recreation.

4.1 Ordination and classification

Just as we need to be able to understand English in order to read this book, so we also need to understand the differing approaches to ordering plant communities to understand fully the aims of a particular worker. For this reason I now summarize two major approaches used by ecologists in the study of plant communities. First there are two, possibly extreme, views of species composition. On one hand there is the view that each species is totally independent of any other and it only occurs at a site by a combination of the chance of it getting there and of its ability to survive once it has arrived. This leads to the dynamic view of vegetation as continuously changing in space and time (Gleson, 1926; Curtis 1959), and to the idea of gradients of species quantity or population size, responding to a combination of environmental factors. Workers using this philosophy will often present their results ordered according to occurrence (for example, Ingelog, Olsson and Bodvarsson (1977) grouped plants growing in a trampled Scots pine stand in Uppsala, central Sweden into three classes, or points on a continuum of tolerance to wear (Table 4.1)) or as an ordination based on a mathematical analysis of differences within a data set (Fig. 4.1).

The opposite view is that plants grow in communities which are similar in nature to organisms, and which may be classified according to a system similar to that produced by Linnaeus for naming plants and animals. The Braun–Blanquet (1928, 1932) system of phytosociology is perhaps the most widely used in continental Europe and utilizes a hierarchy of systematic units based almost exclusively on floristic data (Table 4.2; Muller-Dombois and Ellenberg, 1974). There are, of course, many variants and intermediates between the two extremes mentioned above (see Shimwell (1971) or Muller-Dombois and Ellenberg (1974)) or for surveys of these techniques, their methodology and discussions of their relative merits).

CONTINENTAL SYSTEM OF CLASSIFICATION OF TRAMPLED COMMUNITIES

An example of the result of this technique occurs in Tuxen (1950) where he gives a classification of communities of the Euro-Siberian region. The class, which includes a large number of trampling communities, is the Plantaginetea maioris which has five character species: great plantain (*Plantago major*); rye grass (*Lolium perenne*), annual meadow grass (*Poa annua*), fiorin (*Agrostis stolonifera*) and hairy buttercup (*Ranunculus sardous*). Most of these species will occur in any community which is placed in this class. There is only one order, the Plantaginetalia maioris which has the same character species. There are then three alliances: one is halophytic (salt loving), occurring in salt marshes and intertidal situations, and one alliance occurs in freshwater situations. The remaining alliance, Polygonion avicularis, occurs in trampled situations and has great plantain, rye grass, annual meadow grass, Bermuda grass (*Cynodon dactylon*), swine cress (*Coronopus squamatus*) and knotgrass (*Polygonum aviculare*) as character species.

There follow nine associations, these are the most important units in this type of classification, and it is the characteristics of an

Table 4.1 The tolerance of different field layer and bottom layer species to regular perennial trampling of Scots pine stands in Uppsala (only species displaying clear tendencies included) (from Ingelog, Olsson and Bodvarsson, 1977)

| | Degree of tolerance | | |
	Very low	Low	Moderate to high
Range	Species which, on the whole, are only to be found growing on slightly trampled ground	Species occurring first and foremost on slightly trampled ground but which can also be found growing on moderately trampled ground, although with a reduced biomass and low frequency	1. Species with approximately the same frequency irrespective of whether the ground in question is trampled to a slight, moderate or intensive extent. Frequency begins to decrease only in case of intensive trampling. 2. Species which are only or primarily to be found on moderately or intensively trampled ground
Field layer: tree and shrub species over 0.6 m	*Picea abies* *Pinus sylvestris*	*Acer platanoides* *Betula pubescens* *B. verrucosa* *Populus tremula* *Salix caprea* *Sorbus aucuparia*	
Dwarf shrubs		*Vaccinium myrtillus* *V. vitis-idaea*	
Herbaceous species	*Athyrium filix-femina* *Dryopteris carthusiana* *D. cristata* *D. filix-mas* *Geranium sylvaticum* *Gymnocarpium dryopteris* *Hepatica nobilis* *Lactuca muralis* *Lapsana communis* *Lycopodium annotinum* *Melampyrum pratense* *Orthilia secunda* *Oxalis acetosella* *Polygonatum odoratum* *Pyrola rotundifolia* *Solidago virgaurea* *Thelypteris phegopteris*	*Anemone nemorosa* *Campanula rotundifolia* *Convallaria majalis* *Galium boreale* *Maianthemum bifolium* *Rubus saxatilis* *Viola canina* *V. riviniana* *Taraxacum* spp. (2) *Trifolium repens* (2)	*Achillea millefolium* (2) *Alchemilla* sp. (2) *Fragaria vesca* (1) *Galium verum* (2) *Leontodon autumnalis* (2) *Polygonum aviculare* (2) *P. viviparum* (1–2) *Potentilla erecta* (1)

Table 4.1 *continued*

| | Degree of tolerance | | |
	Very low	Low	Moderate to high
Graminaceous species	*Melica nutans* *Poa nemoralis*	*Calamagrostis arundinacea* *Deschampsia flexuosa* *Luzula pilosa* *Poa angustifolia*	*Agrostis tenuis* (2) *Carex pallescens* (1) *Deschampsia cespitosa* (2) *Festuca ovina* (2) *F. rubra* (2) *Poa annua* (2) *P. supina* (2) *Sieglingia decumbens* (2)
Bottom layer (mosses)	*Hylocomium splendens* *Mnium* spp. *M. longirostrum*	*Dicranum* spp. *Pleurozium schreberi*	*Rhytidiadelphus squarrosus* (1–2)

Table 4.2 The naming system of Continental phytosociological communities from Muller-Dombois and Ellenberg (1974)

Rank	Ending	Example
Class	-etea	Molinio-Arrhenatheretea
Order	-etalia	Arrhenatheretalia
Alliance	-ion	Arrhenatherion
Association	-etum	Arrhenatheretum
Subassociation	-etosum	Arrhenatheretum brizetosum
Variant	No ending	Salvia variant of the Arrhenatheretum brizetosum
Facies	-osum	Arrhenatheretum brizetosum bromosum erecti

association that the practised phytosociologist will recognize in the field. Table 4.3 describes the location of each association and its characteristic species.

ORDINATION OF VEGETATION INFLUENCED BY RECREATION ACTIVITIES

Ordination, apart from its ecclesiastical associations, means arrangement in ranks or classification (*Oxford English Dictionary*). However, in vegetation science it has come to signify a process whereby data from each stand (or quadrat) are compared or contrasted with all the other

Table 4.3 Trampling and associated communities. The character species of the class Plantaginetea maioris and the alliance Polygonion avicularis from Tuxen (1950)

Class:	Plantaginetea maioris
Order:	Plantaginetalia maioris
Character species:	
Plantago major	*Agrostis alba*
Lolium perenne	*Agrostis alba* var. *prorepens*
Poa annua	*Ranunculus sardous*
Alliances:	
1. Polygonion avicularis	
Character species:	
Plantago major	*Poa annua*
Lolium perenne	*Cynodon dactylon*
(anthropogenic trampled community)	
2. Agropyro–Rumicion crispi	
Character species:	
Potentilla anserina	*Agropyron repens*
Agrostis alba	*Leontodon autumnalis*
Rumex crispus	*Carex hirta*
(wet, saline, nitrophyllous community)	
3. Beckmannion–Eruciformis	
Character species:	
Beckmannia eruciformis	*Glyceria fluitans*
Rumex odontocarpus	*Cardamine parviflora*
Rorippa kerneri	*Melilotus dentatus*
(freshwater, nitrophyllous community)	

Associations of the alliance Polygonion avicularis and their character species:

1. Sagineto–Bryetum argentei	
Sagina procumbens	
Spurgularia rubra	
Bryum argenteum	
2. Myosuretum minimi	
Myosurus minimus	
3. Lolium perenne–Plantago maior	
Plantago major	*Matricaria matricarioides*
Lolium perenne	*Cichorium intybus*
Coronopus procumbens	*Coronopus didymus*
4. Sclerochloa dura–Coronopus procumbens	
Regional characteristic species – S. France, Greece and Rumania	
Sclerochloa dura	
Coronopus procumbens	
5. Trifolium fragiferum–Arachnospermum canum	
Regional character species	
Trifolium fragiferum	
Cynodon dactylon	*Arachnospermum canum*

Table 4.3 *continued*

6. Cynodon dactylon–Plantago coronopus
    Regional character species: Quarnero Islands
    Lolium perenne   *Lepturus incuruatus*
    Coronopus procumbens   *Plantago coronopus*
7. Coronopus procumbens–Plantago coronopus
    Regional character species: W. coast Britain
    Coronopus procumbens   *Spergularia salina*
    Plantago coronopus
8. Agropyron littorale–Cynodon dactylon
    Regional character species: Vintschgau alps
    Agropyron littorale
    Cynodon dactylon
9. Junceto–Trifolietum
    Regional character species: Mediterranean
    Juncus compressus   *Poa annua*
    Trifolium fragiferum   *Cichorium itybus*
    Cynodon dactylon
    Lolium perenne   *Plantago major*
    Polygonum aviculare   *Rumex crispus*

stands. There are usually between 10 and 100 stands in each data set. The resulting comparisons are then aligned on an axis which is selected to show the greatest variation existing in the data set. The remaining variation is then reanalysed in the same way to produce a second axis. The coordinates on the two axes are then used to plot the relative position of each stand in two dimensions. This produces a scatter of points or stands, with the most similar being near to each other and the least similar being furthest apart (Fig. 4.1). The ordinated data may then be compared in relation to factors that were not included in the data set, and inferences are then drawn about the relationships between the plants or stands and the environment. In the example, a set of vegetation records was collected from 1 m^2 quadrats in a sand dune area. The quadrats were in associated groups of two or three. In the sets of two, one came from the centre of the footpath and the second from a nearby undisturbed area of vegetation. The sets of three were made on tracks in a similar way, with one being in the centre of a rut, one from partially disturbed edge vegetation and one from nearby undisturbed vegetation. If the assumption is made that the vegetation was originally uniform at each area where a set of two or three quadrats were recorded, then the sets indicate the changes that take place as a result of trampling or use by vehicles. At the same time as records of the presence or absence of plant species in the quadrats were made, records were also made of the soil compactness and soil water, and notes were made on the location of the quadrat. The vegetation data were then ordinated, in this case by the principal components method, and the relationships between the quadrats displayed in two dimensions (Fig. 4.1a). The environmental characteristics of soil bulk density, water content and penetration resistance for each quadrat were then plotted on similar graphs (Fig. 4.1b,c,d). It is now possible to relate soil and use characteristics and associated changes with changes in the species composition of the vegetation. In this case the worn stands are more central in Fig. 4.1a than their associated unworn partners. These are shown connected by solid lines in the case of the sets of three from tracks and dotted lines from the sets of two from footpaths. The most used stands have the highest soil bulk density and the lowest soil penetrability (Fig. 4.1b,d). The vegetation trend across the graph also appears to be associated with soil water but the most used stands are wetter than expected if it is assumed that the contours would have run smoothly across the figure before the paths were created. All of these statements are hypotheses which need to be tested by further work (see Chapter 9 for further discussions of soils). So this technique is best referred to as a hypothesis-generating technique (Gittings, 1969).

Taylor, Reader and Larson (1993) used an alternative ordination technique, multiple discriminant analyses (MDA), to display the degree of similarity in species composition of quadrats subjected to different trampling levels (Fig. 4.2).

In this case they were investigating the effects of different temporal and spatial scales of experimental trampling on cliff-top vegetation in Ontario. The arrows connecting the centres of the two sets of temporal data are pointing in different directions, and this may be interpreted as indicating two different processes. The first is that of immediate impact and the second of long-term adaptation, a subjective deduction from an objective analysis. Their additional ordination (not shown here), utilizing plots on the paths and plots extending 4 m from the path centre, did not show any great difference. However, analysis of the calculated distances used in the MDA and significance levels showed significant differences between all levels of treatment except for the 50 passes over 1 year (Table 4.4). The temporal differences can be ascribed to the differential survival of some species over the long period compared to survival for 1 year (see also Fig. 3.16). The spatial differences accord with intuitive expectations but they do emphasize the need for a standard procedure, as discussed in Chapter 8.

Ordinations of ski-run vegetation data in Tsuyuzaki (1990) and Puntieri (1991) do not show sequential effects associated with different levels of wear but combined effects with other variables such as soil erosion (Chapter 15). The recovery of track vegetation is analysed in Charman and Pollard (1993) by ordination using canonical correspondences analysis (see Fig. 8.14). From this evidence, and the fact

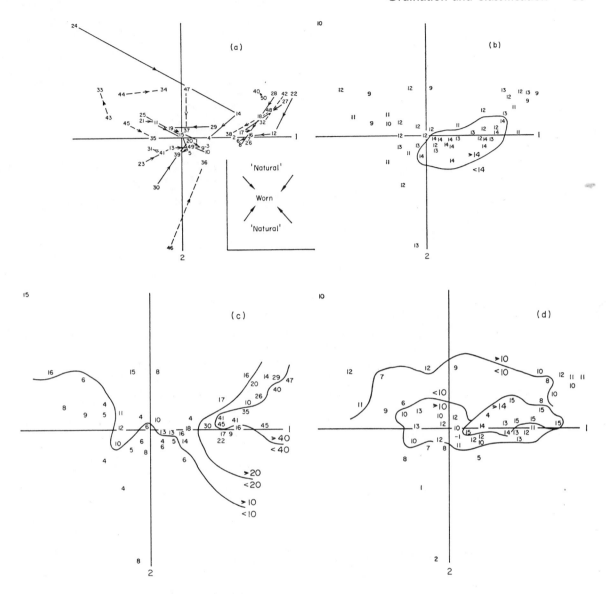

Fig. 4.1 Ordination and interpretation of a vegetation impact survey. (a) Stand numbers, the lines connect stands that were situated at the same site and arrows point towards the stands subject to most wear, ——— track sites; ----, footpath sites. (b) Soil bulk density (g cm^{-3}). (c) Soil water content as percentage volume. (d) Soil penetration resistance (log penetration index from an impact penetrometer × 10). Stand numbers 1–10, tracks; 11–20, track margins; 21–30, adjacent undisturbed vegetation; 31–40, footpaths; 41–50, adjacent undisturbed vegetation. (From Liddle and Greig-Smith, 1975b.)

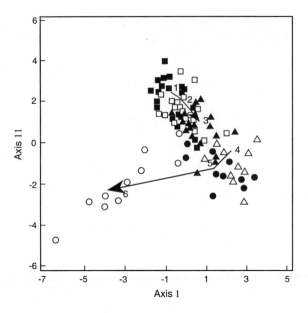

Fig. 4.2 Comparison of species composition of quadrats adjacent to path centrelines classified by trampling level for the canonical multiple discriminate analysis. The arrows trace the direction of changes in species composition through centroids from the least trampled (1) to most trampled paths (6). Short-term trampling over 1 year: ■, zero passes; □, 50 passes; ▲, 500 passes. Long-term trampling over 18 years: △, 100 passes per season; ●, 5000 passes per season; ○, 25 000 passes per season. (Modified from Taylor, Reader and Larson, 1993.)

Table 4.4 Statistical differences among trampling levels in MDA[a] (Taylor, Reader and Larson, 1993)

Treatment	Short term			Long term		
(passes per year)	0	50	500	100	5000	25 000
Short term						
0	–	3.10	3.37***	4.84***	4.72***	4.75***
50		–	3.33**	4.16***	3.98***	4.62***
500			–	4.40***	4.77***	4.65***
Long term						
100				–	4.52***	5.53***
5 000					–	5.50***
25 000						–

[a] Mahalanobis distances, and levels of statistical significance for pairs of the six trampling intensity/duration combinations examined in the canonical multiple discriminant analysis of all quadrats. Blank = NS; ** = $P < 0.01$; *** = $P < 0.001$.

that trampling communities can be identified by subjective methods (see Tuxen, 1950), it is clear that mechanical wear not only has a selective effect on the survival of members of the trampled community, but also that some species are favoured by this situation, provided that wear or the other environmental factors are not too extreme.

CLASSIFICATION BY OBJECTIVE ANALYSIS

One of the fundamental differences between Continental (Zurich–Montpellier) school and the Anglo-American approaches to phytosociology is the level of subjective judgement that is acceptable. The Continental classification scheme discussed above demands personal judgement at most stages of the work from the choice of site, which is common to most systems, through the placing of quadrats to the organization of the data, the analysis and the decision on the final placing of each stand within the classified system. In contrast, the ordination discussed above, while relying on judgement for the field work, hands the calculations of the relationships over to an objective system which is usually computerized. The final interpretation is, of course, a matter for judgement.

There are several alternative classification systems that produce different types of results. One that has proved to have some application in recreation ecology is a polythetic divisive method developed by Hill, Bunce and Shaw (1975). In this technique the stand data are separated into groups at various levels of difference so that the first division separates the stands according to their greatest dissimilarity. This particular system also selects indicator species that together typify each group of stands. Thus the example in Fig. 4.3 from Anglesey, North Wales, has at first division, in the first group, wild thyme (*Thymus drucei*), lady's bedstraw (*Galium verum*) and marram grass (*Ammophila arenaria*), and in the other group silverweed (*Potentilla anserina*) and jointed rush (*Juncus articulatus*). It is therefore possible to go into the field, at least in areas which are similar to the one from which the data was derived, and by searching for these five species in each small area, of say 0.5 m^2, to identify the vegetation as belonging to one or other group. As environmental data were collected at the same time as the vegetation data, it is possible to allocate environmental characteristics to the groups. In this case the soils, where members of the first group occurred, had lower penetration indices (0.94) and volumetric soil water content (9%) than the second group (1.21 and 27%, respectively). Thus stands with the first three species will be softer and drier although the soil bulk densities were similar. At the third level of division it becomes possible to assign predominant habitat categories to the groups and therefore to have a set of path indicator species for this type of site. In this case the plants

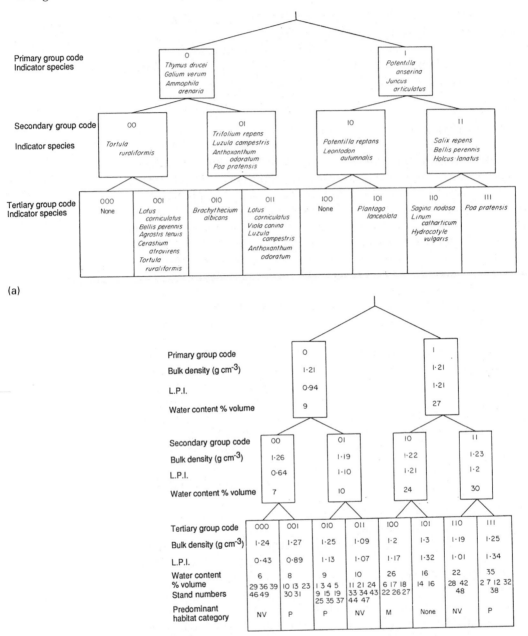

(a)

(b)

Fig. 4.3 Classification of vegetation impact survey data. (a) The principal indicator species for each group in the classification. (b) Soil data and stand numbers (see caption to Fig. 4.1) for each group in the classification. Habitat categories: P, path or track; M, path or track margin; NV, 'natural' vegetation. (From Liddle and Greig-Smith, 1975b.)

growing in the stands with high-density soils (000, 001, 010, 101 and 111) were bird's-foot trefoil (*Lotus corniculatus*), daisy (*Bellis perennis*), bent grass (*Agrostis tenuis*), mouse-ear chickweed (*Cerastium atrovirens*), *Tortula ruraliformis* (moss), *Brachythecium albicans* (moss) and meadow grass (*Poa pratensis*).

Three species – daisy, meadow grass and ribwort plantain (*Plantago lanceolata*) – are generally recognized as trampling-resistant plants occurring in lightly to fairly heavily trampled areas which are moist and quite fertile (Bates, 1935; Perring, 1967; Burden and Randerson, 1972). The other species are perhaps more local in their occurrence on paths, and at least two, marram grass and the moss *Tortula ruraliformis*, are largely restricted to the sand dune habitat. Another interesting point to arise from this study is that measurements of soil bulk density and soil penetrability are not closely linked, and species may occur on soils that have a high bulk density but low penetration index. The ordination and classification together show that the vegetation is strongly associated with the two soil characteristics of bulk density and water content. That differences in soil bulk density are principally the result of wear by vehicles and human trampling is suggested by the fact that the contours on the ordination diagram agree with the levels of wear at each stand, but the variations in soil water content are not arranged in this way and are likely to be due to natural causes (Ranwell, 1959; Willis *et al.*, 1959).

Objective classification of experimental trampling data showed that the differences between north- and south-facing aspects of the paths in a South Wales dune system were the primary factors affecting the species composition (Williams and Randerson, 1989) (Fig. 4.4). The diagnostic species was (*Ononis repens*). The second division separated level and south-facing paths in one grouping and trampled other types of path in the other. Type of treatment, degree of trampling and season were the environmental factors associated with the third-level division. This classification used TWINSPAN – a hierarchical system using dichotomous divisions (Hill, 1979).

These types of classification are, like ordination, useful as hypothesis-generating techniques, but at the same time the indicator species can be used in a predictive system within the area from which the samples were taken, providing the samples were representative of the range of conditions occurring there.

Classification of vegetation data may also be carried out on the basis of the environmental data collected at the same time. This may take the form of a direct measurement of use (Bayfield, 1971b; Frederiksen, 1976; Crawford and Liddle, 1977), surrogate measurements of soil characteristics, such as bulk density or penetrability, that generally respond quantitatively to different levels of use, or (more doubtfully) of characterization of the sample area as a 'track', a 'pathway', 'edge

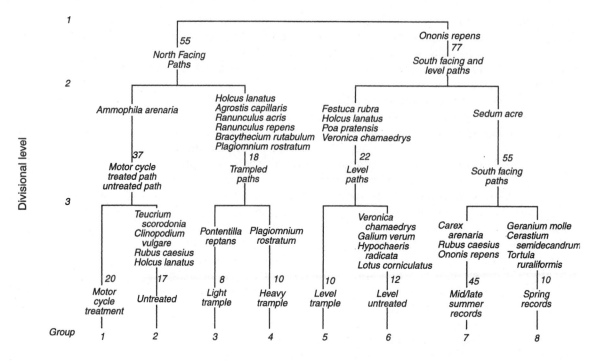

Fig. 4.4 Classification using TWINSPAN showing the effect on species composition of repeated trampling over 3 years (from Williams and Randerson, 1989).

vegetation' or 'undisturbed areas'. I say doubtfully because the subjective criteria used to detect use are usually bare ground, vegetation height or the visual bulk of vegetation, and this leads to the circular logic of saying that 'This is a pathway because the plants are like this (morphology or species)' and then from the data, that 'The plants are like this because they are on the pathway'. However, this technique is much used and, since the judgement of most experienced workers is good, it has proved to be reasonably reliable, especially when checked against other workers' results and/or by other methods.

In a study of human trampling effects on a rocky intertidal community, Beauchamp and Gowing (1982) based their classification of three communities on the surrogate measurements of the differences in accessibility of the three sites. These ranged from 'easily accessible and used by educational tours as well as casual visitors' to 'a small island surrounded by a deep channel at low tide and difficult to climb on to'. This classification was then backed up by a short-term count of actual numbers of people visiting the three sites at low tide on 70 occasions over a period of 21 days and, importantly, they state that 'we have observed over several years that the relative numbers of people visiting the sites are reflected in our counts' (Chapter 18). This kind of composite information is often the most practical approach, especially for management problems where there is not enough money, or time, for a full count of visitors over a period of years.

An early study by Neil Bayfield (1971b), who stimulated much of the later work of the 1970s, utilized the direct technique of measuring visitor numbers by the use of thin-wire tramplometers (Bayfield, 1971a). With this tool he was able to estimate the relative numbers of people crossing a transect which was placed at right angles to main paths (Figs 4.5 and 4.6). The visitor numbers were recorded in summer when the paths were conspicuous, from 13 August to 10 October 1969. However, he also made records during October and June when the paths were more obscure and found that while 81% of the walkers stayed on the paths in summer, only 51% and 41%, respectively, were on the paths in the autumn and spring. This illustrates the point that even when direct measurements of visitor numbers are made, the period of measurement is usually restricted, whereas the present condition of a pathway is often the result of long-standing use by visitors. However, Bayfield's measurements of damage and of bare ground do show a good association with the pattern of use in summer except on the lower (right-hand) slope (Fig. 4.5) (see Chapter 10 for further discussion). The detailed transect across the main path shows how the cover of heather (*Calluna vulgaris*) and deer grass (*Trichophorum cespitosum*) declined towards the centre and the proportion of bare ground and of totally damaged vegetation rose, to nearly 100%, in the area where the level of summer use was also the highest (Fig. 4.6).

SUBJECTIVE CLASSIFICATION

I turn now to the results of subjective classifications by location, ranging from the totally subjective (Bates, 1935; Chappell *et al.*, 1971) to subjectively positioning one end of the line of samples and either randomly or systematically sampling along that line (Dale and Weaver, 1974; Dawson, Hinz and Gordon, 1974). Bates (1935), one of the pioneers of recent trampling research, wrote about two zones and defined them in terms of the species present: 'In the centre of the path, if much used, is an area of bare earth. This is adjoined on either side by zone of smooth stalked meadowgrass (*Poa pratensis*) followed by zones of perennial rye grass (*Lolium perenne*) and wild white clover (*Trifolium repens*), crested dogstail (*Cynosaurus cristatus*), cocksfoot (*Dactylis glomerata*) and plantain species are also fairly constant.' On dry, sandy heaths in districts of low rainfall he observed that 'the dominant species of the path are usually sheeps fescue (*Festuca ovina*),

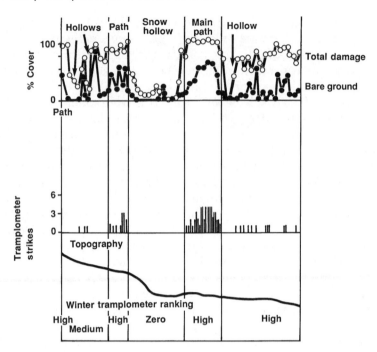

Fig. 4.5 A section of a transect showing the percentage of plants damaged and bare ground across an area which has winter snow cover and is subject to winter and summer wear. The different use levels in summer and winter are shown by the tramplometer strikes and estimates of winter use. (After Bayfield, 1971b.)

Galium spp., fiorin (*Agrostis stolonifera*), *Plantago* and *Heiracium* species ... The lateral boundaries are not sharply defined, but merge gradually into the surrounds, the treading is not evenly distributed and is more concentrated in the centre. The boundary usually occurs where the perennial rye grass and wild white clover reaches its outer limit.' His observations are as valid now as they were over 60 years ago and both highlight the subjective or arbitrary nature of the decisions on the location of the divisions between zones and the fact that much of our perception of zones is based on observations of the plants themselves. However, in support of the subjective approach, I would like to quote another part of Bates' (1935) paper, 'the arrangement of species into zones is so patent to everyday observation that elaboration in the form of statistical data is scarcely necessary'. The question is really one of judgement of 'When does a subjective observation become an unacceptably biased observation?'.

Some more recent examples will help to round out this picture. Chappell *et al.* (1971) divided their site into three areas, 'Zone 1 –

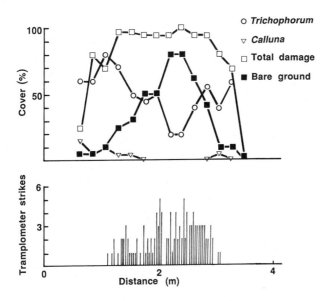

Fig. 4.6 An example of damage to vegetation on a transect across a pathway in the Cairngorm Mountains and the associated distribution of people, showing the percentage of damage to heather (*Calluna*), sedge (*Trichophorum*) and all plants, the percentage of bare ground and the number of tramplometer strikes which indicate the intensity of wear on the same transect. (After Bayfield, 1971b.)

lank tussocky grass with some scrub, accumulating litter and subject to minimal wear; Zone 2 – short sward up to 5 cm high with no significant accumulation of litter, subject to trampling but showing no signs of soil surface disturbance and Zone 3 – areas, regularly used by vehicles as well as being subject to trampling, excessively but not deeply rutted' – a subjective basis to a most interesting study of chalk grassland plants and animals. In this case there were the same number of plant species in Zone 1 and Zone 3 but salad burnet (*Poterium sanguisorba*) which occurred in Zones 1 and 2 was replaced by daisy (*Bellis perennis*) which occurred in Zones 2 and 3.

A more objective approach was used by Dawson, Hinz and Gordon (1974) where quadrats of 20 × 50 cm were regularly placed from a subjectively located centre quadrat on the path and then at distances at 1 m, 2 m and 3 m from the path centre. In this study they found that species numbers increased from a mean of about two in the centre of the path to between four and six at 3 m distance (Fig. 4.7).

Underwood (1992) has presented a detailed critique of the sampling methods used in impact studies and proposed a rigorous approach to the problem. However, while it is desirable to carry out the full sampling programme that he recommends, there are occasions when resources do not allow a fully replicated approach.

A final example of subjective plot location is taken from a study of the effects of traffic associated with oil drilling on the subarctic forest and tundra of the Mackenzie Delta region in Canada (68–74°N) (Lambert, 1972). Lambert divided his site into five subjective categories, each with its own characteristics. First, undisturbed vegetation at some distance from the road; then a band 14 m wide which had 'less disturbance' and a central section 16 m wide that carried the majority of the traffic. The central section was divided into three categories of 'scoured' (where the surface was removed), 'tracks' and, finally, the most severely damaged 'ruts'. There was a decline in vascular plants to about 5% cover, and mosses and lichens (which, because of their importance in this type of vegetation, were recorded separately) were eliminated (Fig. 4.8). The cover of bare ground and water-covered ruts increased with increasing severity of wear.

So, in summary, species have been recorded in many habitats where they have been subject to wear. Each location has its own special characteristics and many species occur only in the one study. However, generalities can be discerned in the species characteristics and these are discussed in Chapter 6.

4.2 Diversity and diversity indices

The diversity of plant or animal species occurring in a site is another general way of viewing survey data collected from recreation areas. At its simplest, diversity may be measured as simply a count of the number of species present in a known area. Dawson, Hinz and Gordon (1974) showed that the number of species on their 3 m transects

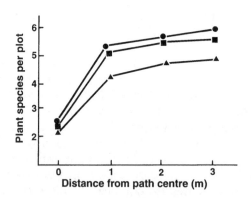

Fig. 4.7 The influence of recreational use (as a function of distance from path centre) upon the number of species of plants on the forest floor: ●, north-facing forest slope; ■, south-facing forest slope; ▲, floodplain forest (after Dawson, Hinz and Gordon, 1974).

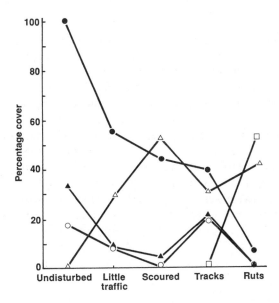

Fig. 4.8 The effects of vehicles on the dwarf shrub community of the tundra of the Mackenzie Delta region of Canada (68–74°N), showing the percentage cover of vascular plants (●), mosses (○), lichens (▲), bare ground (△) and standing water (□) in a badly rutted section of the Gulf Oil ice road. (After Lambert, 1972.)

increased as they sampled further away from the path centre (Fig. 4.7). However, by using surrogate or direct measures, to define a continuous gradient of trampling pressure, rather than just placing the quadrats into a few wear classes, the gradient of change in the species number can be examined. The data listed by Trew (1973) may be arranged for an examination of this gradient, although he did, in fact, divide trampling intensity into five classes (from 250 to 4000 people per year). In the Welsh sand dunes near the city of Swansea he found that there was a rise in species numbers to over 20 per unit area as trampling increased to 1000 people per year (Fig. 4.9a). After this point the plant numbers declined to about five species per unit area. This phenomenon of an initial rise in numbers followed by a steady decline has been noted by a number of workers (Grime, 1973; Trew, 1973; Bayfield and Brookes, 1979) and appears to fit in with the ideas of both Grime (1973) and Connell (1978) (see below). Species numbers have also been used, among other measures, to examine recovery of an experimental track in the Hubbard Brook Experimental Forest, 5 years after closure, where there was a mean of only 3.7 species in 3 m² plots in the recovering area compared to 8.5 in the adjacent controls (Kuss and Hall, 1991).

A general model that summarizes the effects of trampling and environmental management on species of high competitive index, species of high resistance to the stresses imposed by trampling and management, and other species was proposed by Grime (1973) (Fig. 4.9b). He proposed that competitive species dominated at low levels of stress, these are replaced by stress-tolerant species when stress is intense but, at intermediate levels of stress, large numbers of 'remaining' species may join the community. The data of Trew (1973) fits into this pattern, and one of his vulnerable species (the stinging nettle, *Urtica dioica*) fits Grime's category of competitive species. On the other hand, red fescue (*Festuca rubra*), rye grass (*Lolium perenne*) and sweet vernal grass (*Anthoxanthum odoratum*), which all occur in Trew's most trampled site, are moderately productive species

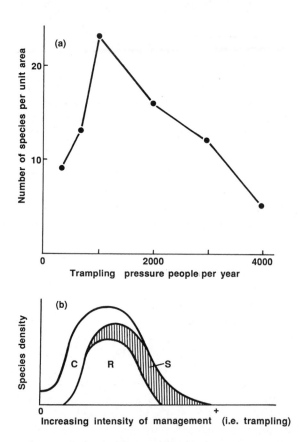

Fig. 4.9 (a) The number of plant species per unit area as a function of use in the sand dunes at Swansea, Wales (data from Trew, 1973), and (b) the distribution of competitive (C), stress-tolerant (S) and remaining (R) species (from Grime, 1973).

and intermediate between competitive and stress tolerant in Grime's triangular ordination (Grime, 1974).

The intermediate disturbance hypothesis (Connell, 1979) offers an explanation of the mechanisms involved. He divides plants and animals into three groups, early colonizers, intermediate colonizers and late colonizers, and suggests that the degree and frequency of the disturbance will determine which of these groups will predominate in any situation. He suggests that the diversity will be highest when the disturbance is intermediate in character (Fig. 4.10). He also postulates that 'the highest diversity ... should occur either at an intermediate stage in succession,' (that is, the progression from bare ground to a fully developed theoretically stable climax forest community) 'after a large disturbance or with smaller disturbances that are neither very frequent nor very infrequent' – this is Connell's well-known 'intermediate disturbance hypothesis'. He defines the situation as an open non-equilibrium, that is 'a site which is open to sources of immigration – new influx of seeds of species that may be locally extinct and a site which is changing – either as a result of succession as early colonisers die out and are replaced by others or as a result of moderate levels of disturbance'. The edges of a well-used footpath, or the centre of one that is lightly used, fulfil these conditions as disturbance by trampling is occurring and seeds or other propagules are often transported on people's footwear or clothing.

More complex diversity indices may be calculated to include some measure of the way the species share the space or dominate a community (discussed by Hill, 1973); although they may appear complex, the diversity indices can give a real insight into the way the biotic communities respond to the effects of wear. The differential sharing of space,

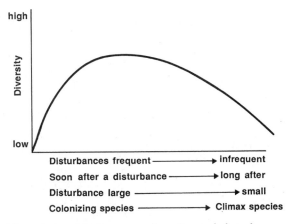

Fig. 4.10 The relationship between diversity and disturbance characteristics, including successional stage (after Connell, 1978).

as reflected by the numbers of individuals of each species that are present in the community (or in other words their dominance), is important in the understanding of likely interactions, and therefore of the causes or processes of change.

The Shannon and Weiner formula,

$$H = \Sigma \; Pi \; \log Pi$$

where Pi is the decimal fraction of total individuals belonging to the ith species, has been used by a number of investigators. Kuss and Hall (1989) found that the trailside vegetation was more diverse (mean index 2.3) than the undisturbed vegetation (mean index 1.83) in the Shenondoah National Park, Virginia. They ascribed these results to changes in competition for light and differential resistance to trampling between the plants of this habitat.

In a recovery study by Kuss and Hall (1991), a comparison was made between worn and unworn areas using this index, and there was strong evidence for a soil compaction effect in the recovering areas after a period of 5 years since use ceased (Table 4.5).

The index proposed by Hill (1973) was diversity (N^2), which equals the sum of all the individuals in the community squared, divided by the sum of the squares of the numbers of individuals of each species:

$$N^2 = \frac{(\Sigma x)^2}{\Sigma x^2}$$

Table 4.5 Comparisons of species diversity and soil penetration resistance by use level and zone of influence. All differences were significant (above 0.01) except the species diversity measure in the untrampled area (after Kuss and Hall, 1991)

Trampling level	Undisturbed (Zone 1) X	SD	Disturbed (Zone 2) X	SD	Percentage change
Species diversity					
0	0.913	0.403	1.000	0.229	±0
100	0.985	0.232	0.658	0.359	−33
400	1.123	0.194	0.726	0.172	−35
800	1.020	0.326	0.615	0.319	−39
X	1.010		0.749		−35.7
Soil penetration resistance (kg cm2)					
0	0.723	0.196	1.212	0.355	+68
100	0.809	0.191	1.744	0.230	+116
400	0.757	0.126	2.309	0.329	+205
800	0.817	0.234	3.000	0.271	+267
X	0.765		2.066		+164

X, mean; SD, standard deviation

This is simple to calculate and, in Fig. 4.11, gives an extra insight into the data of Chappell *et al.* (1971). These graphs suggest that the plants are changing very little if the species number or number of individuals are considered separately, but when the diversity index is examined we can see that the proportions of the species are changing and becoming less uniform as a result of increased trampling damage. On the other hand, while the number of species and numbers of individuals of the soil animals from the same site show a coincident decline, the diversity index rises suggesting that the numbers of each species are becoming more even. At first sight the diversities of organisms in this site do not appear to fit the intermediate disturbance hypothesis (Chapter 8) very well, as the plant diversity is slightly higher in Zone 2 while the soil animals have a higher diversity in Zone 3; however, this brings us to the question of exactly what is intermediate disturbance and whether all groups of organisms respond in the same way so that their diversity is always highest at the same level of disturbance. I would suggest that this is not the case and that each ecological group of organisms will have its peak diversity at a point which relates to their life span, colonization rate and resistance, survival and recovery responses for that particular impact.

The fact that trampling-induced diversity will vary in different communities is generally supported by the calculations of Jurko (1983). He found that Hill's (1973) N^2 index varied little with increased levels of observed damage (a surrogate trampling measure) in grassy communities (ranging, for example, from 8.2 to 5.9, a decrease of 38%, in a sedge and grass (*Carex sempervirens–Calamagrostis villosa*) community), whereas the index rose from 1.5 to 3.8, an increase of 250%, over the same difference in observable damage in a bilberry–dwarf pine (*Vaccinium myrtillus–Pinus mugo*) community of the forest floor. This is in contrast to a reduction calculated by Jurko (1983) of 80% and an increase of only 162%, respectively, in Kostrowicki's index, *I*, for the same communities. Kostrowicki's index:

$$I = \frac{pg}{100}$$

uses only *p*, total cover of species, and *g*, the number of the species (Kostrowicki, 1970).

All of which leads to the following points when considering diversity in trampled or otherwise damaged communities:

- It is essential to understand which elements or characteristics of the community are emphasized by the index concerned.
- It is important to understand what is not being taken into consideration. For example, one site discussed by Jurko (1983) showed a reduction of 20% in species number and 33% in cover; however,

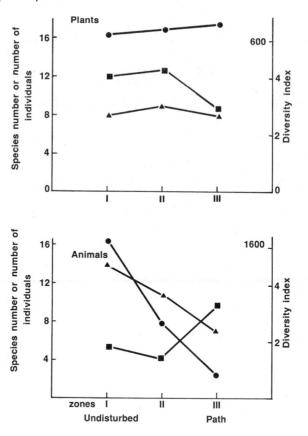

Fig. 4.11 The relationship between the number of individuals (●), species number (▲) and diversity index (■) and level of wear of (a) plant and (b) animal communities in chalk grassland in southern England (data from Chappell *et al.*, 1971).

the sum of the percentage cover of the individual species decreased fivefold (because cover repetition was measured in the first case?) and productivity in terms of biomass was only 10% of that of the control area.

- Some components of the vegetation may be much more vulnerable than others; for example, crustose and fructicose lichens and mosses (Jurko, 1983).
- The species interactions may also change, for example if the dominant species had a low resistance, the immediate response of the community may be domination by the most resilient species.

Indices may be calculated to emphasize selected characteristics of vegetation or other data sets. A number of indices have been developed to reflect the level of damage or disturbance caused by humans. The disturbance index of Wynn and Loucks (1975) was developed specifically to deal with the effects of trampling disturbance at Parfrey's Glen in the Baraboo Hills, Wisconsin, which is a 'designated scientific area' and a recreation site. They utilized the concepts of 'decreaser' and 'increaser' species (Dyksterhuis, 1957) and 'invading' species (Fig. 4.12) (Wynn and Loucks, 1975). Decreasers are at maximum densities in an undisturbed plant community and decrease in density with increasing intensity and duration of disturbance. These species are similar in character to the competitive species of Grime (1973) (Fig. 4.9), and the 'climax' species of Connell (1978) (Fig. 4.10). Increasers are species that were original members of the community and which initially increase in density under light disturbance, apparently in response to lessened competition (Dyksterhuis, 1957). However, these species decrease under conditions of severe disturbance, as must all species when the paths are used so much that only bare ground is left. The increaser species are usually less competitive than the decreaser species which inhibit them in undisturbed communities, and they have some of the characters of stress-tolerant species (Grime, 1973). They would, however, still be classed as members of the climax community by Connell (1978). Invaders include plants or animals that are not capable of entering the original community but which become established when it is disturbed (Wynn and Loucks, 1975). These are the ruderal or 'remaining species' in Fig. 4.9 and the 'colonizers' of Fig. 4.10. Dale and Weaver (1974) also used the concept of decreasers and increasers, based on trends in quantity along a transect in space rather than trends over a period of time. The occurrence of the species that were allocated by Wynn and Loucks (1975) to the decreaser, increaser and invader categories is shown in Fig. 4.12. They were measured on three successive occasions over a 15-month period and their character determined from their trends in time. To estimate their disturbance index, the presence of any of the indicator species was recorded in 20 quadrats (column 1, Table 4.6) (note: no increasers are recorded in their example but they do appear in the summary graph in Fig. 4.12), and factors of 2 and 1.33 (column 2) applied to increasers and invaders to make them equal to the decreasers in number. A weighting of 5, 3 or 1 was then applied to each class respectively (in column 4), producing column 5. The summed disturbance index components were then divided by the total in column 3 and multiplied by 100 to give a 'disturbance index total' per stand, in this case 441.

The potential range of this disturbance index is from 500, if only decreasers are present in all quadrats, to 100 if only invaders are present in each quadrat. A value of more than 400 can be viewed as a healthy ground cover for this community and values of between 250 and 400 indicate an intermediate degree of disturbance, while values lower than 150 indicate severe disturbance (Wynn and Loucks, 1975). As the authors commented, the data are readily obtained, involving only the occurrence of indicator species (once they have been characterized), and the index could be computed annually to determine a trend in the quality of plant communities. In the study of Parfrey's Glen, the 'undisturbed' index ranged from 345 to 475 but the index dropped as low as 164 in a very disturbed part of one area (Table 4.7).

An alternative index that aims to demonstrate 'deformation' of the plant life of a particular community as represented by the phyto-

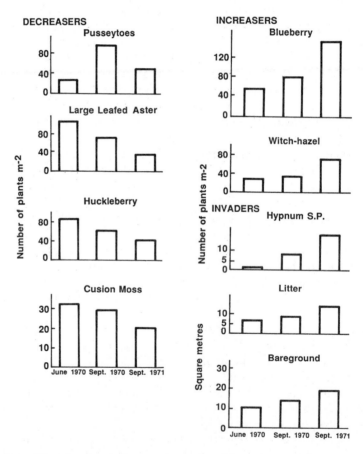

Fig. 4.12 The change in the numbers of indicator species from June 1970 to September 1971 at Parfrey's Glen, Wisconsin. Each decreaser species was reduced by one-third to one-half of its original numbers. The response of the increaser and invader species was more irregular, depending on initial conditions. (After Wynn and Loucks, 1975.)

Table 4.6 Sample calculation showing use of the indicator species and computation of the disturbance index (DI) (Wynn and Loucks, 1975)

	No. of occurrences	Equivalence factor[a]	Column 1 × column 2	Weighting term[b]	DI component	DI sum[c]
Decreasers						
Aster	5	1	5	5	25	
Antennaria	1	1	1	5	5	
Gaylussacia	3	1	3	5	15	
Cushion moss	14	1	<u>14</u>	5	<u>70</u>	
Subtotal			<u>23</u>		<u>115</u>	
Increasers						
Vaccinium		2		3		
Hamamelis		2		3		
Invaders						
Moss	1	1.33	1.33	1	1.33	118.9/26.99 × 100
Litter						
Bare area	2	1.33	<u>2.66</u>	1	<u>2.66</u>	
Subtotal			<u>3.99</u>		<u>3.99</u>	
Total			26.99		118.99	441

[a] The equivalence factor is required because different numbers of species are used for increasers, decreasers and invaders.
[b] Weighting terms were arbitrarily assigned to create a spread of values, high values representing undisturbed vegetation, low values disturbed vegetation.
[c] DI sum = DI component/equivalence factor × 100.

sociological records, or species lists, collected according to the Braun–Blanquet system (Table 3.1) was proposed by Zielski (1978). As the reader will realize, this is based on the Continental approach and utilizes a comparison of the total of the mean percentage cover of the characteristic species of the association (or alliance, order or class), *P*ch, and the total of the mean percentage cover of all the accompanying species *P*t. The first is expressed as a percentage of the second:

$$\Sigma Pch \frac{100}{\Sigma Pt}$$

The stands, or relevés, may then be placed in order of declining coverage of characteristic species in relation to the cover of accompanying species (Fig. 4.13). The association studied was Carici elongatae–Alnetum, elongated sedge (*Carex elongata*) and alder (*Alnus glutinosa*), which occurs in the Brodnica Lake District, Poland, as a woodland in oligotrophic mossy depressions with poor drainage, or in fairly fertile places on lake shores. Zielski subdivided his relevés into three degeneration phases according to the missing elements or steps in degradation (Fig. 4.13) and, on the basis of his extensive analysis of a range of communities, he recommended that parts of the area which have 'Great natural beauty and are important for research and didactic purposes' should come under partial or strict protection.

In 1973 I adopted a slightly different approach to the characterization of sand dune species (Liddle, 1973b). A 'preference index', more accurately called an occurrence index, was based on a subjective assessment of whether the site was a track (A), track edge (B) or adjacent undisturbed area (C). Data from footpaths (D) and their adjacent areas (E) was also included (Table 4.8). The index indicates the occurrence of each species on the worn areas, with a double weighting for the highly worn tracks, (2A) + B + D, compared to its overall occurrence:

Table 4.7 Comparison of values of the vegetational disturbance index over a 15-month period from June 1970 through September 1971 in upland habitats at Parfrey's Glen, Wisconsin (from Wynn and Loucks, 1975)

	Area I			Area II	
	Undisturbed	Part disturbed	Very disturbed	Part disturbed	Very disturbed
June 1970	475	270	178	440	260
Sept 1970	370	270	260	280	232
Sept 1971	345	245	164	212	245

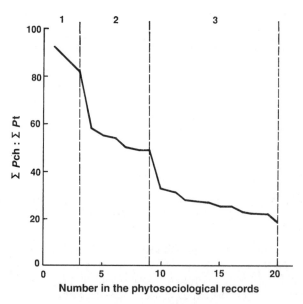

Fig. 4.13 The relationship between the diversity index $\Sigma Pch \times 100/\Sigma Pt$ (see text) and the position of the relevés in the records of the Carici elongatae–Alnetum community in Poland (from Zielski, 1978).

$$\text{Occurrence index} = \frac{2 \times A + B + D}{2 \times A + B + C + D + E}.$$

This figure was calculated for each species and, in order to spread the index from 0 to 1, the lowest value (0.18) was subtracted from each of the others, and the remainders were divided by the highest value (0.65) (Table 4.8). The occurrence index has some significance, particularly as the common path species (daisy (*Bellis perennis*), meadow grass (*Poa pratensis*) and jointed rush (*Juncus articulatus*)) come out at the top of the list (Table 4.8). The advantage of this scheme is that the subjective stage is that of site designation and is not directly related to each species. It would then be possible to designate sites within the area according to the mean occurrence index multiplied by the species quantities (cover or number of individuals), although this is not as simple as the qualitative measures required for the disturbance index of Wynn and Loucks (1975). However, the occurrence index scheme was developed to evaluate the behaviour of each species rather than to characterize the sites.

Boorman and Fuller (1977) devised vegetation vulnerability classes (high, medium and low) by measuring paths crossing each of three vegetation types at Winterton Dunes, Norfolk. The percentage of paths worn to bare sand when crossing each type of vegetation was then used to place the vegetation types in each class. They defined 'highly vulnerable' as a vegetation type with more than 50% of the paths worn bare; 'medium vulnerability', between 10 and 50%; and 'low vulnerability', less than 10%. This was then used to construct a vulnerability map of the dune system from aerial photographs on which the vegetation types could be defined. The vulnerability map was used to characterize sites for management purposes.

4.4 Invader species and dispersal

The fact that Wynn and Loucks (1975) found sufficient numbers of invader plants in their disturbed recreation area to erect a new class and to make this class an important element of their disturbance index, indicates that invasion by new species can be an important event in worn areas. Classical studies of seed dispersal emphasize the adaptations that seeds have undergone for animal dispersal (zoochory) (Ridley, 1930; van der Pijl, 1972), and several investigations have reported seed carriage by people.

Seeds may be carried in mud on peoples' shoes as well as attached to clothing and in pockets and bags. In about 1910 R.L. Prager, visiting Clare Island in County Mayo, wondered how some invader species had come to be on the island (Prager, 1911–15). He observed that mainland mud was adhering to his boatman, Pat Grady's, boots. He collected this and found seeds of chickweed (*Stellaria media*), hemlock

Table 4.8 Occurrence index of species common in survey (Liddle and Greig-Smith, 1975b)

Species	No. of occurrences in each stand group					Uncorrected ratio[a]			Occurrence index (adjusted to scale 1–0)
	A	B	C	D	E			−0.18	
Bellis perennis	5	8	3	7	5	25/33	0.83	0.65	1.0
Poa pratensis	8	7	4	9	3	32/39	0.82	0.64	1.0
Juncus articulatus	4	5	3	3	2	16/21	0.76	0.58	0.9
Potentilla anserina	3	4	3	1	1	11/15	0.73	0.55	0.9
Brachythecium albicans	2	3	1	3	4	10/15	0.67	0.49	0.8
Taraxacum laevigatum	3	1	1	4	4	11/16	0.69	0.51	0.8
Agrostis tenuis	2	4	4	3	3	11/18	0.61	0.43	0.7
Carex flacca	3	4	6	5	4	15/25	0.60	0.42	0.7
Festuca rubra	7	7	7	10	9	31/47	0.66	0.48	0.7
Galium verum	4	4	6	6	6	18/30	0.60	0.42	0.7
Taraxacum officinale	2	5	4	3	4	12/20	0.60	0.42	0.7
Trifolium repens	2	7	5	5	5	16/26	0.61	0.43	0.7
Agrostis stolonifera	2	6	7	5	6	15/28	0.54	0.36	0.6
Leontodon autumnalis	2	4	5	2	3	10/18	0.55	0.37	0.6
L. taraxacoides	3	7	8	10	9	23/40	0.57	0.39	0.6
Thymus drucei	2	5	6	7	6	16/28	0.57	0.39	0.6
Tortula ruraliformis	1	3	2	3	4	8/14	0.57	0.39	0.6
Anthoxanthum odoratum	1	2	2	3	5	7/14	0.50	0.32	0.5
Camptothecium lutescens	1	1	3	4	5	7/15	0.47	0.29	0.5
Carex arenaria	3	4	8	9	9	19/36	0.53	0.35	0.5
Lotus corniculatus	1	4	6	5	5	11/22	0.50	0.32	0.5
Ammophila arenaria	2	2	6	6	8	12/26	0.46	0.28	0.4
Luzula campestris	0	1	3	4	4	5/12	0.42	0.24	0.4
Potentilla reptans	2	1	5	1	3	6/14	0.43	0.25	0.4
Senecio jacobaea	0	4	3	1	4	5/12	0.42	0.24	0.4
Viola canina	1	2	7	6	7	10/24	0.42	0.24	0.4
Poa subcaerulea	0	3	7	4	5	7/19	0.37	0.19	0.3
Prunella vulgaris	0	2	5	3	4	5/14	0.36	0.18	0.3
Viola tricolor	0	1	3	2	4	3/10	0.30	0.12	0.2
Linum catharticum	0	0	3	2	6	2/11	0.18	0.00	0.0

[a] $\dfrac{2A+B+D}{2A+B+C+D+E}$, see text.

(*Conium maculatum*), sow thistle (*Sonchus asper*), scarlet pimpernel (*Anagallis arvensis*), knotweed (*Polygonum aviculare*), toad rush (*Juncus bufonius*), Yorkshire fog (*Holcus lanatus*), various rushes (*Juncus* spp.) and six more indeterminable species. Clifford (1956) also investigated the seeds carried in mud on footwear, in this case from the island of Madeira and many other unspecified places. He placed the mud on pots of sterilized soil in a glasshouse and kept them watered. He was able to identify a total of 43 species that grew from seed in the mud (Table 4.9), including a number of character species of trampled communities (Table 4.3).

In addition to the 'personal' transport of seeds, they may also be carried long distances adhering to motor vehicles. Clifford (1959) also carried out investigations on mud obtained from the underside of vehicles in West Africa and found that early in the rainy season there were about 100 seeds kg^{-1} of dry mud adhering to cars and trucks, and that the number rose to about 180 seeds kg^{-1} of mud early in the dry season. These included 40 different species. He also found that there were fewer seeds in sandy loam (4.4 per sample) than in a loamy sand (mean, 11.7 seeds per sample). He postulated that in the sandy loam, which holds water for a long period, seeds may have germinated and died on the vehicle. However, there is a selective mechanism operating here as he found that the seeds carried in mud were over ten times lighter (mean weight 0.00038 g) than would be expected if they represented the full spectra of roadside vegetation (mean weight 0.00412 g).

A further example, this one from tourists' vehicles in Kakadu National Park, Australia, is the direct consequence of recreation.

Table 4.9 Plants raised from mud off footwear used on the island of Madeira and elsewhere (Clifford, 1956)

Agrostis stolonifera	*Juncus articulatus*
Aira praecox	*J. bufonius*
Anthoxanthum odoratum	*J. effusus*
Atriplex patula	*Lolium perenne*
Bellis perennis	*Matricaria maritima* subsp. *inodorum*
Bromus mollis	*Plantago lanceolata*
Capsella bursa-pastoris	*P. major*
Cardamine pratensis	*Poa annua*
Carex sp.	*Polygonum aviculare*
Cerastium semidecandrum	*Prunella vulgaris*
Chamaenerion angustifolium	*Ranunculus acris*
Chenopodium album	*R. repens*
Cirsium arvense	*Rubus* sp.
Crataegus sp.	*Sagina procumbens*
Deschampsia cespitosa	*Senecio jacobaea*
Euphrasia sp.	*Spergula arvensis*
Festuca ovina subsp. *tennuifolia*	*Stellaria media*
Galium hercynicum	*Taraxacum officinale*
Glyceria fluitans	*Trifolium repens*
Gnaphalium uliginosum	*Urtica dioica*
Holcus mollis	*Viola arvensis*
	Vulpia bromoides

Tourists' vehicles parked at the western and southern entrances to the park were sampled once a month for 7 months (Lonsdale and Lane, 1992). The air intakes and outer bodies were vacuumed and the propagules so obtained identified and counted. A total of 1505 propagules belonging to 84 different species were recovered from 384 cars, but 80% came from just seven species. No propagules of the two major threats to the park, *Mimosa pigra* and *Salvinia molesta*, were found, but 954 propagules were alien to Australia and some were known tropical weeds.

Considering all this evidence, it is not surprising to find that one of the consequences of use of an area for recreation is an increase of ruderal species alongside the paths and tracks, and in some areas the rise in vehicle-borne ecotourism must present a considerable threat to the integrity of the native environment.

4.5 Convergence of species representation and their biogeography

The theories of both Grime (1973) and Connell (1978) (Figs 4.9 and 4.10) postulate that there is an increase in the diversity of species at intermediate levels of management, trampling or other disturbance. This increase in numbers of species requires the immigration of plants into the tracks or paths from areas outside the adjacent undisturbed vegetation, and the analyses of mud on people's shoes and vehicles, and of seeds in domestic animal fur (Liddle and Elgar, 1984), have shown that people who created the paths and hence the environment in which the immigrants may flourish, may also bring in the new species.

This was nicely demonstrated by a species survey of two sand dune areas in The Netherlands. One, the Hoge Veluwe, is an inland dune heathland with an acid podzol, which is very vulnerable to trampling (Fig. 4.14). The other, Meijendel, is a coastal sand dune system with a fairly fertile soil with a high calcium content. The undisturbed parts of these areas had only 3% of their species in common (van der Werf, 1972) (Fig. 4.15). The similarity coefficient used here is simply one where the number of species common to two samples (c) is expressed as a percentage of the number of species in the two samples (a and b).

$$S = \frac{2c \times 100}{a + b}.$$

When comparisons were made between those parts of the two dune systems that had similar and increasing levels of use, a steady decline occurred in the unique species in both areas. There was an associated increase to 17% in the species common to both areas (Fig. 4.15). In other words, the differences between the areas disappeared when they were trampled and the ubiquitous path flora took over.

What is the observed consequence of this immigration of species? In 1759 Benjamin Stillingfleet wrote that he had observed on the

Fig. 4.14 An area of heathland on sandy soils subject to recreation use in the Hoge Veluwe Park in The Netherlands (photograph by M.J. Liddle, 1975).

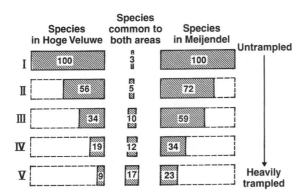

Fig. 4.15 A comparison of the effect of treading on two different types of sand dune areas of The Netherlands, showing reductions in the percentage of unique species and an increase in species common to both sites (after van der Werf, 1972).

Table 4.10 Comparison of total soccer field floras in five ecological regions in Costa Rica by mean similarity coefficients (Frenkel, 1972)

Ecological region	MCT	PW	TWD	TW	LMM
Premontane moist forest–Meseta central type (MCT)	0.73				
Premontane wet forest (PW)	0.48	0.49			
Tropical wet–dry forest (TWD)	0.42	0.36	0.39		
Tropical wet forest (TW)	0.38	0.41	0.30	0.39	
Lower montane and montane forest (LMM)	0.11	0.16	0.05	0.08	0.40

Table 4.11 Selected species percent frequencies representing total quadrat samples and three different treading intensity classes in soccer fields for five ecological regions in Costa Rica, (Frenkel, 1972)

Ecological region and species		Frequencies[a]			
		Total quadrat samples	Treading intensity class		
			A	B	C
Meseta central	No. of samples	115	42	49	24
Eragrostis tenuifolia[b]		83	81	82	83
Eleusine indica		65	78	63	46
Sporobolus poiretii		47	36	55	50
Paspalum notatum		41	14	49	58
Cynodon dactylon		27	26	29	25
Tropical wet–dry	No. of samples	93	45	33	15
Eleusine indica		63	76	58	47
Cynodon dactylon		55	64	61	13
Paspalum notatum		54	44	64	60
Sida rhombifolia		25	9	36	47
Desmodium scorpiurus[b]		24	13	24	53
Euphorbia spp.		16	22	15	0
Tropical wet	No. of samples	136	44	52	40
Cynodon dactylon		64	87	65	38
Paspalum conjugatum[b]		54	23	56	88
Axonopus compressus agg.		47	27	56	60
Eleusine indica		22	57	10	0
Premontane wet	No. of samples	144	40	53	21
Axonopus compressus agg.		62	58	66	62
Sporobolus poiretii		47	43	40	38
Eleusine indica		47	65	43	20
Cynodon dactylon		39	43	36	38
Paspalum notatum		33	18	36	52
Paspalum conjugatum		24	10	28	38
Lower montane and montane	No. of samples	38	6	15	17
Pennisetum clandestinum[b]		98	100	93	100
Trifolium repens[b]		53	0	53	71
Sporobolus poiretii		16	0	13	23

[a]Percentage frequencies are based on the number of 40×40 cm quadrat samples.
[b]Indicates those species which tend to be restricted to a given ecologic region.

Table 4.12 The Malvo–Polygonetum trampled association recorded by Oberdorfer (1960) in Central Chile

Relevé number	2	35	51	65	77	2a
Association character species						
Cynodon dactylon	+2	+	1.2	2.2	2.2	(+)
Malva parviflora	1.2	1.2	+2	+	(+)	●
Amaranthus deflexus	+	+	●	2.2	●	●
Eleusine oligostachya	●	+2	●	●	●	+2
E. indica	+	●	●	●	●	●
Polycarpon tetraphyllum	●	+	●	●	●	●
Sub-order–Class character species						
Polygonum aviculare	3.4	3.4	3.3	2.3	3.2	(+)
Poa annua	●	●	●	+	●	2.3
Modiola caroliniana	●	+2	●	●	●	●
Plantago major	●	●	●	●	●	+2
Coronopus didymus	●	●	●	●	●	+
Accompanying species						
Hordeum murinum	+	●	+2	(+)	+	●
Euphorbia serpens	(+)	●	●	●	●	1.2
Oxalis corniculata	●	●	●	+	●	1.2
Lolium perenne	●	●	(+)	●	●	●
Portulaca oleracea	●	●	●	(+)	+	●
Sagina apetala	●	●	●	(+)	●	●

+ Indicates present at less than 1% cover.

Malvern Hills, England, a spa based on the local mineral springs where 'a walk that was made there for the convenience of water drinkers, in less than a year, was covered in many places with suffolk grass (*Poa annua*) tho' I could not find one single plant of it besides in any part of the hill. This was owing, no doubt, to the frequent treading, which above all things makes this grass flourish' (Stillingfleet, 1759). This grass is commonly found on fertile, moist paths in temperate areas and I have even observed it in subtropical lawns in Brisbane in the winter months. It is one of that select group of plants that can be ranked as widespread 'increasers' on paths and tracks. Frenkel (1972) wrote 'modification of vegetation by treading may be viewed as a complex system of interacting factors in which the trampled habitat presents an extreme environment suitable for colonisation by only a limited number of taxa' (or species). This complex system has common features around the world. This is shown by the evidence gathered in this book, and it is therefore not surprising that similar species do occur on pathways in many widely separated places.

An analysis of the flora of heavily trampled, and presumably unseeded, football pitches in Costa Rica showed that there was a relatively high degree of local similarity (Frenkel, 1972) (Table 4.10), but that there was less similarity on the macro-scale except that two

Table 4.13 Taxonomic relationship among species in trampled vegetation in various phytogeographical region of the world (see Fig. 4.16) (from the author's records with additions from Frenkel, 1972)

		North and East Australian	Hawaiian	North American	Pacific	North American Southern (Subregion)	North American Northern (Subregion)	Andean
Region:		North and East Australian	Hawaiian	North American	Pacific	Southern (Subregion)	Northern (Subregion)	Andean
Number:		34	22	10	10	9	8	30
Location:		Brisbane area	Honolulu	Western Oregon	Montana	Illinois	Canada Great Lakes area	Central C
Habitat when from single site:			Roadside		Prairie edge	Prairie		
Genus:	Poa (Winter annual)	P. annua	P. pratensis	P. annua	P. pratensis	P. pratensis	P. alsodes P. pratensis	P. annua
	Sagina	–	–	S. procumbens	–	–	–	S. apetela
	Spergularia	–	–	S. rubra	–	–	–	–
	Capsella	–	–	C. rubella	–	–	–	–
	Coronopus	–	–	–	–	–	–	C. didym
	Lepidium	–	–	L. draba	–	–	–	–
	Plantago	–	P. major	P. crassifolia	–	P. major	P. major	P. major
	Lolium	–	–	L. rigidum	–	–	–	L. perenn
	Polygonum	–	–	P. aviculare	–	P. aviculare P. hydropiper	–	P. avicula
	Matricaria	–	–	M. matricarioides	–	–	–	–
	Trifolium	–	–	–	–	T. repens	T. repens	–
	Taraxacum	–	–	–	T. officinale	T. officinale	T. officinale	–
	Eleusine	E. indica	E. indica	–	–	–	–	E. indica E. oligos
	Cynodon	C. dactylon	–	–	–	–	–	C. dactyl
	Digitaria	D. didactyla	D. adscendens	–	–	–	–	–
	Juncus	–	–	–	–	–	J. bufonius	–
		–	–	–	–	–	J. spp.	–

species, Bermuda grass or green couch (*Cynodon dactylon*) and goose grass (*Eleusine indica*), occurred on all except the montane and lower montane examples in Frenkel's survey (Table 4.11).

Bermuda grass and goose grass also occur in the trampling association Malvo–Polygonetum which is a member of the trampling class Plantaginetea maioris recorded in Central Chile (Oberdorfer, 1960) (Table 4.12). Thus there are regional trampling floras which occur within restricted areas; in this case part of the Caribbean region (Frenkel, 1972), generally containing some elements that occur throughout the Neotropical phytogeographical kingdom (biogeographers divide the world into phytogeographical kingdoms, which are subdivided into regions; Fig. 4.16). Some species or taxa occur on paths in more than

Patagonian 38 Southern Chile	European (Subregion) 2 North-west maritime Lowlands	European (Subregion) 2 North-west maritime Mountains	European (Subregion) 2 Czechoslovakia	European (Subregion) 2 Southern Sweden	European (Subregion) 2 Russia nr Moscow Pasture and birchwood	Mediterranean 6 Lowlands	Sino-Japanese 4 Japanese islands
P. annua	*P. annua* *P. pratensis* *P. trivalis*	*P. supina*	*P. annua* *P. arenastrum* *P. trivalis* *P. pratensis*	*P. annua* *P. pratensis*	*P. annua* *P. pratensis*	*P. infermis*	*P. annua* *P. pratensis*
–	*S. procumbens* *S. maritima*	*S. saginoides* –	– –	*S.* sp.	– –	*S. apetela* –	*S. japonica*
S. rubra	*S. rupicola* *S. marina*	–	–	–	–	*S. rubra*	–
–	*C. bursa-pastoris*	–	–	–	–	*C. rubella*	
C. didymus	*C. didymus*	–	–	–	–	–	–
L. spicatum	*L. ruderale*	*L. densiflorum*	–	–	–	*L. draba*	*L. virginicum*
P. major *P. hirtella*	*P. major* *P. coronopus* *P. lanceolata*	*P. alpina*	*P. media* *P. major*	*P. major*	*P. major*	*P. crassifolia*	*P. asiatica*
L. perenne	*L. perenne*	–	*L. perenne*	–	–	*L. rigidum*	–
P. aviculare	*P. aviculare* *P. heterophyllum*	–	*P. aviculare*	*P. aviculare*	*P. aviculare*	*P. aviculare*	*P. aviculare* *P. aequale*
M. matricarioides	*M. matricarioides*	–	*M. matricarioides*	–	–	*M. matricarioides*	*M. matricarioides*
–	*T. repens*	–	*T. repens*	*T. repens*	*T. repens*	–	*T. repens*
–	*T. leavigatum* *T. officinale*	–	*T. officinale*	*T. officinale*	–	–	*T. officinale*
–	–	–	–	–	–	–	*E. indica* *E. oligostachya*
C. dactylon	–	–	*C. dactylon*	–	–	–	*C. dactylon*
–	–	–	–	–	–	–	*D. violascens*
J. imbricatus –	*J. bufonius*	*J. squarrosus*	–	–	*J.* spp.	–	*J. tenuis*

one kingdom around the world, for example *Plantago* species grow in such relatively far apart and isolated places as Australia, Hawaii and Europe, the first two of which have a strongly endemic flora. Some of the truly cosmopolitan path taxa, such as the genus *Plantago*, are listed with some of their reported occurrences in Table 4.13. In much of the world, they owe their presence to activities of humans, and a world-wide floristic convergence on pathways in the tropical zones and within the temperate zones is evident. Frenkel (1972) also suggested that there is a 'close floristic communality between trampled vegetation throughout the tropics'. However, he added that 'tropical trampled vegetation bears but slight floristic similarity to temperate trampled vegetation and that within a given region, few [native] species prevail'.

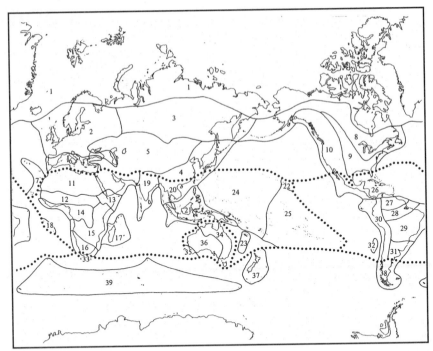

Fig. 4.16 Phytogeographical regions of the world (Lincoln, Boxshall and Clark, 1982).

4.6 Summary

Various approaches, from totally subjective to as objective as possible, have all shown that plant species vary in their resistance to trampling. Certain species are recognized as trampling resistant in well-studied areas such as northern Europe and the USA. In general, diversity has been shown to increase with slight wear and to decrease as the intensity of use rises. The evidence of change fits both Connell's (1979) intermediate disturbance hypothesis and Grime's (1973) prediction based on plant strategies.

Various indices have been utilized to study the effect of trampling, as well as the concept of increaser, decreaser and invader species. There is considerable evidence that some species of the path flora are readily transported by walkers, vehicles and animals. That this transport is on a world scale is demonstrated by the fact that a common path flora exists in nearly all temperate regions, and a widespread tropical path flora has also been recognized.

Plant populations and mechanical wear 5

An approach to the study of plants in their environment that has gained strong momentum is the mapping and counting of individuals, and in this way following changes that take place in plant populations. The work of Vollmer *et al.* (1976) in the Mojave Desert is an example of counting, in which annual plants were counted in 50×50 cm quadrants in the ruts and on the central area between the ruts of tracks made by four-wheel-drive vehicles. They also counted the number of annual plants in an area which had been driven over in random directions, and in undamaged control areas. They found that the numbers of annual plants were only significantly reduced in the ruts compared with all the other situations, although numbers were reduced in all of the driven sites (Fig. 5.1). The same authors also counted the numbers of shrubs in a total of 25 m² that had received different degrees of damage, on the track and in the randomly driven areas (Fig. 5.2). There were more slightly damaged shrubs (0–20% damage) in the randomly driven area and more heavily damaged shrubs (81–100% damage) on the track. About 30% of the shrubs on the track suffered 100% damage, with all the above-ground parts broken off.

However, the study of populations involves more than just recording the density of individuals by counting the numbers per unit area. First, the processes that lead to a given density of plants need to be understood; these are birth, death, immigration and emigration (Fig. 5.3). A particular area is selected in which all the individuals are recorded; this may be a hectare of forest or as little as one decimetre (10 $\times$ 10 cm) of grassland. Births are usually recorded on the appearance of seedlings, although if grass tillers or some other vegetative unit is the subject of study, then the first appearance of the tiller will be recorded as the time of birth. Death is clearly the time at which the organism appears to be dead. This may present some problems, especially if the plants are perennials which spend some time as underground organs such as bulbs, corms, tubers, rhizomes or parts of roots. Immigration may occur as seedlings are moved over the surface of the ground by rain (Liddle, Budd and Hutchings, 1982) or, more often,

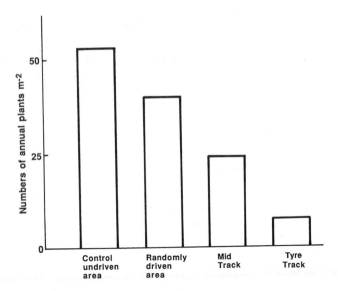

Fig. 5.1 The number of annual plants surviving after various levels of use of a four-wheel-drive vehicle in the Mojave Desert (data from Vollmer *et al.*, 1976).

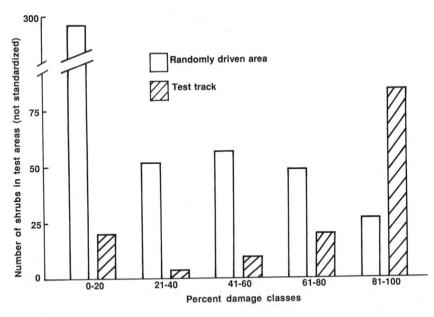

Fig. 5.2 The number of shrubs surviving in the randomly driven area and the test track of an experiment in the Mojave Desert (data from Vollmer *et al.*, 1976).

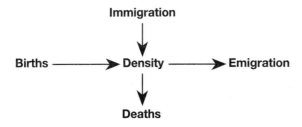

Fig. 5.3 The processes affecting the density of plant or animal populations.

as vegetative portions are moved or grow into the plot or as seeds are carried in from plants outside the plot (Liddle and Elgar, 1984). Emigration occurs in a similar manner, except that the movement is in the opposite direction. Other features, such as the morphology of the plants, the number of vegetative daughter plants that are produced and the number of seeds produced, as well as many environmental variables, such as disturbance or trampling, are commonly studied in search of a greater understanding of population processes.

There have been many studies of the effect of trampling on the number of plants growing in defined areas (Blom, 1977; Kendal, 1982; Thyer, 1982) but I have been unable to locate any longer-term records resulting in the production of life tables showing how trampling changes the population processes, although Fetcher and Shaver (1983) have constructed life tables for cotton grass (*Eriophorum vaginatum*) on the basis of historical records. In a study carried out on turfs transplanted to a glasshouse, Blom (1977) found that the survival of ribwort plantain (*Plantago lanceolata*), buck's-horn plantain (*P. coronopus*) and hoary plantain (*P. media*) seedlings was reduced by simulated heavy trampling on compacted soil when compared to the effect of light trampling on loose soil, while survival was increased for great plantain (*P. major*) (Fig. 5.4). In contrast, the number of dead seedlings was higher in the compacted, heavily trampled soils for all species. Notice here that as new seedlings appeared they were counted into the population, so it is possible for both the percentage survivorship and the number of dead seedlings to rise together. The emergence of seedlings was stimulated where seeds of the four species were sown in the field on to paths trampled by people (Fig. 5.5a), although their percentage death was also increased (Fig. 5.5b). There were more plants on the trampled paths, particularly of great plantain, than any of the other sites examined (Blom, 1977). It is also evident that great

5.1 Plant numbers

plantain and hoary plantain did not germinate well on the untrampled sites. So there are clear differences between species in their ability to germinate and survive in trampled situations, with great plantain being the superior species. We would expect this result on the basis of the other information on these species already discussed in Chapter 4. Blom commented on the importance of separating the indirect effects of soil compaction and soil moisture content from the direct effect of trampling.

Looking now at the survival of mature plants, it is interesting to find that the distribution of trampling treatments in time has a major effect at low intensities of trampling. Kendal (1982) showed that narrow-leaved species were much more vulnerable when trampling was spread over 16 weeks than when the same total intensity (192 or 768

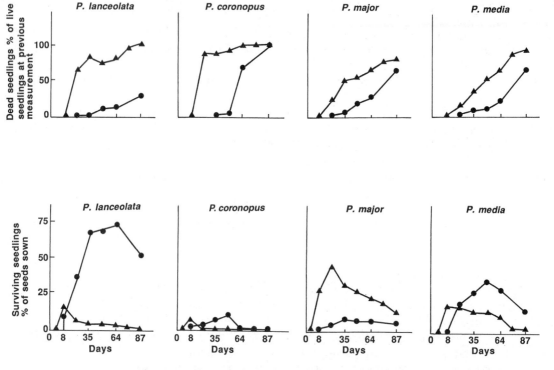

Fig. 5.4 The establishment of *Plantago* species sown in artificially trampled dunes. Top row, the percentage of dead individuals calculated with reference to the number of seedlings which emerged in the preceding period. Bottom row, the percentage of living plants in relation to the numbers of seeds sown. ●, Turves from loose, untrampled soil given light trampling treatments; ▲, compacted turves from the middle of a path given heavy trampling treatments. (Data from Blom, 1977.)

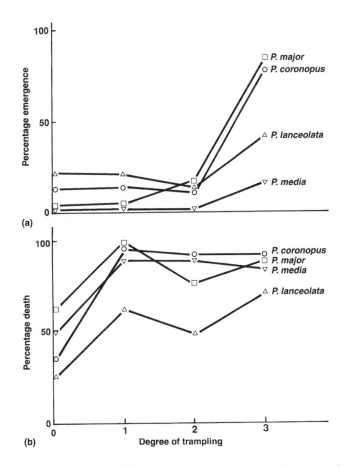

Fig. 5.5   (a) The percentage of four *Plantago* species seedlings which emerged and (b) their percentage death when the seeds were sown on to regularly used paths in sand dunes at Oostvorne, The Netherlands (data from Blom, 1977).

tramples) was spread over 24 weeks or compressed to 8 weeks (Fig. 5.6). Broad-leaved species did not show this difference to the same extent. (The different durations of the experiments necessitated different control measurements for each treatment and the results are expressed as numbers relative to the numbers in the control.)

The effect of recreation on populations of narrow-leaved species seems to have received more attention from ecologists, partly because of the general interest in orchids and the fact that they are susceptible to wild plant 'poaching' (picking when prohibited by law). Bratton (1985) found that the number of shoots of showy orchis (*Orchis spectabilis*) actually decreased with increasing distance from the edges

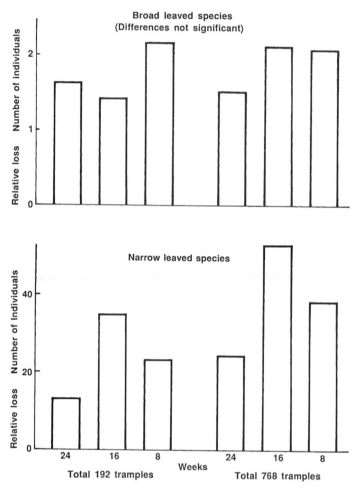

Fig. 5.6 The relative loss of broad-leaved (dicotyledonous) and narrow-leaved (monocotyledonous) species from experimental pathways trampled at two intensities over 24, 16 and 8 weeks. The site was a dense blady grass (*Imperata cylindrica*) ground cover in a eucalyptus woodland, Brisbane, Australia. (Data from Kendal, 1982.)

of trails in the Great Smoky Mountain National Park in eastern USA (Fig. 5.7). However, this effect was not related to the intensity of trail use and the increase in numbers adjacent to trails did not occur with the ladies slipper orchid (*Cypropedium calceolus*) growing in the same area. There were very few blooms of ladies slipper orchid at the trail edge. Again, we see a marked difference in population responses between closely related species, which indicates the difficulty of making general statements in this kind of work.

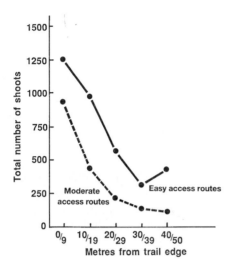

Fig. 5.7 The total number of showy orchis (*Orchis spectabilis*) shoots found at increasing distances from the road or trail edge, whichever is closer, along easy and moderate access routes in the Great Smoky Mountain National Park (after Bratton, 1985).

The European bluebell (*Hyacinthoides non-scriptus*) is particularly robust, and picking the flowering stem just above ground level or pulling the whole of the stem from the bulb so that it broke off underground, each year for a period of 6 years had little effect on flower numbers (Peace and Gilmour, 1949). However, when the leaves and inflorescences were picked off together just above ground level there was a marked and significant decline to about 50% of the original number of plants. These experiments indicate that in this case, visitor education could mean the difference between a well-managed and a destroyed resource. This result can be readily understood as 'the prevention of fruit formation would, if anything, tend to increase the vigour of the bulbs, while damage to the leaf would adversely affect the production of food by photosynthesis and its storage in the bulb' (Peace and Gilmour, 1949).

The historical reconstruction of the life history of cotton grass (*Eriophorum vaginatum*) showed differences between plants growing in undisturbed tundra, plants growing in an area scraped by a bulldozer 8 years before the measurements were made, and those growing in a 15-year-old vehicle track subject to occasional use at Eagle Creek, Alaska (65°26'N) (Fetcher and Shaver, 1983). Visually, tussocks of cotton grass were smaller in the disturbed area (11–22 cm diameter compared to 15–33 cm in the undisturbed tundra), but the number of tillers per tussock was similar in all three sites. The analysis of age distribution of the tillers and their vegetative reproduction was made by separating the tillers of collected tussocks in the laboratory, noting their vegetative connections and dating them on the basis of their number of leaves and the observed leaf production rate. Tillers in undisturbed areas produced a mean of 2.52 leaves yr^{-1}, and in the scraped site and the track site, 3.04 and 3.02 leaves yr^{-1}, respectively. The tillers growing on plants in the track and scraped sites had less than half of the maximum life span of the tillers on plants in the undisturbed area (Fig. 5.8).

It is interesting that, in this environment, changes that were created 8 years previously had nearly as much effect as continuous intermittent disturbance, probably because arctic and alpine habitats have short growing seasons, and are notoriously slow to recover from disruption. There were differences between the disturbed sites in the net reproduction rate (the average number of daughter tillers that lived to average age), with tillers in undisturbed tundra producing a mean of 1.74 daughter tillers, while in the scraped site they produced an average of 1.34 daughter tillers and those on the track produced only 0.53 daughter tillers. Daughter tillers were also produced at an earlier age in the disturbed areas, with peak production in the third year, while peak production was in the fourth year and continued to the tenth year in the undisturbed area (Figure 5.9). This change to higher mortality and earlier vegetative reproduction in disturbed situations is suggested by the authors to be a response to an increase in the quantity of nutrients that are available in disturbed tundra. A similar conclusion was reached by Lamont *et al.* (1994) to explain the increased

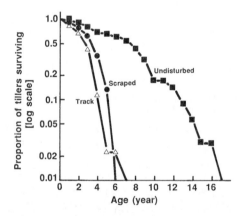

Fig. 5.8 Survival curves for vegetative tillers of cotton grass (*Eriophorum vaginatum*) on areas of Alaskan tundra, some of which had been disturbed 8 years before measurements were made (after Fetcher and Shaver, 1983).

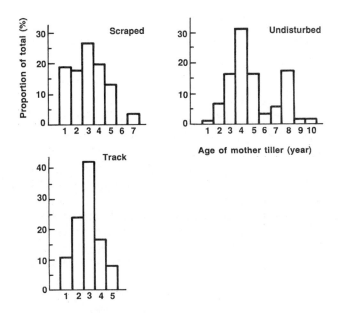

Fig. 5.9 Distribution of daughter tillers of cotton grass (*Eriophorum vaginatum*) by age of mother tiller on areas of Alaskan tundra, some of which had been disturbed 8 years previously. The older tillers show less or no reproduction in the scraped and track areas. (After Fetcher and Shaver, 1983.)

success of roadside plants of *Banksia hookeriana* compared to those growing some distance from the road. The roadside plants were 2.3 times larger, produced 2.5 times more flower heads and had a seed store 4.7 times larger than their non-edge counterparts. In this case the roadside plants had greater access to water and mineral nutrients because of the runoff from the road.

There are many examples of changed morphologies, and some of changed reproductive characteristics, of plants growing in trampled areas. These include, among others, such features as lower height, shorter life span and younger age at reproduction. But the more involved question of whether these are plastic responses that do not remain in the plant when it is no longer trampled, or whether there has been a genetic shift in the population so that the size or other feature continues to be expressed as a definite ecotype, has only occasionally been answered.

5.2 Evolution in response to wear and cutting

Differences were shown to exist between members of three populations of dandelions (*Taraxacum officinale*) growing in:

- a dry area which was mowed and trampled;
- a similar area which was occasionally mowed and less trampled; and
- a wet area which was mowed only once a year at 20 cm height and only infrequently visited by people (Gadgil and Solbrig, 1972).

The whole area had been heavily grazed until 10 years before the observations were made, when the present regime commenced. Four biotypes were identified within these populations on the basis of isozyme patterns (the identification of the enzyme contents of plant cells which vary with genetic variation) and plant morphology. When grown on their own in loam soil with unlimited water, types A and D (Fig. 5.10) had 8.1 and 5.6 leaves, respectively, but this difference was reduced to 2.6 and 2.5 leaves in sandy soil with limited water. The leaf lengths were respectively 13.5 cm and 17.2 cm in the well-watered loam soil, and 12.9 cm and 13.4 cm in the drier sand soil. These two biotypes were in different proportions in the three areas. Biotype A made up 73% and 53% of the population in the trampled areas 1 and 2, but only 17% of the population in the least-trampled area, 3. Biotype D did not occur at all in area 1 and only at 7% of the population at site 2, but it made up 64% of the population in the least disturbed area, 3. The remainder of the population was made up of the intermediate biotypes.

The two biotypes A and D also differed in their reproductive strategies, with type A (from the trampled area) producing an average of

Fig. 5.10 Leaf shapes of two biotypes of dandelions (*Taraxacum officinale*) grown without competition under identical water, light, temperature and soil conditions. A occurred mainly on a dry, trampled and mown area, while D was almost confined to a moist, slightly trampled and only occasionally mown area in Michigan, USA. (From Gadgil and Solbrig, 1972.)

between 2.6 and 3.8 flowers per plant, and type D (from the undisturbed area) producing only 1.2 flowers per plant. Thus the type A can be classed as more of an 'r' selected biotype with a high innate capacity for increase and smaller size (leaf length), while biotype D was more of a 'K' selected biotype with lower reproductive capacity and larger size. However, the difference in leaf numbers does not entirely fit in to this categorization, although the observed strategies are in line with the general idea that 'r' selected species are characteristic of variable or unpredictable environments while 'K' selected species are characteristic of relatively constant or predictable environments. These strategic ideas only hold as long as trampling and mowing are considered to be variable or unpredictable environmental factors.

An extensive study of genetic changes in the morphology of plants growing in lawns and adjacent areas was made by Warwick and Briggs (1978a,b, 1979, 1980a,b,c,d). They followed the same basic technique as that used by Gadgil and Solbrig (1972). In this case the plants used were annual meadow grass (*Poa annua*), great plantain (*Plantago major*), daisy (*Bellis perennis*), ribwort plantain (*Plantago lanceolata*), yarrow (or milfoil) (*Achillea millefolium*) and self-heal (*Prunella vulgaris*). They used four types of habitat:

- lawns, some of which were very lightly trampled (D. Briggs, personal communication);
- grazed areas;
- grasslands seasonally reduced in height by cutting and/or grazing; and
- grasslands which were not cut or grazed.

In general all lawn populations were prostrate in their growth form (Fig. 5.11), whereas plants that were growing in seasonally cut or uncut habitats were erect and considerably taller. The latter were often flowering more profusely. The response of the species when grown under controlled conditions was variable.

Annual meadowgrass (*Poa annua*), which is a common component of lawn and path flora in temperate zones, and in winter in subtropical areas, was collected from two bowling greens and adjacent flower beds, and graded for vegetative and floral erectness according to the criteria set out in Table 5.1 (Warwick and Briggs, 1978b). When grown in controlled conditions, plants from the two bowling greens, almost exclusively, continued to grow in the prostrate form, while two-thirds of those from flower beds were erect (Fig. 5.12). Grains derived from open pollination from some of these plants were sown, the grass was grown to maturity and again scored for vegetative and floral erectness, as well as the number of days to the opening of the anthers (anthesis) (Table 5.2). From this work and enzyme (isozyme) analysis

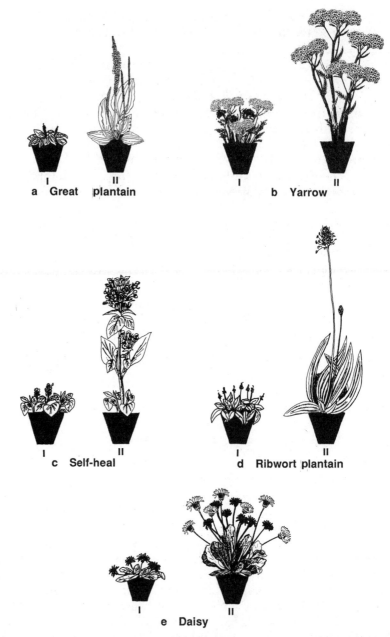

Fig. 5.11 Examples of plants collected from mown lawns (I) and the same species from unmown situations (II), all grown without trampling and cutting in standard conditions. (a) Great plantain (*Plantago major*); (b) yarrow/milfoil (*Achillea millefolium*); (c) self-heal (*Prunella vulgaris*); (d) ribwort plantain (*Plantago lanceolata*); (e) daisy (*Bellis perennis*). (After Warwick and Briggs, 1979.)

the authors concluded that 'growth form and flowering behaviour are genetically determined'. In other words, the variation between lawn and flower-bed populations of annual meadow grass is because they are different ecotypes, and the prostrate habit 'might be at a selective advantage' in the closely mown bowling green habitat (Warwick and Briggs, 1978b). It is also worth noting that there were four prostrate plants from the bowling greens that grew into erect individuals when they were no longer mowed, suggesting that two strategies are involved – genetic adaptation supplemented by a few plants with a high degree of phenotypic plasticity. It is not unreasonable to suggest that similar kinds of adaptation may occur when annual meadow grass (*Poa annua*) is growing on pathways subject to trampling.

Differential genotypic selection of annual meadow grass was also found to have occurred in different parts of golf courses in California (Wu, Till-Bottraud and Torres, 1987). The summer climate is too dry for the survival of this grass without additional water. The germination percentage at 25°C was compared between plants growing on greens which were watered throughout the year, on fairways and in the unwatered rough. The seeds from green populations germinated quickly at 25°C (over 80% in 12 days), while the fairway and particularly the rough populations were slow to respond and only showed

Table 5.1 Growth forms of annual meadow grass (*Poa annua*), as defined by vegetative and floral grades of erectness (Warwick and Briggs, 1978a)

			Vegetative erectness grade scale	Floral erectness grade scale
Growth forms	(a)	Prostrate	1. Prostrate, mat-like (2–3 cm in height)	1. Short-horizontal; mat-like (2–3 cm)
	(b)	Erect	2. Intermediate (7–8 cm)	2. Long-horizontal (4–10 cm)
			3. Erect (9–12 cm)	3. Intermediate (c. 45° angle, 4–10 cm)
				4. Erect (70–90° angle, 10–20 cm)

Table 5.2 Graded growth forms[a] of annual meadow grass (*Poa annua*) for vegetative and floral erectness and days to anthesis (from Warwick and Briggs, 1978a)

Parental growth form	Vegetative erectness	Floral erectness	Days to anthesis
Erect (275 plants)	2.99	3.98	55 ± 2
Prostrate (200 plants)	1.21	2.15	80 ± 3

[a] The scores were based on the number of offspring that conformed to the particular grade of erectness (see Table 5.1); e.g. 2.99 indicates that 274 plants out of a total of 275 fell into class 3 (Erect, 9–12 cm) and 1.21 indicates that 160 plants out of 200 fell into class 1 (Prostrate, 2–3 cm).

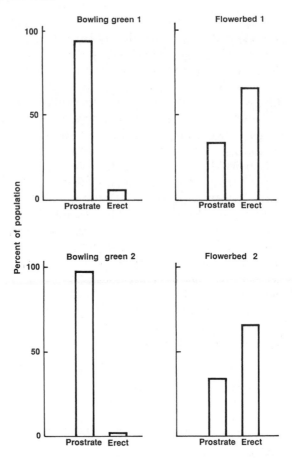

Fig. 5.12 The percentage of prostrate and erect growth forms of annual meadow grass (*Poa annua*) in transplants, from mowed and trampled bowling greens or uncut and untrampled flower beds, to uniform conditions (data from Warwick and Briggs, 1978a).

49% and 5% germination, respectively (Fig. 5.13a). That the dormancy of the latter populations was temperature enforced was demonstrated by the much higher germination rates achieved at 12°C (Fig. 5.13b). Thus the rough populations would mainly germinate in the cooler winter season when the rainfall was higher (up to 400 mm month^{-1}) compared to the drier summers (0.20 mm month^{-1}).

Plants of great plantain (*Plantago major*) also maintained their smaller prostrate growth forms in those individuals collected from lawns, and to a lesser extent from pathways, when grown in cultivation (Warwick and Briggs, 1979) (Figs 5.14 and 5.15). In the great plantain there was very little variation in form when the selection pressures of

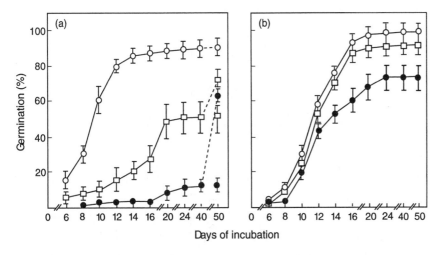

Fig. 5.13 Rate of seed germination of the golf course *Poa annua* populations from three areas of different irrigation and mowing practices. ●, Rough population; □, fairway population; ○, golf green population. Disconnected symbols indicate values of percentage seed germination recorded only after 50 days' incubation from duplicate seed samples. Broken lines indicate transfer of seeds from 25°C (a) to 12°C (b). (From Wu, Till-Bottraud and Torres, 1987.)

trampling and mowing were removed by transferring the plants to uniform conditions. This indicates that the populations were genetically adapted to the trampling conditions and were able to survive because of their genetically fixed small size and prostrate inflorescences rather than because of phenotypic plasticity allowing this form to occur. This species appears to be the only example investigated so far of a path plant which has little phenotypic plasticity within its path populations. Its worldwide distribution on paths, on suitable soil types within suitable climatic zones, testifies to the success of this strategy.

The other species investigated, daisy (*Bellis perennis*), ribwort plantain (*Plantago lanceolata*), yarrow (*Achillea millefolium*) and self-heal (*Prunella vulgaris*) all had dwarf and/or prostrate forms when growing in lawns or heavily trampled and grazed habitats (Fig. 5.16) (Warwick and Briggs, 1979). When these were transplanted, the overall pattern of distinct phenotypic differences associated with the different types of habitat was not maintained in cultivation. Meerts and Vekemans (1991) also found that small size in trampled annual knotweed was due to phenotypic plasticity. There were, however, some genetically fixed dwarf, prostrate plants in samples collected by Warwick and Briggs (1979) from many of the lawn and heavily grazed populations. Like the work on annual meadow grass discussed above, this suggests

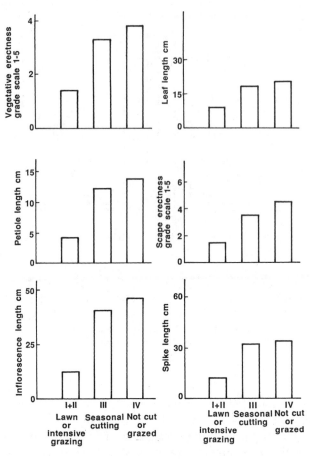

Fig. 5.14 Cultivation trial of great plantain (*Plantago major*) from different habitats. I, lawns, especially old, established turf in the courts and gardens of several Cambridge (England) colleges; II, intensively grazed grasslands; III, grasslands seasonally reduced in height by cutting; and IV, grassy areas not seasonally reduced in height by grazing and/or mowing. All parameters shown were significantly different (0.1%) between habitat types, and the plants from trampled and mown situations were generally smaller and more prostrate. (Data from Warwick and Briggs, 1979.)

the importance of both genotypic differentiation and phenotypic plasticity as adaptive strategies in short-turf habitats (Warwick and Briggs, 1979). It has been pointed out that factors other than trampling and cutting, such as SO_2 and copper pollution, may have been significant factors in these studies (D. Briggs, personal communication). The studies of Law (1975) (see Law, Bradshaw and Putwain, 1977) throw

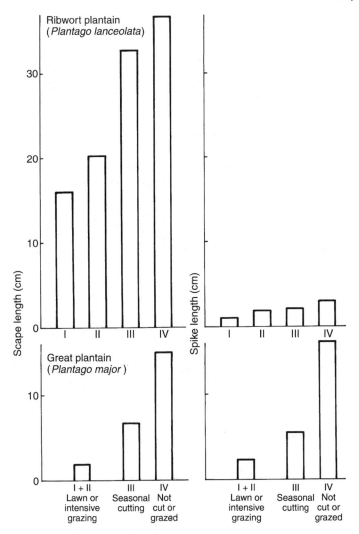

Fig. 5.15 Scape (flowering stem) and spike (inflorescence) length of ribwort plantain (*Plantago lanceolata*) and great plantain (*Plantago major*) in field populations (from Warwick and Briggs, 1979).

some light on the genetics of *Poa annua* but he was primarily concerned with the effects of density rather than trampling.

In addition there is considerable literature on the response of turf grass species and cultivars in cultivated situations which would require a separate treatise to do full justice to that body of information. Journals carrying papers on this research are: the *Journal of the Sports*

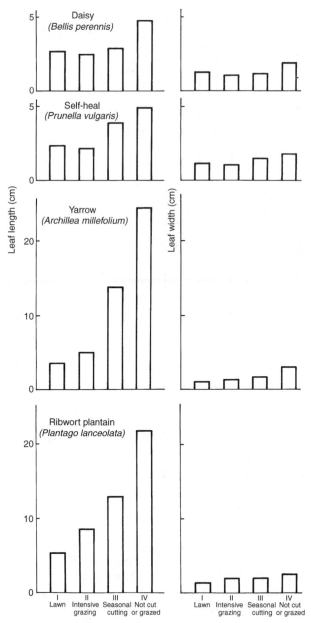

Fig. 5.16 Leaf length and width of four species of plant growing in four different habitats. I, lawns, especially old, established turf in the courts and gardens of several Cambridge (England) colleges; II, intensively grazed grasslands; III, grasslands seasonally reduced in height by cutting; and IV, grassy areas not seasonally reduced in height by grazing and/or mowing. In all species there were significant differences (P = 5% or less) between at least two of the habitat means. (Data from Warwick and Briggs, 1979.)

Turf Research Institute; *Rasen-Turf-Gazon*; *Zietschrift für Vegetations-technik*; and the *Proceedings of the International Turf Grass Research Conferences*.

Trampling has been suggested to differentiate selectively between cyanogenic (the capacity to produce hydrogen cyanide in the leaves when damaged) and acyanogenic forms of bird's-foot trefoil (*Lotus corniculatus*) which are distinct genotypes (Keymer and Ellis, 1978). These authors suggest that trampling may act by causing the release of hydrogen cyanide from the cyanogenic plants as a result of physical damage to the leaves. This will reduce the vitality of the plants so effected.

In conclusion, it can be said that there appear to be two genetic strategies for survival of plants in trampled situations, and that they both allow the production of small, often prostrate individuals. One is to have small, prostrate forms which are genetically fixed and the other is to have sufficient phenotypic plasticity so that the adult plant can exist and flower as a small, reduced individual. Populations may consist of only small genotypes, but given the small size of the environmental mosaic in which intense trampling often occurs (e.g. narrow paths or bowling greens) it is necessary for the plants to be self-fertilizing or cross only with other members of its own genotypic form to maintain the genetic integrity of the population. These conditions appear to be fulfilled only by the great plantain (*Plantago major*). A more usual situation is where the population is a mixture of adapted genotypes and phenotypically small individuals. But it is worth noting that the small genotypes would not appear with such regularity if there were no selective advantage to be gained from this adaptation.

5.3 Survival value of small size?

At first sight it seems self-evident that smaller plants will survive better than larger ones in trampled areas. However, there are other features that contribute to survival, such as flexibility and tough leaves, so it is pertinent to ask the question 'Is small size a real advantage?'

Warwick and Briggs (1980a,b,c,d) investigated this question with three of the species used in their 'genetic work', by means of transplant experiments where roadside individuals were transplanted to lawns, and by subjecting the different forms of annual meadow grass and great plantain to experimental trampling.

Daisy plants were transplanted from trampled and untrampled habitats into a closely mown lawn. Most plants survived and approximately 60% produced fruiting capitula. The only indication of previous selection for small size giving an advantage was that three plants that were previously growing in a lawn and one that was growing in a grazed meadow produced approximately 50% of the capitula which fruited (Warwick and Briggs, 1980a). Similar experiments with yarrow had essentially similar results.

Great plantain was subjected to a reciprocal transplant experiment between tall roadside vegetation and a mown lawn. The major feature to emerge was that the lawn individuals produced more seeds than did the roadside individuals, although some of the tall scapes of the latter group did escape due to their flexibility (Warwick and Briggs, 1980b). Coefficients of selection (Table 5.3) were calculated as follows for the roadside plants in mown plots and for lawn plants in tall grass plots. In mown plots:

$$1 - \frac{\text{dry weight (or numbers) of roadside plants}}{\text{dry weight (or numbers) of lawn plants}},$$

this gives an estimate of the force of selection acting against roadside plants; and in tall grass plots:

$$1 - \frac{\text{dry weight (or numbers) of lawn plants}}{\text{dry weight (or numbers) of roadside plants}},$$

this gives an estimate of the force of selection acting against lawn plants. The differential fitness did not act against genetically tall roadside plots grown on a lawn with respect to survival or vegetative growth, but reproduction was strongly influenced (Table 5.3). However, except in terms of survival and seed set, selection acted strongly against lawn plants in tall grassland. This experiment indicated that the small lawn plants do have a reproductive advantage in the lawn situation. When cutting at 2 cm above the soil was carried out in another experiment the roadside plants were at a much greater disadvantage (Warwick and Briggs, 1980c). This suggests that although the smaller plants are at an advantage, the general features such as flexibility of leaves and scapes of the great plantain also confer an advantage in trampled and mown situations. The authors also commented that, in contrast to annual meadow grass, the lawn variants of great plantain appear to be behaving as 'r' strategists while erect plants are apparently 'K' strategists (Warwick and Briggs, 1980b).

In an artificial trampling experiment on great plantain and annual meadow grass (Warwick and Briggs, 1980d) also showed that these two important path species apparently had different reactions to trampling. Under heavy trampling treatment the erect forms of great plantain suffered more damage than the prostrate forms, particularly with respect to the reproductive structures (Table 5.4). In contrast, both forms of annual meadow grass showed rather similar reductions under the heavy trampling treatment (Table 5.5). The changes in resource allocation patterns under heavy trampling are also different between the two species. A greater proportion of the resources go into shoot growth in both species, under all treatments (Table 5.6), but only in annual meadow grass does shoot growth maintain a reasonably high level (39%) (see also section 7.7).

Table 5.3 Calculated forces of selction acting against roadside and lawn plants (from Warwick and Briggs, 1980b)

Character	Force of selection acting against roadside plants in mown plots	Force of selection acting against lawn plants in tall grass plots
Survival	0.06	0.00
Vegetative dry weight	0.00	0.76
Number of individuals setting seed	0.82	0.00
Spike dry weight	0.88	0.74
Reproductive dry weight	0.66	0.77
Total dry weight	0.05	0.76

Table 5.4 Experiment simulating trampling stress on great plantain (*Plantago major*) (from Warwick and Briggs, 1980d)

Character	Growth form	Trampling stress Control	Light	Heavy	Percentage of control[a]	Source of variation: Between growth forms / Between treatments / Growth form × treatments
1. Total dry wt per plant (g)	Erect	11.27 >	3.39 >	0.94	(8.3)	**
	Prostrate	7.65 >	2.46 >	1.09	(14.2)	***
		***	ns	ns		***
2. Shoot dry wt per plant (g)	Erect	7.21 >	2.50 >	0.71	(9.8)	***
	Prostrate	4.23 >	1.53 >	0.72	(17.0)	***
		***	**	ns		***
3. Root dry wt per plant (g)	Erect	1.14 >	0.38 >	0.18	(15.8)	**
	Prostrate	0.80 >	0.25 >	0.17	(21.3)	***
		**	ns	ns		***
4. Spike dry wt per plant (g)	Erect	2.02 >	0.30 >	0.30	(1.5)	ns
	Prostrate	2.04 >	0.46 >	0.14	(6.9)	***
		ns	ns	ns		ns
5. Reproductive dry wt per plant (g) (spike and scape)	Erect	2.92 >	0.51 >	0.05	(1.7)	ns
	Prostrate	2.62 >	0.68 >	0.20	(7.6)	***
		*	ns	ns		***
6. Number of individuals setting seed (out of 40)	Erect	40		34	7	
	Prostrate	40		40	35	

[a] Growth form means and form values obtained under heavy trampling stress, percentage of control dry weights are indicated in brackets.
Results of analyses of variance are included, except for character (6) and the statistical results of between growth form comparisons are given for each treatment.
Treatment means followed by > are significantly greater than the following means; ns, not significantly different; *, $P < 0.05$; **, $P < 0.01$; ***, $P < 0.001$.

Table 5.5 Experiment simulating trampling stress on annual meadow grass (*Poa annua*) (from Warwick and Briggs, 1980d)

Character	Growth form	Control		Light		Heavy	Percentage of control[a]	Source of variation Between growth forms Between treatments Growth form × treatments
1. Total dry wt per plant (g)	Erect	5.28	>	2.92	>	1.65	(31.3)	ns
	Prostrate	5.90	>	2.90	>	1.76	(29.8)	***
		*		ns		ns		***
2. Shoot dry wt per plant (g)	Erect	3.13	>	2.08	>	1.22	(39.0)	***
	Prostrate	3.67	>	2.24	>	1.42	(38.7)	***
		*		ns		ns		*
3. Root dry wt per plant (g)	Erect	1.47	>	0.63	>	0.38	(25.9)	ns
	Prostrate	2.04	>	0.57	>	0.31	(15.2)	***
		*		ns		ns		***
4. Reproductive dry wt per plant (g)	Erect	0.68	>	0.21	>	0.05	(7.4)	Paired *t*-tests due to
	Prostrate	0.19	>	0.09	>	0.03	(15.8)	unequal sample sizes
		***		***		ns		
5. Number of individuals flowering (out of 60)	Erect	53				49	38	
	Prostrate	29				22	19	

[a] Growth form means and form values obtained under heavy trampling stress, percentage of control dry weight indicated in brackets.
Results of analyses of variance are included for characters 1, 2 and 3.
Statistical results of between growth form comparisons are given for each treatment.
Treatment means followed by > are significantly greater than the following means; ns, not significantly different; *, $P < 0.05$; **, $P < 0.01$; ***, $P < 0.001$.

In general, small prostrate plants of great plantain are less damaged by simulated trampling than erect individuals, but there appears to be no evidence that small prostrate growth forms of annual meadow grass have any advantage in these conditions. In both species reproductive features are more affected under heavy trampling, with relatively more energy being diverted into vegetative structures and less into flowering. This is in contrast to the distribution of biomass by annual knotweed (*Polygonum aviculare* ssp. *aequale*) in which the allocation to reproduction increased as a response to trampling while the size decreased as a result of phenotypic plasticity (Meerts and Vekemans, 1991).

5.4 Summary Recreation impacts ultimately reduce the numbers of plants in a given area. In the case of picking bluebells (*Hyacinthoides non-scriptus*) for recreation, the way in which they are picked has a major influence on

Table 5.6 Resource allocation patterns: growth form means and results of analyses of variance (except reproductive dry wt of *Poa annua*). Statistical results of between growth form comparisons are given for each treatment (from Warwick and Briggs, 1980d)

Taxa	Character	Growth form	Control		Light		Heavy		Source of variation: Between growth forms / Between treatments / Growth form × treatments
Plantago major	1. Shoot dry wt per plant	Erect	64	<	79		72	Increase	**
		Prostrate	55	<	64		67		*
			*		**		ns		ns
	2. Root dry wt per plant	Erect	10		13	<	24	Increase	*
		Prostrate	10		10	<	17		***
			ns		ns		**		*
	3. Reproductive dry wt per plant	Erect	26	>	8	>	4	Reduction	**
		Prostrate	35	>	26	>	16		**
			*		***		**		*
Poa annua	1. Shoot dry wt per plant	Erect	61	<	72		75	Increase	***
		Prostrate	64	<	80		80		***
			ns		*		*		ns
	2. Root dry wt per plant	Erect	27	>	22		22	Reduction	ns
		Prostrate	34	>	19		19		***
			*		ns		ns		***
	3. Reproductive dry wt per plant	Erect	12		6		3	Reduction	Paired *t*-tests due to unequal sample sizes
		Prostrate	2		1		1		
			*		*		ns		

Treatment means followed by < are significantly less than the means preceded by it, P 0.05; treatment means followed by > are significantly greater than the means preceded by it, $P < 0.05$; other adjacent-treatment means are not significantly different from each other, $P < 0.05$; ns, not significantly different; *, $P < 0.05$; **, $P < 0.01$; *** $P < 0.01$.

the outcome. The time over which the impact occurs is also important. However, the changes brought about by soil compaction may favour seed germination or some aspects of vegetative growth.

Since trampling impacts often occur in a linear form, with adjacent untrampled plants with which outcrossing may occur, it is not surprising that phenotypic plasticity is the main adaptive mechanism by which small-sized path individuals are produced. However, annual meadow grass (*Poa annua*) does appear to have evolved a path genotype, and genotypically determined small individuals do occur in populations of other species.

While small-sized individuals are often found on paths, with the exception of the great plantain it has been hard to demonstrate that their small size aids in their survival. Trampling generally inhibits the development of flowering and directs relatively more energy into vegetative structures.

It should also be noted that comparison with the effects of grazing can be misleading as in the latter situation nutrient cycling through the animal is an important part of the interaction (cf. Westoby, 1989) and selection against palatability may also occur. The argument that herbivory (which includes the effect of trampling in the case of mammals) can act mutualistically on plants, enhancing their growth, was dismissed by Belsky *et al.* (1993). However, the literature on the effects of grazing can provide some insights into the ecology of trampling and should not be ignored in the context of trampling studies.

Mechanical wear and plant form and function 6

One area of knowledge in which ecologists have looked for explanations of plant responses to various factors in the environment, is that of plant form and function. In this chapter I explore the various morphological patterns that have been observed in plants of trampled areas and discuss their associated explanations in terms of resistance of plants to trampling or other damage. In this case, we are looking at the differences between species or groups of species rather than the subtle level of within-species variation that may indicate the short-term processes of evolution, discussed in the previous chapter.

The higher plants (angiosperms) can be divided into the broad-leaved plants (dicotyledonous species) and narrow-leaved plants (monocotyledonous species). The first group includes herbs, such as dandelions and daisies, and most deciduous trees, while the second contains, amongst other groups, the grasses, sedges, lilies and orchids. The two major groups differ in their response to trampling, and this difference throws some light on the processes involved. If we imagine an untrampled, lightly grazed temperate grassland as our starting point, it would be dominated by tall grasses such as cocksfoot (*Dactylis glomerata*) or false oat grass (*Arrhenatherum elatius*) and perhaps blady grass (*Imperata cylindrica*) in subtropical situations. With the commencement of light trampling, the tall grass stems (culms) and leaves are bent and broken and the vegetation becomes more open. The lower-growing broad-leaved species receive more light and invader species can also join the community. At this stage the dominance has switched to the broad-leaved species of low growth form, such as great plantain (*Plantago major*), daisy (*Bellis perennis*) and dandelion (*Taraxacum* spp.); this dominance is only lost when the trampling intensity increases and narrow-leaved species are the survivors. With the increasing level of trampling, the total biomass declines until plants

6.1 Taxonomic groups (ratio of broad- and narrow-leaved species)

can no longer survive at all. This pattern was illustrated by Liddle and Greig-Smith (1975b) in the Welsh sand dunes, using the surrogate measure of soil hardness (Chapter 3) and comparing sites that were assumed to have different levels of trampling (Fig. 6.1). The whole sequence was found on the car-track sites, where there was bare ground in the tracks, and relatively undisturbed grassland at the edges. Footpaths and cattle tracks only showed the first and last parts of the sequence, respectively.

The first stage of this process was investigated at Shillong, India, using the grass (*Paspalum dilatatum*) and the herb wild white clover (*Trifolium repens*) (Pradhan and Trepathi, 1980). They noted that the total leaf area and the above-ground biomass of the grass was greater than that of the accompanying clover in the untrampled area, but that the proportions were reversed in the trampled sites. The two species were then grown by themselves and together in pots. These were given two levels of artificial 'trampling' and one set was kept as untrampled controls. As the total intensity and frequency of 'trampling' increased, the proportion of the grass declined, while the proportion of clover increased (Fig. 6.2a,b). The species in the 'trampled' mixtures had

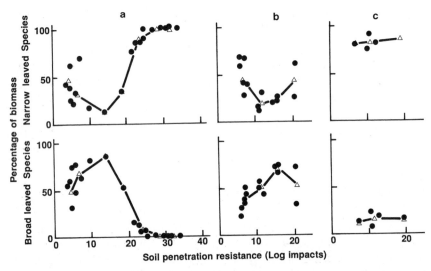

Fig. 6.1 The relationship between narrow-leaved (monocotyledonous) and broad-leaved (dicotyledonous) species and the intensity of wear, as indicated by the surrogate measure of soil penetration resistance between 6 and 12 cm depth. The biomass of plants was recorded as proportion of the total biomass at each measurement. (a) Car track; (b) footpath; (c) cow track. ●, Quadrat records; △, mean of all occurrences in each group of five increments in penetration resistance. (From Liddle and Greig-Smith, 1975b.)

greater leaf areas and leaf numbers than when grown and 'trampled' by themselves, although the grass biomass was reduced. This work shows that the clover responds favourably to trampling when mixed with the grass; the grass was more severely affected. The authors attributed the stimulation of the clover in the mixture to the greater damage caused to the grass which normally offered intense competition to the legume. It is suggested that the competition is for light. Trampling reduced the leaf area and other parameters of the clover in monoculture but not in the mixtures; the grass was reduced in both situations (Fig. 6.2c). They also suggested that the grass was protecting the clover, at least at the intermediate levels of artificial trampling used in this experiment.

In the second stage of this process, both groups decline in biomass but the broad-leaved species are proportionally more vulnerable and are reduced and eliminated first. This differential can be attributed to the greater resistance of grasses, which can survive considerable levels of trampling, particularly in the form of small tussocks. For example, the percentage reduction per passage calculated by Kellomaki (1973) for his woodland ground flora was 2.5% for wavy hair grass (*Deschampsia flexuosa*) and 10% for all the herbs grouped together. This change in ratio between narrow- and broad-leaved plants will not follow the first part of this pattern in situations where the dominant plant is broad-leaved, such as heather (*Calluna vulgaris*) or some other woody species. I would expect, however, that the broad-leaved species would be replaced by narrow-leaved species unless the trampling was very intense.

Trampling by horses in montane habitats in the Eagle Cap Wilderness, Oregon appears to have induced the first part of this process (Cole, 1981), as graminoid cover was replaced by forbs (broad-leaved species), but even in the most used area total cover was still 63% (Table 6.1).

Table 6.1 Characteristics of lightly and heavily grazed meadows of the montane valley-bottom and moist subalpine types used by recreational pack stock (horses) (Cole, 1981)

Meadow type	Montane valley-bottom			Moist subalpine			
Intensity of use	Light	Heavy	Heavy	Light	Light	Heavy	Heavy
Total cover (%)	104	73	63	105	103	101	102
Graminoid cover (%)	81	30	19	90	80	35	40
Forb cover (%)	23	43	44	15	23	66	62
Forb cover/graminoid cover	0.28	1.43	2.32	0.17	0.29	1.89	1.55
Deschampsia caespitosa cover (%)	23	–	–	8	40	5	4
Carex scopulorum cover (%)	–	–	–	70	4	5	8
Annual species cover (%)	–	6	2	–	–	–	–
Alien species cover (%)	2	1	20	–	–	–	–

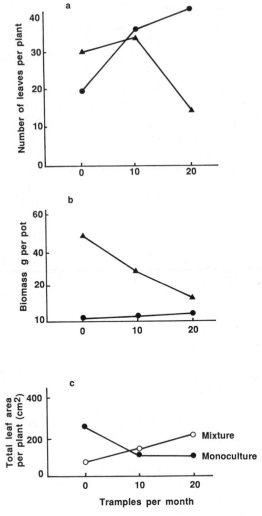

Fig. 6.2 The relationship between paspalum (*Paspalum dilatatum,* △) and white clover (*Trifolium repens,* ●); at three levels of artificial trampling, after 24 weeks of treatment: (a) the number of leaves per plant; (b) the biomass per pot; and (c) the leaf area per plant of clover when grown with paspalum and in monoculture. (From data of Pradhan and Trepathi, 1983.)

6.2 Life form   The life form of plants provides another, more differentiated, way of grouping plant species according to their morphology. The system of Raunkier (1934) is the most commonly used, and this is based on the position of the vegetative buds or persistent stem apex, that survive over winter or periods of drought. These range from the phanero-phytes, which have their buds 25 cm or more above the ground surface, through to the geophytes, with subterranean perennating organs (Fig. 6.3). These classes are often subdivided, and subdivisions relevant to the study of trampling are listed in Table 6.2.

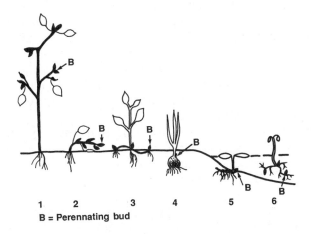

Fig. 6.3   The six main life forms of plants according to Raunkier (1934): 1, phanerophyte; 2, chamaephyte; 3, hemicryptophyte; 4, geophyte; 5, helophyte; and 6, hydrophyte. B, perennating bud. (After Shimwell, 1971.)

Table 6.2   Life forms (Clapham, Tutin and Warburg, 1962)

1. Phanerophytes – woody plants with buds more than 25 cm above soil level
2. Chamaephytes – woody or herbaceous plants with buds above the soil surface but below 25 cm
3. Hemicryptophytes – herbs (very rarely woody plants) with buds at soil level
4. Geophytes – herbs with buds below the soil surface
5. Helophytes – marsh plants
6. Hydrophytes – water plants
7. Therophytes – plants which pass the unfavourable season as seeds.

Some subdivisions and abbreviations are as follows:

Phanerophytes
   N Nanophanerophytes, 25 cm–2 m high
Hemicryptophytes
   Hp Protohemicryptophytes, with uniformly leafy stems, but the basal leaves usually smaller than the rest
   Hs Semi-rosette hemicryptophytes, with leafy stems but the lower leaves larger than the upper ones and the basal internodes shortened
   Hr Rosette hemicryptophytes, with leafless flowering stems and a basal rosette of leaves

In 1938 Bates published a short discussion in which he regarded perennial meadow grass (*Poa pratensis*), rye grass (*Lolium perenne*) and white clover (*Trifolium repens*) as the dominant components of footpath vegetation. He also pointed out that their cryptophytic life form may be demonstrated by cutting a vertical cross-section through the vegetation and soil at right angles to the axis of the footpath, and extending out into the surrounds. He also observed that the annuals (therophytes), knot grass (*Polygonum aviculare*), rayless mayweed (*Matricaria suaveolens,* syn. *matricarioides*), swine cress (*Senebiera cornopus* syn. *Coronopus squamatus*) and seedlings of great plantain (*Plantago major*) are squeezed out or buried in the mud during the winter, and only establish in the outer trampled zones where there is less disturbance (see also Fig. 3.11).

It might be expected that the generalization 'that cryptophytic life form survives on footpaths and therophytes are confined to the edge zones' would have some variation. In Welsh sand dunes, Liddle and Greig-Smith (1975b) found that the semi-rosette and rosette hemicryptophytes (see Fig. 6.3 and Table 6.2) had the highest representation on tracks and paths, with the generally taller protohemicryptophytes having a track occurrence index of 0.7 (Table 6.3). Woody chamaephytes (track occurrence index 0.9) and herbaceous chamaephytes

Table 6.3 Track and path occurrence index of different life forms occurring in the Aberffraw sand dune system (Liddle and Greig-Smith, 1975b)

	No. of occurrences in each stand group					Track occurrence index
	A	**B**	**C**	**D**	**E**	
Nanophanerophytes	0	0	3	3	4	1.0
Woody chamaephytes	2	5	6	7	6	0.9
Herbaceous chamaephytes	5	13	18	15	17	0.7
Protohemicryptophytes (Hp)	13	27	45	40	48	0.7
Semi-rosette hemicryptophytes (Hr)	28	49	55	60	51	1.0
Rosette hemicryptophytes (Hr)	21	42	41	31	37	1.0
Geophytes	0	2	8	3	6	0.3
Therophytes	0	9	21	16	23	0.3
Acrocarpous mosses	1	5	5	4	9	0.5
Pleurocarpous mosses	5	11	18	13	21	0.6
Liverworts	0	2	1	4	7	0.5
Lichens	0	2	5	4	2	0.6

A, track; B, adjacent to track; C, associated natural vegetation; D, footpath E, associated natural vegetation. Note that plants with occurrence index of 1 are most common and 0.1 least common on paths (see Table 4.8 for details of calculation).

(track occurrence index 0.7) both have surprisingly high occurrence in the worn areas, considering that their buds are above the soil surface. The woody chamaephyte was wild thyme (*Thymus drucei*) which can withstand light trampling, probably due to its close-growing prostrate habit, and was, I believe, a perfumed component of Elizabethan lawns, which were hand mown. The geophytes, which might be expected to have a high survivorship, on paths were represented by the genus *Equisetum* which has very brittle stems. In another location woody phanerophytes (25+ cm high) were found to decrease markedly along trails in the mountainous Shenandoah National Park, Virginia, while the cover of hemicryptophytes and a graminoid (grass-like) therophyte and similar geophyte increased (Table 6.4) (Hall and Kuss, 1989). Sun and Liddle (1993a) also found that the two woody species in their simulated trampling experiment did not survive even light trampling and that in eight tropical and subtropical field sites there were no woody plants in trampled areas (Sun and Liddle, 1993c).

In general, the more protected the bud or apex is from direct damage and/or detachment from the plant, either by uneven ground, burial or protective structures, the more likely it is that the plant will be able to survive the effects of trampling and vehicles.

Table 6.4 Increasers, decreasers and unchanged plants along the trail transect (Hall and Kuss, 1989)

Plants	Life form	Growth form
Increasers		
Panicum latifolium	Hemicryptophyte	Graminoid semi-rosette without runners
Poa pratensis	Geophyte	Graminoid with rhizome
Juncus bufonius	Therophyte	Graminoid
Stellaria pubera	Chamaephyte	Perennial herb
Potentilla canadensis	Hemicryptophyte	Semi-rosette with runners
Fragaria virginiana	Hemicryptophyte	Rosette with runners
Viola papilionacea	Hemicryptophyte	Rosette herb with runners
Decreasers		
Dennstaedtia punctilobula	Geophyte	Delicate herb with rhizome
Quercus rubra	Phanerophyte	Deciduous tree
Parthenocissus quinquefolia	Phanerophyte	Woody sprawler
Cornus florida	Phanerophyte	Deciduous tree
Vaccinum vacillans	Phanerophyte	Deciduous shrub
Viburnum acerifolium	Phanerophyte	Deciduous shrub
Unchanged		
Agrostis sp.	Hemicryptophyte	Graminoid rosette without runners
Agrostis alba	Hemicryptophyte	Graminoid rosette without runners
Lonicera japonica	Phanerophyte	Woody sprawler

6.3 General morphology   In this section the interactions between trampling or vehicles and different plant morphologies are considered at the species level, starting with higher plants (angiosperms), separated into broad-leaved species and grasses, and then the mosses and liverworts (bryophytes) and finally the lichens. The treatment of the higher plants considers stems, broad leaves, flowers, the underground organs, the grasses and, finally, plant anatomy.

STEMS OF HIGHER PLANTS

The most obvious effect of mechanical wear is a breaking down of taller structures. The only protection that these structures have is their apparent resistance to mechanical forces. Herbs up to 50 cm tall are readily trampled, and anyone wearing shoes or boots may walk through this type of vegetation, although those with bare feet are likely to avoid any plants that may appear sharp or rough (Chapter 5). Taller vegetation, such as blady grass (*Imperata cylindrica*) or reed beds (*Phragmites australis*), may be penetrated by people wearing some protection, and waders or waterproof boots in the wet reed bed. Woody shrubs are pushed down or broken by four-wheel-drive vehicles, provided they do not appear to be able to damage the car. The general principle of personal protection and visual selection of routes (the alpha stage) is discussed in Chapter 2. The other major cause of damage to stems of woody plants is cutting or breaking for firewood. This has been recorded particularly from the USA and Australia, where the climate is favourable for camp fires and afternoon barbeques. This is considered later in this section.

The reduction in height of heather (*Calluna vulgaris*) shoots was recorded in an area used for student field exercises by a mean of 440 students per year over a period of 8 years (Bayfield and Brooks, 1979). They divided the area into five zones of different intensities of use, on the basis of exact knowledge of where students had been working. In the most used area, the cover was reduced to about 55%, and the height of the heather stems, recorded in four classes, was between 0 and 10 cm for nearly 50% of the records (Fig. 6.4). The tallest class, over 30 cm, occurred in 13% of the records from the unused area but this was reduced by the heaviest use to 1.5% of the records. It appeared that the taller stems of heather were particularly susceptible to trampling, as there was a substantial reduction in the two tallest categories, even in the zone that had the lightest use.

Woody stems are often cited by authors as being a major reason for vulnerability (see phanerophytes in the previous section). Greller, Goldstein and Marcus (1974) noted that plants that survived winter snowmobile traffic along a track at an altitude of 11 400 ft (3474 m)

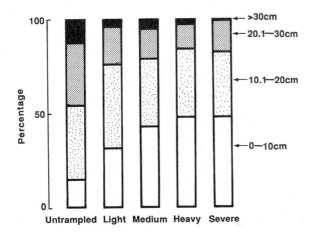

Fig. 6.4 The relationship between height of heather and intensity of trampling of the study area at Kindrogen, Scotland. The percentage of total point-quadrat strikes of heather shoots is shown in each of four height classes. Individual pins could strike shoots in more than one height class. (From Bayfield and Brooks, 1979.)

in the Colorado Mountains had, among other features, little woodiness, and, in the Denali Highway region near Mount McKinley, Alaska, Sparrow, Wooding and Whiting (1978) found that off-road vehicles used in summer did more damage to tall shrubs than to other plants. Older, and therefore more woody, heather growing on the Kinder Scout upland on the Bleaklow Plateau, Derbyshire, was found to be more vulnerable to trampling than young shoots of the same plant (Beeching, 1975). Conversely, greater survival of (*Acamptopappus shockleyi*) and (*A. sphaerocephalus*) in areas used by four-wheel-drive vehicles in the Mojave Desert was suggested to be the result of having 'less brittle' stems than other species which were more heavily damaged (Vollmer *et al.*, 1976). Very detailed experiments, in which 18 species of plants were selectively measured, were carried out in summer in an upland area of the Yosemite National Park (Holmes and Dobson, 1976). The plants were arranged in a sequence of their survival rate, which was taken as the remaining cover after 10 passages. The presence of various morphological features affecting their survival were recorded (Fig. 6.5). Woodiness appeared only once above the 50% 'survival rate' and the authors commented that the plants most sensitive to trampling were of two types:

- plants with woody parts above ground; and
- plants with tall, entirely non-woody (herbaceous) stems and having all leaves on the above-ground part of the plant. (Fig. 6.5).

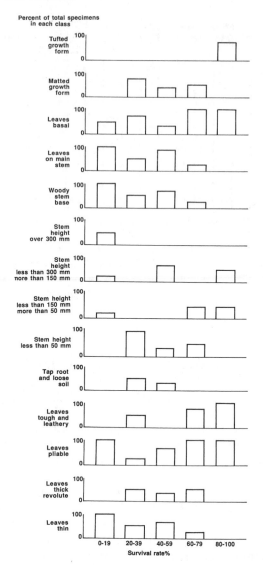

Fig. 6.5 The relationship between various morphological characteristics and survival of plants with those features in a trampling experiment. Species in each group: 0–19% survival, *Castilleja lemmonii, Senecio triangularis, Lupinus covillei, Polygonum bistortoides*; 20–39% survival, *Vaccinium nivictum, Eriogonum incanum, Mimulus primuloides, Calyptridium umbellatum*; 40–59% survival, *Salix eastwoodiae, Phyllodoce breweri, Lupinus lyallii*; 60–79% survival, *Senecio scorzonella, Carex scopulorum, Antennaria alpina,* moss; 80–100% survival, *Juncus parryi, Poa epilis* and *Calamagrostis* sp., *Oryzopsis kingii, Aster alpigenus* (see Fig. 6.6 for illustrations of some species). (Data from Holmes and Dobson, 1976.)

The first type included *Salix eastwoodiae*, *Vaccinium nivictum* and *Phyllodoce breweri*. The second type was characterized by *Lupinus covillei* and *Senecio scorzonella* (Fig. 6.6). Holmes and Dobson (1976) considered that herbaceous plants with basal leaves were the most resistant, in part because the upper leaves protect the lower ones. Also pliable and leathery or thick basal leaves were 'usually essential for a high survival rate', for example *Aster alpigenus* (Fig. 6.6). Dense clumping of stems and leaves and tough foliage, as shown by *Oryzopsis kingii*, was also an advantage. Herbaceous stems which are easily crushed and flattened are a disadvantage, and the authors noted that 'after the first 15 to 20 steps, 100% of the stems of *Senecio trangularis* were broken or flattened' (Fig. 6.6). However, it should be noted that where stems are bent flat with the ground but not detached, it is possible that layering might take place and new plants produced at each node of the original damaged stem.

Horizontal stems above the ground (runners and low-growing stolons) are subject to compression and abrasion (Fig. 6.7); they may also be subjected to shearing forces where they pass over a stone or other hard-edged object, or if the ground erodes away beneath them, as on dry sand dunes. Common wild thyme (*Thymus drucei*) has this form and a path occurrence index of 0.6 (Table 4.8) (Liddle and Greig-Smith, 1975b). I noted that this species occurred on footpaths where the leaves may be stripped from the branches, but it does not survive so well on tracks. This may be due to the softer soil surface on the drier areas of the Aberffraw sand dunes where wild thyme occurs, cushioning the branches from compression damage, but not providing protection from the horizontal torque forces produced by motor cars, 'it thrives where competition for light is reduced, and trampling pressures are not extreme' (Liddle, 1973b).

Two related species, silverweed (*Potentilla anserina*) and creeping cinquefoil (*Potentilla reptans*), also occurred in the survey, with path occurrence indexes of 0.9 and 0.4, respectively. The main morphological difference between these two species appears to be the longer leaf petioles of the creeping cinquefoil and the folded young leaflets of the silverweed. There may be differences in the flexibility of stolons, but the silverweed occurs on the tracks mainly as small, isolated rosettes without stolons, so a change in form and possibly physiological differences are implicated here. The other species in this category that occurs commonly on paths and picnic areas is white clover (*Trifolium repens*), preference index 0.7 (Table 4.8). Again, this was able to survive treading much better than wear by cars, and it has been noted on paths and trampled areas in many other situations (e.g. camp grounds (La Page, 1967), trampling by birds (Gillham, 1956), sheep paths (Davies, 1938) and path edges (Bates, 1935; Klecka, 1937)). This species occurs with a reduced leaf size in trampled areas.

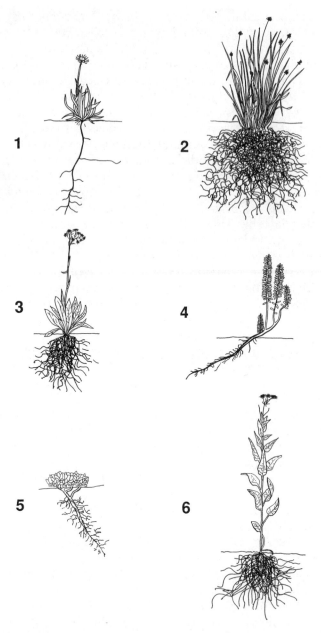

Fig. 6.6 Growth form of six species of plants tested for trampling resistance in Yosemite National Park, California. 1, *Aster alpigenus* (87%); 2, *Juncus parryi* (80%); 3, *Senecio scorzonella* (60%); 4, *Phyllodoce breweri* (50%); 5, *Vaccinium nivictum* (20%); 6, *Senecio triangularis* (5%). Survival rate (%) is shown in brackets. (Redrawn by courtesy of D.O. Holmes and H.E.M. Dobson.)

Fig. 6.7 Abraded and broken runners of spinefex grass (*Spinifex hirsutus*) on a car track in sand dunes on Fraser Island, Queensland (photograph by M.J. Liddle).

So again we have some indication of altered plant form (Goryshina, 1983).

The last three species discussed above have much more flexible stolons than common wild thyme, but a vertical stem shortened to a few millimetres and protected by a rosette of leaves, such as that of daisy (*Bellis perennis*), occurrence index 1, appears to confer a greater potential for survival, at least under moderate levels of trampling. The other protective feature that merits discussion here is the protection of nodes and axillary buds by the presence of stipules or modified stipules (ochreae) which occur notably on common knot grass (*Polygonum aviculare*) (Fig. 6.8). It is apparent that stems in the form of hollow tubes do not survive as well as those that are solid, and that flexibility and high tensile strengths are advantageous.

LEAVES

It is also evident that one of the adaptations to the trampling environment is small size, and a reduction in leaf size is a common plastic response to the effect of trampling. The area of leaves of common white clover (*Trifolium repens*) was reduced by 57% when it was

Fig. 6.8 Stipules and ochre of *Polygonum* species (after Harder *et al.*, 1965).

growing on a car track (Liddle, 1975a) and Goryshina (1983) noted that the leaf areas were significantly reduced on white clover, great plantain and knot grass growing in trampled situations. The ratio of stem biomass to leaf biomass is also significantly increased by 'stamping' on bilberry (*Vaccinium myrtillus*) in September and harvesting in the following July when a normal ratio of 3.58 rose to 17.2 after treatment (Anderson, 1961). This effect was thought to have been brought about by soil changes rather than direct mechanical bruising of leaf buds as no visual signs could be detected by the author. However, Goryshina (1983) commented that the effect may be due to direct mechanical damage but, in addition, trampling may also inhibit leaf growth.

Leaf length, width, thickness and number per plant were also reduced under a regime of simulated trampling (Sun and Liddle, 1993d) (Table 6.5). The species included five tussock grasses, two prostrate grasses, three herbaceous species and one woody species that survived the treatments. In this experiment light trampling had a major effect on leaf length of mat grass (*Axonopus compressus*), Rhodes grass (*Chloris gayana*) and green panic (*Panicum maximum*). The reduction in leaf length of green couch (*Cynodon dactylon*), elastic grass (*Eragrostis tenuifolia*), rye grass (*Lolium perenne*), common sida (*Sida rhombifolia*) and white clover (*Trifolium repens*) was much less and apparently linearly related to trampling intensity, suggesting that a different process may be operating than with the first group. The reduction in width was partly due to the loss of the mature leaves and was greater for species with wider leaves, as was the reduction in leaf thickness of the species with thicker leaves. Direct damage to the plant was the main cause of the reduction in leaf number – either detaching the leaves or breaking the stems and thus reducing the size of the plant. The same results were generally recorded in the field except that in some cases leaf thickness slightly increased (Sun and Liddle, 1993c). The increase was only statistically significant for *Axonopus compressus*. Leaf length may also be affected by other factors, such as flooding which may cause the leaves to elongate (Engelaar, 1994). In his experiments with great plantain (*Plantago major*), common sorrel (*Rumex acetosa*) and curled dock (*R. crispus*) the combined effect of flooding and trampling produced leaves longer than the controls, but trampling alone generally produced leaves shorter than the controls (Fig. 6.9).

The advantage of basal leaves that are pliable, leathery or thick has already been noted (Holmes and Dobson, 1976). However, the leaves of heather (*Calluna vulgaris*) and of crowberry (*Empetrum nigrum*) are thick and small and yet do not survive trampling. In the case of heather, Bayfield and Brookes (1979) noted that high water potential deficits are developed in shoots following trampling, and these can persist for many months. This may be due to detachment of leaves,

Table 6.5 Relative morphological parameters of both lightly and heavily trampled plants (%) of each species measured before the last trampling treatment, and results of statistical analysis (Sun and Liddle 1983d)

Species	Leaf length	Leaf width	Leaf thickness	Total number of live leaves	Percentage of live leaves that were broken per plant	Plant height
Axonopus compressus						
Light trampling	58.5	99.7	95.7	66.6	25.3	38.1
Heavy trampling	40.1	83.0	73.9	35.8	40.6	16.3
Chloris gayana						
Light trampling	54.2	94.5	95.2	68.6	63.6	20.0
Heavy trampling	44.5	82.2	81.0	51.0	48.9	16.3
Cynodon dactylon						
Light trampling	67.8	81.0	92.9	53.6	23.5	65.8
Heavy trampling	59.5	70.0	76.9	12.5	24.7	58.7
Eragrostis tenuifolia						
Light trampling	74.5	98.6	100.0	76.1	22.0	53.3
Heavy trampling	70.4	97.6	100.0	57.1	53.9	41.5
Hypochoeris radicata						
Light trampling	49.5	49.7	96.3	129.2	21.5	66.0
Heavy trampling	28.9	39.3	59.9	26.4	95.1	44.0
Lolium perenne						
Light trampling	87.3	98.3	100.0	65.2	18.3	54.1
Heavy trampling	84.3	94.1	100.0	56.0	39.7	54.8
Panicum maximum						
Light trampling	49.8	81.8	98.4	122.5	82.6	12.5
Heavy trampling	35.7	46.0	85.9	16.6	100.0	4.6
Sida rhombifolia						
Light trampling	57.5	59.2	92.3	71.8	27.5	45.7
Heavy trampling	30.0	36.1	53.9	15.8	73.2	8.2
Sporobolus elongatus						
Light trampling	71.5	95.2	100.0	87.3	28.2	49.5
Heavy trampling	65.9	94.0	100.0	71.4	46.5	47.2
Trifolium repens						
Light trampling	74.8	98.0	92.4	28.4	10.8	51.5
Heavy trampling	34.6	16.3	68.3	9.7	9.1	16.7
F test[a]						
Light trampling	11.84**	31.9**	7.6**	45.8**	23.9**	41.3**
Heavy trampling	9.8**	26.4**	61.7**	47.2**	8.5**	36.1**

[a] **Significant at 0.025 level.

which appears to happen fairly easily in both species, or to damage to the leaves which are also fairly rigid. In either case, I suggest that rigidity of both the leaves and the more mature stems is likely to increase vulnerability to wear. It is self-evident that thin leaves are delicate and

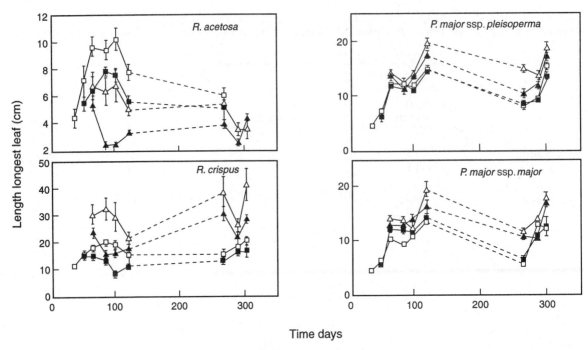

Fig. 6.9 Mean length of longest leaf of *Rumex acetosa, R. crispus, Plantago major* ssp. *pleiosperma* and *P. major* ssp. *major* plants that were trampled, flooded or both, compared to an untrampled, non-flooded control series. □, Control; ■, trampled; △, flooded; ▲, trampled and flooded. (From Engelaar, 1994.)

easily damaged by trampling, and that long leaf stalks (petioles) are also a weakness when compared to sessile leaves. Perhaps the best indication of resistant leaf qualities is given by the premier broad-leaved trampling plant, great plantain (*Plantago major*) which has thick, strong-veined and flexible leaves of very variable size, which form the base of the rosette (Fig. 6.10). A range of leaf sizes was found in four other broad-leaved species by Warwick and Briggs (1979). They found that leaves from daisy (*Bellis perennis*), self-heal (*Prunella vulgaris*), yarrow (*Achillea millefolium*), ribwort plantain (*Plantago lanceolata*) and greater plantain (*Plantago major*) were all shorter and narrower when the plants were growing in lawns or intensely grazed grasslands, a bit larger in seasonally mown grasslands, and considerably greater in grasslands not seasonally reduced in height by mowing or grazing (Fig. 5.16). Except for the great plantain, these differences were not consistently maintained in cultivation without mowing or grazing, although there were no dwarf or prostrate plants in the populations from the ungrazed or unmown habitats (see also Chapter 5).

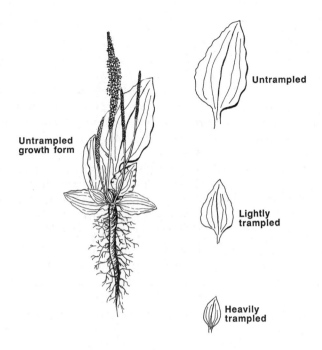

Fig. 6.10 (a) Great plantain (*Plantago major*); (b) leaf forms of great plantain subject to various levels of wear (from Goryshina, 1983).

An examination of the range of leaf shapes was made by Jurko (1983) who classified the leaves by their form, predominant habitat and persistency (Table 6.6). He compared data from a number of forested and unforested subalpine sites that had been used by tourists in the Tatra Mountains, Czechoslovakia. He found that under the protecting tree canopy hairy and three-dimensional leaves were almost absent when compared to a sheltered subalpine community, and that there was a higher proportion of microphyllous and mesophyllous leaves. There was also a higher proportion of hygromorphic and fewer scleromorphic forms. He found that there was a rise in the proportion of three-dimensional leaves (filiphyllous) on all trampled sites except one; and one species with felty leaves, the bell flower (*Campanula alpina*), was able to expand on some sites (Fig. 6.11). Graminoid (grass-like) leaves 'strongly receded' in the trampled communities except in the dwarf pine (*Pinus mugo*) communities, where they increased as the canopy 'was thinned'. Similarly, graminoid leaves increased their share of the biomass within the other forest site, the Adinostylo–Piceetum alliance, especially the woodrush (*Luzula nemorosa*), wavy hair grass (*Deschampsia flexuosa*) and the grass *Calamagrostis villosa*. Nanophyllous leaves generally declined in proportion, especially in the mossy and forest sites.

Microphyllous leaves, which were in high proportions in the forested sites, exhibited variable changes, reducing in the Adinostylo–Piceetum and rising in the Calamagrostio villosae–Piceetum. The proportion of mesophyllous leaves rose in both forested sites, but was relatively unchanged elsewhere. The group of mesomorphic leaves generally declined with trampling, with the notable exception of those in the mossy site where the scleromorphic leaves declined; otherwise sclero-morphic forms tended to rise on most trampled sites.

In summary there was a tendency for three-dimensional scleromor-phic forms to predominate in the trampled subalpine areas and for graminoid and mesophyllous forms to increase in trampled areas of the lower forests. The number of deciduous leaves declined as a result of trampling in the two spruce forest sites, while the proportion of overwintering green leaves rose, as did the proportion of evergreens; there were no marked changes in the other sites. The same author also calculated Vareschi's (1980) leaf diversity coefficient (Cd) for the differently trampled areas in these communities (Cd = Af, where A = the number of species and f = the number of leaf categories) and found that it fell as the intensity of trampling increased in all types discussed here (Fig. 6.12).

Simulated trampling of sea plantain (*Plantago maritima*) plants grown singly in pots in a glasshouse experiment demonstrated that light trampling, the equivalent of one step for 5 seconds per day, can produce a 'positive' effect on plant growth. There were about double the number of leaves on the trampled plants when compared to the

Table 6.6 Leaf characteristics used in Jurko's (1983) calculation of leaf diversity coefficient

Size and shape
TD	Three-dimensional (filiphyllous)
F	Felty leaves (hairy on one or both sides)
G	Graminoid leaves (grass-like)
N	Nanophyllous – up to 2 cm^2 leaf area
M	Microphyllous – from 2 cm^2 to 6 cm^2 leaf area
MS	Mesophyllous – from 6 cm^2 to 100 cm^2 leaf area
m	Macrophyllous – over 100 cm^2 leaf area

Predominant habitat type
m	Mesomorphic
sc	Scleromorphic
hg	Hygromorphic

Persistency
I	Evergreen
W	Green over winter
d	Deciduous

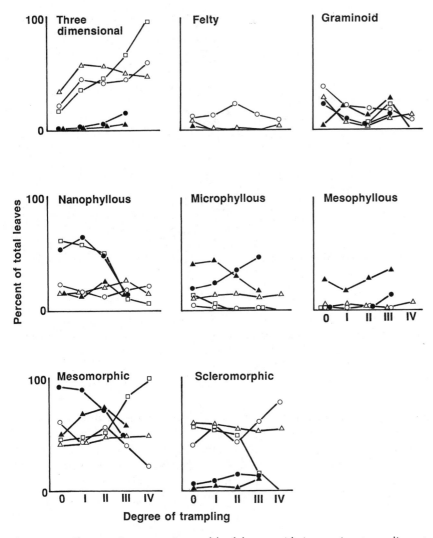

Fig. 6.11 Changes in proportions of leaf forms with increasing trampling at five sites in the Tatra Mountains. ○, △, Subalpine and □, mossy 'grasslands'; ▲, ●, 'dwarf pine forest' (From Jurko, 1983.)

controls, although the trampled leaves were about half the length of the leaves on the control plants (Blom, 1983) (Fig. 6.13). There was also an increase in the biomass and the number of daughter rosettes of the trampled plants. However, when the same treatments were carried out on sea plantain grown in turfs which had been cut and taken to the glasshouse, the trampled plants had fewer and shorter leaves, and they also produced only about one-quarter of the biomass

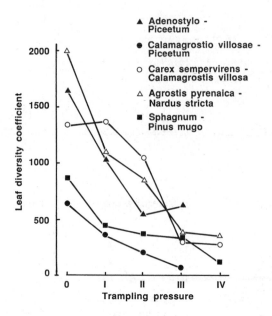

Fig. 6.12   The relationship between leaf diversity index and increasing trampling pressure in worn communities' areas of the Tatra Mountains, Czechoslovakia; open symbols are 'grasslands' and closed symbols are 'dwarf pine forests'. (After Jurko, 1983.)

of the untrampled controls (Fig. 6.13). Blom commented on his results: 'apparently the competitive effects of the surrounding vegetation together with the effects of trampling caused the reduced growth of the trampled plantago plants in the turfs'. However, these results do show that under certain conditions it is possible to stimulate an increase in vegetative growth by trampling.

FLOWERS

A pathway across a temperate meadow or along an open way through a broad-leaved woodland may often be picked out by the parallel lines of daisies (*Bellis perennis*) growing along each side. These are the delightful visual result of reduced competition between grasses and broad-leaved plants. This same site is the location for many other flowering species, so the general statement that light trampling can visually enrich a grassland community is certainly true. However, this effect is indirect, working through the initial reduction of grasses, so the direct and indirect effects of trampling or other recreational activities on flowering should be examined.

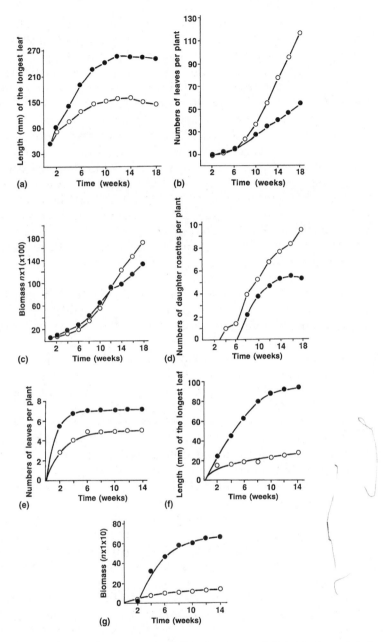

Fig. 6.13 The effects of trampling on the length, number of leaves and, for one treatment, numbers of daughter rosettes, and biomass, per plant of sea plantain (*Plantago maritima*). (a), (b), (c) and (d) plants grown separately in pots in the glasshouse; (e),(f) and (g) plants grown with other species in turfs taken to the glasshouse; ●, control plants; ○, trampled plants. (From Blom, 1983.)

Measurements of the percentage of orchid shoots which bloomed in the Great Smoky Mountain National Park showed that showy orchis (*Orchis spectabilis*) flowered significantly more frequently (25.4% flowering) by easy access trails than in less used difficult access situations (20.7%) (Bratton, 1985). However, the numbers of pink lady's slipper (*Cypripedium acaule*) were completely reversed, with 2.5% and 23.5% flowering along easy and difficult access routes, respectively. The mean number of shoots per unit area of showy orchis was 703 on easy and 362 on difficult access routes, so this also increased the number of blooms for visitors to see. But the highest number of pink lady's slipper shoots occurred on the moderate access routes (mean 902 and 10.5% blooming), with 233 and 212 on the easy and difficult routes, respectively. In this case, the visitors could see more blooms on the moderate access routes.

The mechanism of these changes is uncertain, as there were correlations between the percentage of pink lady's slipper plants blooming and the distance to trails on which only foot and horse traffic were permitted, and roads where visitors could drive their vehicles. These correlations have been combined in a predictive model which suggests that maximum blooming, at about 22%, will occur in a position about 600 m from the trails and about 6500 m from a road (Fig. 6.14). Bratton considered that a 'lack of large individuals [of pink lady's slipper] is the primary reason for the low flowering percentages in the easy access sites'. The leaf length was reduced from a mean of 14.61 cm in the difficult access sites to a mean of 8.74 cm in the easy access situations, and leaf size had a very close correlation with blooming success. Since 'road or trail effects extend at most 10 metres to 50 metres to the road edge' (Cole, 1978), direct effects of trampling or other factors relating to roads or trails are difficult to implicate as causal factors. However, since these are discernible blooms, that can be seen from some distance, Bratton (1985) suggested that 'the much lower proportion of large, blooming plants in the high access areas suggests that poaching (picking or removing by visitors) removes the largest individuals', although this does not explain how the effect was reversed for the showy orchis, with more blooms appearing by the easy access routes.

A more direct experiment was carried out on sown plants of sea plantain (*Plantago maritima*) in ungrazed dune-edge sites and an area grazed by cattle (Blom, 1983). After the second year, there were no flowers on the plants on the grazed area and only a maximum of 22.2% flowered after the third year where they were grazed, in contrast to a maximum of 83.3% in an ungrazed site. While an experiment with grazing only implicates trampling as a part of the mechanisms of damage, it does fit in with intuitive expectations that flowering will be inhibited by direct mechanical damage.

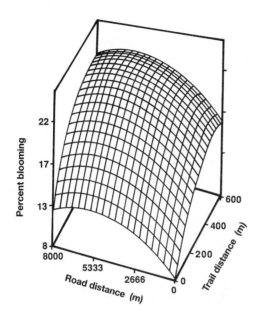

Fig. 6.14 A model to predict the percentage of blooming plants of lady's slipper orchid (*Cypripedium acaule*) populations at various distances from the nearest road and the nearest maintained trail in the Great Smoky Mountain National Park (Bratton, 1985).

The length of the flowering stems of the great plantain (*Plantago major*) and ribwort plantain (*Plantago lanceolata*) were measured in populations from lawns and ungrazed habitats by Warwick and Briggs (1979). The plants from lawns and grazed areas had considerably shorter flowering stems and spikes (Fig. 5.15) although the authors also observed that the lawn plants of great plantain had prostrate flowering stems, often protected by some of the rosette leaves.

UNDERGROUND ORGANS

It is rather difficult to separate the direct effects of mechanical forces on roots and rhizomes from the indirect changes caused by soil compaction or erosion (see Chapters 11 and 15). However, it can be observed that roots exposed by trampling-induced soil erosion are easily abraded and have their outer 'bark' removed. If the erosion is severe enough to leave a large space beneath the root, then it is quite likely to be broken on footpaths and will almost certainly be severed on vehicle tracks, unless it is so large that it is avoided by walkers and drivers.

This interaction with erosion was noted by Holmes and Dobson (1976) in the sparsely covered dry flats of the lodgepole pine (*Pinus contorta*) subalpine ecosystem: 'The root system of plants growing in the well drained areas such as *Eriogonum incanum*, tend to consist mostly of tap roots, with little lateral branching or spreading to hold the soil in place' (see Fig. 6.5). This plant had only a 20% survival rate after it had been walked on 100 times; 'As the soil was removed from under the plants the subterranean woody stems were crushed and broken'. The authors commented that in a moist, peaty soil with a dense root content, a stable surface was resilient to trampling pressures and thus cushioned the plants. In general, the fibrous, dispersed roots of the narrow-leaved (monocotyledonous) plants appear to confer a greater survival potential in trampled situations than do the single tap roots of the broad-leaved (dicotyledonous) plants; however, in moist, nutrient-rich, temperate situations, broad-leaved plants which are well adapted in other respects, such as great plantain (*Plantago major*), daisy (*Bellis perennis*) and common knot grass (*Polygonum aviculare*), do very well.

Some broad-leaved species such as (*Erythronium grandiflorum*) are also able to survive as they are active in early spring and spend the season when the trails are in frequent use as bulbs (Dale and Weaver, 1974). Bulbs, however, are affected by trampling, especially as the intensity of trampling increases; a weekly treatment reduced the dry weight of the bulbs of bluebell (*Hyacinthoides non-scriptus*) to nearly one-third of that of untrampled bulbs (Blackman and Rutter, 1950) (Fig. 6.15).

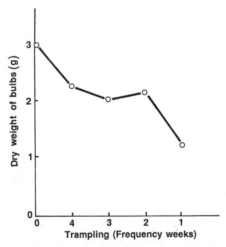

Fig. 6.15   The relationship between the dry weight of bluebell (*Hyacinthoides non-scripta*) bulbs and trampling after various intensities of trampling of the growing plant (data from Blackman and Rutter, 1950).

They also found that postponing trampling until 3 weeks after the leaves had appeared in spring allowed an almost full development of bulb dry weight. Although grasses are discussed below, it is worth adding Edmond's (1964) comment that 'it is interesting to note that timothy (*Phleum pratense*) with its tuberous shoot base, was relatively tolerant of treading'.

The effects of trampling on root surface area appeared to be slightly greater through mechanical damage to the leaves than through soil compaction, but the combined effect on two plantain species was not much greater than for each of the factors separately (Fig. 6.16) (Engelaar, 1994). Only the most affected species, common sorrel (*Rumex acetosa*), had a severe reduction in root growth, while the plantain species were only reduced to about 70–90% of the controls and the shoot:root ratios remained unchanged.

Rhizomes may be damaged by the effects of soil movement. Sand sedge (*Carex arenaria*) growing in the Aberffraw sand dunes was found

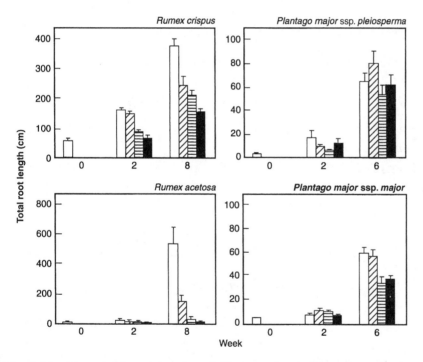

Fig. 6.16 Mean total root lengths of *Rumex crispus, R. acetosa, Plantago major* ssp. *pleiosperma* and *P. major* ssp. *major*. Plants which remained untrampled (open bars) or had their leaves trampled (diagonal hatched bars), surrounding soil trampled (horizontally hatched bars) or both (closed bars) twice a week for 2, 6 or 8 weeks. (From Engelaar, 1994.)

to flourish in tyre ruts of an experimental track after driving had ceased, but excavation of a rhizome that had been present when the driving occurred showed that it had an unusually high number of dead growing points just underneath the rut (Fig. 6.17). The rhizomes of beach grass (*Ammophila breviligulata*) respond to disturbance in the fore dunes by increasing their proportion of horizontal to vertical growth. In undisturbed fore dunes the biomass of the horizontal growth was 19% of the whole and comprised 36% of its length. After disturbance by vehicles the proportion of the horizontal biomass rose to 65% and 80% of the length was growing horizontally (Brodhead and Godfrey, 1977). This response was not as marked in the less fertile high dune.

As implied above, a rhizome growing beneath the soil surface with its adventitious buds offers the potential for regrowth after the above-ground parts of the plants have been damaged. In the case of *Phyteuma hemisphaericum* (Fig. 6.18), responding to the slicing action of the metal edges of skis on pistes, 'The characteristic which has evolved for entirely different reasons provides the plant with an advantage

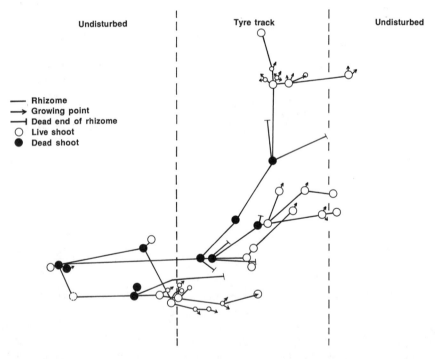

Fig. 6.17 A rhizome of sand sedge (*Carex arenaria*) that had been growing under an experimental tyre track at Aberffraw, Wales. Note that the dead growing points of the rhizome occur only under the tyre track.

enabling it, to some extent, to survive the new 'habitat stress' factor' (Grabherr, 1985).

An unusual plant strategy that provides a mode of recovery following grazing is that of cryptogeal germination of the Australian plant *Jedda multicaulis* (Fig. 6.19) (Clarkson and Clifford, 1987). In this seedling the plumule is initially positively geotropic, burying the cotyledons below the ground surface. Several centimetres of stem become buried and the associated auxiliary buds provide a 'capacity for vegetation regeneration following destruction of the above ground parts', more often by fire but also occasionally by trampling.

GRASSES

The grasses have been separated from the other plants in this discussion partly because of their different morphology and partly because they are the subject of extensive research for both lawn and sports turf use as well as for cattle and sheep grazing.

Grass morphology is generally more consistent than that of the broad-leaved species (Fig. 6.20). The features that are of special significance in trampled situations are the rhizomatous or stoloniferous main stem, with growing points at or near the surface of the ground, its frequent branching, especially in tussock forms, the persistent meristems that occur at the base of the leaves, and the way in which the

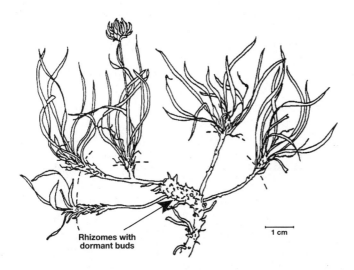

Fig. 6.18 *Phyteuma hemisphaericum* present as a 'stable' species on the eroded parts of ski pistes, where the metal edges of the skis slice off the natural vegetation cover. Dashed lines indicate soil surface (Grabherr, 1985).

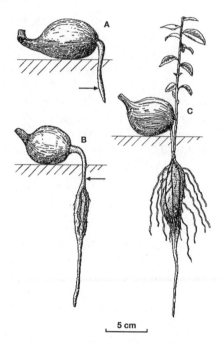

Fig. 6.19 Sequence of events in the germination of *Jedda multicaulis*. Arrows indicate the position of the plumule. (a) The pseudo-radicle is positively geotropic and grows downward into the soil. (b) Once the pseudo-radicle has penetrated the soil to a depth of 5–10 cm the root rapidly develops. (c) The shoot bursts through the fused colytedonary petioles and grows to the surface. The lower part of the stem with its adventitious buds remains below ground. (Clarkson and Clifford, 1987.)

apical meristem of the stem is shielded by the sheathing leaf bases, except when the flowering culm is developed. A survey of eight tropical and subtropical sites (Sun and Liddle, 1993c) showed that, of the 45 species recorded, the tussock forms comprised 72% of those that occurred on heavily trampled areas (Table 6.7). Interestingly, Balph and Malecheck (1985) found that cattle tended to avoid trampling on tussocks of crested wheatgrass (*Agropyron cristatum*) and that the higher the tussocks, the less they were trampled. This stresses the importance of the alpha stage in animal (or human)–plant interaction and its potential evolutionary consequences (Chapter 8).

Trampling greatly alters the morphology of grass plants, reducing tiller and leaf length, leaf number and the dry weight of aerial parts, but in one case, red fescue (*Festuca rubra*) growing in a sand dune area showed no change in the number of tillers per plant (Liddle, 1975a) (Fig. 6.21). Light trampling treatments in a simulated trampling experiment on five 'tussocky' grasses (Sun and Liddle, 1993b)

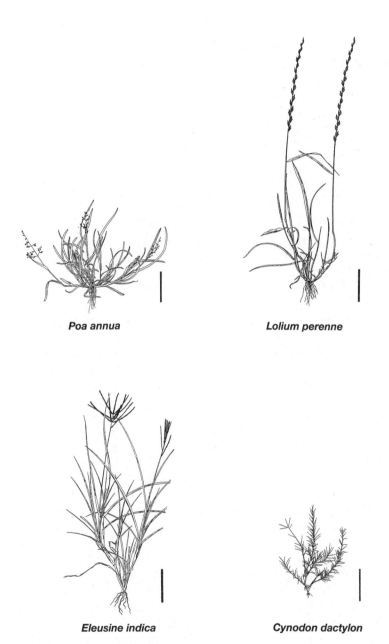

Poa annua

Lolium perenne

Eleusine indica

Cynodon dactylon

Fig. 6.20 Four species of grass commonly found on paths or tracks. Annual meadow grass (*Poa annua*) and rye grass (*Lolium perenne*) are temperate species which may extend to the subtropics; crowsfoot grass (*Eleusine indica*) and green couch (*Cynodon dactylon*) are widespread tropical species. (Note: there are alternative common names for these species.) All have 'worldwide' distribution. Bars indicate 10 cm for each species at maximum size. (From Tothill and Hacker, 1973.)

Table 6.7 Percentage distribution of four morphological types in four wear class areas at each site (Sun and Liddle, 1993c)

	Untrampled	Slightly trampled	Moderately trampled	Heavily trampled
Prostrate	36.6	66.6	66.7	27.9
Upright herbaceous	45.6	10.8	0.0	0.0
Upright woody	13.4	22.6	0.0	0.0
Tussock	4.4	0.0	33.3	72.1

caused a small increase in the number of tillers in green panic (*Panicum maximum*), elastic grass (*Eragrostis tenuifolia*), rye grass, (*Lolium perenne*) and slender rat's tail (*Sporobolus elongatus*), and a slight reduction in rhodes grass (*Chloris gayana*) (Table 6.8). Heavy simulated trampling produced smaller increases in rye grass and slender rat's tail. Apart from green panic, all the species had the longest tillers in the controls and the shortest under heavy trampling, and both treatments also reduced tiller diameter. The mutual protection was clearly demonstrated by the fact that in plants that had five or fewer tillers in the controls almost all were broken by the end of the experiment but very few of the others were damaged. O'Connor (1956) found that in addition to a reduction in the average leaf length of cocksfoot (*Dactylis glomerata*) from 9.1 cm to 5.8 cm by six passages of a tractor, the number of tillers was also reduced from 7.3 to 1.1 per plant (see pp. 148–151 for a discussion of leaf reduction in grasses and other plants).

> It appears that there are two types of processes involved in the reduction of tiller length, one involving injury and the second occurring in intact tillers. Elastic grass, rye grass and slender rats tail had uninjured reduction, while rhodes grass and green panic had injured reduction in tiller length. Flexibility of tillers also plays an important role in determining which of these two responses will occur when plants are subjected to trampling (Sun and Liddle, 1993b). Elastic grass, rye grass and slender rats tail had the most flexible tillers while rhodes grass and green panic had the least flexible tillers.
>
> (Sun and Liddle, 1993a)

It is also worth noting that the leaves of elastic grass and slender rat's tail had high tensile strengths, approximately 140 g mg^{-1}, while the other three species had strengths below 100 g mg^{-1} (Sun and Liddle, 1993a).

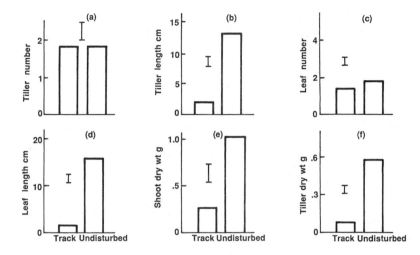

Fig. 6.21   Some aspects of the morphology of 20 plants of red fescue (*Festuca rubra*) from one 10 × 10 cm quadrat at two sites in a sand dune area at Aberffraw, Wales. (a) Number of tillers per plant; (b) length of live tillers; (c) number of live leaves per tiller; (d) length of live leaves; (e) mean dry weight per shoot; (f) mean dry weight per tiller; (I) two standard errors. (From Liddle, 1975a.)

Human trampling has also been shown to reduce the stem (culm) length of crested dog's tail (*Cynosurus cristatus*) in proportion to the intensity of trampling received, as measured using Bayfield's (1971a) tramplometer system (Young and Pendlebury, 1969) (Fig. 6.22). Trew (1973) compared the flowering culms of rye grass and red fescue in trampled and untrampled areas (Table 6.9). Height and reproductive parameters were apparently reduced to a greater extent in rye grass than in red fescue. Edmond (1964) also noted that the yield of cocksfoot was particularly sensitive to trampling, and suggested that it 'may possibly have been due to an elevation of its growing points in the absence of grazing animals during the period before grazing' in his experiment. Branson (1953) classified some of the grasses growing in Montana according to the presence and height of their growing points during the grazing season (Table 6.10, Fig. 6.23). He found that those with vegetative growing points (Fig. 6.24) above the soil (Class I) decreased, while those with the vegetative growing points at or below the soil surface (Class II) increased under grazing. The third class (III) declined because most of their growing points rose above the surface as they produced flowers, leaving few or none in vegetative condition. The height of the growing points above the ground may also depend upon the amount of trampling received by the grasses. The growing points of rye grass were 44.7 mm and 62.2 mm above the ground at two successive harvests in untrampled plots, but in plots that had been trampled by sheep, but not grazed, the heights were 27.2 mm and 34.7 mm, respectively (Edmond, 1974). The angle at which the tillers grow may also be affected by compaction, with high pressures causing shallower angles; for example, a compaction pressure of 200 lb (91 kg) reduced the tiller angle of perennial rye grass from 49° to 35° (Edmond, 1958).

Table 6.8 The mean tiller number, length and diameter of five grasses after light and heavy simulated trampling (Sun and Liddle, 1993b)

Species	Control	Light trampling	Heavy trampling	ANOVA (F-test)
Panicum maximum				
Tiller number (per plant)	2.5	3.4	2.1	21.9**
Tiller length (cm)	26.0	2.3	8.0	59.9**
Tiller diameter (mm)	11.8	8.0	5.3	219.0**
Chloris gayana				
Tiller number (per plant)	5.0	4.5	3.3	117.0**
Tiller length (cm)	26.5	9.8	7.4	18.7**
Tiller diameter (mm)	8.9	7.8	5.8	21.2**
Eragrostis tenuifolia				
Tiller number (per plant)	8.0	8.7	9.3	65.0**
Tiller length (cm)	23.7	13.1	12.1	28.9**
Tiller diameter (mm)	7.4	6.7	6.0	7.7*
Lolium perenne				
Tiller number (per plant)	13.0	14.1	13.5	8.0*
Tiller length (cm)	4.4	3.6	2.9	21.0**
Tiller diameter (mm)	2.4	2.2	2.2	4.9*
Sporobolus elongatus				
Tiller number (per plant)	7.0	9.3	8.5	53.7**
Tiller length (cm)	16.8	13.3	12.7	5.6*
Tiller diameter (mm)	7.3	6.1	5.1	19.4**

ANOVA tests: ** significant at $P \leq 0.1$; * significant at $P \leq 0.05$.

Table 6.9 Changes in the physiognomy of two grasses with trampling (Trew, 1973)

Character (average)	*Lolium perenne* Trampled	*Lolium perenne* Untrampled	*Festuca rubra* Trampled	*Festuca rubra* Untrampled
Height above ground (cm)	7.35	23.76	5.1	9.72
Number of seed pods	12.61	21.00	13.00	18.00
Length of seed head (cm)	3.10	8.60	1.75	3.20
Seed pods per cm of seed head	0.25	0.41	0.14	0.18
Number of leaves on seed stem	2.9	3.1	–	–
Length of leaves (cm)	2.1	5.3	–	–

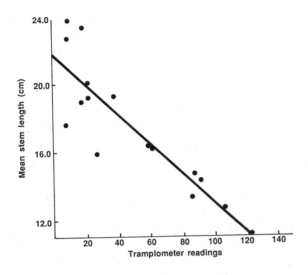

Fig. 6.22 The relationship between the height of the flowering stem (culms) of crested dog's tail (*Cynosurus cristatus*) and intensity of trampling (from Young and Pendlebury, 1969).

Table 6.10 A tentative classification of grasses based on the position of the growing points in relation to the ground during the growing season, and on the ratio of fertile to vegetative stems (Branson, 1953)

Class I Vegetative growing points above the soil	Class II Vegetative growing points at the soil surface	Class III Few or no vegetative growing points
Switchgrass (*Panicum viergatum*)	Kentucky bluegrass (*Poa pratensis*)	Little blue stem (*Andropogon scoparius*)
Western wheatgrass (*Agropyron smithii*)	Buffalo grass (*Buchloe dactyloides*)	Canada wild rye (*Elymus canadensis*)
Big bluestem (*Andropogon gerardii*)	Blue grama (*Bouteloua gracilis*)	Plains muhly (*Muhlenbergia cuspidata*)
	Side-oats grama (*Bouteloua curtipendula*)	Marsh muhly (*Muhlenbergia racemosa*)

	June grass (*Koeleria cristata*)	Scribner panic grass
		  (*Panicum scribnerianum*)
		Wilcox panic grass
		  (*Panicum wilcoxianum*)
		Annual grasses (including: hairy chess
		  (*Bromus commutatus*), little barley,
		  (*Hordeum pusillum*), six weeks fescue
		  (*Festuca octoflora*) and others)

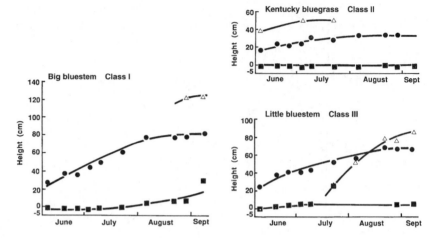

Fig. 6.23 Average heights of growing points (■), leaves (●), and flower stalks (△) over part of the growing season. Examples of three classes are shown (see Table 6.10). (After Branson, 1953.)

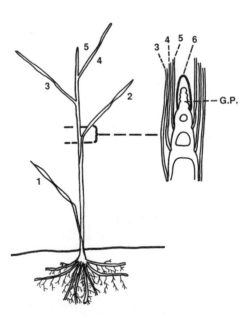

Fig. 6.24 Diagram of young western wheatgrass (*Agropyron smithii*) stem and enlarged longitudinal section of growing point (GP). The leaves are given the same number in both figures. (After Branson, 1953.)

The plant stem and leaves may be subjected to bending, compression, stretching and abrading by surface friction forces, as a result of trampling. Compression of plant stems will obviously result in major damage if they are hollow or very soft (Fig. 6.25). Rigid, lignified, suberized or thickened stems will withstand considerable compression, and Shields and Dean (1949) found that the small, thick-walled cells in a young southern blue gum eucalyptus (*Eucalyptus globulus*) stem enabled it to withstand compression from a sectioning microtome blade much more readily than the less dense tissue of a large-celled stem of wandering Jew (*Zebrina pendula*) (Fig. 6.26).

6.4 Plant anatomy

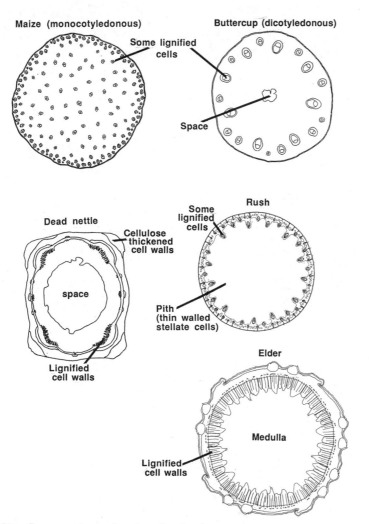

Fig. 6.25 Cross-sections of various kinds of stem (based on drawings in Vines and Rees, 1964).

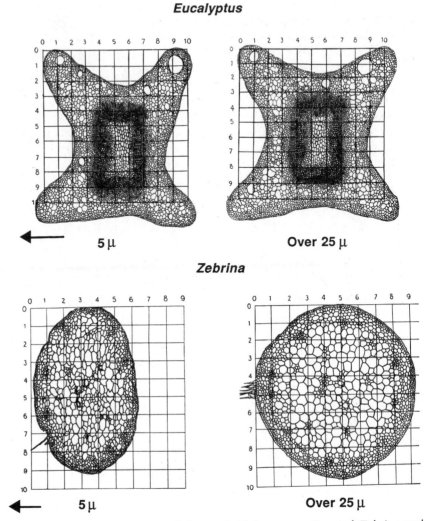

Fig. 6.26 Relative distortion of thin and thick cross-sections of *Zebrina* and *Eucalyptus* when cut with similar force. At a thickness of 5 μm the smaller cells of the *Eucalyptus* reduce distortion to 5%, compared to 30% in the larger celled *Zebrina*. (From Shields and Dean, 1949.)

Although the area of the leaves of white clover (*Trifolium repens*), great plantain (*Plantago major*) and knot grass (*Polygonum aviculare*) may be markedly reduced by trampling (compression and other forces), the structure of the assimilatory (photosynthetic) tissues and their cell dimensions appears not to change (Goryshina, 1983). Also, no signs of the development of features that prevent water loss were observed by Goryshina, although trampling-resistant plants are often said to show these features at the morphological level. Goryshina's (1983) main observation was that trampling inhibited the processes of cell division during leaf blade formation, but there was very little effect on the growth of leaf cells to full size once they had been formed. The final leaf size was reduced by the reduced number of cells.

Bending of stems and leaves of small plants commonly occurs in areas used for recreation, either because the trampling foot forces the leaf or stem to the ground or because it is bent over a stone or twig on uneven ground. Hollow structures are liable to collapse under bending forces unless there is a flexible internal support such as the 'mole' used by plumbers when bending metal pipes. Thus hollow stems are liable to collapse and break, as are brittle, solid stems such as those of broad-leaved plants in the early stages of secondary thickening (Fig. 6.25). Those stems that are oval in cross-section will withstand a greater degree of bending at right angles to their smaller measurement than in the other direction. Leaves and stems which have flexible strengthening tissues such as suberin and hemicellulose, as found in the plantain family, are probably the most resistant of all the higher plant structures, being strong and yet flexible. The force required for breakage will depend on the intrinsic strength of the stem and its cross-sectional area.

The ability of a structure to withstand stretching forces depends on its tensile strength. Tensile strength depends on the cellulose cell walls and the strengthening materials such as hemicellulose and lignin (Evans, 1967a). Longitudinal orientation of the microfibrils making up the cellulose wall may also increase its tensile strength (Preston, 1955; Probine, 1963). Evans (1967b) tested young leaves of a number of grasses for strength (S), expressed as

$$S = \frac{\text{breaking load (g)}}{\text{dry weight (mg) of 5 cm length of leaf}}.$$

These are listed, in order of decreasing strength, together with their cellulose and sclerenchyma (lignified strengthening tissue) content, expressed as a proportion of the cross-sectional area of the leaf, in Table 6.11.

Although Evans found that within species there were positive correlations between cellulose and strength, and sclerenchyma content and strength, his data show no consistent relationship between these features when the species are compared. This suggests that although these features are significant once the general pattern of tissues is established, the detailed layout of the tissues in the leaf is also very important. Studies by Shearman and Beard (1975c) demonstrated that leaf tensile strength of seven species of grass was correlated with their wear tolerance, as defined in Shearman and Beard (1975a) (Table 6.12). There was also a high correlation $(r = 0.88)$ between the proportion of the dry weight of grass made up of cell wall material and wear tolerance when expressed as weight per unit area of turf, but not when it was expressed as a percentage of the dry weight of each plant (Shearman and Beard, 1975b) (Table 6.13). Further analysis of the cellulose, hemicellulose and lignin contents of the same seven species (Table 6.14) all showed a higher correlation with wear tolerance when expressed on the basis of a weight per unit area of ground (four significant correlations) than when expressed as a percentage of the dry weight of the plant (no significant correlations). This suggests that the way the plants make up the turf may be more important for wear resistance than the details of their individual morphology.

The horizontal component of trampling or vehicle forces is responsible for the surface abrasion of above-ground plant parts. Evenness of the leaf or stem surface and the thickness and toughness of the outer waxy cuticle would appear to be the main protective features,

Table 6.11 Strength, cellular content and sclerenchyma area of various grasses in New Zealand (data from Evans, 1967b)

Species or variety	Strength	% Cellulose by weight	% Sclerenchyma
Perennial rye grass (*Lolium perenne* var. *ruahui*)	112.8	16.1	2.72
Cocksfoot (*Dactylis glomerata*)	105.4	17.2	2.57
Timothy (*Phleum pratense*)	84	17.2	3.25
Manawa rye grass (*Lolium perenne* x *Lolium multiflorum*)	82	15.6	1.88
Prairie grass (*Bromus unioloides*)	81.8	17.2	3.35
Brown top (*Agrostis tenuis*)	62.6	15.3	1.92
Rough blue grass (*Poa trivialis*)	62.4	14.4	–
Yorkshire fog (*Holcus lanatus*)	50.8	15.6	1.43

Table 6.12 Leaf tensile strength defined as grams required to break the leaves of seven turf grasses (Shearman and Beard, 1975a,c)

	Wear tolerance[a]	Tensile strength (g/leaf)
Manhattan perennial rye grass (*Lolium perenne*)	2.5	635
Kentucky 31 tall fescue (*Festuca arundinacea*)	2.4	722
Merion Kentucky bluegrass (*Poa pratensis*)	2.3	635
Pennlawn red fescue (*Festuca rubra*)	1.3	305
Italian rye grass (*Lolium multiflorum*)	1.0	696
Cascade chewings fescue (*Festuca rubra* var. *commutata*)	0.5	269
Rough bluegrass (*Poa trivialis*)	0.2	412

[a] Mean of wheel and slide injury ratings subtracted from 5, which equals 100% bare soil.

Table 6.13   Total cell wall (TCW) content of various species of turf grass on a weight per unit area (mg dm^{-2}) and on a percentage of total dry weight basis (% dry wt) (data from Shearman and Beard, 1975b)

	TCW	
	mg dm^{-2}	% dry wt
Manhattan perennial rye grass (*Lolium perenne*)	726	48
Kentucky 31 tall fescue (*Festuca arundinacea*)	805	51
Merion Kentucky bluegrass (*Poa pratensis*)	739	45
Pennlawn red fescue (*Festuca rubra*)	507	51
Italian rye grass (*Lolium multiflorum*)	500	46
Cascade chewings fescue (*Festuca rubra* var. *commutata*)	482	51
Rough bluegrass (*Poa trivialis*)	414	40

Table 6.14   Cellulose, hemicellulose and lignin content as percentage of dry weight of seven species of turf grass, 10 weeks after seedling emergence (Shearman and Beard, 1975b)

	Cellulose	Hemicellulose % of dry weight	Lignin
Manhattan perennial rye grass (*Lolium perenne*)	23.0	21.5	4.37
Kentucky 31 tall fescue (*Festuca arundinacea*)	24.5	21.1	6.05
Merion Kentucky bluegrass (*Poa pratensis*)	24.0	17.5	4.33
Pennlawn red fescue (*Festuca rubra*)	24.7	21.6	6.15
Italian rye grass (*Lolium multiflorum*)	22.8	19.3	5.85
Cascade chewings fescue (*Festuca rubra* var. *commutata*)	26.9	20.0	6.22
Rough bluegrass (*Poa trivialis*)	18.5	19.4	2.63

although a felt of hairs and leaf flexibility may also help to dissipate abrading energy.

In summary, the limited evidence available suggests that the percentage and distribution of cell wall materials both contribute to a plant's resistance to wear.

Mosses have been recorded as a part of the trampled flora, or at least as affected by trampling, by a number of authors (for example, Bayfield, 1971b; Liddle, 1973b; Liddle and Greig-Smith, 1975b; Studlar, 1980, 1983; Pentecost and Rose, 1985; Jónsdóttir, 1991). Surprisingly, a detailed consideration of the effect of trampling on the life forms of mosses was not published until Studlar (1983). As there are few generally used common names for mosses, their Latin names will be used in this section.

6.5 Mosses

Mosses have been placed in two groups strictly according to their morphology: acrocarpous, where the stems stand vertically up from the substrate, and pleurocarpous, which are much branched, generally horizontally growing forms. However, a more ecologically meaningful classification, which not only takes into account the growth form but also includes a consideration of the way individuals are assembled and the influence of external factors, has been put forward by Magdefrau (1982). This has 10 forms, as set out in Fig. 6.27. It would be instructive to analyse the mosses recorded by Studlar (1980) according to this scheme, which incidentally is also applicable to liverworts.

Mosses, like higher plants, appear to have a range of responses to trampling pressures. The pleurocarpous *Brachythecium albicans* flourished in the centre of a sand dune car-track whereas *Camptothecium lutescens*, also a pleurocarpous moss, was absent from the track but present in quantity in the adjacent undisturbed vegetation (Fig. 6.28). *Camptothecium lutescens* had a similar distribution on a footpath, at the edges of which two acrocarpous mosses occurred.

Bryophyte (mosses and liverworts) cover, frequency and species richness was greater within the trails examined by Studlar (1980) in the Appalachian Mountains, Virginia, than it was in the vegetation adjacent to the trails. Bryophytes are an important, sometimes trample-resistant component of trail vegetation and thus probably important in preventing soil erosion on trails (Studlar, 1980), although Jónsdóttir (1991) reported that *Rhacomitrium lanuginosum* was very sensitive to trampling in the Icelandic highlands. Mosses growing in Alaskan tundra also appeared on a track in quantity, although they were much reduced in scoured areas and ruts (Fig. 4.8) (Lambert, 1972). It is also interesting to note that in Studlar's (1980) study all the trail species were present in the adjacent forest and those species normally associated with disturbance in that area, *Bryum argenteum*, *Ceratodon purpureus* and *Funaria hygrometrica*, were absent from the trails. Studlar (1980) commented that 'Bryophytes may be expected to thrive on trails, providing there is sufficient rainfall, since some species are known to do well in unstable, temporary, severe or bare habitats'. The removal of plant litter, which can inhibit moss growth, as a result of trampling, may also be an important factor.

Fig. 6.27 Life forms of mosses. (a) Annuals: i. *Diphyscium foliosum*, ii. *Phascum curvicollum*; (b) short turf: *Trichostomum brachydontium*; (c) tall turf: *Dicranum undulatum*; (d) cushion: *Grimmia pulvinata*; (e) mat: *Hookeria lucens*; (f) weft: *Thuidium delicatulum*; (g) pendants: i. *Papillaria deppei*, ii. *Phyllogonium viscosum*; (h) tail: *Prionodon densus*; (i) fan: *Neckeropsis undulata*; (j) dendroids: i. *Mniodendron divaricatum*, ii. *Rhodobryum roseum*. (a)i and ii × 4; (b–j) × 2/3. (From Magdefrau, 1982.)

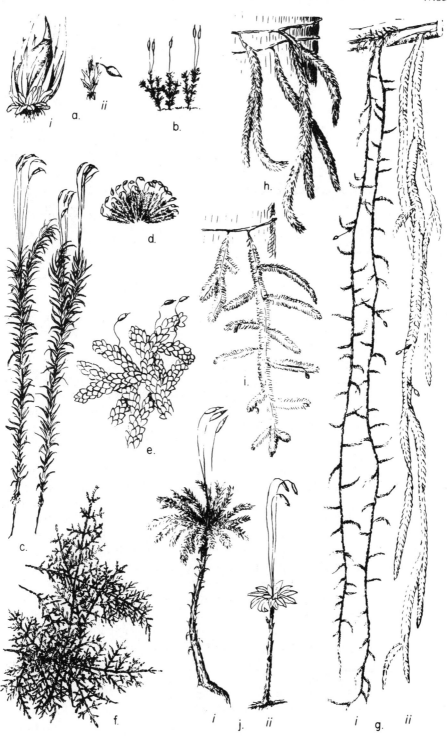

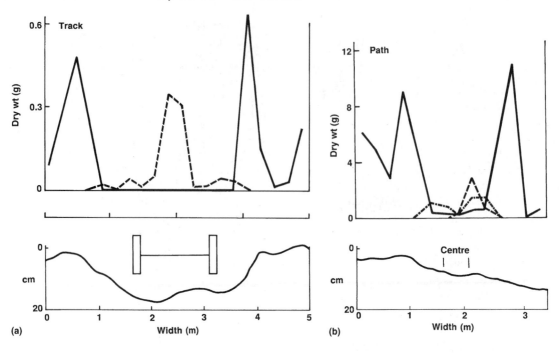

Fig. 6.28 The distribution of various mosses across (a) a car track and (b) a footpath at Aberffraw, Wales. Solid line, *Camptothecium lutescens*; dashed line in (a), *Brachythecium albicans*; in (b), *Bryum species* (annuals). Dashed and dotted line, *Tortula ruraliformis* (annual). Lower graphs, soil profiles vertical exaggeration × 5 (From Liddle, 1973b.)

The survival of *Polytrichum commune*, an acrocarpous moss, studied in a separate experiment, was attributed to the low growing points (meristems) and the wiry, horizontal, buried or surface-growing 'rhizomes' which are formed from leaf fragments attached to broken stems (Studlar, 1980; personal communication, 1988). *Ditrichum pallidum* was also found to be resistant because of its protected meristems, but this has a short turf growth-form where growing points, stems and leaves were protected by adjacent plants in the turf, and by being pushed into the soil (Studlar, 1980). A parallel is also drawn by Studlar between the boat-shaped leaves of *D. pallidum*, with their thick midribs, and the short, tough concave leaves cited by Bates (1935) as characteristic of resistant grasses.

The height of both *Polytrichum commune* and *Sphagnum recurvum* was also reduced by experimental trampling, with the sphagnum being the more vulnerable of the two (Fig. 6.29) (Studlar, 1980). Further trampling experiments (Studlar, 1983) were carried out on a *Sphagnum palustre* and *S. henryense* (tall turfs) mixed stand, *Polytrichum*

commune (tall turf), *Ditrichum pallidum* (short turf), and on two pleurocarpous mosses, *Thuidium delicatulum* (wefts) and *Hypnum impo-
nens* (smooth mat) (Fig. 6.30).The author ranked the species, from
greatest to least damage, as *Sphagnum* (both species), *Polytrichum*,
Thuidium and, lastly, the two horizontal species, *Ditrichum* and
Hypnum. The *Sphagnum* species were reduced by only 1600 passages
in the case of *S. recurvum*, to a peat made up of isolated leaves,
detached heads or capitula, and stem and shoot fragments 1–3 cm
long. Upright shoots of the turfs of *Sphagnum*, *Polytrichum* and
Ditrichum, and ascending primary stems of the wefts of *Thuidium*,
were fragmented and knocked down by trampling. Short turfs of
Ditrichum and wefts and smooth mats of *Hypnum* were relatively
protected from damage to the fragments by soil encrustation. The
cover of the tall turfs was sometimes increased by damage as the stems
were then lying horizontally on the soil surface.

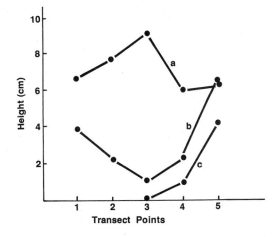

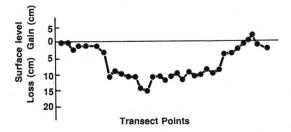

Fig. 6.29 Top, change in height of a *Polytrichum* moss colony in Jefferson
National Forest, Virginia; (a) before trampling, (b) after 1600 passages made
over 12 days, and (c) after 4200 passages made over 28 days. Bottom, change
in level of a *Sphagnum* moss colony after 1600 passages. Horizontal line indicates the original level of the moss. (From Studlar, 1980.)

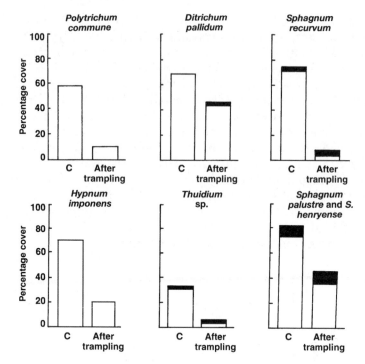

Fig. 6.30 The effect of trampling on the percentage cover of six species of moss. They all received 4200 passages over 28 days, except the *Sphagnum* which received 1600 passages on one day. The shaded portion indicates species other than the dominant one. (After Studlar, 1983.)

There has been little detailed work on the relationship of moss anatomy to trampling effects. Studlar (1980) noted that the presence of many non-living cells in the *Sphagnum* leaf and the fact that its stems rest on the soft remains of the previous year's growth may contribute to its vulnerability. As in the higher plants, the processes of leaf cell formation in *Dicranum scoparium* and *Polytrichum piliferum* were inhibited by trampling during leaf blade formation but there was almost no effect on cell extension (Goryshina, 1983).

6.6 Lichens   In some alpine and polar habitats lichens form an important part of the ground cover and they are also ecologically significant in arid lands (Rogers, 1977). They are primarily plants of harsh environments and are able to withstand extremes of temperature and desiccation, although they may be present in almost any habitat. Lichens are susceptible to considerable damage from trampling, recorded from sites in the Rocky Mountains in Colorado (Willard and Marr, 1970), the

Coastal Ranges, Washington (Bell and Bliss, 1973) and Mt McKinley, Alaska (Stelmock and Dean, 1979), as well as Scotland (Bayfield, 1971b), the Austrian Alps (Grabherr, 1985), low-lying northern forest in Sweden (Kardill, 1974), coastal dune areas of Denmark (Alstrup, 1978) and sheep-trampled arid areas of Australia (Rogers, 1977). Like mosses, there are few generally used common names for lichens and Latin names are used in this section.

Lichens are unusual plants, being composed of a symbiotic union of photosynthetic algae and saprophytic fungi. The two components may be of various species but they are always constant so that 'species' and growth forms of lichen can be defined. Three essential forms are usually recognized: crustose, foliose and fruticose (Fig. 6.31) (Hale, 1983). These are, however, 'in no sense a natural division' but 'points on a scale of continuous differentiation from primitive to highly structured thalli with many intermediates' (Hale, 1983). So these intermediates are likely to be found in the field. The soil lichens of the genus *Cladonia* studied by Bayfield, Urquhart and Cooper (1981) have hollow finger-like vertical growths called podetia (Fig. 6.32). A part of a podetia wall is made up of closely cemented fungal strands (hyphae) which form a strengthening tissue (pseudoparenchyma or sclerotic thickening). Like all lichens they are soft and flexible when moist and become brittle and crumble easily when they dry out.

Bayfield, Urquhart and Cooper (1981) reported that on the Cairngorm Plateau (1000 m) in Scotland, which is used extensively by walkers in summer, damage to lichens, particularly the large foliose species, is immediately apparent in the form of broken or crushed thalli, a feature that I have also observed in coastal sand dunes at Aberffraw, North Wales and reported by Grabherr (1985) from the Austrian Alps. A detailed survey of the amount of damage suffered

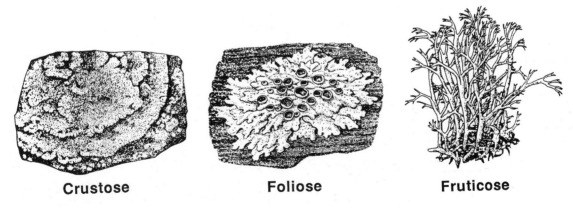

Crustose   **Foliose**   **Fruticose**

Fig. 6.31 Growth forms of lichens: (a) crustose; (b) foliose; (c) fruticose (from Hale, 1983).

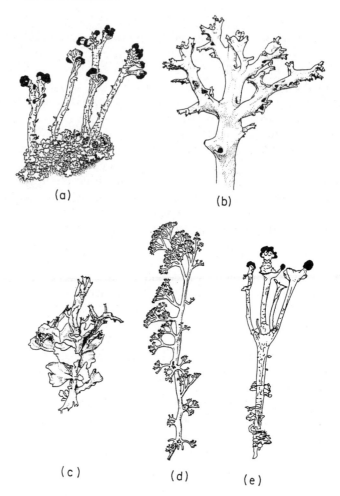

Fig. 6.32 Podetia of lichens: (a) *Cladonia cristatella*; (b) *C. perforata*; (c) *C. turgida*; (d) *C. rangiferina*; (e) *C. gracilis* (from Hale, 1983).

by four species of fruticose, ground-growing *Cladonia* (Bayfield, Urquhart and Cooper, 1981) revealed that adjacent to the most heavily used paths *C. uncialis* received the highest damage (22%), followed by *C. arbuscula* (16%), *C. rangiferina* (10%) and *C. impexa* (6%) (Fig. 6.33). In many cases the mean damage resulting from widespread low levels of use was itself very low, for *C. rangiferina* and *C. arbuscula* it exceeded 2% only at 1 and 5 m from the most popular paths and for *C. impexa* this level was reached only at 1 m from the path. By comparison, *C. uncialis* exceeded 2% damage at 6 out of 12 combinations of distance and use shown in Figure 6.33. At higher levels of use, in field experiments (Bayfield, 1971b), lichen cover was reduced on the Cairngorms from 17% in undamaged conditions to 5% after 240 passages (Fig. 6.34), and a similar level of reduction occurred on sand dunes by Lake Huron, Ontario, where 300 passages reduced lichen cover from 100% to about 30% (Bowles and Maun, 1982).

The detailed effects of trampling on lichens at different levels of moisture content were studied in the four species mentioned above, in the laboratory (Bayfield, Urquhart and Cooper, 1981). Turfs of the four species of *Cladonia* were collected, cut into 10 cm² blocks and the water content adjusted by a mist spray and various periods of drying. These blocks were then set in the ground and subjected to a single footfall from a 55 kg person wearing smooth-soled shoes. All species suffered much more damage when their water contents dropped below about 25% of the oven-dry weight (ODW) of the lichen thallus (Fig. 6.35). The damage to *C. uncialis* in the moist, elastic state was

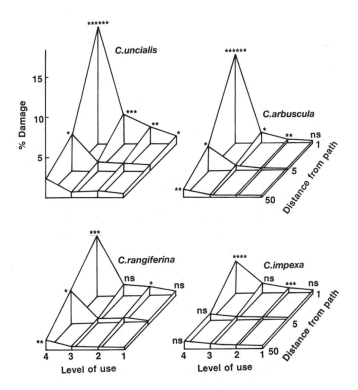

Fig. 6.33 Comparison of damage to four species of *Cladonia* at various distances from the path and various levels of use, in the Cairngorm Mountains, Scotland. Use level 1, no visitors; 2, 1–6.1 people per day; 3, 6.2–18.8 people per day; and 4, over 18.8 people per day. Significance (analysis of variance) shown by *, 0.05; **;0.01; ***, 0.001; ns, not significant. (Bayfield *et al.*, 1981.)

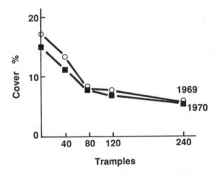

Fig. 6.34 The effect of various levels of trampling on the cover of lichens, Cairngorm Mountains, Scotland. ○, 1969 one year after trampling; ■, 1970, after the second year of 'recovery' (from Bayfield, 1971b).

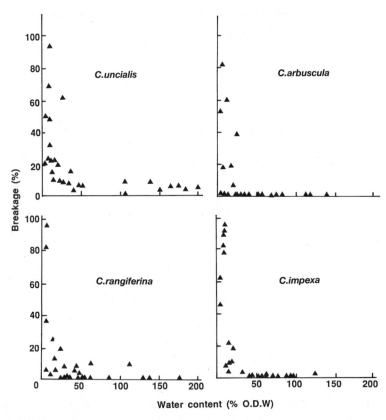

Fig. 6.35 The effect of lichen water content (as % oven dry weight) on the percentage breakage of *Cladonia* podetia when trampled (from Bayfield, Urquhart and Cooper, 1981).

often more than 5% but it only occasionally exceeded 2% for the other three species. In all cases the detached fragments were mostly less than 1 cm long and over three-quarters of the fragments were less than 5 mm long, when the thalli were dry.

A detailed examination of cross-sections of the podetia revealed that those of *C. uncialis* had thinner walls, a narrower layer of sclerotic tissue and a greater podetium diameter than the other three species (Table 6.15). The authors commented that this may, to some extent, explain the relatively high fragmentation of moist *C. uncialis*, which they also demonstrated by experiment (Table 6.15). Crawford and Liddle (1977) remarked that 'the relative trampling tolerance of different species (of higher plants) cannot be considered in isolation from the habitat in which they are growing'. This is also true for lichens, and Bayfield, Urquhart and Cooper (1981) observed that *Cladonia uncialis* is, in contrast to the other three species, often found as a colonist of patches of bare ground where it may tend to dry out more thoroughly (and become more brittle) than the other three species, and where it could be more liable to human trampling. In conclusion, it is also worth noting that Rogers and Lange (1971) found that the frequency of occurrence of four species of desert lichens declined rapidly around waterholes heavily trampled by sheep.

6.7 Summary

There are many summaries in the literature of qualities that contribute to trampling resistance, or vulnerability, of plants. There are clearly qualities such as flexibility, toughness, protected meristems and low growth habit that confer on plants varying degrees of resistance to trampling. There are also features, such as suberized and hemicellulose tissues, lignified tissues, overlapping leaf sheaths and rhizomatous growth, that contribute to the resistant qualities. It is therefore advantageous to study and be aware of these qualities and features when seeking an understanding of the trampling process, for academic or

Table 6.15 Comparison of the range of dimensions of transverse sections of podetia of the four species of *Cladonia* (mm) and the mean fragmentation of 10 moist (> 30% moisture) and 10 dry (< 30% moisture) samples after a single footfall. Sections of podetia were cut from about the middle of mature well-grown specimens (Bayfield, Urquhart and Cooper, 1981)

	C. uncialis	C. impexa	C. arbuscula	C. rangiferina
Thickness of podetium wall	0.1–0.15	0.15–0.2	0.2–0.25	0.23–0.35
Width of sclerotic thickening	0.05–0.06	0.08–0.1	0.1–0.11	0.1–0.18
Diameter of podetium	1.1–2.3	1.1–1.3	1.2–1.9	1.1–1.16
Fragmentation (% of total biomass)				
Moist samples	6.9	1.5	0.5	2.8
Dry samples	30.9	38.4	26.5	29.3

management purposes. However, as with all biological material, it is infinitely varied, and a degree of caution is recommended when interpreting these features. The following comments are therefore not meant to devalue the considerable level of knowledge that we have achieved, but are given as an aid to the interpretation in situations where there is no direct information.

When resistant features of a species of plant are being considered, it must be borne in mind that a single disadvantageous feature can be enough to reduce drastically the resistance of the whole plant, even if all the other characteristics appear favourable. If I might paraphrase, 'the trampling resistance of a plant is determined by its weakest feature'.

Resistant features can perhaps be placed in two groups. First, is the type of feature, such as protected meristems, where any increase in protection will add to the plant's resistance, unless too much of the plant's reserves are utilized. The second type of feature depends very much on having an optimum level; strengthening of the stem will increase resistance up to the point where the stem starts to become inflexible. It is then more likely to break when trodden on or driven over, and further lignification will only lower its resistance, at least until it reaches a large enough size to resist the forces imposed on it, or so that it looks strong and is avoided by walkers and drivers.

The emphasis on different features will also shift, depending upon whether comparisons are being made within a group of very similar plants or among widely dissimilar species. In the study of grasses, the total amount of cellular material appears to be positively correlated with resistance, but in a comparison between a grass and an upright broad-leaved plant the general morphology will obviously be more important.

Turning to more practical levels of consideration, a population of plants growing together, such as a single-species grass sward, or a short turf of mosses, will often have relatively different resistance than would be shown by a comparison of single individuals of the same species. This was demonstrated by the work of Shearman and Beard (1975b), who found that there was a good correlation between the amount of lignin per unit area of turf and its resistance to wear, but that the relationship did not hold for individual plants. As noted above, the relative trampling resistance of different species of plant may change according to the habitat in which they are growing.

The reaction of humans or trampling animals to the plant or situation in which it is growing (for example, rush tussocks inviting use as 'stepping stones' across a peat bog) must also be considered when full understanding of any situation is sought. Finally, a full understanding of any ecological situation will usually require a knowledge of at least the recent history of the site.

The physiological reactions of plants to damage 7

The information presented in this chapter is almost all derived from experiments which were not designed to test the effects of recreation. However, the examples I have selected do have a bearing on the processes that are likely to take place when people, animals or vehicles come into contact with plants. It is clearly desirable that more direct work be done to examine plants' responses to the actual processes of recreation impact.

The physiological reactions to a range of stimuli, from gentle touching through a series of increasing force and more severe damage to a consideration of the effect of breaking branches or stems (Fig. 7.1), will be considered in this chapter. Where possible the physiological reaction is indicated and the ultimate morphological change is then presented. However, the linkages shown in Fig. 7.1 are not complete or fully researched so the details of the logical structure must be regarded as tentative.

Walkers in vegetated areas probably touch far more plants in passing than they actually tread on, especially if they are on little-used paths or in longer vegetation. It follows that a widespread effect of outdoor recreation is the plants' reaction to this gentle stimulus which may be a very occasional touch, or repeated brushing in more frequently used locations.

It is perhaps not widely known, although long established, that plants respond to contact with a bioelectric change that is essentially the same as that which occurs in an animal's nervous system. This bioelectric impulse takes the form of an action potential (about 50–100 mV) similar to that of an axon, induced according to an all-or-none reaction (Roblin, 1985). This impulse originates even when stimulus is not injurious and was called 'middle wave conduction' by Sibaoka (1950) (Fig. 7.2). It occurs in most plants although it may be absent in some species and it can be modified by biological conditions in others (Pickard, 1973).

7.1 Touching

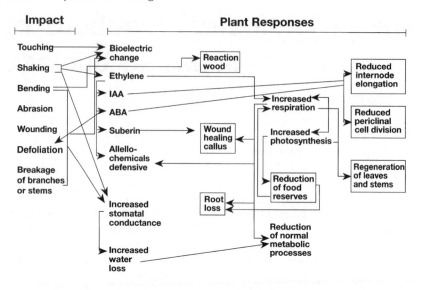

Fig. 7.1 Some of the relationships between types of impact and physiolog-ical response. Note that impacts are placed in order of increasing severity and, for example, that the bioelectric change will also take place under the more severe impacts. Square boxes indicate morphological or tissue changes. IAA, indole acetic acid; ABA, abscissic acid.

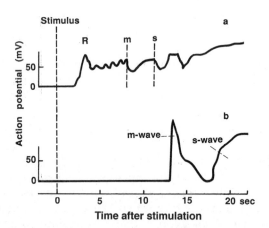

Fig. 7.2 Action potential changes recorded a few millimetres from the point where the plant (*Mimosa pudica*) was treated (a) and about 5 cm from the stimulation site, (b). R, rapid response phase; m, a longer-term, more drastic change in potential; s, an even slower longer-term change in potential. (After Sibaoka, 1950.)

Repeated contact by brushing has been shown to affect the growth of seedlings, reducing the stem or petiole length (Biddington and Dearman, 1985). This kind of mechanical stress may also strengthen the stems and petioles of some species (Heuchert, Marks and Mitchell, 1983) and incidentally induce resistance to frost and drought (Jaffe and Biro, 1979). All of these responses may be advantageous to plants growing in trampled areas, increasing their mechanical resistance and their survival in the more extreme microclimate which may be induced by a reduction in vegetation cover (Liddle and Moore, 1974) (p. 260).

Experimental brushing of seedlings with paper for 1.5 minutes each day for three or more days also increased the leaf thickness in cauliflower, lettuce and celery, and the weight of chlorophyll per fresh weight of leaf tissue increased in lettuce and celery but declined in cauliflower seedlings (Biddington and Dearman, 1985). The petiole and hypocotyl thickness was also reduced by brushing cauliflower seedlings, indicating that some species are adversely affected by this kind of contact. The leaf area was reduced by over 20% in all three species at some stage of their growth, although the authors did not report whether this was due to a reduction of cell replication or of cell elongation (Chapter 6) (see also Goryshina, 1983). Reduction of leaf area is also a common feature of plants in trampled areas.

Root length was reduced by 10–20% and the number of lateral roots by 10–25% when seedlings were brushed (Biddington and Dearman, 1985). The effect on the root : shoot ratio was variable, being increased in lettuce, reduced in celery, and unaffected in cauliflower seedlings. Again this suggests that some species may be disadvantaged by these effects of contact.

Finally, it is worth noting that growth responses to brushing were seen several days after brushing had ceased, noticeably in leaves which were barely visible at the time of brushing (Biddington and Dearman, 1985). Thus even very occasional use of the pathway may still have a longer-term effect on the morphology of a plant touched by the visitors.

7.2 Shaking

Shaking of plants has usually been carried out in an attempt to simulate the effects of wind. However, walkers or vehicles also shake plants when passing through vegetation, and Jaffe (1973) pointed out that adaptations designed to protect plants from stresses produced by high winds may be similar to those protecting them from the stresses induced by moving animals. The results of the shaking experiments are therefore likely to be relevant to recreation ecology.

Shaking tends to be a more intense treatment than touching, although there is still no disruption of plant tissue. Treatments ranged from 'rubbing the first internode (of stems less than 2 cm in length) about 10 times gently between forefinger and thumb while slightly bending it from side to side' (Biro *et al.*, 1980), applying treatment with a gyratory shaker at 260 r.p.m. for 10 minutes each day for 16–35 days, to shaking 20 cm high plants with a 4 cm plastic collar around the stem so that the whole plant was shaken continuously at 2–3 Hz, deflecting it 20° from the vertical at the end of each stroke (Grace *et al.*, 1982).

Plants subjected to these more robust treatments show a reduction in most size parameters when compared to unshaken control plants. For example, the stem length of alaska pea (*Pisum sativum* cv. Alaska) was reduced by a mean of 28% in four different treatments (Akers and Mitchell, 1984) and plants of maize (*Zea mays*) were reduced by 50% in height (Neel and Harris, 1972). Leaf numbers (Neel and Harris, 1972), fresh weights, dry weights and areas were also reduced, with 38% of the area being 'lost' from shaken leaves (Akers and Mitchell, 1984). The rates of extension of leaves of rye grass (*Lolium perenne*) and sand fescue (*Festuca arundinacea*) were also reduced by about 40% when the plants were grown in wind tunnels at high wind speeds of 7.4 m sec^{-1} (26.6 km h^{-1}), compared with the rate of extension when plants were grown at low wind speeds of 1 m sec^{-1} (3.6 km h^{-1}) (Russell and Grace, 1978) (Fig. 7.3). The work on the alaska pea also showed a reduction in the fresh and dry weights of fruit as well as a reduction in mean number of seeds per fruit (from 3 to 1.9) (Akers and Mitchell, 1984).

However, some parameters, which might be allied with plant strength, showed an increase. Akers and Mitchell (1984) found that in one case the stem's specific weight (mg dry weight cm^{-1}) increased from 2.4 to 2.9, and the specific weight of leaves (mg dry weight cm^{-2}) increased from 2.2 to 2.8, suggesting that the plants were becoming more robust. Normal elongation of the internode and epidermal cells

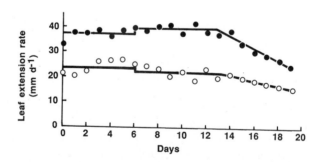

Fig. 7.3 The effect of high wind speeds (26.6 km h^{-1})(O) and low wind speeds (3.6 km h^{-1}) (●) on the rate of extension of rye grass (*Lolium perenne*) leaves. Lines indicate means over 6-day periods. (After Russell and Grace, 1978.)

of bean (*Phaseolus vulgaris* cv. Cherokee wax) plants was shown to be reduced by mechanical stress (Biro *et al.*, 1980; see p. 165–6 for details of treatment), while the radial enlargement of the stem internode increased (Fig. 7.4).

Microscopic examination of longitudinal sections of the stem tissues revealed that although cell elongation was reduced in the epidermal and cortical tissues, there was, by inference, a decrease in anticlinal cell divisions in the vascular tissues and pith where the cell length remained the same as in the controls. Examination of radial cross-sections showed that the increase in diameter of the stems was a function of both increased cell diameter and cell number. The increased size of the cortex was primarily due to increased radial cell enlargement, while the vascular tissues expanded as a result of increased cambial activity, producing higher numbers of secondary xylem and pith cells (Table 7.1 and Fig. 7.5) (Biro *et al.*, 1980). Whitehead and Luti (1962), Walker (1960) and Grace and Russell (1982) all recorded reduced cell expansion as a consequence of mechanical stress from wind. These results contrast with the results of Goryshina (1983) who examined the cells of trampled leaf tissues from three species of plants and attributed the smaller leaf sizes to a reduced number of cell divisions. However, Grace *et al.* (1982) also found that the epidermal cells from shaken leaves of sand fescue (*Festuca arundinacea*) were no smaller than those of the control leaves so that the

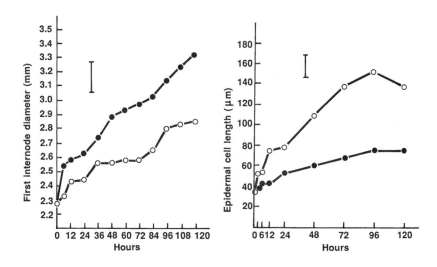

Fig. 7.4 The time course of radial enlargement of the first internode, and epidermal cell elongation of kidney bean (*Phaseolus vulgaris*). ○, Control plants; and ●, plants which were shaken daily; I, maximum standard deviation. (After Biro *et al.*, 1980.)

authors concluded that the reduced rates of growth were due to a reduced rate of division, rather than a reduction in cell expansion. So it seems that a reduction in both cell division and/or elongation may occur as a result of trampling or mechanical stress on leaves and stems, but cell size changes have been recorded only in stems.

The thickness of cells was also increased by wind activity (Venning, 1948; Walker, 1957, 1960; Larson, 1965) and Heuchert, Marks and Mitchell (1983) found that shaking increased the elastic strength of tomato stems while enhancing the cellulose content of their fibre.

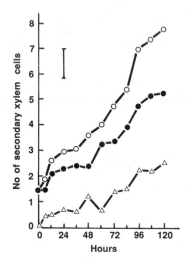

Fig. 7.5 The time course of the production of secondary xylem cells at the first internode of kidney bean (*Phaseolus vulgaris*) plants. ●, Control plants; ○, plants which were shaken daily; △, difference between treatments; I, maximum standard deviation. (After Biro *et al.*, 1980.)

Table 7.1 A comparison of cell numbers in control and shaken bean internodes. Each datum is followed by its standard deviation (Biro *et al.*, 1980)

Tissue	Number of cells along section radius		Radial thickness of tissue type (μm)	
	Control	Rubbed	Control	Rubbed
Pith	6.8 ± 0.5	8.8 ± 0.4	283.7 ± 2.6	313.0 ± 4.8
Secondary xylem	5.1 ± 0.3	8.3 ± 0.3	70.8 ± 1.4	88.6 ± 1.7
Secondary phloem	6.6 ± 0.4	6.8 ± 0.2	58.6 ± 0.9	57.7 ± 1.3
Primary phloem	3.0 ± 0.2	3.2 ± 0.2	23.2 ± 0.6	12.9 ± 0.8
Cortex	4.2 ± 0.3	4.7 ± 0.3	108.5 ± 1.0	131.7 ± 2.0

Turning to the physiological changes induced by mechanical stress, a number of authors have observed changes in water content. Specific water content (mg H_2O mg dry weight^{-1}) was reduced in the stems and leaves of alaska pea in some experiments when it was subjected to shaking (Akers and Mitchell, 1984). The stomatal conductance was almost doubled by the effect of shaking sand fescue (*Festuca arundinacea*) and the total water potential was reduced (Grace *et al.*, 1982) (Fig. 7.6). The authors suggested that shaking may have caused the stomata to open as leaf water potential fell, thus causing a higher transpiration rate which, in turn, caused a lower water potential and consequently reduced extension growth. Although, as the authors pointed out, 'this hypothesis is not consistent with all the observations' and it seems to have an element of circularity in the logic, it does seem likely that shaking by walkers or vehicles could affect the water potentials of some plants.

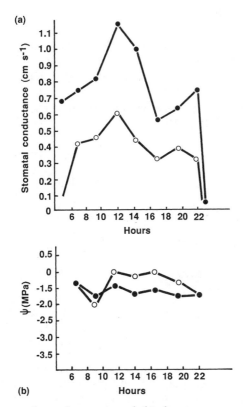

Fig. 7.6 (a) Stomatal conductance and (b) the water potential in unshaken (O) and shaken (●) sand fescue (*Festuca arundinacea*) plants (from Grace *et al.*, 1982).

7.3 Bending   Walkers and vehicles also bend plants, often repeatedly, when passing through and over vegetation in areas used for recreation. In addition to triggering the physiological reactions discussed above, more extensive tissue changes may be observed when a plant is repeatedly bent. Most of the research into this phenomenon has examined the production of 'reaction tissue' in woody plants, usually in connection with forestry. Reaction wood is a variation of normal wood structure, and is assumed to be a response to natural or artificial reorientation of a woody axis from its usual position.

Reaction wood is characterized in conifers by thick-walled, heavily lignified, rounded tracheid cells, and in dicotyledons by fibres with an inner unlignified, more gelatinous, layer of the secondary wall in addition to the usual lignified layer (Fisher and Stevenson, 1981). These fibres are assumed to shorten at maturation and, as they occur in suitable positions within the trunk or branch, are able to bend the displaced woody axis to its initial 'normal' position within the tree crown (Fisher and Stevenson, 1981). Therefore, reaction wood occurs as a response to an external influence.

The evidence suggests that the primary inducing factor in reaction wood formation is the force of gravity, presumably working through the auxin hormone systems, to bring the branch or leader back to its normal position (Fig. 7.7a). Where there has been no displacement with respect to gravity, stress (compression or tensile) apparently produces the same effect, 'correcting' the architecture of the affected plant (Fig. 7.7b) (Fisher, 1985).

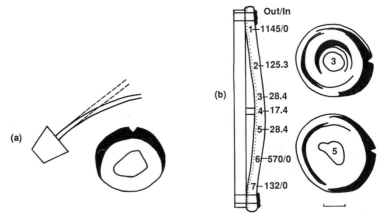

Fig. 7.7 Distribution of reaction wood (solid areas) in (a) a sapling tilted after initial growth without support, and (b) the stem of a vertical sapling of *Terminalia catappa* bent sideways with a brace. Cross-sections were taken at seven sites and the ratio of the amounts of reaction wood on the outer and inner sides of the stem is also shown. In (a) reaction wood was formed only on the upper side of the stem. (From Fisher, 1985.)

In the light of the evidence given above, it appears that any displacement of branches or stems due to recreation activities is likely to induce the formation of reaction wood in the affected areas, and similar changes may well occur in many kinds of plants.

Here again, no measurements on plants from recreation areas appear to have been made. However, even a cursory examination of plants growing on paths and tracks will show evidence of surface abrasion and the subsequent lesions which develop. This is not very surprising considering the magnitude of the horizontal vector of walking forces. The abrasions may be the result of direct contact between the plant and shoe or wheel, probably exacerbated by the presence of sand or soil particles. In addition, a path or track which has low vegetation cover and patches of bare ground will allow a greater flow of air on windy days, and thus allow the wind to pick up particles which may also abrade the plants (Skidmore, 1966; Mackerron, 1976).

Abraded leaf surfaces have their waxy cuticle either deformed or removed, the epidermal cells may also be removed, and with further abrasion the damage may extend to underlying palisade and mesophyll cells (Black and Mack, 1984) (Fig. 7.8). Apart from the physiological responses discussed in previous sections, abrasion may also cause a reduction in leaf turgor, which may be permanent, and a decrease in stomatal conductance (Black and Mack, 1984) (Fig. 7.8).

Crowberry (*Empetrum nigrum*) has been observed to lose most of its leaves up to 1 week after being trampled (Hylgaard and Liddle, 1981): 'in 3 to 4 days the colour of the leaves gradually changes from green to yellowish-brown due to wilting, probably induced by transpiration losses'. 'Microscopic examination of the leaves showed that the ericoid (rolled) structure of the leaves is destroyed by trampling' (Hylgaard and Liddle, 1981). The leaf loss may have been the result of the release of wound-induced ethylene, which was implicated in the experiments of Black and Mack (1984). They observed extensive leaf fall from poplars (*Populus* spp.) after abrasion by volcanic dust from the 1980 eruption of Mt. St. Helens, Washington State and, after the series of abrasion experiments mentioned above, they suggested that ethylene production may have been induced by the consequent damage to the leaves. Abeles (1973) reported that ethylene-mediated leaf abscission may occur without any visible evidence of senescence. It therefore seems possible that trampling which has abraded the leaves, perhaps coupled with more extensive damage, has also stimulated the production of ethylene and the leaf abscission process in crowberry.

7.4 Abrasion

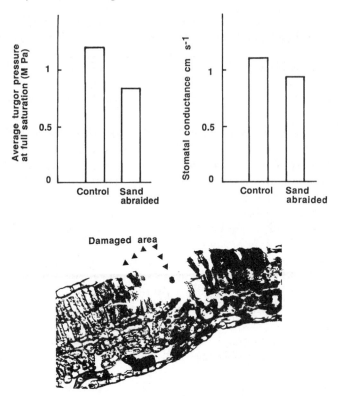

Fig. 7.8 Cross-section of abscissed leaf of lombardy poplar (*Populus nigra* var. *italica*) after ash abrasion. Removal of the epidermis and damage of the mesophyll was localized. Intact epidermal cells (extreme left side of upper leaf surface) surround all damaged areas. Histograms show average turgor pressure (MPa) at full saturation and stomatal conductance (cm s⁻¹) of control and sand-abraded plants of lombardy poplar. (After Black and Mack, 1984.)

7.5 Wounding   Trampling, vehicles, wood collection and many recreation-related activities cause mechanical damage and possibly kill plants growing in amenity areas. The plants' response to damage or wounding is therefore of direct relevance to understanding the ecological effects of recreation. It is not surprising to find that wounding stimulates responses that are similar in nature to, but more extensive than, the phenomena discussed above.

Bioelectric changes are the plants' first response to wounding; this is perhaps not surprising as 'research during the last few years suggests that bioelectric changes triggered by environmental modification are early events in the metabolic processes which direct the whole plant or some of its organs towards a particular metabolic development' (Roblin, 1985). The wounding response appears to be more long-lasting although of a similar magnitude to that produced by mechanical stimulation (Fig. 7.9a). Examination of the variation in amplitude of the response over time shows that the response to mechanical stimulation is followed by a much slower response, suggesting that two mechanisms may be involved in the wounding response (Roblin, 1985) (Fig. 7.9b). Roblin suggested that both responses are linked to the release of a 'stimulating substance' by wounding.

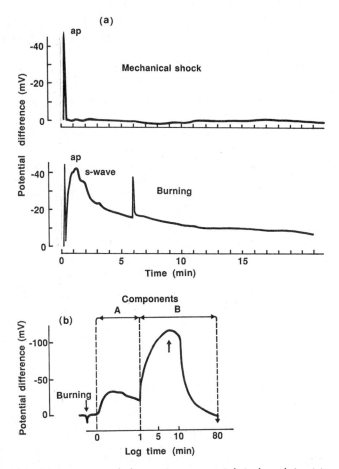

Fig. 7.9 The time course of the action potential induced in (a) mimosa (*Mimosa pudica*) petiole when the leaf was stimulated mechanically or 'gently' burnt, and (b) a kidney bean (*Vicia faba*) stem following burning of the leaf blade. (a) Shows the different potential produced by the two stimuli, with a large slow wave reaction to burning. The two phases of a burning stimulus are clearly differentiated when time is plotted on a log scale (b). (After Roblin, 1979, 1985.)

There are many reports of trauma-induced ethylene production (Williamson, 1950; Cooper, Rasmussen and Waldon, 1969; Cooper 1972). Stresses of many kinds induce increased ethylene production and Yang and Pratt (1978) suggested that 'any condition deviating from the normal environment of the intact healthy plant may lead to such an increase'. For example, gentle stroking of the leaf blades caused lily plants to produce more ethylene, resulting in reduced growth (Hiraki and Ota, 1975). It has been suggested that wounding overcomes whatever factor limits ethylene production (McGlassen and Pratt, 1964). When bark was peeled from the wood of tangerine terminal shoots, 5–10 times as much ethylene was produced in the peeled bark wood as in the intact stems and when the stems were completely macerated about 13 times as much ethylene was produced by the macerate.

The chemical pathways by which ethylene is produced in wounded tissue were not generally understood in 1978 at the time of Yang and Pratt's review. While the physiological functions of ethylene production were also unknown, Yang and Pratt (1978) suggested that it may activate a system to enable plants to cope with initial wound stress. An example with direct relevance to compacted soils in recreation sites was reported by Goeschl, Rappaport and Pratt (1966). Young seedlings responded to a load which was applied in a manner designed to oppose growth, with an increase in ethylene production. There was also a decrease in stem elongation and increased radial expansion, resulting in a sturdier axis capable of exerting and withstanding forces required for emergence from compacted or crusted soils. Ethylene may also be involved in wound healing as it produces swellings in apple twigs, callus formation and even root formation in stem cuttings of tomatoes (Zimmerman and Wilcoxen, 1935).

One of the more important wound-healing substances is suberin, a normal constituent of bark. Suberization is a general process involved in wound healing, and cut tissues gradually become covered with a layer of suberin which increases their diffusion resistance to water vapour. This process takes some time, up to 7 days in the case of potato tuber tissue (Kolattukudy, 1978) (Fig. 7.10). The rate of suberization rises with temperature up to 25°C and falls at higher temperatures, and high humidity generally has a favourable effect, as does oxygen up to 20% concentration. Abscissic acid can promote suberization and wound healing and, as may be expected, it acts more quickly in younger tissues.

Other chemicals may also be produced when a plant is wounded and some of these may have a defensive function (allelochemicals – generally, 'chemicals by which organisms of one species affect the growth, health, behaviour or population biology of organisms of other species' (Whittaker and Feney, 1971)). For example, trypsin

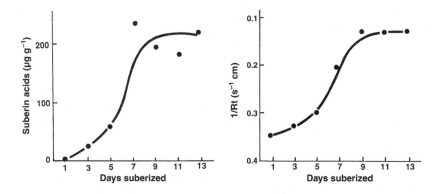

Fig. 7.10 Time course of suberization in potato tuber tissue. (a) Suberin acids laid down on wound; (b) development of diffusion resistance of tissue surface to water vapour. (Redrawn from Kolattukudy, 1978.)

(a proteolytic enzyme inhibitor) was released from 17 of 23 plant species, of 10 genera, from which the leaves were detached and supplied with a 'wound hormone' prepared from other leaves which had been ground to a powder (Walker-Simmons and Ryan, 1977). The trypsin inhibitor could act as a poison, or at least give indigestion to an animal that swallowed it. This would, if known by the animal, protect the plant from further damage.

Scots pine (*Pinus sylvestris*) seedlings responded to the removal of 50% of their bark at the base of their stems by increasing the production of resin acids (Gref and Ericsson, 1984). Two and a half times the normal concentration of acids was found in the bark adjacent to the wound. This included dehydroabietic acid, which may be a deterrent to larval feeding and provide increased decay resistance against white rot and brown rot fungi.

Several other compounds are released from wounded tissue, possibly induced by an increase in the production of IAA (indole-3-acetic acid) (Tanaka and Uritani, 1979). Earner and Jaffe (1982) suggested that 'mechanical perturbation of bean internodes induces ethylene evolution which in turn induces the accumulation of high levels of IAA and the production of abscissic acid (ABA) both of which contribute to the retardation of elongation of the internodes'.

The biochemical responses of plants to wounding are clearly complex and much of the work has been carried out on young growing plants and excised plant tissue; nevertheless, it is clear that these responses will also occur as a result of recreation-associated wounding. However, the most serious consequence of wounding may be the exhaustion of the plant's food reserves, especially by repeated damage.

7.6 Food reserves   The production of wound tissue and new organs must be carried out at the expense of some energy. If the plant has its photosynthetic area significantly reduced, it will have to mobilize food reserves which may have been required for further growth.

The utilization of food reserves is accompanied by production of CO_2 from respiration, and ethylene production is usually accompanied by a rise in respiration rate, which may increase sevenfold (Yang and Pratt, 1978). There is also little doubt that, in some cases, ethylene plays a hormonal role in the induction of an increased respiration rate (Lee, McGlasson and Edwards, 1970), thus utilizing stored food material. It is clear therefore that the consequence of wounding is the utilization of stored or readily available food by the plant.

Van Sambek and Pickard (1976) found that a localized wound 'has a dramatic effect on the gas exchange of an entire shoot if substances from the damaged cells gain entry to the transpiration stream'. They found that in the five species tested there was an initial burst of CO_2 production (from increased respiration), reaching a peak about 5 minutes after the plant was damaged. They suggested that the initial burst utilizes carbon compounds present within the cell and, by implication, subsequent material is obtained from elsewhere in the plant. When the tillers of Italian rye grass (*Lolium multiflorum*) were defoliated, leaving the main shoot intact, more carbon products went to the tillers, although in this case the roots continued to receive the same amount as previously (Marshall and Sagar, 1968). The other change was an overall increase in the export of assimilates from the main shoot. The authors suggested that the rate of photosynthesis may also have increased in the remaining tissues following defoliation. Damage by trampling or vehicles may then alter the 'carbon economy' of the whole plant.

In higher plants with vascular conducting systems (Tracheophyta), carbohydrate reserves are mainly stored in rhizomes, roots, tubers and stem bases, so they are relatively protected from the initial impact of trampling. In bryophytes, particularly mosses, starch and oil are both abundant and widespread, although starch storage organs (in the sense that they occur in higher plants) are rare among bryophytes (Watson, 1967). From regeneration studies it is evident that there are sufficient food reserves for wound healing in quite small particles, at least of some mosses (Bayfield, 1976). Lichens often have starch granules in the lower part of the thallus, which forms the 'base' of the plant, but even this part is still very vulnerable to trampling, especially when dry.

Energy-rich carbohydrates are stored when there is an excess of photosynthesis over respiration and body building. Storage is most likely to occur after periods of rapid growth (for example in late spring in seasonal climates, this may reduce the resources immediately available to produce stems, leaves and flowers). As soon as the aerial parts are fully developed, carbohydrates are produced by grass leaves in excess of current requirements. Storage of reserves is generally most active in the autumn (Weinmann, 1952). However, in the subtropical Bermuda or couch grass (*Cynodon dactylon*) total available carbohydrates were lowest in the roots and rhizomes at the early growing stage and replenishment took place at the time of active shoot growth (Hall *et al.*, 1948). This highly trample-resistant grass produces numerous rhizomes which are high in starch content, and shoot growth is found to consume only a relatively small fraction of the large carbohydrate reserve. Bermuda grass is one of the warm-climate grasses having the very efficient C4 photosynthetic system and this explains to some extent why production of carbohydrates nearly equals utilization even at periods of maximum growth.

Temperate grasses may have a second period of increased growth in early autumn and this may reduce storage of carbohydrates, which is further depleted by winter respiration. Species of plants are very variable in their responses, and climate, especially in the tropics, will also have a major influence on these processes so that generalizations will need to be interpreted according to local conditions.

7.7 Defoliation

Damage caused to plants by trampling and vehicles is not the same as defoliation by grazing or in experimental situations, as the wounded parts may often be left attached to plants damaged by trampling. The damaged organs may cause a drain on food reserves for wound healing or they may have sufficient photosynthetic tissue left to be self-sufficient, or even to export carbohydrates. Damage may also stimulate the production of new leaves or tillers, and this may also deplete food reserves (Liddle and Greig-Smith, 1975a; Kendal, 1982; Sun and Liddle, 1993d).

There have been very few investigations of the effect of trampling on food reserves, so the grazing and turf grass literature will be referred to here as the nearest approximation to the effects of recreation, although the selective nature of grazing and the recycling of nutrients must be borne in mind when considering the effects on various species.

In general, the effects of repeated defoliation are cumulative and the weight of, or percentage of carbohydrate in, the storage organs will gradually decrease. In grazing terms, the depletion of carbohydrates is accompanied by a reduction in the herbage yield (biomass) (Weinmann, 1952). This is equivalent to a reduction in vegetation cover in recreation areas. Weinmann (1952) also noted that 'there is evidence that plants depleted of carbohydrate reserves are more susceptible to adverse conditions, such as frost, heat and drought'. Severe defoliation may also lead to a reduced root system, which will impair the plant's capacity for absorbing water and nutrients.

As pointed out by Grace (1977) there are exceptions where defoliation appears to have little effect on final yield (Jones, Dunning and Humphries, 1955; Carlson, 1966; Grant and Hunter, 1966) and, as mentioned above, defoliation can stimulate an increase in the rate of photosynthesis in the remaining tissue (Waring, Khalifa and Treharne, 1968). A possible example of this is Bermuda grass which was subjected to 91 close cuts with a lawn mower in one season, and showed no significant lowering of the percentage of carbohydrate in the roots and rhizomes; 'However almost complete exhaustion of carbohydrate reserves was brought about by complete defoliation by means of scissors repeated at weekly intervals for less than a season' (Weinmann, 1952). It is also evident that plants which have trampling-resistant stems capable of photosynthesis, e.g. knot grass (*Polygonum aviculare*), have a greater survival potential in trampled areas.

Carbohydrates are utilized for any metabolic activity that requires energy and are an essential part of the growth and maintenance of any plant. For example, Clement *et al.* (1978) observed that the rate of nitrogen uptake by perennial rye grass (*Lolium perenne*) fluctuates diurnally in response to photosynthesis as there are considerable costs involved in nitrogen uptake (Veen, 1981; Johnson, 1983). Nitrogen uptake fell by 90% following defoliation of perennial rye grass (Clement *et al.*, 1978) so we may expect comparable disruptions of metabolic processes in plants damaged by recreation activities. The interaction between species and environment moderates or increases the direct effect of damage; for example, Kentucky bluegrass (*Poa pratensis*) ceases all photosynthesis at 32°C (Younger and Nudge, 1976) while Bermuda grass (*Cynodon dactylon*) continues to grow happily!

7.8 Summary   Damage to plants by recreation activities will have graded physiological and biochemical consequences, depending upon the degree of disruption of the plant body, the species concerned and the environment in which it is growing. Some of the linkages between the actions and their various consequences are summarized in Fig. 7.1.

There is no doubt that the physiological reaction of plants to human contact is an extremely important element of their responses. Even light brushing has consequences for the plant which often manifest in a change to the morphology of the new growth. More robust impacts, such as shaking, have been shown to increase the strength of some stems and change the cell water potentials. Reaction wood may develop when plants are bent and held in a distorted or new position. Abrasion may also damage the surface layers of plant structures, probably resulting in an immediate loss of water and hence turgor. This can lead to the loss of leaves which, in the case of crowberry (*Empetrum nigrum*), may occur some considerable period after the impact. Depending upon its severity, wounding leads to extensive and complex physiological reactions, and reserves are utilized in callus formation and replacement growth. Food reserves are reduced by almost all of these impacts and, depending upon the growth phase, they may take some time to replace.

The indirect effects of recreation, such as soil compaction, on plant physiology have not been discussed in this chapter, partly because they have not been studied in the recreation context and partly because the extension of this account into all possible areas of knowledge would require a considerable series of volumes. For those wishing to follow this subject, the sophisticated experiments of Engelaar (1994) on nitrification and nitrate acquisition by the roots and shoots of great plantain (*Plantago major* ssp. *major*) and common sorrel (*Rumex acetosa*) in compacted soils would make a good starting point. The physiological consequences of recreation impacts have hardly been researched and further investigation of the nature and effect of recreation damage at the physiological level is required urgently.

8 Plant strategies (resistance, survival and recovery) and regeneration of communities

The term strategy, used to refer to a suite of morphological and physiological characteristics of plants, has received some criticism for its anthropomorphic connotations but it is convenient and has entered the ecological vocabulary (Grime, 1974, 1979; Lincoln, Boxshall and Clark, 1982). Grime (1979) defines strategies as 'groupings of similar or analogous genetic characteristics which recur widely among species or populations and cause them to exhibit similarities in ecology'. The definition of Lincoln, Boxshall and Clark (1982) is 'the plan, method or structure utilized by an organism or group of organisms to meet a particular set of conditions'. Both definitions apply to single organisms as well as to groups of organisms, presumably populations, species or communities. In experimental practice only small areas of plant communities are subject to wear. The responses are measured in terms of characteristics that may be applied to either the community or to individual species (Bayfield, 1979b; Sun and Liddle, 1991). All or any characters described in the preceding chapters may form a part of the suite of characteristics that comprises a plant strategy.

8.1 Plant strategy

Perhaps the best known approach to plant strategy is that of Grime (1974, 1977, 1979). He considered that species could be characterized according to their relative competitive ability, stress tolerance and degree of ruderal characteristics. He developed an objective experimental system in which relative growth rate and a competitive (or morphology) index, based on height, lateral spread and amount of litter produced, were used to place plants in a triangular ordination (Fig. 8.1a). This has been used to place plants in positions that relate to survival under trampling stress as well as under other environmental features, such as disturbance and grazing (Fig. 8.1b), and allows predictive statements about the species measured by Grime in relation to the trampling environment.

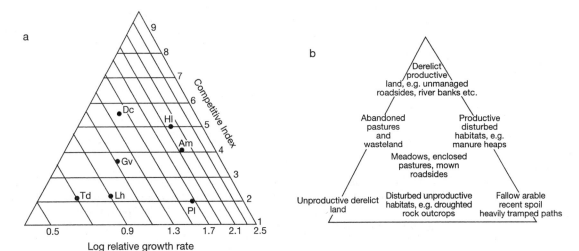

Fig. 8.1 (a) Triangular ordination diagram of herbaceous species in the study area at Malham according to their competitive indices and relative growth rates (Rmax = grams per gram per week). Positions on the diagram are obtained from Grime (1974). Species: Am, *Achillea millefolium*; Dc, *Deschampsia cespitosa*; Gv, *Galium verum*; Hl, *Holcus lanatus*; Lh, *Leontodon hispidus*; Pl, *Plantago lanceolata* and Td, *Thymus drucei*. (b) The interactions between competition, stress and disturbance, and the locations of selected habitat types in the ordination. (Modified from Aspinall and Pye, 1987.)

This scheme has been utilized by a number of recreation ecologists. Aspinall and Pye (1987) related the qualities defined by Grime (1974, 1977) to the positions of species on the axis of an ordination of species occurring in trampled limestone grassland in the Malham area of North Yorkshire, UK. This provided some degree of explanation of the processes occurring in their sites. Engelaar (1994) used a modified scheme derived from Grime (1979) to place six species that he had been examining in great detail in relation to flooding and trampling (Fig. 8.2). He equated flooding with stress and trampling with disturbance in a well-reasoned argument but, as he points out, one aspect of trampling (soil compaction) should be considered as a stress factor. This leads to his final conclusion that 'it is not always possible to classify a process in the field simply as stress or disturbance'. This is a major problem with predictions based on the analysis of strategy and should be considered in relation to the following discussion.

The concept of a strategy as a response to trampling pressure has been applied to different phases of the human (or other animal)/plant interaction. This process can be divided into three stages and these should be regarded as a sequence of events that result from a single impact, for example a step on one place (Liddle, 1984) (Fig. 8.3).

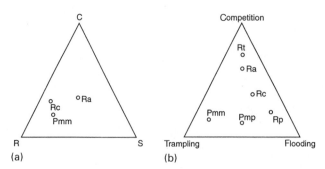

Fig. 8.2 Classification of the species studied according to the C-S-R model of Grime. (a) The calculated classification of *Rumex* and *Plantago* species as proposed by Grime, Hodgson and Hunt (1988). (b) The estimated classification based on their occurrence in relation to flooding and trampling. Rt, *R. thyrsiflorus*; Ra, *R. acetosa*; Rc, *R. crispus*; Rp, *R. palustris*; Pmm, *P. major* ssp. *major*; and Pmp, *P. major* ssp. *pleiosperma*. (From Engelaar, 1994.)

Symbol	Event	Quality
Alpha	Visual/Audio signals	Inviting or repulsive
Beta	Contact	Resistance (vulnerability)
Gamma	Death or survival	Survival
	Regrowth	Recovery

Fig. 8.3 The general scheme proposed for animal/plant interactions. The outcome of the reaction may be summarized for a single plant as: outcome = $\alpha \times \beta \times \gamma$, where $\alpha = 0$ or 1; $\beta = 0$ to/or ± 1 or higher if in terms compatible with γ; $\gamma = \pm$ some plant characteristics in the terms of the outcome (e.g. cover, height, tiller no., etc.). (Liddle, 1984.)

THE ALPHA PROCESS

This includes all of the signals (e.g. sight, scent, etc.) that are received by the human from the plant before physical contact occurs between that human and that specific plant. The duration of this stage is from the time the first signal is perceived by the human until either contact is made or the human stops reacting to the signals. The outcome of this stage is that the human either moves away from the plant or makes contact with it ($\alpha = 0$ or 1). It is also possible for the human to make contact with the plant without perceiving it, in which case the interaction commences at the beta stage.

THE BETA PROCESS

This starts when the human first contacts the plant and ceases when the human is no longer in contact with it. There are two possible patterns in time. One, the contact is short and the plant does not have time to react during the period of contact. In the second case, the contact period is longer and the plant has time to react while contact continues. The outcome of both of these interactions is that the plant may be unaffected or gain or lose some material substance, or it may be changed in form.

THE GAMMA PROCESS

This commences when the plant starts to respond to the beta process, and the beta and gamma processes may overlap in time. The plant may die, or change its form or its growth rate as a consequence of the beta processes. The ends of the gamma processes are difficult to define but they may be considered to continue until the plant returns to the condition existing before the interaction started or until it dies ($\gamma = d$, Δr or Δf, where d is death, r is the growth rate and f is a quantitative measure of its form). This is equivalent to a change in plant fitness as described by Lubchenco and Gains (1981). In nature, one plant is likely to be subject to repeated events and, in practice, the measurements must record the sum of the consequences of these events.

The scheme provides an analytical basis for comparison of the various interpretations of the common terms associated with the impact process (Fig. 8.4). The alpha stage has hardly been investigated. There are just a few comments in the literature about walkers

Symbol	Event	Various terminologies						
Alpha	Visual/audio/ scent signals	Pre-contact signal (f)						Vulnerability (a,c)
Beta	Contact	Resistance (c,b,g,d)	Vulnerability (f)		Damage (a)			
Gamma	Death or survival	Survival (h,d)	Tolerance (b)	Resilience (b,h)	Recovery (a)	Tolerance (c,b,e)		
	Regrowth	Recovery (h,d)	Resilience (g,c)					

Fig. 8.4 Terminologies used by various authors to define the responses of vegetation to human trampling. (a) Bayfield (1979b); (b) Cole (1993); (c) Cole and Bayfield (1993); (d) Kay and Liddle (1987); (e) Liddle (1975a); (f) Liddle (1984); (g) Liddle and Thyer (1986); (h) Sun and Liddle (1993a, b, c).

choosing side paths, short cuts or tussock plants and avoiding 'sharp' plants when barefooted (e.g. Nickerson and Thibodeau, 1983) but virtually nothing that refers to specific plants. The concept of increasing protection of the walker or driver leading to less sensitivity to the nature of the vegetation deals with the effects of the alpha stage on human behaviour rather than on the vegetation itself (Chapter 2). The beta and gamma stages are, however, much discussed, either generally or within the total concept of trampling.

However, one problem that has bedevilled the discussions of strategy has been the varied definitions of the terms adopted by different authors (Fig. 8.4). The definitions used in this discussion are taken from *The New Shorter Oxford English Dictionary* (Brown, 1993) as follows.

- **Recovery**: 'Possibility or means of recovering or being restored to a former, usual or correct state.'
- **Resilience**: '(1) The action or act of rebounding or springing back. (2) The ability to recover readily from, or resist being affected by, a setback, illness, etc.'
- **Resist**: 'Withstand the action or effect of . . .'
- **Resistance**: 'Ability to withstand something.'
- **Survival**: 'The action or fact of continuing to live after some event.'
- **Tolerance**: 'The ability of an organism to withstand some particular environmental condition.'

Three other terms that occur in the ecological literature have a bearing on this discussion:

- **Stability**: 'The capacity of a physical system to resist disintegration, decomposition or decay. Resistance to displacement.' (This appears to be equivalent to resistance but has been used for the total effects of impacts.)
- **Inertia**: 'The tendency to continue in the same state, to resist change.' This also is equivalent to resistance but it depends on when it is measured as sometimes there has been time for recovery, or recovery is part of the process.
- **Durability**: 'Able to withstand change, decay, or wear.' This appears to be equivalent to resistance.

Perhaps the least contentious term from this group is **resistance**, used here to indicate the immediate response of a plant or community to trampling impacts. Resistance is frequently measured as the amount of cover or biomass remaining immediately after experimental trampling treatments and it may be expressed either directly or in relationship to the cover or biomass, either before treatment or in comparison with control plots (Bayfield, 1979b; Sun and Liddle, 1991, 1993a; Cole, 1993; Cole and Bayfield, 1993). Relative resistance may be expressed either as the amount of damage after a given level of trampling (Cole, 1987a) or the level of trampling needed to cause a given level of damage (Liddle, 1975a). In one case it was defined as the amount of force or number of impacts required to produce breaks in the coral skeleton or a specified amount of polyp damage and death (Kay and Liddle, 1987). There is an excellent discussion of the meaning of this and some other terms in Cole and Bayfield (1993).

Vulnerability has also been used to define the immediate effects of trampling (Cole, 1987a); this is the reciprocal of resistance. More recently, Cole and Bayfield (1993) used vulnerability to refer to the combined result of resistance and recovery. Cole (1987a) plotted relative loss of vegetation cover as a percentage (in various vegetation types) on the y axis, the resultant curves were inverted compared to those in Figs 3.3 and 3.4. While the data can be readily interpreted, the rising curve appears to this author to be intuitively reversed as it is increasing while actually recording the loss of plant material or species. It does seem correct, however, for increases in mineral soil exposure or penetration resistance.

Durability is rarely used today but it occurs in the landmark paper of Wager (1964) to denote the cover remaining after simulated trampling.

Survival is not often considered as a separate characteristic but it has considerable significance when plants such as crowberry (*Empetrum nigrum*) or cotton grasses (*Eriophorum* spp.) show a delayed response to trampling damage (Bayfield, 1979b; Hylgaard and Liddle, 1981; Hylgaard, 1982; Cole, 1993). Cole (1993) recorded that the following four species and two groups declined in relative cover 1 year after trampling: mountain boxwood (*Pachistima myrsinites*), mountain heather (*Phyllodoce empetriformis*), grouse whortleberry (*Vaccinium scoparium*), shining clubmoss (*Lycopodium lucidulum*), and mosses and lichens in Washington's Cascade Mountains, USA. Survival was defined as 'the probability of survival of a fragment or colony (of coral) after a given amount of damage' (Kay and Liddle, 1987). The same authors defined recovery as the ratio of the growth rate of a (coral) colony or fragment after a given amount of damage to the growth rate of an undamaged colony or fragment.

Two central questions need to be answered in this discussion:

- Is the collection of certain characteristics appropriately called a strategy?
- Do plants actually exhibit the supposed collection of attributes?

The Oxford English Dictionary defines strategy as 'The art or skill of careful planning towards an advantage or desired end'. Plants have

8.2 Validation of the strategy concept

evolved, as a result of selection pressures, in such a way that their genetic pattern predetermines their growth and morphology within certain ranges and kinds of expression. These ranges and kinds of expression constitute the characteristics that we describe as composing that particular species. If these characteristics have, for example, the qualities that allow the plants to survive trampling, then the genetic plan is an example of a plan that conveys a certain advantage. I maintain that this is in sufficient agreement with the definition of strategy for the term to be used in this context. (It is worth noting that Begon, Harper and Townsend (1990) used both the terms 'strategy' and 'tactics' in a similar context.) Whether there is a consciousness that foresees or desires a particular end, becomes a metaphysical question that we each have to answer for ourselves.

The best way that I can approach the second question with the data that are presently available is to assess the recorded responses of species to trampling. Cole (1993) lists the response levels as high, medium or low in terms of resistance, 'resilience' (recovery) and tolerance of 127 example species. The response levels in terms of resistance and recovery show that only four species out of 127 examples had both high resistance and high recovery rates (Fig. 8.5). Forty-six species had either high resistance or high recovery and the remaining 81 were not adapted to survive trampling.

The lack of a combination of both high levels of resistance and of recovery (four examples) means that most plants show either high

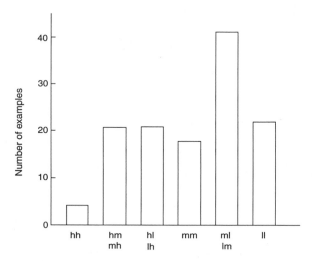

Fig. 8.5 The number of species of plants that have particular combinations of responses leading to high (h), medium (m) or low (l) resistance (first letter) or recovery rates (resilience, second letter). Total 127 species. (Data from Cole, 1993.)

resistance or high recovery rates, or do not have a trampling survival strategy. Plants do not all react in the same way to trampling stress and some do have certain combinations of characteristics which enable them to survive. In my opinion it is reasonable to call these combinations of characteristics strategies as they constitute a plan that achieves a particular end.

Further discussion of the terminology of change independent of humans or as a consequence of human action may be found in Connell and Slatyer (1977), Holling (1973), Westman (1978) and in many ecology texts, such as Whittaker (1975), Begon, Harper and Townsend (1990) and Krebs (1994).

A number of workers have measured the immediate effects of trampling on resistance of the vegetation or species, usually by subjecting untrampled vegetation to various levels of wear. Sun and Liddle (1993d) subjected 13 species to experimental trampling in the glasshouse and measured, among other parameters, the above- and below-ground biomass immediately after the treatment finished (Fig. 8.6). One interesting aspect of these results was that the roots appeared to be more resistant to trampling than the shoots. Three woody or erect herbaceous plants had very low resistance and did not survive the trampling treatment.

Two species, elastic grass (*Eragrostis tenuifolia*) and green couch (*Cynodon dactylon*) were subjected to experimental trampling at 10-day intervals in glasshouse conditions over a period of 80 days (Sun and Liddle, 1991). The biomass was recorded after each simulated trampling. The resistance index of green couch declined towards the end of the experiment while that of elastic grass rose, albeit somewhat irregularly (Fig. 8.7a). The recovery index of both species rose at first and then that of elastic grass fell back to the starting level while that of green couch continued to rise (Fig. 8.7b). The authors concluded that elastic grass had a resistant strategy while green couch had a recovery strategy, although it should be noted that the biomass of green couch was very low immediately after the last five treatments (Sun and Liddle, 1991 – their Fig. 5c, not 5b as indicated in the caption).

Some montane heath communities on the Cairngorm Mountains, Scotland, were subject to various levels of experimental trampling and their recovery recorded over a period of 8 years (Bayfield, 1979b). This much longer time course showed that some species were very vulnerable and slow to recover (*Sphagnum rubellum*, Fig. 8.8a), some sustained quite heavy damage with a delayed response but were fully recovered at the end of the 8-year period (*Eriophorum* spp., Fig. 8.8b) and some sustained relatively less damage (except at the highest level

8.3 Resistance of single species

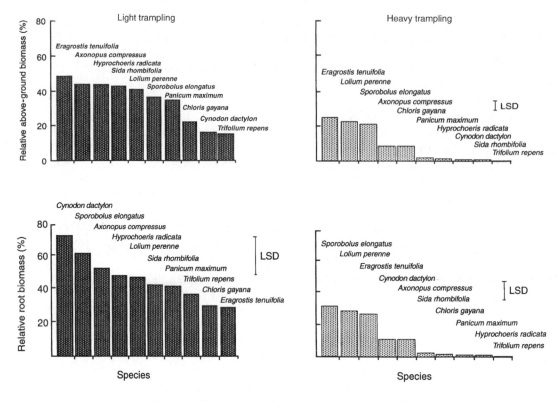

Fig. 8.6 The relative above-ground biomass of both lightly and heavily trampled plants of the 10 species that survived trampling, calculated as a percentage of controls (from Sun and Liddle, 1993d).

of wear) and rapidly replaced lost cover (*Trichophorum cespitosum*, Fig. 8.8c). These strategies could be classed as vulnerable, recovery and resistant/recovery, respectively. The dieback of cotton grasses (*Eriophorum* spp.) after 2 years was apparently due to the death of bruised tissues and frost damage (Bayfield, 1979a).

An alternative presentation of recovery data in the form of an ordination allows a quick assessment of the performance of each species (Fig. 8.9). From this it is clear that only sedges (*Carex* spp.) are both resistant and able to recover well, but cotton grasses (*Eriophorum* spp.), heaths (*Erica* spp.) (at two sites) and deer grass (*Trichophorum*) (at two sites) all made a good recovery. Bog moss, bear berry, crowberry and ling (*Sphagnum, Arctostaphylos, Empetrum* and *Calluna*) are all vulnerable and slow to recover. From single-species data it appears that some species have a clear resistant or recovery strategy, rarely both, and some species are not in any way adapted to withstand trampling.

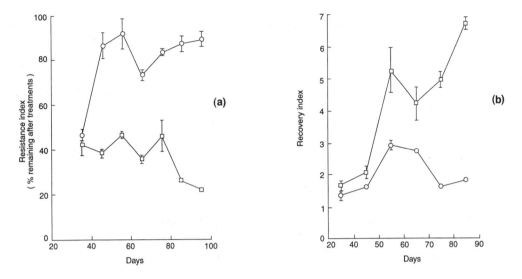

Fig. 8.7 The resistance index (a) and recovery index (b) of *Eragrostis tenuifolia*
(○) and *Cynodon dactylon* (□), calculated as the percentage of the biomass
remaining after each simulated trampling. I, two standard errors. (From Sun
and Liddle, 1991.)

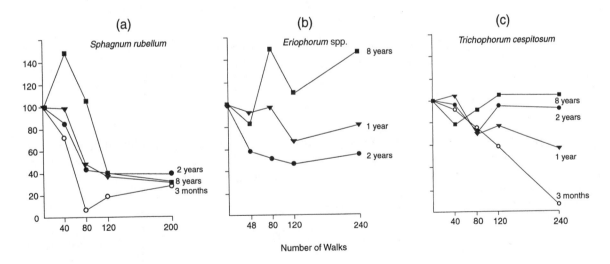

Fig. 8.8 Relative intact cover of species in the lichen-rich *Calluna–
Trichophorum* heath on Cairngorm at various times after disturbance by
trampling (from Bayfield, 1979b).

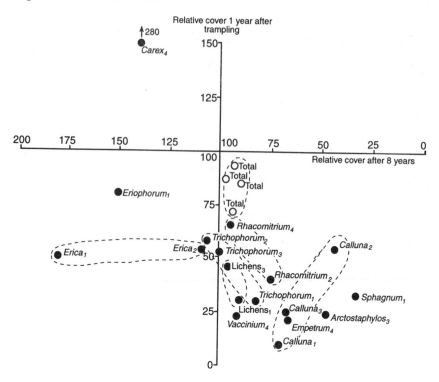

Fig. 8.9 Relative intact cover of individual species, and total vegetation cover, at the four sites on Cairngorm after 1 and 8 years. Site numbers are shown after each species. (From Bayfield, 1979b.)

Single species of coral were also tested separately for trampling resistance (Kay and Liddle, 1984, 1987) and Bayfield used a conceptually similar approach on four species of lichen (Bayfield, Urquhart and Cooper, 1981).

8.4 Resistance, recovery and tolerance of communities of plants

A survey of the resistance, tolerance and resilience of alpine vegetation in four locations in the USA (Cole, 1993) used the strategic terminology in an accurate manner. Cole measured the amount of cover immediately after trampling treatment and expressed it relative to the cover of the control plots:

$$\text{relative cover} = \frac{\text{surviving cover on trampled plots}}{\text{initial cover on trampled plots}} \times \text{cf} \times 100\%$$

where

$$\text{correction factor (cf)} = \frac{\text{initial cover on control plots}}{\text{surviving cover on control plots}}$$

This quantity was used as the x axis in his composite figures (Fig. 8.10). The y axis was the relative cover 1 year after trampling. This defines the state the vegetation was able to maintain when subjected to these particular levels of trampling and rest periods. If these areas were repeatedly subject to this regime, there may be cumulative effects on vegetation and soils that could modify these results.

Cole (1993) applied this 'experimental protocol' to 16 sites in four locations and analysed the data according to two aspects of plant life form (Fig. 8.11a,b), growth form (Fig. 8.11c) and degree of canopy closure (Fig. 8.11d). The abundant species in each of the four locations and the elevation zone and region of his 16 samples are shown in Fig. 8.11e,f.

The life-form data (Fig. 8.11a) show that some examples of each type had low resistance but that the hemicryptophytes generally had the highest tolerance and the greatest recover rates (distance above the diagonal line). The second type of life-form analysis (Fig. 8.11b) showed that forbs had the lowest resistance while the shrubs had the lowest tolerance and recovery rates, two points showing negative recovery. The erect growth forms (Fig. 8.11c) had generally very low resistance and the caespitose (tufted or clumped) forms high resistance. Caespitose forms also had the highest tolerance but many

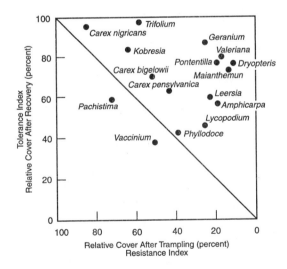

Fig. 8.10 Resistance, tolerance and resilience of the 16 vegetation types in four alpine regions in the USA. Resilience is indicated by the perpendicular distance from the diagonal line of equal resistance and tolerance. (From Cole, 1993.)

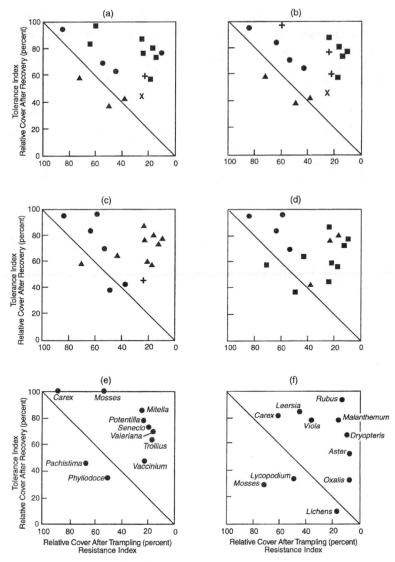

Fig. 8.11 Resistance, tolerance and recovery (resilience) of different plant groups from four locations in the USA. (a) Vegetation types dominated by chamaephytes, hemicryptophytes and cryptophytes. Chamaephytes have perennating buds above the soil surface; hemicryptophytes have perennating buds at the soil surface; cryptophytes have perennating buds below the soil surface. ▲, Chamaephytes; ■, hemicryptophytes; ●, cryptophytes; +, hemicryptophytes and cryptophytes; ×, chamaephytes and cryptophytes. (b) Vegetation types dominated by graminoids, shrubs and forbs. ●, Graminoids; ▲, shrubs; ■, forbs; +, graminoids and forbs; ×, forbs and clubmoss. (c) Growth form of dominant species. ●, Caespitose or matted; ▲, erect; +, graminoids and forbs. (d) Vegetation types with varying degrees of canopy closure. ●, Open; ▲, partially forested; ■, forested. (e) Abundant species in four vegetation types in Washington's Cascade Mountains. (f) Abundant species in four vegetation types in New Hampshire's White Mountains. (From Cole, 1993.)

of the erect forms had high recovery rates. It is interesting to note
negative relationship between
1 the lowest resistance having
sa. This supports the views of
Liddle (1987) that there is a
ant or to have good recovery
both characteristics. However,
e or matted species tended to
esistance and recovery, so the
simple division into resistance
)pen sites (Fig. 8.11d) had the
; on some of the forested and

analyses from the Washington
the New Hampshire White
ne forms, such as mosses, may
)very rates in one place and a
This is probably because there
wo locations. The other group
s and these both had similar
that the abundant species in
strong negative relationship
itrast to the White Mountains
species had a wide range of
eneric level, the concept of an
e idea of strategies provides a

te the long-term (or supposed
, etc. (Liddle, 1975b; Bayfield,
ibines all the other qualities
) the final outcome. A tolerant
resistance, recovery, or both.
cations of paths, campsites or
juantitative measures of vege-
)r long periods of time is an
es present, to the level of use
they have experienced. This is often the input parameter when sites
are being monitored for 'limits of acceptable change' management
criteria.

Resistance, resilience (recovery) and tolerance were calculated in a
number of ways by Cole and Bayfield (1993) for three vegetation types
(Table 8.1). The calculation of the mean result after a number of
passes from 0–500 has the advantage that anomalous data cannot have
a major effect on the result. This method was used to prepare Figs
8.10 and 8.11, from Cole (1993).

Table 8.1. Indices of resistance and tolerance for three vegetation types (Cole and Bayfield, 1993)

Index	Vegetation type		
	Valeriana	*Vaccinium*	*Carex*
Resistance			
Minimum number of passes that cause a 50% cover loss	25	200	650
Mean relative cover after 0–500 passes	16	49	85
Resilience			
Percentage increase in cover 1 year after 50% loss	100	−32	86
Mean increase in cover 1 year after 0–500 passes, as a percentage of the damage caused by trampling	76	−22	80
Tolerance			
Maximum number of passes that leave at least 75% cover 1 year after trampling	300	75	> 700
Mean relative cover 1 year after 0–500 passes	80	38	97

8.5 Regeneration of communities

Regeneration is of primary importance in land management and follows closely behind controlling or preventing damage to the vegetation and soils as a management priority. The sequence of plants and animals that occur on a regenerating site is a successional process. Probably the most widely accepted discussion and definitions of models of succession are those of Connell and Slatyer (1977) (Fig. 8.12). The crucial question for management is whether there are propagules of the original vegetation left in the soil, in which case secondary succession can take place, or whether the substrate has been denuded and seeds or other propagules will have to be introduced. Secondary succession allows the regeneration process to start some way down the sequence in Fig. 8.12.

An example of this process occurred in military camps abandoned 40 years previously in the Mojave Desert, California (Prose, Metzger and Wilshire, 1987). Here they found that the dominant woody shrub, creosote bush (*Larrea tridentata*), had recolonized the hand-cleared tent sites and parking lots, but not the graded roads where up to 20 cm of the top soil had been removed (Fig. 8.13). They surmised that although the vegetation had been hand-cleared to ground level, the root crowns of the creosote bush had remained and provided a basis for secondary succession, a similar process to that described by Puntieri (1991) in Argentina where *Alstroemeria aurea* recovered in areas cut for ski runs. The higher total cover recorded by Prose, Metzger and Wilshire (1987) in the parking lot was apparently due to the colonizer (*Ambrosia dumosa*) also being present in greater numbers than in the control area, so that there was an earlier successional process coincident with the regrowth of the dominant species. The soils were also compacted in the used sites, especially on the abandoned roads,

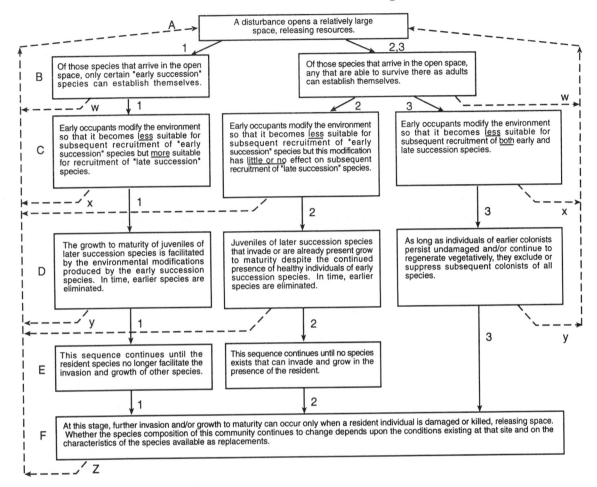

Fig. 8.12 Three models of the mechanisms producing the sequence of species in succession. A to Z, the sequence in time. The dashed lines represent interruptions of the process, in decreasing frequency in the order w, x, y and z. 1, The facilitation model; 2, the tolerance model; 3, the inhibition model. (From Connell and Slatyer, 1977.)

and this is considered to be an important factor in these successional sequences. Two mining camps in the Great Basin Desert, Nevada, abandoned 70 years previously, also showed disturbance effects on the vegetation (Knapp, 1992). The used areas were dominated by an introduced annual, *Bromus tectorum*, and the author considered this to be a permanent change, in this case the processes of succession were diverted by a permanent change in the environment caused by the introduced plant.

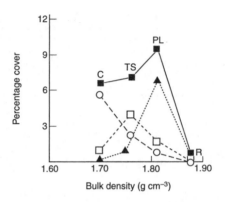

Fig. 8.13 Relation of the mean soil bulk density and vegetation cover at Iron Mountain abandoned military camp, Mojave Desert. ■, Total percentage cover; ○, long-lived perennials; □, long-lived opportunistic; and △, short-lived species. C, control; TS, tent site; PL, parking lot: R, road. (From Prose, Metzger and Wilshire, 1987.)

Another study of regeneration on previously used military tracks, this time on Dartmoor, south-west England, showed that moorland/blanket bog vegetation had rather poor recovery over a 20-year period, and in one case the direction of succession was toward a grassland community (Charman and Pollard, 1993) (Fig 8.14). This contrasts with a track through a grassland site that showed very little difference from the off-track vegetation after the same period of recovery.

The morphology of species is very important in the recovery process, and in stressed environments recovery from seed is not always possible. Grabherr, Mair and Stimpf (1987) showed that it was possible to reproduce the stolons of *Agrostis schraderana* and alpine meadow grass (*Poa alpina*) in large quantities, and these were used to revegetate high alpine areas on Pfannhorn (2660 m) in South Tyrol, Italy. Grabherr, Mair and Stimpf (1988) also recorded that the spread of some species in alpine situations can be extremely slow, for example *Carex curvula* rhizomes only advanced at the rate of 1 mm yr^{-1}.

Recovery after various intensities of trampling suggests that there is a level of impact above which the recovery process proceeds at a constant rate, presumably as long as propagules are present (Cole, 1987a) (Fig. 8.15). In this example recovery was approximately the same for all levels of use above 800 passes per year, although below that level recovery was proportional to the amount of impact. A similar response was recorded in the Hubbard Brook Experimental Forest, New Hampshire, where recolonization of impacted areas that received 100 trampling passes, as measured by plant cover, dominant indices,

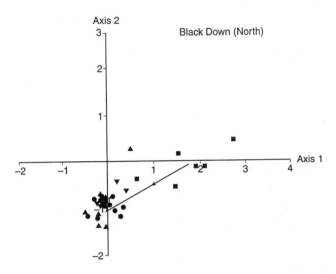

Fig. 8.14 Canonical correspondence analysis of track vegetation (▼, 24; ▲, 14; and ■, 3 years after closure) and adjacent 'off-track vegetation'. ●, grassland. The arrow indicates the suggested time sequence for vegetation change. (From Charman and Pollard, 1993.)

floristic dissimilarity and species diversity, was similar to that of areas receiving four and eight times more trampling, despite major differences in soil penetration resistance (Kuss and Hall, 1991).

The comparison between levels of wear at the same site makes the assumption that environmental conditions were similar in all plots. An earlier paper by Cieslinski and Wagar (1970) emphasized the interaction of aspect and slope on the recovery of damaged vegetation in the Cache National Forest, Utah. They found that survival was greatest on the coolest north-eastern slopes and least on slopes with the hottest south-western aspect. They also found that there was greater 'end of season stocking' on the steeper slopes. This does not imply that impact of wear was less on the higher gradient but, in their opinion, it was related to soil coarseness or some other factor associated with slope rather than slope itself. There is, of course, also an interaction with the species, some of which may show further damage (in one case up to 8 years after the experimental trampling ceased) (Bayfield, 1979b, his Fig. 2).

A successional sequence for revegetation of ski grounds on the highlands of Hokkaido, northern Japan, showed a range of strategies (Tsuyuzaki, 1993) (Fig. 8.16). Groups G and E appear to follow the facilitation model (see Fig. 8.12). Groups C, D and F are closest to the inhibition model, while A_1, A_2 and B are somewhere between the inhibition and tolerance models.

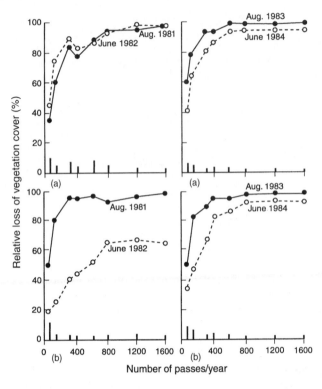

Fig. 8.15 Recovery after the first and third seasons of trampling in the (a) *Clintonia–Vaccinium* and (b) *Clintonia* vegetation types. August measurements immediately followed trampling; June measurements preceded trampling. The difference provides a measure of recovery. Selected mean values are plotted; bars at bottom show the median standard error for the two measurement periods (From Cole, 1987a.)

As is evident from the examples discussed throughout this chapter, there has been little work on the recovery of tropical vegetation after recreation impacts, but one example comes from the rainforests of Costa Rica (Boucher *et al.*, 1991). They found that an old trail abandoned for 32 months had diversity indices of 2, compared to 3 in the forest and 0.44 on trails still in use. Cover was still reduced but seedlings, 'platanillos' and herbs were favoured in this environment. These same types are also favoured in light gaps and adaptation to natural disturbances may give them an advantage in recovery from low level human disturbance. This suggests that trail closure for a few years may be sufficient to allow vegetation recovery in tropical rainforests, in contrast to desert and more boreal situations. This may be an example where the productivity theory (Fig. 3.15) does hold.

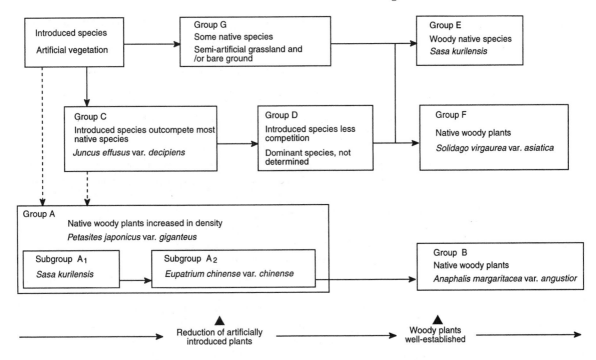

Fig. 8.16 Proposed model for successional sere on ski grounds in the highlands of Hokkaido, northern Japan (from Tsuyuzaki, 1993).

In conclusion, it should be realized that restoration of vegetation to its original condition may not be possible for the following reasons (adapted from Cairns, 1990):

● There is no description of the original condition to serve as a management goal.
● Some of the species may have been lost.
● The biological or physiochemical environment may have changed (from pre-disturbance conditions).
● The basic ecological knowledge is not available.
● The cost may be prohibitive.

Professor Bradshaw (1987) has said 'Restoration ecology is the acid test for ecological theories' and at this stage of the science we are still very much engaged in description, with tested predictions of genuine causality few and far apart.

8.6 Summary   The concept of strategy has been accepted by most ecologists in spite of its anthropomorphic connotations. It may be defined as a suite of morphological and physiological characteristics utilized by an organism or group of organisms to meet a particular set of conditions.

Many terms dealing with strategy occur in the literature and Liddle (1984) has suggested an analytical scheme categorizing the different stages of trampling impacts and responses. Resistance, tolerance and vulnerability are perhaps the most frequently used.

Strategy has been used to characterize the responses of plants to trampling. Sun and Liddle (1991) suggest that they may in general be either resistant or resilient. Resistant species are relatively less damaged by impacts but tend to have slow recovery rates, resilient ones are easily damaged but recover quickly.

Strategy concepts have been applied to communities as well as individual species and this is probably the most useful application of the concept for management purposes.

Regeneration of communities is also an important management concern and it is clear that the time scales are very varied. In most cases the time for full recovery has been underestimated. In some cases, deserts and high alpine habitats for example, it is doubtful if complete recovery to original conditions will ever occur. The reasons for this conclusion were clearly stated by Cairns (1990) and are set out above.

Introduction to soils 9

Soils, plants and animals may be viewed as an interacting system, supported and derived from the underlying rocks, reacting with and limited by the air and climate that surrounds them. These components are all subject to change at different rates. The temperature, humidity and wind speed may alter from one second to the next, animals may be grazing one minute and flying or galloping from a predator in the following one, plants may be in bud one day and in bloom shortly after the following sunrise, soils build, leach or wash away each season, and rocks that were deposited in one millennium may be eroded in the next. In our study of the effects of recreation in different parts of the earth's cloak, it is as well to remind ourselves of the interacting and changeable nature of the material we subject to our scrutiny. This short introduction to soils is intended for the person who has no prior knowledge of the subject and as a reminder for others; fuller descriptions may be found in Corbett (1969), FitzPatrick (1971), Russell (1973), Brady (1974), White (1979) or McLaren and Cameron (1996).

9.1 Soil types

Soils are products of the interactions between climate, underlying rocks, the animals within the soils and the plants growing on their surface. It is therefore not surprising to find that in each place around the world there are different soils, and that in places with similar rocks and climatic conditions, similar soil types are to be found. The different soil types are often characterized by comparison of the different layers that occur within them. In some cases the difference between the soil layers is dramatic and self-evident (as in podzols) while in others it is very subtle and only detected by close visual inspection and perhaps physical and chemical analysis (as in a kraznozem).

The podzols which form in the humid temperate parts of the world and the kraznozems which are found under humid tropical conditions illustrate here some of the terms used for the layers of the soil profile

(Fig. 9.1). A brief description of the processes of formation of each layer is given in Fig. 9.2. In both cases litter layers on the surface are the primary source of the organic content of the soil, and as it sinks down to the lower layers it becomes macerated and the nutrients pass gradually into the water that is held in the soil. The base rock is slowly broken up by acid from the litter layer, the action of water flow, and changes and extremes of climate. There is a gradual and continual mixing of the organic and inorganic material by the action of water flow, soil movement and soil animals, particularly earthworms which, according to one estimate, may take only 80 years to pass the entire top 15 cm of the soil through their guts (Barley, 1959). The nature and components of the upper soil layers are most immediately affected by trampling, and the rest of this chapter briefly discusses these properties in an undisturbed soil.

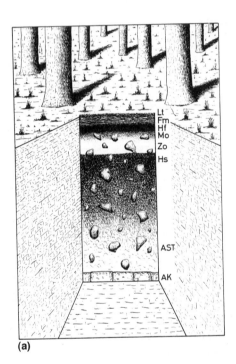

(a)

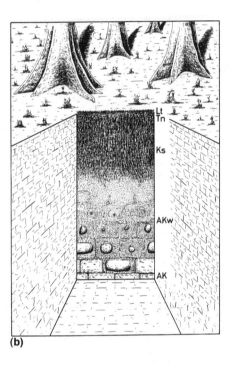

(b)

Fig. 9.1 A podzol (a) and (b) a krasnozem soil profile. Lt, freshly fallen plant litter; Fm, partially decomposed litter; Hf (Humifon), amorphous organic matter; Mo (Modon), black organic matter and mineral grains; Zo (Zolon), white weathered minerals, Hs (Husesquon), deposited organic material leached from the zolon; AST, acid drift; AK, acid rock; Tn (Tannon), organic and mineral material; Ks (Krasnon), bright red clay; AKw, chemically weathered acid rock. (From FitzPatrick, 1971.)

Horizon			Description	
O	Ol or H	A_{00}	Litter layer; loose fragmented organic matter in which original plant structures are still readily discernible Fermentation layer; partially decomposed organic matter	Horizons of biological activity and eluviation
	Of or H			
	Oh or H	A_0	Humified layer; heavily transformed organic matter with no discernible macroscopic plant remains	
A		A_1	Dark-coloured horizon, with high organic content intermixed with minerals. Ap denotes ploughed layer	
E		A_2 or E_a	Light-coloured horizon, with low organic content due due to high eluviation	
		A_3	Transitional layer, sometimes absent	
		B_1	Transitional layer, sometimes absent	
B		B_2	A dark layer of accumulation of transported silicate, clay, minerals, iron and organic matter, showing the maximum development of blocky and/or prismatic structure	Horizons of eluviation; accumulations of clay, humus or iron oxides leached or translocated from the upper layers
		B/Cor B_2	Transitional layer	
C			Weathered parent material, comprising mineral substrate with little or no structure; containing a gleyed layer or layers of accumulated calcium carbonate or sulphate in some soils	
R		O	Underlying rock	

Fig. 9.2 Soil profile and horizon nomenclature; *, horizons formed *in situ* (from Lincoln, Boxshall and Clark, 1982).

9.2 Soil composition

The skeleton of the soil consists of mineral particles which vary considerably in size, from stones over 2 cm in their largest dimension to clay with particles of less than 0.002 mm (Fig. 9.3). The proportions of the particles also vary greatly, from silt soils with over 70% of the particles being less than 0.02 mm to sands with over 85% of the particles being greater than 0.05 mm, according to the United States Department of Agriculture Classification System. Notice that the system that originated in what was the Soviet Union has more classes, each with a smaller size range (Fig. 9.3). The mineral particles are slowly weathered and release nutrients into the soil solution, these may form secondary minerals such as clay, which affect the chemistry of the soil through their ion-exchange properties. In a natural soil the different sized particles are all mixed together to form a matrix, and the spaces between them contain the other soil constituents (Fig. 9.4).

The organic solids occur between the mineral particles, in varying degrees of decay, from complete dead leaves, twigs or animals through

Size (mm)	USSR	United States system	International system	approximate number of particles per gram	approximate surface area, cm² per gram	visibility of individual particles	physical character	mineralogical composition
3.0								
2.0	gravel							
1.0	coarse sand	very coarse sand	coarse sand	5.4×10^2	21	visible to naked eye	loose and single-grained; not sticky or plastic	mainly quartz with some rock fragments
0.5	medium sand	coarse sand						
0.25		medium sand						
0.1	fine sand	fine sand	fine sand	5.4×10^5	210	visible to naked eye	loose and single-grained; not sticky or plastic	mainly quartz and feldspar with some ferromagnesians
0.05	coarse silt	very fine sand						
0.01			silt					
	medium silt	silt		5.4×10^8	2,100	visible under microscope	smooth and floury; only slightly cohesive	mainly quartz and feldspar with some ferromagnesians, mica and clay minerals
0.005	fine silt							
0.001	coarse clay	clay	clay	7.2×10^{11}	23,000	invisible under microscope except in upper range; many particles resolved by electron microscope	sticky and plastic when moist; hard and cohesive when dry	mainly clay minerals with some quartz
0.0005	fine clay							
0.0001	colloidal clay							

Fig. 9.3 Soil particle sizes, their terminology according to the systems of the former USSR, USA and the 'International System' and some of their characteristics. (Modified from Bradshaw and Chadwick 1980.)

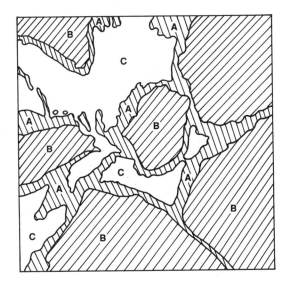

Fig. 9.4   Drawing from a photomicrograph of a thin section of a sandy soil showing how colloids and clay may accumulate in soil as coatings on the grains of sand and as bridges between the grains: A, colloids and clay; B, large mineral particles; C, soil pores. Scale not given on original photomicrograph. (After Black 1968.)

an organic colloid to individual ions in the soil solution. This dead organic matter is partially fibrous and water retentive and its physical properties help to bind the soil together into aggregates. The smallest particles of organic matter, colloids, ions and clay also occur on the surface of the mineral particles as well as forming bridges between the grains (Fig. 9.4).

The pores between the solid particles and the organic material are filled either with air or water. Water in the soil is part of one of the major cycles of material which take place in and near the earth's surface. The water falls as rain and some is retained in the soil, the rest drains to streams, lakes, rivers and eventually to the sea (Fig. 9.5). It continuously evaporates from open surfaces, the moisture is transpired through plants, and so rejoins the earth's atmosphere, ready to fall again as rain. Oxygen is essential for aerobic respiration of the soil's micro-organisms and passes from the air into solution in the soil water. But if the soil is completely waterlogged, either at the surface or some depth below, oxygen will only diffuse through the water at a slow rate. Oxygen will then be in short supply, and conditions will become anaerobic, slowing or stopping the growth of many plant roots. These conditions do not occur in well-drained or sandy soils.

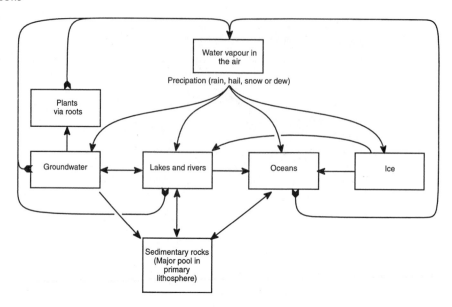

Fig. 9.5 The water or hydrologic cycle.

The soil also contains fungi which live saprophytically on the dead remains of the plants and animals and contribute to their mineralization. Fungal filaments may also contribute to binding soil particles into aggregates and so improve its structure. Bacteria contribute to the breakdown of dead material, and, particularly the genus *Rhizobium*, are the main fixers of atmospheric nitrogen. They often occur as small colonies in their own gummy secretions.

Many minute soil organisms, including protozoa, mites and springtails, contribute to the breakdown of organic remains. However, the major soil animals, usually the earthworms, pass large amounts of soil particles through their guts, macerating the organic material and releasing nutrients with their acid digestive juices. Worms also loosen the soil, creating pores through which water may drain and oxygen enter the soil.

Desert soils deserve a special mention as they have very fragile surface layers and may take thousands of years to mature. The surface crusts may be formed by biological activity and composed of living plants (lichens and algae), or they may be the result of chemical weathering, which is extremely slow in arid environments as this requires water to take place, or they may sometimes be the consequence of mechanical sorting, leaving larger stones on the surface. As with other soils, biological activity is the main force influencing the rate of soil formation (Dregne, 1983).

The soil acts as a store of nutrients, ideally releasing them at the rate required by the plants. Nutrients are continuously cycled from inorganic states through living organisms and, by various routes and avenues, back to the stores in the soil, rocks, oceans or atmosphere. The amount of macro- and micro-nutrients required by plants varies for each element and for different plants (a general indication is given in Table 9.1), as does the nature of their mineral cycles.

Nitrogen is perhaps the most important single nutrient for plant growth, and it is nearly always in short supply (Bradshaw and Chadwick, 1980). It is an important component of proteins and therefore of protoplasm. It is usually taken up by plants in the form of nitrate (NO_3^-) which is very soluble and easily leached from the soils (Fig. 9.6). The major store of nitrogen is in the atmosphere, and this is made available to the plants by nitrogen-fixing organisms. These are able to reduce gaseous nitrogen to ammonium (NH_4^+), for example *Rhizobium* which is symbiotically associated with roots of leguminous and other plants. Bacteria are instrumental in oxidizing the ammonium through a series of steps to nitrate. Since organic nitrogen is returned to the soil in plant litter, it, and animal dung and bodies, become the primary supply of nitrogen (as NH_3 or NH_4^+) in mature soils. Some of the leached nitrate may be reduced to less soluble forms in the lower layers of the soil and be held there for deeper rooting plants.

Phosphorus, the second macro-nutrient, is essential for plant enzymes and is taken up from the soil as the H_2PO_4 or HPO_4^{2-} phosphate ion. The main source of phosphate in the soils is the underlying rock and derived mineral particles, often apatite and fluorapatite.

9.3 Soil chemistry and the requirements for plant growth

Table 9.1 Some of the ions or elements required by plants (from various sources)

	Cations		Anions	
In large quantities	K^+	Potassium	$H_2PO_4^-$	Phosphate
	Ca^{2+}	Calcium	SO_4^{2-}	Sulphate
	Mg^{2+}	Magnesium	NO_3^-	Nitrate
	NH_4^+	Ammonium		
In small quantities	Fe^{3+}	Iron		Availability reduced under alkaline conditions
	Mn^{2+}	Manganese		
	Zn^{2+}	Zinc		
	Cu^{2+}	Copper		
	B^{3+}	Boron (occurs as BO_3^{3-})		
	Mo^{6+}	Molybdenum (occurs as MoO_4^{2-})		Availability reduced under acid conditions

Some of the ionic valencies may vary according to the conditions in which they are found.

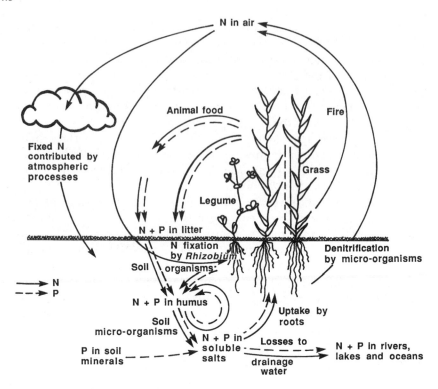

Fig. 9.6 Nitrogen and phosphorus cycles at the air-and-earth's surface (from Bannister, 1976; Bradshaw and Chadwick, 1980).

Many soils are derived from minerals containing little phosphorus and it is usually removed from tropical soils by leaching, so is often in short supply. In addition, phosphate will combine with many components of the soil, forming complex, insoluble compounds from which it is only slowly released. Over half of the soil phosphorus may be contained in organic materials. Because of the way phosphorus is complexed, it will travel only up to 2 mm in the soil solution (Bradshaw and Chadwick, 1980) and plant roots have to grow into the precise locations where it is available. Fungal associations (mycorrhizas) may assist the higher plants to obtain phosphorus. The phosphorus cycle is also illustrated in Fig. 9.6.

Potassium and other cations occur in quantity in the soil, usually in close association with the clay particles because of the latters' surface negative charge. The availability of cations depends upon the cation exchange in clay soils, but where there is little clay or organic matter, potassium is readily lost, especially in areas of high rainfall or in hot, wet climates where the clays are destroyed.

Soil acidity (pH) depends upon the concentration of hydrogen ion, H^+, in the soil solution. The pH may range from 3 in base-deficient acid soils through neutral (pH 7) to pH 9 where there is an excess of calcium or sodium. The acidity of soil affects the amount of nutrients which are available to plants, the biological activity, chemical weathering and the solution of toxic materials such as aluminium (Fig. 9.7).

Water is crucial to plant growth as it is not only the medium in which all the soil 'goodies' are dissolved and transported to be taken up by the roots, but it is also an essential component of all cytoplasm. Plants lose vast amounts of water by transpiration as they have to expose a large area of moist cell surfaces to the air within the photosynthetic tissues, in order to absorb carbon dioxide from the atmosphere. According to Bradshaw and Chadwick (1980) most plants use at least 250 kg of water for each 1 kg of dry matter they produce, and in drier regions this may amount to as much as 800 kg per 1 kg of dry matter.

Soil provides a store of water that buffers the intermittent, and sometimes insufficient, supply from the rain. The amount available to plants is the difference between the total quantity present in the soil and that fraction of water which is held by such strong capillary, hydration and other forces that the plant cannot extract it. This water potential (Ψ, psi) is measured by the force required to extract it from the soil. The amount held by the soil against gravity is called field capacity, usually taken as $\Psi = -0.33$ bars, and the level at which plants wilt is

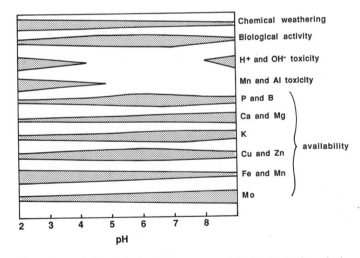

Fig. 9.7 The effect of the pH of soil on some of its biological and chemical characteristics. The breath of the band is related to the degree of activity or availability (from Bannister, 1976, after Larcher, 1973).

considered to be $\Psi = -15$ bars. These values are largely constant from one soil to another, but soils differ in the percentage moisture retained (Table 9.2).

Most herbaceous plants root to a depth of 60 cm and more vigorous species can root to twice this depth. Trees may root to 3 m or more (Bradshaw and Chadwick, 1980). This provides access to a reasonable store of water for plants to draw upon for most of the year, except in very dry climates where the native species often have physiological mechanisms for avoiding drought. In drier regions, the growth of native plants may also be restricted to times of sufficient water supply.

9.4 The effects of recreation on soils

At first sight the consequences of recreation activities are compaction and erosion of the soil as a result of the mechanical forces, and increased levels of nutrients from litter and human waste. However, as with most ecological events there is a series of changes in soil condition as a result of recreation impacts. Some of the changes involve feedback loops, either accentuating the change or stabilizing the situation (Fig. 9.8). The general relationships between soil characteristics were summarized by Cole (1987b) (Table 9.3). Each of these qualities is considered separately in Chapters 10–16.

Plant roots are also a part of the soil environment and the direct effect of trampling and vehicles on plants is either to damage or kill them. This results in a reduction of cover and biomass of living tissue and an initial increase in plant litter. The reduction in living material

Table 9.2 The available water capacity in various soils which differ considerably in their retention of water (from Bradshaw and Chadwick, 1980, after Salter and Williams, 1965)

Soil type	Field capacity (% water)	Permanent wilting point (% water)	Available water capacity (cm)[a] (assuming rooting depth of 30 cm)
Sand	6.7	1.8	1.98
Sandy loam	19.8	7.9	4.75
Fine sandy loam	25.6	9.5	6.4
Silty loam	35.3	12.7	7.05
Clay loam	30.1	16.3	4.15
Clay	39.4	22.1	4.83
Peat	156.8	70.6	9.05

[a] This is the difference between field capacity and permanent wilting point calculated for 30 cm of soil and expressed as the depth of available water.

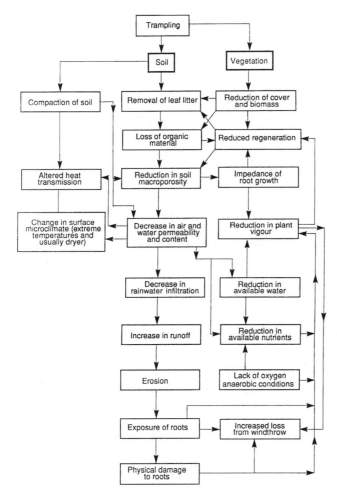

Fig. 9.8 Changes in soil condition resulting from trampling.

Table 9.3 Relationships between soil characteristics and susceptibility to impact (Cole, 1987b)

Soil property	Level of susceptibility		
	Low	Moderate	High
Texture	Medium	Coarse	Homogeneous; fine
Organic context	Moderate	Low	High
Soil moisture	Moderate	Low	High
Fertility	Moderate	High	Low
Soil depth	None	Deep	Shallow

may also result in a reduction of root growth. This will, in turn, cause a destabilizing of the soil as the old roots decay and no new ones take their place. The penetration of roots and root hairs and their subsequent decay normally contribute to the porosity of the soil so, in the long term, there will be a reduction in soil porosity as the surface vegetation is reduced.

Trampling also results in the maceration and physical removal of litter from the path surface as well as a reduction of the depth of organic soil layers. This will also lead to a reduction of porosity as does the direct force of compaction. The reduction in porosity means that there is less space for air and water, and a consequent reduction in the rate of diffusion of oxygen required for the respiration of soil fungi, bacteria, soil animals and plant roots, leading eventually to a drastic reduction in the living processes of the soil. The reduction in soil water also reduces the flow rate and the movement of dissolved mineral elements from their source to plant roots. The recycling processes are also inhibited by the reduction of bacterial and other activity. The resources available to plants are therefore limited by a number of different processes, which all tend to lead to a further reduction of plant growth.

The physical action of feet or wheels may also loosen or displace some of the particles at the soil surface, and this, together with the reduction in plant cover, leads to soil erosion. This is accentuated by the fact that rainfall cannot readily penetrate the compacted soil and a greater proportion flows over the surface. This increased runoff is a powerful erosive force, especially on sloping ground.

Desert sand dune soils may be destroyed by only one pass of a vehicle (Wilshire, 1983). The microfloral (algal) and mechanical crusts may be disrupted and the pattern of surface runoff may also be changed, with tracks forming channels resulting in accelerated erosion (Wilshire, 1983).

Exposed soil surfaces also lead to a drier microclimate and one in which the temperature ranges are more extreme (Liddle and Moore, 1974) and, combined with water shortages, temperatures are often beyond the lethal limits of very young seedlings. However, the heat conducting properties of the soil are also improved, and this may slightly ameliorate some of the extremes of temperature.

Recreation can therefore set in train a series of processes leading to, at best, a more vigorous growing environment for plants and, at worst, ugly scars on a previously natural and undisturbed landscape.

Litter and organic material 10

The litter discussed in this chapter is dead plant material. This may still be attached to a living plant, detached but still a whole leaf or stem, or even a complete plant. More usually it consists of dead detached portions of plants on the soil surface. Trampling increases the cover of litter on the ground by detaching dead and live materials from the plants. The litter already lying on the surface will be broken up and may be either kicked or scuffed off the pathway, blown away by the wind or washed away by surface water flow. In addition, it is reduced by the natural decomposition processes (Fig. 10.1). The balance between death of surface plants and loss of litter will determine the top cover of litter recorded in the field (Cole, 1982) (Fig. 10.2). Should the area of plants that die be greater than the area of litter that is lost from the site, then the recorded top cover of litter will increase as a result of trampling (Fig. 10.2b). This might occur with widespread light use. If, however, the worn area is more restricted, then the litter will be broken into fine pieces and easily lost from the path, while the cover of plants remains higher, so the recorded top cover of the litter will be reduced (Fig. 10.2c). This balance needs to be considered when examining the various, rather inconsistent, reports of the effect of trampling on the cover of litter.

10.1 Measurements of litter

Cole (1982) recorded the cover of litter in 26 campsites, mostly beside subalpine lakes, in the Eagle Cap Wilderness, Oregon. He found there was no significant difference in the cover of litter between light-use, moderate-use and heavy-use sites and the range was between 49% to 56% cover. However, when the cover of litter was expressed as a percentage of the adjacent undisturbed control conditions, the ratios between cover of litter on the impacted site and the controls were 0.9, 2.3 and 2.4 on the light-, moderate- and heavy-use sites, respectively – a marked increase in the quantity of litter under the higher levels of use. He also commented on two M.S. theses which reported no significant difference in cover of litter between light- and heavy-use sites (Coombs, 1976; Fichtler, 1980) and one publication (Legg and

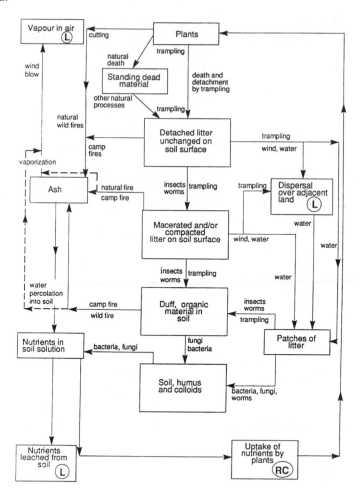

Fig. 10.1 The processes that may take place in the litter layer as a result of human trampling; L, lost from impacted area; RC, recycled within impacted area.

Schneider, 1977) in which a significant increase of litter cover in light-compared to heavy-use sites was recorded. In contrast, Brown, Kalisz and Wright (1977) reported a reduction of the mean cover of litter from 64% in undisturbed sites to 55% in disturbed sites of picnic and camping areas near sea-level in mixed oak (*Quercus alba*, *Q. coccinea* and *Q. velutina*) and white pine (*Pinus strobus*) forests of Rhode Island and southern New England.

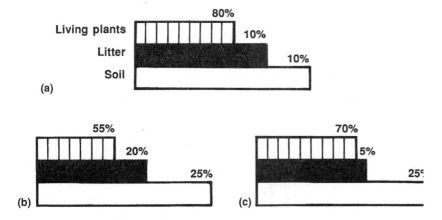

Fig. 10.2 The effect of plant death and litter loss on the recorded cover of litter. (a) Original condition; (b) widespread light use – a reduction in plant cover but apparent increase in litter; (c) heavy confined use – actual and measured decrease in litter. Note that an actual reduction in litter quantity may not be recorded if a greater reduction in plant cover has exposed or created more litter (b).

It is clear that greater accuracy in the records of changes in the amounts of plant litter can be achieved by the more laborious method of biomass sampling (Chapter 3). Different parts of a heather moor at Kindrogen, Scotland, were used at different intensities for teaching purposes. As the level of use increased, there was a steady increase in the biomass of litter from about 20% of the total biomass in the control to about 70% in the heavy-use areas, but this dropped to about 40% under severe use (Fig. 10.3) (Bayfield and Brooks, 1979).

This is probably a true recording of the effects of trampling on litter, with the initial increase followed by a decrease in the amount of dead plant material on the soil surface, as the intensity of use increased.

A triangular figure was presented by Jim (1987), plotting the cover of vegetation, bare ground and litter (Fig. 10.4). From this it is evident that, even in the same environment, sites may range from nearly 100% cover of vegetation to nearly 100% cover of litter with only small amounts of bare ground, although a few moderately used sites had over 80% bare ground, as did all but one of the severely used sites.

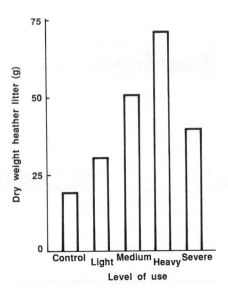

Fig. 10.3 Mean dry weights of heather (*Calluna*) litter (separated from total litter weights) and associated level of use in a study area at Kindrogan, Scotland (from Bayfield and Brooks, 1979).

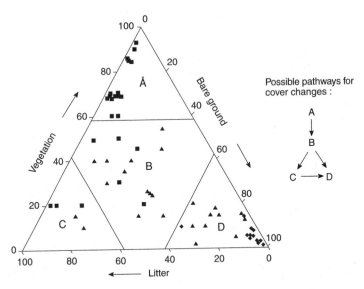

Fig. 10.4 The relative proportions of three cover types in 60 recreational sites in the Shing Mun Country Park, Hong Kong. The four regions (A to D) classify sites by the stages of degradation. The symbols indicate aggregate scores of erosion indicators: ■, slight; ▲, moderate; and ◆, severe degradation. (From Jim, 1987.)

Where light trampling occurs on litter which is not blown or washed from the site there is a gradual compaction of the litter mass. Duffey (1975) enclosed hay in nylon mesh bags measuring 20 × 20 × 8 cm. These were then pegged to the ground and some were lightly or heavily trampled at 5 or 10 treads per month for 12 months. The depth of the untrodden litter gradually declined from an initial 8.9 cm to 5.5 cm after 12 months. The depth of the lightly trodden litter bags was reduced to 2.4 cm and the heavily trodden litter was 1.7 cm deep (Fig. 10.5). The dry weight of the control showed a loss of 34% (from 125 g to 82.3 g) after 12 months while a litter bag receiving heavy treading had lost 94% (from 125 g to 7.7 g) of its initial weight. Duffey was not able to weigh more of the trodden bags because of the extreme difficulty in separating the litter from adhering soil particles. The trodden litter was fragmented by the physical impact of the human foot. Measurements made on sections shown in Fig. 10.5 showed that the untrodden litter had lost 55% of its volume after a 12-month period, while heavy trampling increased this loss to 81% (Table 10.1). The percentage of the volume that was air space fell from 63% in the control after 12 months to 38% in the litter that was heavily trodden (Table 10.1). In summary, heavy trampling approximately halved the volume of litter and the proportion of air spaces within it when the material could not be blown or washed from the site.

Compaction of litter and the reduction of contained air spaces had an unexpected consequence in a eucalyptus forest on a sandy soil approximately 100 km north of Brisbane, Australia. It was observed that after a small ground fire, lightly used pathways around an experimental plot still had unburnt, compacted litter lying on their surface while the adjacent uncompacted litter had been burnt away. Measurements made after the fire showed a mean of 446 g m^{-2} of litter on the path and 711 g m^{-2} in the unburned and untrodden areas; there was, of course, no litter left in the untrodden, burned area. It is possible

10.2 Compaction of litter

Table 10.1 Change in volume and proportion of air space in litter bags and sliced sections before and after treading (Duffey, 1975)

	Pre-treatment	Controls (after 12 months)	5 treads/month (after 12 months)	10 treads/month (after 12 months)
Volume (cm^3)	3560	1600	1128	708
Percentage fall from pre-treatment	–	55	68	81
Percentage of section consisting of air spaces	–	63	54	38

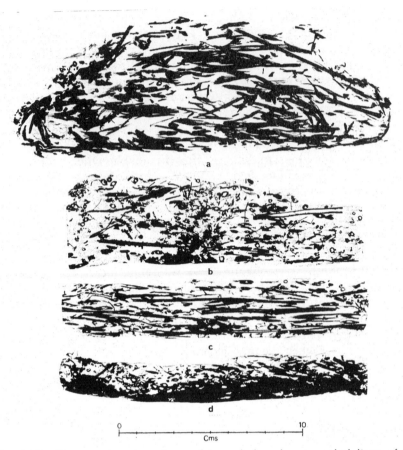

Fig. 10.5 Photographs of sections of trampled and untrampled litter after embedding in gelatin: (a) at the beginning of the experiment; (b) control (untrampled) litter after 12 months in the field; (c) after 60 treads during the 12-month period; (d) after 120 treads during the 12-month period. (From Duffey, 1975.)

that besides having less air space the compacted litter may have held more moisture which could also have inhibited its flammability. Litter may also be actively collected to act as kindling for campfires and any litter that underlies a fire will naturally be burnt away, together with some or all of the organic material in the surface layers of the soil. Fenn, Gogue and Burge (1976) (see Cole and Dalle-Molle, 1982) found that a large fire using 63.5 kg of wood burning over 50 hours not only destroyed all litter but altered organic matter to a depth of 10 cm or more, with 90% being lost from the top 2.5 cm of soil. Impacts were less in fine-textured and moist soils and where softwood fuels were burned.

Evidence for the dispersion of litter by vehicles comes from work on the impact of air-cushion vehicles on the arctic tundra of northern Alaska where there was a steady reduction in the cover of litter from 40% in undisturbed tundra to zero after 50 passages of the vehicle (Rickard and Brown, 1974) (Fig. 10.6). Photographs in a separate report on this work vividly demonstrate that larger amounts of organic material were blown into the air as the vehicle passed, leaving a ridge of litter 10–15 cm high adjacent to the track (Rickard, 1972). It is likely that the direct horizontal forces exerted by feet and wheels will also disperse litter but in lower quantities than those dispersed by hover-craft.

There is evidence for increased surface flow of water (Chapter 15) on impacted areas and this would carry away any fine particles of litter that may be lying on the surface of a path or track, especially during heavy temperate rains or the more frequently heavy tropical showers. The removal of surface layers of litter and organic material is apparent in a comparison of podzol profiles from impacted and adjacent undisturbed areas (Fig. 10.7) (Leney, 1974). The organic humus in the untrampled profile is replaced by a layer containing sand in the trampled profile. This suggests that not only have the surface layers been lost from the impacted area but that the lower B horizon appears to have been impregnated with more organic material, or duff, to create a new A horizon (duff is a product of litter decomposition in which the original structure is no longer discernible). This is now

10.3 Dispersion of litter

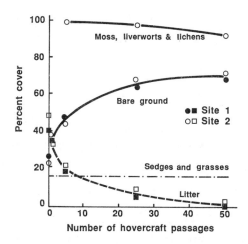

Fig. 10.6 The effect of repetitive passes of an SK5 air-cushion vehicle on the plant cover at Barrow, Alaska. Site 1, drained lake basin; site 2, drier upland area. ○ ●, bare ground; □ ■, litter. (After Rickard and Brown, 1974.)

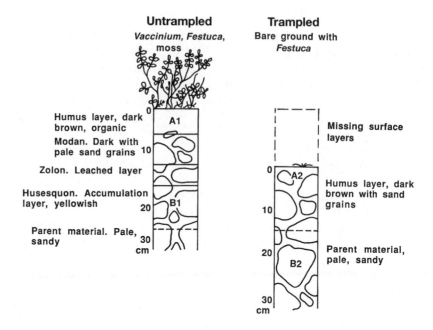

Fig. 10.7 The effects of trampling on the soil profile at Rothiemurchus, Scotland. The original profile descriptions by Leney are retained in the figure. (From Leney, 1974.)

immediately on top of the parent material. Interpretations of processes of trampling on soils need to take into consideration the changed soil profiles.

10.4 Incorporation of litter into the soil

The depth of duff on campsites in the Eagle Cap Wilderness, north-east Oregon, was very variable, with shallower layers in the least used and heavily used sites (Fig. 10.8) (Cole, 1982). However, when these were considered in relation to the depth of duff in adjacent control areas, there was a clear relationship between percentage loss and increasing use (Fig. 10.8). The highest-use areas had lost a mean of over 70% of the organic layer of the soil profile.

Cole (1982) found no correlation between soil organic content and use, but Dotzenko, Papamichas and Romine (1967) did find that there were negative correlations between these two parameters. Their samples were taken from 39 tent sites in each of three campgrounds in the Rocky Mountains National Park, Colorado. In each campground 13 samples were taken from each category of heavily, moderately and lightly used tent sites. The mean organic matter contents of the soils

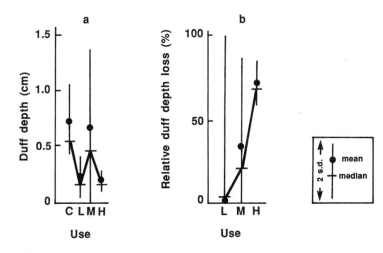

Fig. 10.8 Decrease in duff depth in relation to the amount of use: (a) duff depth and (b) relative reduction in duff depth, on controls (c), lightly (L), moderately (M) and heavily used (H) campsites. (From Cole, 1982.)

of three types of site were 3.6%, 3.5% and 5.1%, respectively. Willard (1971) also recorded a halving in soil organic content of used areas of two Texas state parks. Since the results are not always clear and there are at least two papers where increases have been recorded (Dotzenko, Papamichas and Romine, 1967; Cole and Fichtler, 1983) it seems unwise to conclude that use of areas for walking or camping will always lead to an immediate reduction in the organic content of the soil.

Larger pieces of the litter in the form of branches and logs for campfires are also collected from campsite and picnic areas. The range of collection may be considerable but most will occur within 100 m of the camp. As most of the nutrient cycling takes place through leaves and twigs, and soil organic material comes from the same source, collection of wood for fires is not usually a problem in this respect (Cole and Dalle-Molle, 1982). However, these authors pointed out that the decayed wood has a high water-holding capacity and can form significant long-term stores of nitrogen, phosphorus and sometimes calcium and magnesium, as well as being a significant site of nitrogen-fixing organisms. Dead wood may also form a substrate for seedling establishment and some important fungi which usually live in association with plant roots (ectomycorrhizal fungi). Dead wood may also form a habitat for macrofauna. For example, a 'shelter wood'

10.5 Other effects of recreation on dead plant material

from which all of the larger cut wood over 7.5 cm diameter was removed, had fewer spiders and other arthropods than an area where the larger wood had been left on the ground (Fellin, 1980). Dead wood either standing or lying on the ground may also form a nesting site or a substrate for the food supply of small mammals and birds (Cole and Dalle-Molle, 1982). The removal of dead wood also has an effect on the aesthetic value of the site and may allow visitors to penetrate further into the surrounding, previously undisturbed vegetation. The author has observed a rainforest campsite where cut wood was available but at a small charge – the forest floor had been 'cleared' up to 200 m in one direction where visitors had removed the fallen wood, thus extending considerably the area of disturbance to wildlife.

10.6 Summary   The amount of litter (or dead plant remains) lying on the soil surface may increase as a result of trampling impacts that kill and break plants, or decrease as the litter already present is broken up by trampling and the small particles are scuffed, blown or washed away from the trampled area. Because of the two processes involved and problems associated with expressing the amount of litter as a percentage of area covered by dead plant remains, various workers have reported both increases and decreases in the amount of litter present after an area has been trampled. Bayfield and Brooks (1979) (Fig. 10.3) suggested that it increases at first and then declines, and Jim (1987) considered that 'the conversion from vegetation to litter was faster than that of litter to bare soil initially, but the converse is true at advanced stages of degradation'. These are likely to be realistic assessments of the normal changes that occur. Light trampling may also compact litter, and in one instance this reduced its flammability.

Litter may be dispersed by vehicles, especially hovercraft, and the increased surface water flow associated with a reduction in vegetation cover will also carry away the finer particles of litter (duff). There is also some evidence that litter may move down soil profiles as they are changed by trampling impacts. As there is conflicting evidence, it seems unwise to conclude that use of an area for walking or camping will always lead to a reduction in the organic content of the soil, although where campfires are permitted branches and logs are likely to be removed from a considerable area around the site.

Soil bulk density, porosity and penetrability **11**

Many studies of the impact of recreation activities on the environment have utilized measurements of soil bulk density or penetrability, as indicators of soil compaction. 'A change in the state of compaction of a soil results in a change in its volume. A change in volume is caused by forces that may originate either from vehicles, trampling, by human or other animals or from natural causes such as drying and wetting' (modified from Harris, 1971). There is a clear relationship between the amount of use, either by vehicles or walkers, and the degree of soil compaction, as shown by measurements made over a period of 3 days in sand dune soil (Liddle and Greig-Smith, 1975a) (Fig. 11.1). Bulk density and penetration measures are also clearly related to each other, although there is a suggestion that the relationship might be curvilinear (Liddle and Greig-Smith, 1975a) (Fig. 11.2). Had the compaction been continued, the compaction process would have reached a maximum and no further increases in bulk density or penetration resistance would have been recorded. However, soil compaction is a complex process, as might be expected from the highly complex nature of soil.

Much of the following discussion is based upon papers in Barnes *et al.* (1971).

A soil consists of a matrix of solid particles that vary from organic peat to gravel, enclosing voids that may be filled with gas or liquid. When a soil is subjected to a load that causes a volume change, there are four physical changes that can take place in that soil:

11.1 Bulk density and porosity

- rearrangement of the positions of the solid particles;
- change in the volume of the pore spaces;
- compression of the solid particles;
- compression of liquids and gas within the pore spaces.

The first two are the primary changes that occur and the last two only occur to a very minor extent and are not normally considered in recreation studies.

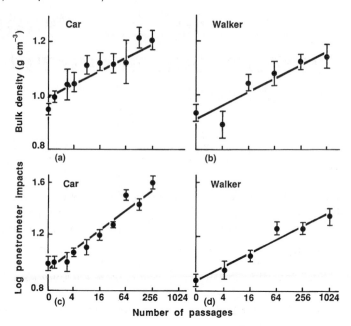

Fig. 11.1 The relationships between the number of passages of a car and a walker and (a) and (b) bulk density and (c) and (d) log penetrometer impacts. ●, means; I, two standard errors. (From Liddle and Greig-Smith, 1975a.)

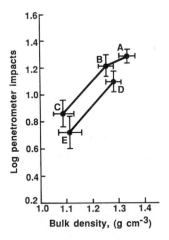

Fig. 11.2 Relationship between bulk density and log of penetrometer impacts. A, track stands; B, picnic area stands; C, adjacent 'natural vegetation' stands; D, footpath stands; E, adjacent 'natural vegetation' stands; ●, means; I, two standard errors of each group. (Liddle and Greig-Smith, 1975a.)

The condition or state of compaction of the soil can be defined in terms of bulk density, porosity and void ratio (Fig. 11.3). The bulk density is defined as the weight of oven dry solid (Vs) per unit of volume (V):

$$Bd = \frac{Vs}{V}$$

Porosity is the total non-solid volume and the void ratio is the ratio of non-solid to solid volume (Fig. 11.3). The calculations for the data in Fig. 11.3 are as follows:

$$\text{Porosity} \quad (n) = \frac{Vv}{V} = \frac{50}{100} = 0.5$$

$$\text{Void ratio} \ (e) = \frac{Vv}{Vs} = \frac{50}{50} = 1.$$

The functional relationship between the various changes that occur when soil is compacted is shown in Fig. 11.4.

The degree of compaction or re-arrangement of solid particles that occurs as a result of forces applied to the soil's surface depends on the extent to which the soil particles can change position. In saturated conditions the water acts as a cushion and the speed of water movement away from the affected area will determine the degree of compaction.

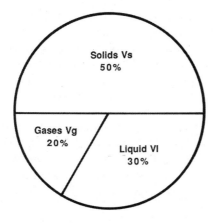

Fig. 11.3 Volume composition of a typical silt loam soil. Total volume (V) = Vs + Vg + Vl. Volume of voids (Vv) = Vg + Vl; porosity (n) = $\frac{Vv}{V}$; V and void ratio (e) = $\frac{Vv}{Vs}$. (After Harris, 1971.)

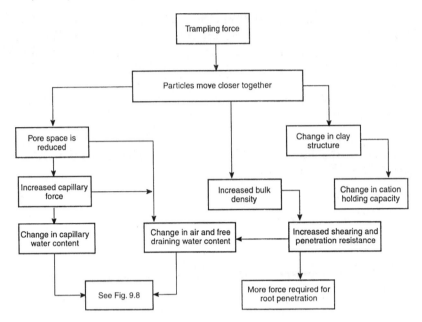

Fig. 11.4 Changes that take place in some soil characteristics as a result of trampling.

The re-arrangement of the soil particles also depends on the structural arrangement of the particles and, in fine-grained soils, on the degree of bonding between adjacent particles. The resulting compaction is due primarily to a change in the volume of the voids. The relationship between load and void volume is logarithmic with increasing forces being required to bring about similar amount of change as the soil compacts (Fig. 11.5). The pore system, especially the interconnectedness of the pores, is also important as it affects the flow of air and water.

In uncompacted, fertile, well-structured soils there are large voids with many interconnecting passages (Fig. 11.6a). These allow a ready flow of air and water and hence good root growth. When these soils become compacted the pore volume is reduced as well as the extent of the interconnecting passages (Fig. 11.6b). However, the surface layers of the soil are also subjected to lateral forces, especially from a rolling wheel, and the soil particles are moved in three dimensions (Fig. 11.7). Thus the pores in surface layers are less linear, although Trouse (1971) suggests that this is due to the lower quantity of undecayed roots in the surface layers of these samples from arable fields (Fig. 11.6c). The most compacted zones have been found at 15 cm depth under car tracks and footpaths (see Fig. 11.19; vegetation records are shown in Fig. 3.2) and at 19 cm under cow paths (Liddle, 1973b; Liddle and Greig-Smith 1975b). The uncompressed surface layers may, in this case, also be the result of lateral forces disturbing the surface layers of the poorly graded sand dune soils. The depth of the most compacted layer is likely to depend upon the ratio of vertical to horizontal components of force and the type and condition of the soil.

Most soils contain sand, silt and clay and a 'well-graded' soil containing both coarse- and fine-grained particles will have more contacts between particles than a poorly graded soil, mainly made up of larger particles. Therefore a well-graded soil will have more resistance to movement between particles and hence compaction will be less, for a

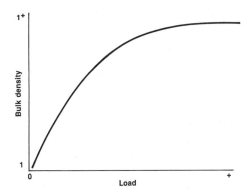

Fig. 11.5 Hypothetical curve showing relationships between soil bulk density and trampling or vehicle loads. Note that after a certain reduction of porosity the soil will resist further compression.

Fig. 11.6 Pore systems of thin slices of soil, the dark areas represent the voids. (a) Uncompressed soil; (b) tilled soil subsequently compressed by traffic – note the small isolated voids typical of the pore system of compressed surface soil; (c) compressed soil from the hard layer formed as a traffic pan during the tillage operation – note the suggested connectedness of the voids. (From Trouse, 1971.)

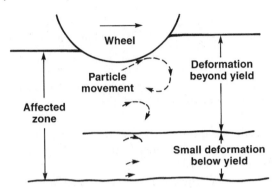

Fig. 11.7 Deformation and displacement patterns in the soil under a moving wheel. Note the horizontal elements of displacement, especially near the soil surface. (From Freitag, 1971.)

similar amount of use, than in a poorly graded soil, although the max-imum bulk density reached under continuous use may well be greater than in poorly graded soils. The water present in the soil will facilitate particle movement up to the point where the pores are full and the water has to undergo compression or move out of the loaded area for compaction to occur (Fig. 11.8) (Chancellor, 1971). As the pressure on the soil is increased the bulk density at first increases and then, as the proportional moisture content rises, the bulk density no longer increases so much, and ceases altogether when the soil is saturated. At that point the water resistance prevents any further compaction unless water is forced out of the pores, like squeezing a wet sponge.

The proportion of three different particle size classes, sand, silt and clay, can be used to classify soils using the nomenclature of the US Department of Agriculture. These are grouped into 12 basic soil classes and presented as a triangular diagram (Fig. 11.9). The interaction between the particle size composition, moisture content and porosity under a standard load of 2 bars (a bar = 0.99 atmosphere) is shown in Fig. 11.10 (Chancellor, 1971). The poorly graded soil with the narrowest distribution of particle sizes (soil C) tended to be forced to acquire greater dry bulk densities than the well-graded soil with a wide range of particle sizes (soil A). The finer-textured soils and clays (soil A) tended to resist further compaction at a lower density for a given force than did the coarser soils (soil C).

Consideration of clay particles shows that their alignment is crucial to their bulk density and water-holding capacity. Clay particles may be considered as planar, and when these are randomly arranged the bulk density and moisture content are low (Harris, 1971) (Fig. 11.11). As the clays are compacted, they pass through a disorganized phase

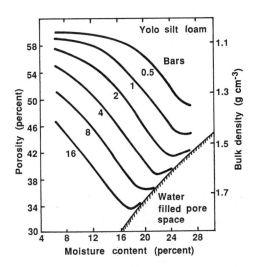

Fig. 11.8 The effect of moisture content on the stress–compaction relationship. The different lines indicate different levels of applied pressure in bars. Note that the percentage pore space and moisture content are volume measurements. (From Chancellor, 1971.)

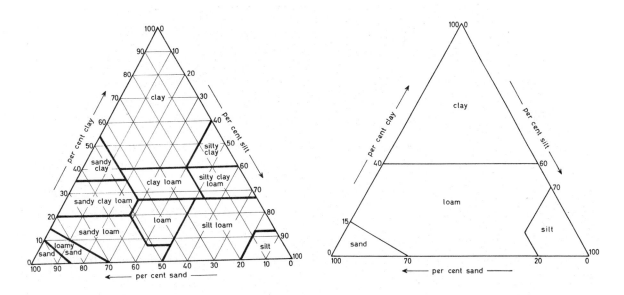

Fig. 11.9 Triangular diagram of texture classes according to the US Department of Agriculture. Sand 2–0.05 mm; silt 0.05–0.002 mm and clay <0.002 mm. The lowest percentage line is read from each point, for example a loam may have 20% clay, 40% silt and 40% sand. (From US Department of Agriculture, undated.)

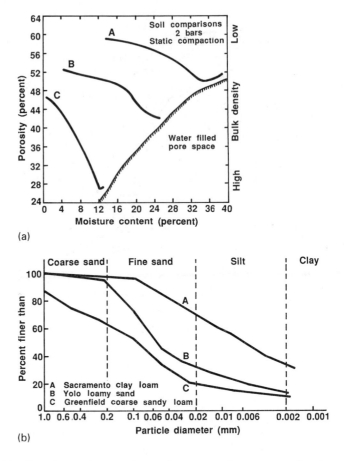

Fig. 11.10 Density–moisture relationships (a) and grain size distributions (b) for three soils. The soils were subjected to a pressure of 2 bars in a static compaction cylinder. Soil A is the finest of the three and soil C has a very broad distribution of particle sizes and assumes the most compact state. Note that in (a) the percentage pore space and moisture content are volume measurements. (After Chancellor, 1971.)

where they are flocculated and have the lowest dry bulk density. In the final state the particles are parallel and have a higher bulk density (Fig. 11.11).

Compression of the solid particles, gases or liquids is such a small effect that it will not be considered in this discussion.

Some changes in bulk density that have been measured in recreation areas range from an increase of $0.08\,\text{g cm}^{-3}$ in a campground soil which was derived from granitic bedrock in Oregon (Cole, 1982) to an increase of $0.56\,\text{g cm}^{-3}$ in a birch woodland soil on pathways near

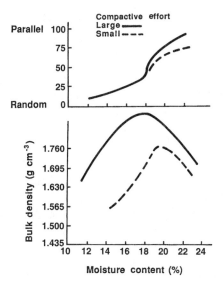

Fig. 11.11 The effect of moisture content on particle orientation for compacted Boston blue clay (Lamb, 1960.)

Moscow (Yur'eva, Matveva and Trapido, 1976) (Table 11.1). The highest mean soil bulk density presented in the series of papers listed in Table 11.1 is 1.51 g cm⁻³, recorded from beneath the soil surface in an agricultural situation (Arndt and Rose, 1966). Other factors will also influence the degree of change in bulk density of soils compacted by recreation activities. The comparative measurements from wet and dry areas of Welsh sand dunes suggest that the wetter soils are compacted to a greater degree than the dry ones, but this could also be associated with differences in soil structure. Wet soils occur naturally in hollows where finer particles of sand and a greater build-up of organic material both occur. The broader distribution of particle size may also be the cause of the greater uncompacted bulk density of the loam soils compared to the sandy soils in the Connecticut State Parks studied by Lutz (1945) (Table 11.1, see also Fig. 11.10). It is clear that, as expected from work in agricultural situations, well-graded soils with a broad distribution of particle sizes generally compact to a greater degree as a result of continuous use for recreation, than poorly graded soils. Russell (1973) quotes an approximate density of 1.6 g cm⁻³ as the maximum for plant growth, although any change in soil compaction changes the root environment to some degree. The change in the bulk density of snow as a result of snowmobile use has quite different implications for the survival of plants and animals, and is discussed in Chapter 24.

Table 11.1 Some examples of bulk density changes in soils subject to recreation or vehicle use

Vegetation and soil type	Site use	Bulk density (g cm⁻³)			Increase as % of original wt	Core depth (cm)	Location	Author
		Uncompacted	Compacted	Change				
Dry sand dune,	Car track	1.16	1.36	+0.20	17	6	Abeffraw,	Liddle (1973b)
pasture	Footpath	1.14	1.29	+0.15	13		Wales	
Wet	Car track	1.06	1.30	+0.24	22			
dune slack	Footpath	1.09	1.29	+0.20	18			
Tippera	Agricultural	1.06	1.30	+0.24	23	2.5	Northern	Arndt and Rose
clay loam	traffic	1.12	1.51	+0.39	35	7.5	Australia	(1966)
Chalk	Trampling and	0.68	1.02	+0.34	50	2.5	Nr Winchester,	Chappell *et al.*
grassland	vehicles	0.96	1.00	+0.04	4	5	England	(1971)
North-facing	Trampling	0.84	0.98	+0.14	17	5	Iowa	Dawson, Hinz and
forest slope								Gordon (1974)
Floodplain forest glacial till		0.95	1.39	+0.44	46	5		
Lebanon silty clay loam, native forest	Campgrounds	1.06	1.31	+0.25	24	not stated	Ozarks Missouri	Settergren and Cole (1970)
Soils derived from granite bedrock subalpine fir, Engleman spruce	Campgrounds	0.88	0.96	+0.08	9	5	Eagle Cap- Wilderness, Oregon	Cole (1982)
New England, mixed oak and white pine forest, granite till or outwash loamy sandstone to sandy loams	Campgrounds and picnic areas	0.54	0.93	+0.39	72	5.1	Washington County, New England	Brown, Kalisz and Wright (1977)
Merrimac sands, coarse texture	Trampling	0.01	1.37	+0.36	36	10	Warton Brook State Park, Connecticut	Lutz (1945)
Sand loam, fine texture, forest park	Trampling	1.06	1.30	+0.24	23	10	Sleeping Giant State Park, Connecticut	
Recreation area, birchwoods	Trampling	0.91	1.47	+0.56	62	not stated	Krasnaya Pakhre, near Moscow	Yur'eva, Matveva and Trapido (1976)
Snow	Snowmobile							
Shade		0.14	0.53	+0.39	279	not	Ottawa Area	Neumann and
Sun		0.19	0.56	+0.37	195	stated		Merriam (1972)

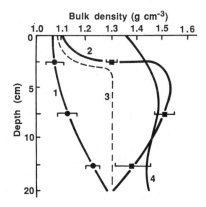

Fig. 11.12 Bulk density profiles in crop bands (1), traffic bands (2), bullrush millet stubble land (3), and a grazed native pasture (4) (from Arndt and Rose, 1966).

The depth to which changes take place will depend upon the soil type and the magnitude of the compaction forces. Bulk density has not often been recorded below the top 5 cm in recreation studies, but Arndt and Rose (1966) did measure it to 15 cm in their studies of the effects of traffic in an agricultural situation. They found the maximum compaction at 7.5 cm, while the surface layers of the tilled soil were relatively unaffected (Fig. 11.12). In contrast, the grazed native pasture shown in the same figure was quite compacted in the surface layers. This may be expected as the greatest compaction forces occur at the soil surface (Cohron, 1971); however, the particles at the surface also show considerable movement (Fig. 11.7), and where the form of use exerts a high horizontal component of force, surface compaction may not always occur.

It is clear that the distribution of compacted soil down the profile is dependent on the nature of the compacting forces, in particular the ratio of horizontal to vertical forces and the depth of penetration of the hoof, foot or wheel into the surface layers. Weaver and Dale (1978) recorded the depth of experimental horse trails as nearly 15 cm, and the author has measured hoof 'prints' up to 40 cm deep in a wet clay soil in Sussex, England.

11.2 Particle size and aggregates

Soil particle distributions and soil aggregate sizes are changed by the effects of compaction and they also have a direct effect on the soil density. There have been very few comparative measurements of soil particle size distribution in soils compacted by recreation use and adjacent uncompacted soils. In his study of state parks in Connecticut,

Lutz (1945) found that in the used areas 0.82% of particles were over 2 mm in diameter, and in the unused soils, 0.33%. He believed that this was the result of sorting by wind, the largest mineral particles having been left behind. There was also a small, statistically significant decrease in the percentage of sand (2.0–0.05 mm diameter) in the soils of the used area (85.75%) compared to the unused area (89.39%). He suggests that these are within the 'suspension competence' of the winds having blown the sand from the used to the unused area where it is trapped by the vegetation which disturbs the laminar flow of air. In contrast, the percentage of silt (0.05 mm diameter) was 7.37% in the 0–10 cm layer of soil from the used area and 4.61% in the uncompacted soils. He again invokes wind transport to explain this difference, suggesting that the finer particles are blown right away from the site, but this does not explain the greater amount of silt in the compacted soils. This could perhaps be due to the movement of silt down the profile, suspended in drainage water, as a consequence of the break-up of the surface soil structure.

A similar relationship was observed in grassland soils at Cape Fear, North Carolina (Hosier and Eaton, 1980). There were more coarser particles in the impacted soils and the finer (mid-range) particles were in greater quantities in the non-impacted soils (Fig. 11.13). There were no measurable quantities of silt or clay. The authors suggested that the greater cover of grasses had trapped the wind-blown sands in the non-impacted grasslands and also reduced the velocity of oceanic overwash thereby promoting the collection of water-borne material. In general, the recorded changes in soil particle size distribution appeared to be associated with changed windflow, soil structure and drainage conditions.

Soil structure can also be described in terms of the coarser aggregates or clods in the soil. A good structure, with aggregates over 1 mm in diameter, is important for maintenance of the larger pores

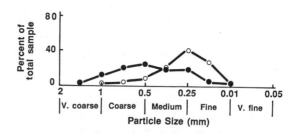

Fig. 11.13 The distribution of sand-sized particles under grasslands at Fort Fisher Beach (impacted) (●) and Bald Head Beach (non-impacted) (○) (from Hosier and Eaton, 1980).

and consequent good aeration and drainage. Aggregates need to be distinguished from clods, which are compressed lumps of soils generally over 0.2 cm in diameter (see Chancellor, 1971). Compaction destabilizes aggregates (which is evident as they disintegrate when they are subsequently wetted) but increases the number of clods in the soil. Arndt and Rose (1966) found that the mean content of clods less than 1.25 cm in diameter fell from 74.6% to 34.5% where soil had been compacted by tractors, but there was a sevenfold increase, from 7.2% to 48%, in larger clods (5–15 cm in diameter) as a result of traffic compaction. In contrast, the weight diameter of aggregates in a chalk grassland soil did not change as a result of trampling and use by vehicles (Chappell et al., 1971). Weight diameter is a means of expressing the relative size range of soil aggregates and was given by:

$$\text{weight diameter} = \Sigma \frac{(a\ b)}{100}$$

for each sample where a = percentage of soil remaining on a sieve of particular size and b = the mid-point of the sieve's aperture range for the same size fraction. However, when the aggregates from the top 2.5 cm of soil were tested for water stability there was a marked drop, from 83.2% of aggregates of 1–2 mm diameter being water stable in soils from untrampled areas to only 50.6% in soils from areas used by walkers and vehicles. Larger aggregates (2–4 mm in size) showed an identical loss in water stability in the trampled surface layer. This loss of structural stability was thought to be more serious than changes in bulk density and pore space, because a loss of stability leads to a lower infiltration of rainwater and possible surface erosion.

11.3 Soil pore space

The changes in soil porosity have been measured directly by some workers and indirectly (as infiltration rates) by others. Total pore space is generally in the region of 50% or more of the soil volume. Chappell et al. (1971) found that the surface layers (0–2.5 cm) of their undisturbed chalk grassland soil had 72% pore space and the layers 2.5–5 cm deep had 60% pore space. The pore spaces of both these layers were reduced to 58% in the area used by walkers and vehicles. Reductions of 14% and 2%, respectively. The pattern of lower pore space at greater depths and less change at greater depths was also recorded by Lutz (1945) in sandy and sandy loam soils, and by Burger (1932) in woodland soils in Germany (Fig. 11.14a,b). The surface pore space was reduced from about 68% to 55% in trampled areas, and under playing fields at forest camps it was reduced still further to about 45% (Fig. 11.14c). In all these cases the porosity declined with depth, a characteristic that was also found by Webb (1982) in desert soils, where there was a reduction in all pore sizes to a depth of 6 cm, more especially those greater than 4.5 μm, as a result of increasing numbers of passes by a motorcycle (Table 11.2).

Infiltration rate is not a direct measure of porosity as it is affected by the soil's suction pressure at the time the measurement is made and because water will not flow evenly through a soil with a layered texture, especially if a less permeable layer overlies a more permeable layer. The platelets in clay soils become oriented when trampled or puddled in the presence of moisture (Fig 11.11). If the soil then dries out, a hard crust of varying thickness is formed and moisture is consequently retained on the soil surface. In spite of these problems, infiltration rates have been found to be a useful measure of soil change as a result of use by walkers or vehicles. The mean infiltration rates were drastically reduced in the loamy sands and sandy loams of the granitic soils of the park and picnic grounds studied by Brown, Kalisz and Wright (1977) in New England. They used a 10.2 cm diameter, single-ring infiltrometer driven 5.1 cm into the soils and found a mean rate of 378 cm h^{-1} in the uncompacted soils. This was reduced to a mean of only 12 cm h^{-1} in the trampled areas. Cole (1982), using a double-ring infiltrometer, also in granitic soils in the Eagle Cap Wilderness, Oregon, divided his readings into instantaneous infiltration

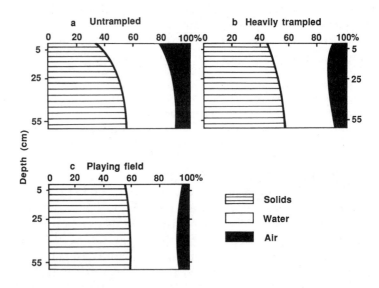

Fig. 11.14 The proportions of solid components, water and air in soils at Vaumarcus, Central Europe: (a) untrampled soil; (b) heavily trampled soil; (c) playing field soil (after Burger, 1932).

Table 11.2 Comparison of the volume of pores with radius greater than 4.5 microns (μm) in motorcycle-compacted and undisturbed soils, Fremont Peak Study Area (Webb, 1982)

	Pore volume (cm³/100 g)				
	Undisturbed	1 Pass	10 Passes	100 Passes	200 Passes
Pore radius (μm) 0–30 mm depth					
150	8.9	6.4	2.9	2.9	2.2
50–150	3.2	2.1	3.2	3.6	3.6
30–50	2.6	1.9	3.1	2.1	2.6
15–30	2.4	2.1	2.6	3.4	2.8
7.5–15	2.4	2.0	2.6	1.9	2.4
4.5–7.5	1.7	1.7	1.2	0.4	0.8
< 4.5	6.4	7.8	5.7	5.7	5.0
30–60 mm depth					
150	2.5	4.5	1.4	1.6	1.8
50–150	4.4	3.7	3.1	3.2	3.5
30–50	3.3	3.4	2.5	2.4	2.6
15–30	3.4	1.5	2.4	2.6	2.1
7.5–15	2.3	2.3	2.1	2.3	1.9
4.5–7.5	1.0	1.1	1.0	0.7	0.6
< 4.5	5.8	6.3	6.0	5.9	5.8

rates (the first 1 cm of water) and saturated infiltration rates (the first 5 cm of water). His light-use sites had mean instantaneous rates of 36 and saturated rates of 14.4 cm h^{-1}. These were reduced to means of 19.2 and 7.8 cm h^{-1}, respectively in the heavy-use sites. The infiltration rates of the compacted areas of the two studies were of a similar order of magnitude, but there is a major difference between values obtained in the 'control' areas of Brown, Kalisz and Wright (1977) and the 'light-use' areas of Cole, indicating that infiltration rate is a very variable soil characteristic. However, many other features such as soil type, water content at the time of the measurement and the time over which the measurements were made (not stated by Brown, Kalisz and Wright, 1977) may also have affected this comparison.

An example of this variation was recorded at the two desert locations in southern Nevada studied by Eckert *et al.* (1979), where there were coppice soils on low, small dunes under shrubs, and 'interspace' – mostly barren soils. The interspace soils had gravel pavements, either of a single layer embedded in the mineral soil or of several pebbles thick forming a gravel mulch over the mineral soil. The coppice soils had similar terminal infiltration rates of about 3 cm h^{-1} (after 30 minutes) at both sites, and 20 passes of a 0.75 tonne pick-up truck reduced these to about 2 cm h^{-1} in both cases. The interspace gravel-mulched soil at Crystal Springs had a terminal infiltration rate of about 2 cm h^{-1} in its undriven state, whereas the gravel-embedded soil at Blue Diamond had a terminal (30 minute) rate of only about 0.2 cm h^{-1} (Fig. 11.15). The infiltration rate of the gravel-mulched soil was reduced by 20 passages of a truck to about 0.75 cm h^{-1}, whereas the infiltration rate of the gravel-embedded soil was almost unchanged, at about 0.2 cm h^{-1}, by the treatment. The sediment loss characteristics responded differently (Chapter 15).

The relationship between the intensity of use (expressed as the number of passes of a motorcycle) and terminal infiltration rate measured with a double-ring infiltrometer was found by Webb (1983) to be curvilinear, with reduced variation at higher levels of use (Fig. 11.16). This is likely to be due to the soil reaching a point where any further compaction would require very high instantaneous pressures, which would break up the soil particles, and suggests that most experiments relating soil compaction to wear by vehicles or walkers have not taken the impact to the point where an equilibrium is reached. The reduction in infiltration rates will generally cause a major increase in the soil surface runoff and consequent erosion during heavy rainfall.

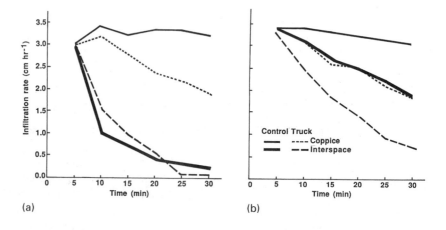

Fig. 11.15 Infiltration curves for coppice and gravel-embedded interspace soils on track and control treatments at Blue Diamond (a) and Crystal Springs (b), two desert sites in southern Nevada (from Eckert *et al.*, 1979).

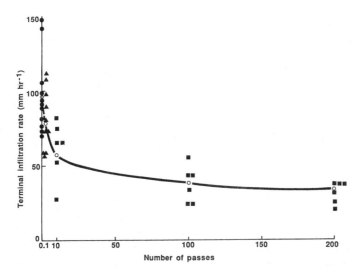

Fig. 11.16 The effect of repeated motorcycle passes on the terminal infiltration rate of Fremont Peak loamy sand. Note the curvilinear response and reduced variation at high levels of use. (Modified from Webb, 1983.)

Penetrometers are easy and quick to use, so a large number of readings can be made in a short time. This produces results that can be displayed with small error terms and an apparent high degree of accuracy when compared with bulk density measurements (Fig. 11.1). Further comparison of the graphs in Fig. 11.1 shows that the number of impacts required to penetrate a standard distance of 5 cm was 16 after 16 passes in a car and 40 after 256 passages, a rise of over 2.5 times in the number of impacts. The bulk density rose only from 1.1 g cm^{-3} to 1.2 g cm^{-3} in the same tracks. This demonstrates the higher sensitivity of the penetrometer. The successful use of the log scale for penetration impacts also indicates that the relationship between penetrability, as measured by impact penetrometers, and bulk density is curvilinear. However, there are many problems associated with the interpretation of these results. For example, as water content increases, the bulk density of a soil subjected to a given amount of compaction energy increases to a peak after which the density declines. In contrast, the penetrometer resistance declines steadily, so the measurements are not strictly comparable (Chancellor, 1971). If the penetrometer measurements are to be related to soil density, it is imperative that the moisture content, soil type and the penetrometer type (shape, size and material of the penetrometer point, as well as the operating procedure) be identical for all compared points. In other words, compara-

11.4 Penetration resistance

tive measurements can only be completely reliable when comparisons are made by the same operator using the same equipment within one soil type at one level of soil moisture. One other consideration is that a good relationship between the penetrometer resistance and percentage root penetration has been demonstrated (to 850 kg cm^{-2}, where no further root penetration occurred) (Taylor, Roberson and Parker, 1966). In so far as a penetrometer can give an indication of the possible growth of plants, it does have an advantage over measurements of bulk density.

The comparison of recreation and control sites in southern New England by Brown, Kalisz and Wright (1977) provides one example of the use of a spring penetrometer, which measures force. Here the penetration resistance had a mean of 1.25 kg cm^{-2} in the control sites and 3.05 kg cm^{-2} in the recreation sites, a rise of 240% (Table 11.3). Bulk density rose by 185% at the same sites, indicating that when penetration is measured in terms of force rather than impacts there is a similar order of response to that of bulk density measurements.

An extensive series of penetration tests was carried out by Wilshire and Nakata (1976) in the Mojave Desert, California, to determine the effects of motorcycles and military practice manoeuvres on the desert soils. Their penetrometer had a 30° graduated cone, and this was loaded by a standard 'full' body weight. The depth of penetration of the cone into the soil was then measured. They investigated a number of soil types, and the results for areas of alluvium derived from limestone are presented here (Fig. 11.17). They commented that 'initial or light impact, results in a considerable lateral displacement of material and loss of cohesion, whereas repeated impact results in compaction'.

The relationship between penetrability and depth can reveal more about the compaction processes than surface readings of bulk density, and readings at depth are more easily made by penetrometer. A survey of tracks of Fort Fisher Beach, North Carolina, showed that at a force of 20 kg cm^{-2} the depth of penetration was about 25 cm in the undisturbed sites, 15 cm between the tracks, and about 7 cm in the tracks

Table 11.3 Characteristics of the soil surface of recreation and control sites (Brown, Kalisz and Wright, 1977)

	Recreation sites	Control sites	Percentage differences
Penetration resistance (kg cm^{-2})	3.05	1.25	240
Bulk density	0.93	0.54	185

Both site differences significant at the 0.01 level.

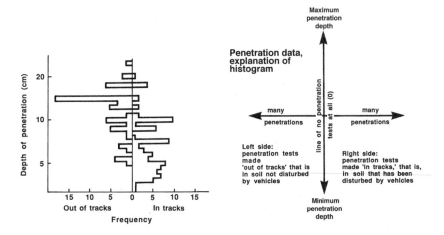

Fig. 11.17 Relationship between soil penetrability and occurrence of compacted soils in and out of motorcycle tracks in the Mojave Desert, California. The histograms show the number of test measurements that penetrated to a given depth. The greater number of bars on the lower part of the right-hand 'in track' histogram indicate more compacted soils. (From Wilshire and Nakata, 1976.)

themselves (Fig. 11.18) (Hosier and Eaton, 1980). It is interesting to note that at an applied force of about 3 kg cm^{-2} the depth of penetration was greatest in the intermediate area, suggesting that here the surface layers were considerably loosened by traffic.

Sequential readings made with an impact penetrometer provided data on which compaction profiles of the soil to a depth of 60 cm, beneath a sand dune track and a footpath, were constructed (Liddle and Greig-Smith, 1975b) (Fig. 11.19). It can be seen that the most compacted layer is at a depth of about 15 cm in both cases. This may be another demonstration of the fact that surface layers are subject to lateral movement, but it is interesting that, although the lateral forces exerted by vehicles are much greater, the depth of maximum compaction is the same in both cases. Apart from the 'displacement' of the zone of maximum compaction to some depth in the soil, the compaction contours show a pattern that would be intuitively predicted from the vertical stress lines measured in a silt loam soil by Reaves and Cooper (1960) (Fig. 2.8). Penetrometer measurements can add to our understanding of the effects of recreation on soils, but, like all other measuring devices used in ecological studies, their potential and their limitations need to be understood when the results are interpreted.

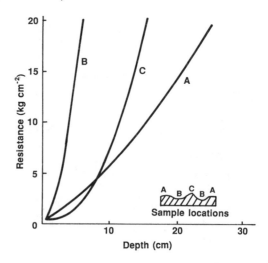

Fig. 11.18 Soil penetration resistance at various depths: (a) in non-impacted areas; (b) in vehicle tracks; and (c) in areas between tracks at Fort Fisher Beach, North Carolina. (From Hosier and Eton, 1980.)

The positive relationships between the soil qualities discussed above and soil moisture, organic carbon, pH and erosion indicators were demonstrated by Jim (1987) in the Shing Mun Country Park, Hong Kong for three levels of use (Table 11.4).

11.5 Summary   Recreation activities such as walking, driving vehicles or riding horses result in soil compaction. This results in changes in volume which have often been measured by the bulk density (the dry weight per unit volume) of the soil or by penetration resistance. The two measures are related.

Soil compaction involves a re-arrangement of the solid soil particles and a change in volume and nature of the soil pore spaces. The depth of the most compacted layer varies according to the nature of the compacting forces and the type and condition of the soil. The amount of compaction that occurs for a given force is related to the range of sizes of the soil particles, a well-graded soil with both coarse and fine particles will compact more slowly than a coarse soil. When all the pores are saturated no further compaction can occur unless water is squeezed out of the affected area. When subjected to pressure, clay particles, which are planar, pass from a random arrangement through a 'disorganized' phase, where they are flocculated and the clay has the lowest bulk density, to a state where they are parallel and have the highest bulk density.

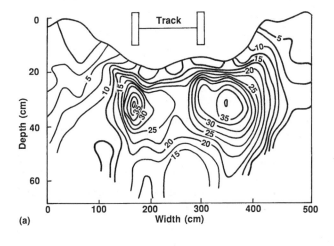

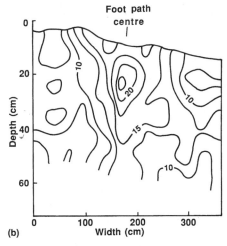

Fig. 11.19 Soil hardness isolines in sand dune soils at Aberffraw, North Wales: (a) under a car track, and (b) under a footpath. Vertical exaggeration 5×. (From Liddle and Greig-Smith, 1975b.)

Measured changes in compaction of soils subject to recreation impacts vary from 0.08 g cm^{-3} to 0.56 g cm^{-3}. A bulk density of 1.6 g cm^{-3} is said to be the maximum for plant root growth. Bulk density has usually been measured in the surface layers in recreation studies; in one agricultural situation the highest density was found at a depth of 7.5 cm.

The recorded changes in soil particle size distribution appear to be due to changed windflow, soil structure and drainage conditions. The larger soil aggregates are often destroyed when compaction occurs,

Table 11.4 Soil properties in three site-erosion-classes, (modified from Jim, 1987)

Soil attributes	Levels of use		
	High	Medium	Low
Duff depth (cm)	0	1.5	6.0
Texture	Loamy sand	Loamy sand	Loamy sand
Stone content (%)	16.5	10.6	17.0
Structure			
shape	Apedal	Apedal	Crumb
size	–	–	Medium
grade	Massive	Massive	Moderate
Aggregate stability (%)			
surface soil	43.6	59.3	69.7
subsurface soil	44.0	66.9	77.4
Penetration resistance (kg)	6.0	5.2	4.5
Bulk density (g cm^{-3})	1.35	1.35	1.29
Pore volume (%)	43.6	44.8	48.1
Infiltration rate (cm h^{-1})[a]			
instantaneous	1.3	11.8	48.0
saturated	0.6	4.8	22.8
Moisture content (%)	16.7	20.9	23.8
Organic carbon (%)	1.9	2.4	3.3
pH	7.5	5.3	6.3
Sheet-erosion indicators (cm)			
platform exposure	2.9	1.6	0.0
root exposure	6.3	3.1	3.7

[a] Instantaneous rate refers to that before the soil is saturated by the added water: saturated rate, which is more stable, refers to that after the soil has been saturated.

and the number of larger, water-stable aggregates was reduced by trampling on chalk grassland soils.

Infiltration rates are affected by the sizes and connectedness of the soil pores, which are in turn affected by compaction. Infiltration rates have been studied at a number of sites and in one case a rate of 378 cm h^{-1} in untrampled soils was reduced to only 12 cm h^{-1} in a used area.

Penetration resistance is a useful, sensitive measure of soil compaction, but the results are not always comparable and there are problems of interpretation. Standard impact penetrometers appear to be more sensitive than those that measure the force required for penetration. Penetrability of the soil is sometimes increased by traffic in the surface layers but it is usually reduced at depth. Penetrometer readings have been made to depths of 60 cm and compaction contours mapped in a vertical profile.

Generally, soils undergo change as a result of use for recreation. The surface layers may be loosened but compaction usually occurs at a level below the disturbed soil. Measurements of soil compaction have been made in various ways but the results must always be interpreted with an understanding of the nature of the soils and the limitations of the techniques used.

12 Soil water, air, temperature and nutrients

Soil water, air and nutrients, are, like most soil qualities, interdependent, especially where the amounts available to plants are concerned. Water and air occupy the pore spaces within the soil. The nutrients, originating from the mineral and organic material or outside the soil system, all eventually pass into solution in the soil water, where they become available to plants or are leached from the soil.

12.1 Soil water

Plants and animals depend on a supply of water to maintain their cellular life processes, and plants also need large quantities for transpiration. An adequate water supply is therefore necessary to maintain soil fertility and plant cover. Since compaction alters the soil structure and the associated pore sizes, shapes and volumes, and since the nature of the pores determines the soil water content, compaction also alters water content and transmission in soils. Compaction usually produces layers of soil with high bulk density and low porosity rather than a uniformly compacted soil (Warkentin, 1971). These layers may form at the surface, due to raindrop action or sliding compaction forces, or they may form beneath the surface, due to rolling loads and consequent movement of surface particles (see Fig. 11.7). It is therefore possible for soils to be altered so that they hold less or more water for a longer period of time.

Measurements of the quantity of soil water may be expressed as a percentage of soil wet weight or dry weight, percent volume, or as soil suction. Soil suction is measured as the suction to which water must be subjected to prevent it from flowing through a permeable membrane into the soil. That is, equal to the suction being exerted by the soil on any water that comes into contact with it. Plants can exert a suction of up to 15 bars (1.5 MPa) to withdraw water from the soil. It can be seen from Fig. 12.1 that at 20% water content by weight a suction of about 0.1 bar is required to remove water from a sandy soil, while a suction of about 10 bars is required to obtain water from a clay soil with the same percentage water content.

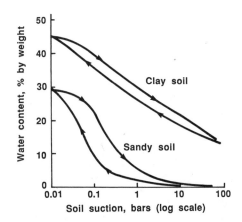

Fig. 12.1 Generalized water retention curves for two soils with different grain-size distributions. The relationship between soil suction and water content are different according to whether the soil is drying out or receiving water. The water content at similar suction pressure is higher on wetting than drying. (From Warkentin, 1971.)

Compacted soils retain less water at low suctions and more water at higher suctions where it is still available to plants – this can be seen where the solid lines cross over below a suction of 0.1 in Fig. 12.2. However, the extra retention is greater in sandy than clay soils. The relationship between the soil bulk density and infiltration rate is shown for three soil types in Fig. 12.3. Bulk densities between 1.1 and 1.5 g cm^{-3} have little influence on infiltration rate, but higher bulk densities eventually result in lower infiltration rates for the soils tested (see Chapter 11 for a discussion of infiltration rates).

Thus the effects of compaction on soil water are: a balance between the reduction of the larger voids and a consequent reduction in the free drainage and infiltration rates in wet periods, and a retention of extra water by capillary action in the smaller pores in dry conditions. The consequences for recreation and maintenance of plant cover depend on the balance between rainfall and soil type.

It is clear that compaction reduces the pore space in all soils (Chapter 11) and in so doing reduces the volume of larger pores and increases the potential for the retention of water by capillary forces. The first question discussed here is, what effect does this have on water content in the field and at what suction pressure is the water being retained? Unfortunately most recreational ecologists have measured total water content only as a percentage of the soil weight or volume. Chappell *et al.* (1971) found that the 'moisture' content of the top 5 cm of soil,

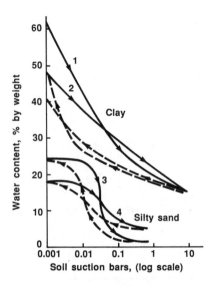

Fig. 12.2 The effect of compaction on water retention curves of a silty sand and a clay soil. 1 and 3 are low and 2 and 4 high dry densities. (From Warkentin, 1971.)

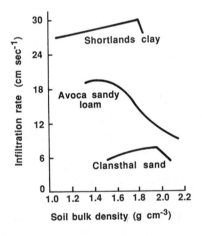

Fig. 12.3 The influence of soil bulk density on water content of three Natal soils at a matric suction of 0.3 bar (from Warkentin, 1971, after Hill and Sumner, 1967).

expressed as percentage of the dry weight of the soil, was reduced from 39% in unused areas to 21% in heavily trampled areas of their chalk grassland site. This occurred with a 14% reduction in volume of the total pore space, but no significant reduction in air space. This indicates that the soil water has a new relationship with the solids in the compacted brown loam soils. In contrast, there was no significant difference between used and unused sites in the survey of the Aberffraw dunes when all the means of all the measurements were taken (Fig. 12.4a) (Liddle, 1973b). However, when the stands were divided into wet and dry groups on the basis of vegetation type, the dry track and path stands had significantly higher water contents than the adjacent unused areas (Fig. 12.4b).

In contrast, Dotzenko, Papamichas and Romine (1967) demonstrated a negative relationship between bulk density (by implication campsite use) and soil water content (Fig. 12.5). In this case the soil was a sandy loam with a finer pore structure which may have retained the higher water content in the uncompacted soil. The soil organic content was also higher in the uncompacted soils (7.6%) and reduced in the campsite soils to 4.2%; this may have retained additional water in the uncompacted soils.

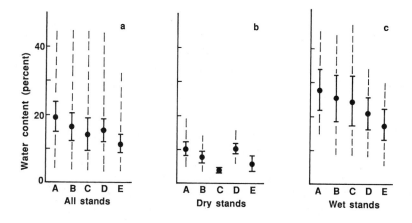

Fig. 12.4 Volumetric water content of sand dune soils in Anglesey, North Wales. (a) Data from all areas in the survey; (b) data from five dry areas; (c) data from five wet areas; A, track stands; B, picnic area stands; C, adjacent 'natural vegetation' stands; D, footpath stands; E, adjacent 'natural vegetation stands'. ●, mean; I, two standard errors; |, total range. (From Liddle and Greig-Smith, 1975a.)

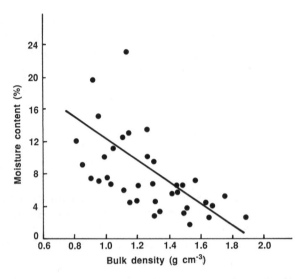

Fig. 12.5 The relationship between bulk density and percentage moisture content in soils of the Aspenglen campground, Rocky Mountain National Park, Colorado (from Dotzenko, Papamichas and Romine, 1967).

12.2 Drainage and water balance

The amount of water in the soil is constantly varying as a result of the changing input from precipitation and varied rates of output by drainage and evaporation. Two points of practical use are defined on the drying curve as field capacity and wilting point. The field capacity is the amount of water retained by a previously saturated soil when free drainage has ceased, usually expressed as percentage dry weight of the soil. This gives an indication of the level of water retention in a soil over a longer time period. The wilting point is the amount of water retained in the soil by such a high suction that it is not available to plants, and therefore permanent wilting occurs. This is generally considered to be about 15 bars (1.6 MPa). This is also expressed by most workers as the percentage dry weight of the soil. The difference between the field capacity and wilting point gives an indication of the water available to plants for their survival. In field measurements of recreation areas and adjacent unused 'control' sites in a silty clay loam in the Missouri Ozarks (Settergren and Cole, 1970), there was a significant difference between the surface layers of two sites for both measures (Table 12.1). However, when the differences between the two measures were calculated, it was found that the percentage of water available to plants was the same in both soils.

When this was adjusted to take account of the different soil bulk densities, marginally more water was shown to be available in the compacted soils (Table 12.1). Lutz (1945) also found higher field capacities in the compacted areas of sandy loam soils and in the 10–20 cm layer of a compacted sand soil. Liddle and Greig-Smith (1975a) tested the availability of soil water to plants by growing them in compacted and uncompacted sandy soil from their field site. Plants grown in the compacted soils survived for a considerably longer period than those growing in uncompacted soils, after water was no longer supplied to the pots.

Water moves in the soil in response to a gradient of potential or suction. In saturated soils the water also moves under a gradient of positive pressure from the force of gravity and this process is usually referred to as soil drainage. This takes place through the larger pores and is called mass flow. In unsaturated soils the force is negative, and the soil suction or soil water potential is created by capillary forces in the smaller soil pores. The rate of water movement depends upon the magnitude of the potential gradient and the transmission coefficient of the soil. This is called its hydraulic conductivity. Water can also move through soils in response to a thermal gradient, usually from hot to cold (Warkentin, 1971), and at higher soil suction pressures and temperatures flow in the vapour phase becomes significant.

The amount of water in the soil at any one time depends on the balance of the processes indicated in Fig. 12.6. From the previous

Table 12.1 Water available to plants in compacted and uncompacted soils in the Missouri Ozarks (Settergren and Cole, 1970)

Depth (cm)	Field capacity (a)		Wilting point (b)	
	Compacted	Uncompacted	Compacted	Uncompacted
0.0–7.5	25.1	31.5	8.8	15.6
7.5–15.0	25.6	25.1	8.3	8.5
15.0–22.5	26.2	23.7	10.5	4.7
22.5–30.0	28.3	22.5	12.9	6.8
	Available moisture by weight (a − b)		Available moisture by volume (a − b × c) when c = mean bulk density	
0.0–7.5	16.3	15.9	21.4	16.9
7.5–15.0	17.3	16.6	22.7	17.6
15.0–22.5	15.7	19.0	20.6	20.1
22.5–30.0	15.4	15.7	20.2	16.6

The mean bulk density in the compacted area was $1.31 \, \text{g cm}^{-3}$ and in the uncompacted area, $1.06 \, \text{g cm}^{-3}$.

discussion it is clear that by removing the vegetation and compacting the soil, recreation can have a considerable impact on the routes by which water moves through the environment. The dynamic nature of the process suggests that longer-term studies of water content of recreation-site soils may be required to give a clearer picture of their true condition. In a New England study, Brown, Kalisz and Wright (1977) followed the changes in soil moisture content of picnic areas, campground areas and adjacent undisturbed areas from 2 June to 10 October 1973. They used neutron probes to record soil moisture to a depth of 1.52 m. They found that the compacted soil in the recreation sites recharged and lost moisture more slowly than did the control-site soils (Fig. 12.7). Compacted soils also contained less moisture than the control sites following storms, and more moisture after extended periods of depletion. As less water entered the compacted soil, it also percolated to shallower depths on the recreation sites. However, the amount of water lost by evaporation and drainage is also partly reduced by higher bulk densities (Arndt and Rose, 1966).

The following five classes of drainage given by FitzPatrick (1971) give a good general guide to the conditions that result from different outcomes in the balance between gain and loss of the soil water:

- **Excessively drained.** Water moves rapidly through the soils, which have bright colours due to oxidizing conditions.
- **Freely drained.** Water moves steadily and completely through these soils with little tendency to be waterlogged.

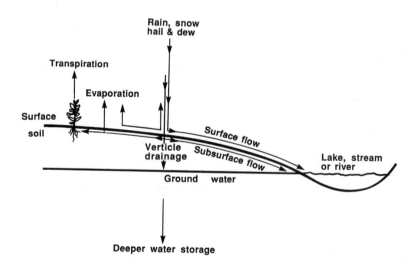

Fig. 12.6 The major routes by which water moves through the environment.

- **Imperfectly drained.** These soils are moist for part of the year, with one or two horizons showing mottling owing to an extended period of wetness and reduction of iron to the ferrous state. The wetness in these soils, as in the next two classes, may be caused by an impermeable horizon, high water table or constantly high precipitation rate.
- **Poorly drained.** These soils are wet for long periods of the year, with the result that many horizons are mottled, with at least one that is blue or grey, resulting from intensely reducing conditions.
- **Very poorly drained.** These soils are saturated with water for the greater part of the year so that most of the horizons are blue or grey (caused by reducing conditions). Peat may also be present as a result of the high degree of wetness.

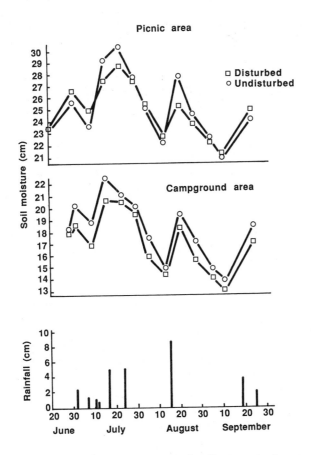

Fig. 12.7 Moisture in the top 152 cm of soil at a picnic area and campground, and the local precipitation record at southern Rhode Island (from Brown, Kalisz and Wright, 1977).

Soils may become wetter because of compaction, inhibition of adjacent drainage or an increased inflow caused by a diversion of normal drainage routes. The wetter soils may then react differently to recreation impact. Experiments with sheep on a salty loam in New Zealand showed that when the water table was at the surface, bulk density decreased and the bearing capacity, as measured by a cone penetrometer, was too low to be recorded (Table 12.2) (Lagocki, 1978). The soil structure also appeared to be destroyed. Bulk density did not increase as a result of the sheep trampling until the water table fell below 20 cm, when the bearing capacity also increased. In the arctic tundra of the Mackenzie Delta region of Canada, tests were carried out on the effects of track vehicles on the tundra soils of varying wetness from 'waterlogged' to 'free-draining'. After one pass, the 'secondary terrain structure' (peat) was nearly 90% damaged, whereas the free-draining soil had only 13% damage (Fig. 12.8) (Bellamy, Radforth and Radforth, 1971). Ten passages caused 100% damage to the structure in the waterlogged area, whereas the free-draining site was only 75% damaged after 100 passages. These experiments clearly show that waterlogged ground is very susceptible to trampling or vehicle damage. Clay soils will become puddled or poached in these conditions, with the platelets lying parallel, or oriented, and capable of forming a very hard surface on drying. Bjorkhem et al. (1974) concluded, after a study of the effects of forestry vehicles on soils and tree growth, that 'the risk of damage is much bigger when the soil is moist than when it is dry'. And they also commented that fine-grained

Table 12.2 The effect of sheep trampling on the bulk density and bearing capacity of the surface soil (0–6cm) in relation to the depth to the water table below ground surface (Lagocki, 1978)

Depth to water table (cm)	Bulk density (g cm^{-3})				Bearing capacity (kPa)			
	Control		Trampled		Control		Trampled	
	Mean	SE	Mean	SE	Mean	SE	Mean	SE
0	0.77	0.079	0.60*	0.063	226	8.7	<10	–
10	0.73	0.078	0.73	0.066	253	6.0	90***	6.9
20	0.77	0.083	–	–	281	5.6	225**	6.4
25	–	–	0.89*	0.065	–	–	–	–
30	0.72	0.078	0.93**	0.060	325	8.2	360**	7.2
40	–	–	1.00	0.110	360	5.6	386**	5.9
50	–	–	1.00	0.110	–	–	392	9.5

Differences, between control and trampled measurements, $P < 0.05$(*), 0.01(**), 0.001(***).

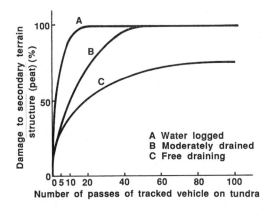

Fig. 12.8 The relationship between damage to secondary terrain structure (peat) and the number of passes of a tracked vehicle on tundra in the Mackenzie Delta region of Canada: A, waterlogged; B, moderately drained; and C, free-draining soils (from data of Bellamy, Radforth and Radforth, 1971).

loam and clay soils are much more sensitive than coarse moraine soils, which have a mixture of bigger material. Coarser soils, even when compacted, will also drain much faster than fine-textured or clay soils.

Inhibited or altered drainage also has considerable effect on soil erosion patterns (discussed in Chapter 15).

12.3 Soil air

Oxygen is essential for aerobic respiration of soil organisms. Since plant roots, soil animals and most micro-organisms depend on aerobic respiration to release energy for growth and movement, a good supply of air is essential for maintenance of soil fertility and plant growth. Air occupies the pore spaces between the soil particles that are not filled with water. Air may move through the soil pores by mass flow (the movement of all the gases together) or by gaseous diffusion (in which the molecules of each gas move along a concentration gradient). The driving force for mass flow is a gradient of total pressure, which may be caused by water flow through the soil or variations in air pressure at the soil surface. The driving force for gaseous diffusion is a partial pressure gradient, caused by different concentrations of a particular gas. For example, utilization of oxygen and release of carbon dioxide within the soil will cause a gradient of oxygen from the higher partial pressure above the surface to the lower partial pressure within the soil, and a gradient of carbon dioxide in the opposite direction. Compaction or high water content may reduce soil porosity enough to stop or limit gaseous transfer. Consequently compaction may alter the air composition of biologically active soils (for a full discussion of this topic, see Grable, 1971).

The effect of different bulk densities of a silty clay loam (see Fig. 11.9 for constituents) on air porosity at different suction pressures was investigated by Grable and Siemer (1968) (Fig. 12.9a). At a water suction of 30 cm, air porosity was over 30% when the bulk density was $0.93\ g\ cm^{-3}$, but when the soil was compacted to $1.23\ g\ cm^{-3}$ the air porosity was reduced to about 12%. The adhesion of soil particles into larger aggregates forms an essential part of the soil structure, and the size of these aggregates also has a direct influence on soil porosity (Fig. 12.9b). Here it can be seen that aggregates in the size range of 3–6 mm have an air porosity over 40% at water suction pressure of 30 cm, but when they are reduced to less than 0.5 mm diameter, the same water suction occurs when the porosity is less than 10%. When the soil becomes saturated and water suction becomes negative, the air porosity is near zero.

In general, air porosities of more than 19% will provide good conditions for plant growth (Grable, 1971). Porosities below 19% may lead to deficiencies of oxygen or a build-up of carbon dioxide within a biologically active soil. The actual oxygen content will also depend on the collective respiration rate of the soil organisms (Grable and Siemer, 1968). Normally the relative humidity of the soil atmosphere is quite high and the increase of carbon dioxide is at the expense of the oxygen content, while gaseous nitrogen remains fairly constant (Trouse, 1971). Recreation impacts can reduce the volume of air within the soil by compaction and by poor drainage (Fig. 12.10). Compaction increases

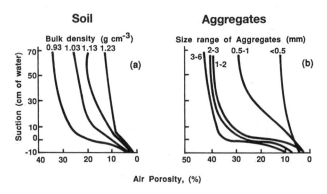

Fig. 12.9 The influence of bulk density (a) and size of aggregates (b) on air porosity after desorption of a silty clay loam at different suctions. Suctions were developed at the bottom of columns about 10 cm high. Bulk densities of the aggregates ranged from 0.84 to $1.07\ g\ cm^{-3}$ in the largest and smallest fractions, respectively. (From Grable, 1971, after Grable and Siemer, 1968.)

the volume of solids at the expense of air space, while water displaces air when the drainage is poor. Chalk grasslands were found to have air porosities of 37% and 45% at 2.5 cm depth in trampled and unused areas, respectively. At depths of 5 cm the porosity in the trampled areas was 33%, well above the limiting amount of 19%. In contrast, the volume of air in the soil in sand dune tracks was about 10%, compared to about 40% in soils at picnic areas, footpaths and natural vegetation (Fig. 12.11a). In this situation, the lack of air space probably led to a reduction of soil oxygen and a consequent inhibition of plant root growth. Measurements of the field capacity of these soils showed that the track soils contained water above their field capacity, while the others were 15% or more below field capacity (Fig. 12.11b). This suggests that in this case the reason the track soils had an inadequate aeration was poor drainage, possibly compounded by compaction.

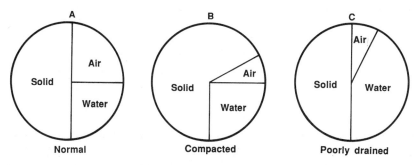

Fig. 12.10   The effect of compaction and poor drainage on the volume of air space in the soil (from Letey, 1961).

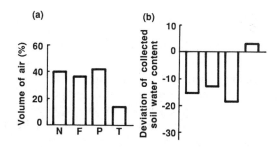

Fig. 12.11   Water content in the field at the end of December of sand dune soils from Aberffraw, Anglesey, North Wales. (a) Percentage volume of air; (b) percentage deviation of the soil water content from soil field capacity (volumetric); N, natural vegetation; F, footpaths; P, picnic areas; T, tracks. Only the track measurements were statistically different from the rest. (From Liddle and Greig-Smith, 1975a.)

12.4 Soil temperature   The temperature of a particular piece of soil will depend upon its thermal characteristics, the temperature and thermal characteristics of the surrounding soil and the temperature and thermal characteristics of the air and vegetation above the soil surface.

The thermal processes in soil may be measured as conductance, diffusivity and convection. Conduction of heat takes place through the soil solids and depends upon their specific heat or thermal capacity (pc)(the amount of heat energy required to heat 1 g by 1°C) and the amount of contact between particles. A compacted soil will have greater contact between particles (Fig 11.6c) and this will increase conductance and hence thermal diffusivity (k), but it may be offset by the greater volume of solids requiring heating in $1 \, cm^3$ of soil. Conductance is measured over a defined distance, usually 1 cm. Diffusion of heat may take place through the soil air and, where a temperature gradient is created in the soil profile, soil air may move by convection through the air-filled soil pores, thus compaction by reducing the air-filled soil pores (Fig. 12.10) will also affect the diffusivity (k) characteristics.

The thermal characteristics are related as follows (Lowry, 1967):

$$\text{thermal diffusivity (k)} = \frac{\text{conductivity } (\lambda)}{\text{specific heat (pc)}}.$$

The air temperature at the soil surface will be governed by the amount of incoming longwave and shortwave radiation, the amount of shortwave radiation intercepted by moisture and other particles in the air, the amount of reflected longwave radiation from the soil and movements of the air itself. The vegetation cover above the soil will intercept some of the longwave and shortwave radiation and restrict the flow of air.

The two processes of vegetation change and soil compaction associated with the creation of paths and other recreation activities will clearly have a direct effect on the temperature and thermal environment of the soil.

The interception by vegetation of incoming heat energy is shown in Fig. 12.12, where the maximum daytime temperature occurred at about 10 cm above the ground surface in the vegetated area and just below the soil surface on the unvegetated track (Liddle and Moore, 1974). The two points just above the track surface were probably under-recorded due to the design of the equipment. The rest of this discussion is based on Liddle and Moore (1974) as I am not aware of any other detailed records of thermal changes due to the impact of recreation.

When the run of temperatures over more than one day in vegetation is compared with those on a path, the effect of exposing the soil

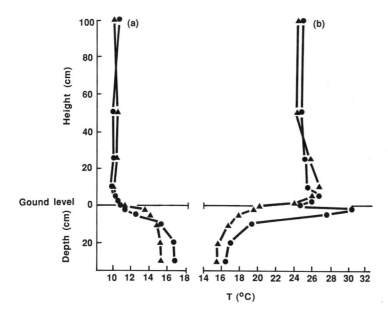

Fig. 12.12 (a) Mean midnight (10.00–02.00 hours), and (b) midday (12.00–14.00 hours) temperatures; ●, dry track; ▲, dry vegetation. The effect of compaction can be seen by comparison of the circle with the triangle. Note the broken horizontal axis. (Modified from Liddle and Moore, 1974.)

surface and compacting the soil can be clearly seen (Figs 12.13, 12.14). The maximum temperature moves from 5 to 10 cm above the soil to 5 cm beneath the track surface. The absolute maximum is also raised from 28° to 30°C, while the minimum soil temperature at 5 cm depth dropped from 14° to 11°C – a 5° difference (25%) over the range of temperatures.

In a second experiment the effects of vegetation removal and soil compaction were measured separately, in the first case by comparing the temperatures of a vegetated area with an area cleared of vegetation for the experiment. To examine the effect of soil compaction the temperatures of the uncompacted, clear area were compared with those of a compacted track. These results were expressed as difference diagrams. The first comparison (Fig. 12.15) showed that the soil in vegetated area was 15°C cooler than the exposed uncompacted area, and at mid-morning the air temperature was up to 5° warmer in the vegetation, a difference of 20°C. The effect of soil compaction was much less and mainly in the opposite direction to the effect of the removal of vegetation (Fig. 12.16). The track soil was 7°C cooler at midday and about 1°C warmer at night than the uncompacted exposed soil.

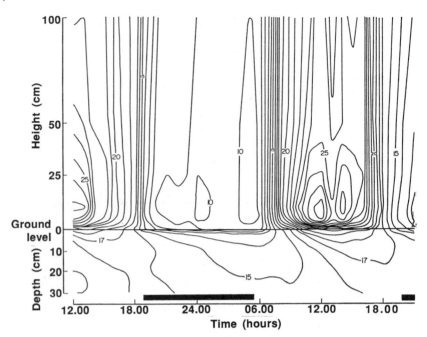

Fig. 12.13 Dry natural vegetation temperatures over a 32-hour period. The temperature is shown by isotherms for every degree C boundary except where they are very close together. The period of darkness is indicated by the solid line below the main diagram. (From Liddle and Moore, 1974.)

The thermal capacity and the thermal conductivity both markedly increased in the track soil compared to the uncompressed soils (Table 12.3). Thus they would store more heat energy in the soil for a given heat input, with a smaller rise in temperature, and there would be a faster conductance of heat to deeper levels away from the soil surface. This would result in less heating at the compacted soil surface during the day and more stored energy to keep it warmer at night.

The combined effect of vegetation removal and soil compaction was that the track soil was 9°C warmer during the day and 1°C warmer at night, while the air temperature over the track was 6°C cooler than that over the vegetation during the day and 2°C warmer at night.

We also investigated the effect of different soil water situations, comparing the temperatures on a dry track with 6% soil water by weight, with a wet track with 15% water content (Fig. 12.17). The wet soil may also have had greater silt content than the dry soil. Here we found that the wet track was 7°C cooler than the dry track in the day and 2°C warmer at night. The thermal constants were all higher in the wet track soils compared to the dry track soils. The

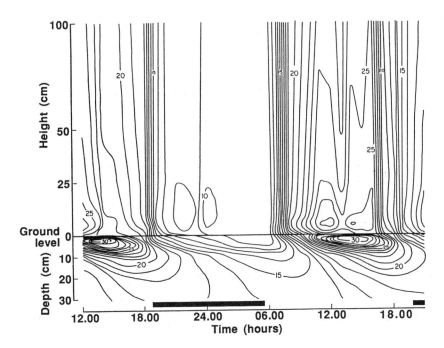

Fig. 12.14 Dry track temperatures over a 32-hour period. The temperature is shown by isotherms for every degree C boundary except where they are very close together. The period of darkness is indicated by the solid line below the main diagram. (From Liddle and Moore, 1974.)

Table 12.3 Thermal properties of soils (Liddle and Moore, 1974)

	Thermal capacity (pc) (cal deg^{-1} cm^{-1})	Thermal diffusivity (k) (cm^2 sec^{-1})	Thermal conductivity (λ) (cal deg^{-1} cm^{-1} sec^{-1})
Track stand (I)	0.3059	0.0036	0.0011
Vegetation stand (II)	0.2524	0.0033	0.0008
Cut vegetation stand (III)	0.2579	0.0031	0.0008

conductivity was twice as high on the wet track and the other factors only increased by 22% (thermal capacity) and 60% (thermal diffusivity), so conductivity would appear to be the major factor influencing this result.

In general, the microclimatic effects of track formation are complex, with the removal of vegetation appearing to be the major factor in dry areas under high incident radiation levels. The situation is different

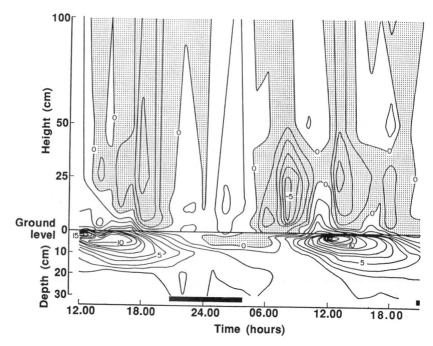

Fig. 12.15 The difference in temperature between the area cleared of vegetation and the area on which the vegetation was undisturbed, showing the effect on soil and air temperatures of vegetation removal alone. The shaded areas indicate times and heights when the site without vegetation was colder than the site with vegetation. (From Liddle and Moore, 1974.)

in wet areas where reduced thermal diffusivity of the track soil compared to wet soil under vegetation may augment the effect of vegetation removal (Liddle and Moore, 1974).

12.5 Soil nutrients Recreation impacts on soils may alter their compaction and drainage characteristics. Compaction and water content influence the two mechanisms, diffusion and mass flow, by which nutrients move to plant roots. The amount of nutrients mineralized from soil organic matter may also be altered by changes in the degree of compaction, soil water and consequent aeration.

As mentioned in Chapter 9, cations are adsorbed on to the surface of clay platelets in the soil. These platelets are agglomerated into packets or tactoids, with exchangeable cations sandwiched between the platelets. In this position the electrostatic forces hold the platelets together and the cations are no longer available for plant growth. Compaction increases the number of platelets per tactoid and this

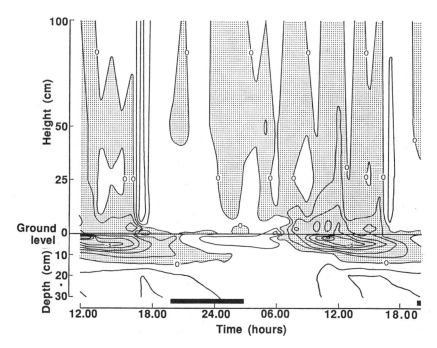

Fig. 12.16 The difference between the track temperatures and those of the area cleared of vegetation showing the effect on soil and air temperatures of soil compaction alone. The shaded areas indicate times and heights when the compressed soil was colder than the uncompressed soil. (From Liddle and Moore, 1974.)

increases the proportion of exchangeable ions that are sandwiched between the two mineral surfaces (Fig. 12.18). While compaction of soil may increase the number of platelets per tactoid by only a small percentage, the external surface, where ions are still mobile, may be appreciably reduced. Other forces that may affect the diffusion of cations are increases of positive charges and the counterflow of positive ions of a different species or diffusion of negative ions in the same direction as the original positive ion flow.

The mass flow of cations in solution is reduced by compaction when soils are relatively saturated, but as the number of small pores are increased, more water is held in compacted soils at medium suctions. The hydraulic conductivity of a clay loam at a water pressure of approximately 8 bars (0.8 MPa) was doubled when it was compacted from $1.1 \, g \, cm^{-3}$ to $1.5 \, g \, cm^{-3}$ (Kemper, Stewart and Porter, 1971). Thus for this soil the mass flow of nutrients in solution was also doubled by this degree of compaction. Movement of nitrates, sulphate and other largely non-adsorbed anions also depends to a larger degree

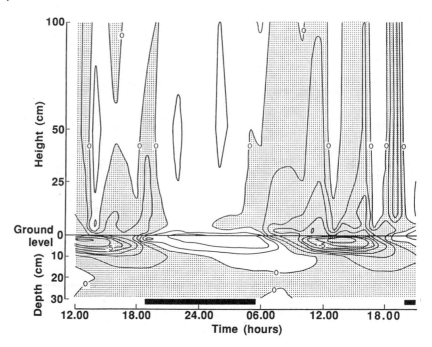

Fig. 12.17 The difference between the temperatures of the wet track and the dry track showing the effect on track temperatures of higher soil water content. The shaded areas indicate times and heights when the wet path was colder than the dry path. (From Liddle and Moore, 1974.)

on mass flow of the soil solution than does the movement of the more tightly adsorbed anions and cations (Kemper, Stewart and Porter, 1971).

Field measurements of the soil nutrient levels from recreation areas often produce non-significant or opposing differences between used and unused areas. For example, Chappell *et al.* (1971) found no significant differences in soil nutrient levels between used and unused areas, and the differences found by Cole (1982) for total nitrogen, potassium and phosphorus were also not significant. However, Leney (1974) found that trampled and untrampled vegetated areas adjacent to a bare pathway at Balmedie, Scotland, (Fig. 12.19) had significantly higher levels of nitrogen and potassium. In this case the bulk densities were 1.2, 1.1 and 1.6 g cm^{-3}, respectively. At Brown's Bay Provincial Park, by the St. Lawrence River in Ontario, Rutherford and Scott (1979) found that the nitrate (NO_3) concentrations were significantly reduced by use – a similar result to that of Leney. Rutherford and Scott suggested that part of the explanation could be reduced microbial action due to poor aeration, leading to reduced breakdown of organic material. A horse track passing through a bracken (*Pteridium aquilinum*) and the heather (*Calluna vulgaris*) site on a southern English heathland had higher levels of nitrogen in the path and the path-edge soils (Liddle and Chitty, 1981) (Fig. 12.20). The authors examined the likely sources of elements at the site and concluded that 'it seems possible that the source of the additional elements on the tracks may be dung deposited by horses whose food is grown outside the heathland ecosystem'.

Liddle and Chitty (1981) also found a higher level of phosphorus on the track at the heather site (Fig. 12.20). Phosphate concentration

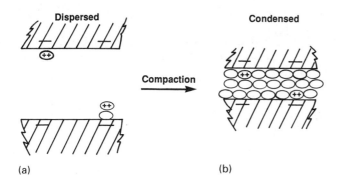

Fig. 12.18   The effect of compaction on the surface charge and retention of water molecules of calcium-saturated 2 : 1 type clays: (a) dispersed, (b) condensed (tactoid) states (from Kemper, Stewart and Porter, 1971).

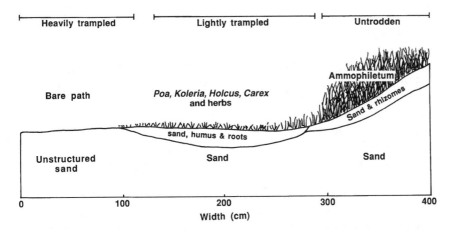

Fig. 12.19   The effect of trampling on the soil profile of the sand dunes and slack (sand dune pasture) habitats at Balmedie, Scotland (from Leney, 1974).

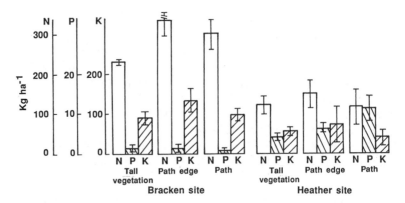

Fig. 12.20 Nitrogen (N), phosphorus (P) and potassium (K) in the soil of a sandy lowland heath at Chobham Common, 42 km south of London, England. The nitrogen in the path and path edge soils of the bracken (*Pteridium*) site and the phosphorus in the path soil of the heather (*Calluna*) site were significantly higher than the quantities in the soils of the undisturbed sites. (From Liddle and Chitty, 1981.)

in the shallower podzolic soils studied by Rutherford and Scott (1979) increased in campsites in grassed areas and decreased in forest campsites. Cole (1982) found significant increases in the cations magnesium, sodium and calcium in similar soils, and Rutherford and Scott (1979) found an increase in chlorine (an anion) in campsite soils.

Hydrogen ion concentration (pH) has a major influence on the chemical and biological processes in the soil. In about half of the examples that I have found in the literature the pH level was about 0.3 higher on pathways. In these cases the pH of the adjacent undisturbed areas was between 4.08 and 5.42. Where the soil pH had dropped in the trampled sites, it was higher (at pH 7.5, 6.45 and 6.38) in the adjacent natural areas, and in these cases it fell to pH 5.3, 5.5 and 6.1, respectively (Jim, 1987; Leney, 1974; Dawson, Hinz and Gordon, 1974). From the limited data available, pH 5.5 would appear to be the point towards which some paths and campsite soils are changed by trampling; however, it is hard to envisage the mechanism that would produce this result.

Campfires may also have a major effect on soil nutrients, depending on their maximum temperature and duration (Cole and Dalle-Molle, 1982). At temperatures above 400°C most nitrogen and sulphur and much phosphorus is lost from the soil. The moisture-holding capacity and infiltration rates are reduced, while soil pH and the amounts of most cations increase. The ash from campfires may also cause changes which are generally favourable to growth and can lead to compositional changes in the vegetation understorey.

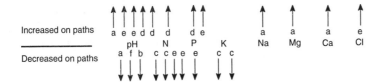

Fig. 12.21 Changes in concentrations of some soil nutrients as a response to trampling. The tendency for various elements to increase or decrease in path soils when compared to adjacent vegetation according to various authors. a, Cole (1982); b, Dawson, Hinz and Gordon (1974); c, Leney (1974); d, Liddle and Chitty (1981); e, Rutherford and Scott (1979); f, Jim (1987). Note there are further examples in Leney (1974) but these have been omitted so that the trends illustrated above are not dominated by measurements from only two habitats.

The changes in soil nutrient status as a result of recreation activities are clearly complex and dependent upon a variety of known and unknown processes which may work together or in opposition. The final outcome will depend upon inputs from dung, urine and litter, natural levels in the soil, breakdown rates of organic material, release rate from adsorbed or mineral sources, leaching in drainage water and uptake by plants. The recorded changes are summarized in Fig. 12.21, but it must be borne in mind that they are from a variety of soils subject to a range of uses at a range of intensities and that the method of determination may also have varied. pH rises more often than it falls, but the 'marginal' sites generally used for recreation tend to have acid soils in their natural states. Soil macronutrients (NPK) do not show any particular trend, while some micronutrients appear to have increased in used areas. No safe generalization can be made about the changes in soil nutrient status as a result of recreation activities on the basis of the papers considered here. At this stage of our knowledge each site needs individual consideration.

12.6 Summary

Water occupies a variable proportion of the soil pore space, the remainder is occupied by air. The suction pressure holding the water in place depends on the size of the pores, with smaller pores exerting greater suction. Since compaction reduces the size of soil pores the water may be held with greater suction pressure during dry periods in trampled soils. The smaller pores also reduce infiltration rates and free drainage in wet periods.

Measurements of soil water content made in recreation areas have shown both increases and decreases as a result of compaction, but in dry soils more water generally appears to be available in path or track soils and in periods of drought. Wet soils with inhibited drainage are more vulnerable to damage from trampling or vehicles, and fine-grained loam and clay soils are much more sensitive than coarser soils. Inhibited or altered drainage may also lead to increased erosion from the greater runoff on the soil surface.

The air content of soils is also reduced by compaction and the consequent reduction in size and proportion of soil pores. Air movement through the soil may take place by mass flow or gaseous diffusion. The utilization of oxygen by respiring organisms may result in a concentration gradient, decreasing down the profile, especially where air movement is inhibited by water. Air porosities of 19% or more will provide good conditions for plant growth. In contrast, the conditions in heavily trampled soils are frequently below the optimum for plant growth.

The temperatures of trampled soils are affected by the removal of vegetation and compaction of the soil. Vegetation removed exposes the soil surface to higher levels of radiation but may also increase the possibility of air movement and convective cooling. Soil compaction tends to increase its conductivity and thus facilitate cooling of the surface layers, at least in dry areas; in wetter areas this effect may be reversed. In general, paths and tracks create a warmer and drier micro-climate than that found at the soil surface beneath vegetation.

Soil nutrient changes do not show any consistent trend in recreation areas, with increases or decreases both recorded in studies made to date. pH tends to rise more often than it falls, but this may be due to the fact that most studies have been made in areas of fairly acid soil. Each site needs to be evaluated separately for these qualities.

The living soil 13

The soil animals and 'plants' discussed in this chapter include the micro-organisms bacteria, fungi, algae and protozoa; the mesofauna, including eel worms (class Nematoda), earthworms (subclass Oligochaeta), insects (class Insecta), mites and spiders (class Arachnida), springtails (order Collembola) and many other less numerous invertebrate groups. For the purpose of this discussion, most of the vertebrates (rabbits, moles, some snakes, lizards, birds, tortoises and turtles) are generally excluded as their responses to recreation or similar impacts do not appear to have been studied.

Considering soil as a substrate for plant growth, the bacteria, fungi and earthworms are probably the most important soil organisms as they play an important or major part in the cycling of nutrients (Fig. 9.6). The earth-worms macerate the larger organic particles, which are further broken down by fungi and finally bacteria. The latter are especially important in the nitrogen cycle, where they may fix the gas from the atmosphere and release it from dead organic material.

13.1 Micro-organisms

Bacteria are the smallest and most numerous of the free-living micro-organisms in the soil, where they number several millions per gram of soil with a live weight variation of between 1000 and 6000 kg ha^{-1} in the top 15 cm of soil (FitzPatrick, 1971). This weight is slightly less than that of the fungi but greater than that of all the other organisms combined (Clark, 1967). Bacteria oxidize nitrogen from the ammonium ion (NH_4^+), which is usually the first nitrogenous breakdown product from organic material, to nitrate (NO_3^-) which is then available to plants; these are the nitrifying bacteria. However, there are also deni-trifying bacteria which can reverse the process even when oxygen is in short supply.

In an experiment to determine the effect of compaction on soil micro-organisms, Hubbell and Gardner (1948) found that the numbers of bacteria were reduced while the numbers of actinomycetes (a class of fungi) and other fungi were hardly affected. The total numbers of organisms in their compacted clay soil, after 90 days of incubation in stable conditions, were 0.975 million g^{-1} of dry soil, while there were 3.408 million g^{-1} of dry soil in the uncompacted control. When the maximum particle size was reduced from 2.38 mm to 0.12 mm, there were 0.783 million g^{-1} and 4.249 million g^{-1} in the compacted and uncompacted clay soils, respectively. The effect of compacting a sandy loam soil was to reduce the total number of organisms from 3.892 million g^{-1} to 0.714 million g^{-1} when the original particle size was a maximum of 2.38 mm. However, when the maximum particle size was reduced to 0.12 mm the difference in numbers was not so large – the numbers were 2.866 million g^{-1} and 0.868 million g^{-1}, respectively. In all cases compaction had reduced the total number of organisms, but the data in Table 13.1 suggest that this was only a reduction of the numbers of bacteria.

Numbers of fungi, actinomycetes and bacteria determined separately for the two original maximum soil aggregate sizes are given in Table 13.1, together with the total pore space and the non-capillary pore spaces. Since aeration takes place largely through non-capillary pore space, it is interesting, if confusing, to note that a non-capillary pore space of 2.5% apparently inhibited bacteria when the soil aggregate size was 2 mm but had little effect when it was 0.12 mm, except when the latter soil was compacted and the non-capillary pore space reduced to 0.5%.

In general, compaction appears to reduce the number of bacteria in the soil, a conclusion which partially agrees with the results of Duggeli (1937). He found that the number of nitrifying bacteria ($NH_4 \rightarrow NO_3$) in trampled soils was reduced by about a factor of 10 when compared with untrampled soils, although the denitrifying bacteria ($NO_3 \rightarrow NH_4$) increased by the same proportion. The effect of compaction on the action of bacteria was measured by Whisler, Engle and Baughman (1965) as a proportion of ammonium and nitrate nitrogen in the soil after 14 days' incubation. They found that nitrogen present as nitrate was reduced by a range of 50–98% by compaction in the soils they tested (Table 13.2). In the two soils where the amount of nitrate

Table 13.1 Effect of compaction and pore space on microbial activity in a Gila clay after 90 days' incubation (Hubbell and Gardner, 1948)

Maximum aggregate size (mm)	Treatment	Actinomycetes $\times 10^4$ (hundred thousands)	Other fungi $\times 10^2$	Bacteria $\times 10^6$ (millions)	Total pore space (%)	Non-capillary pore space (%)
2	Control	3.00	9.00	3.10	63.8	8.0
	Compacted	5.35	20.00	0.42	39.0	2.5
0.12	Control	1.93	11.25	4.02	58.5	2.1
	Compacted	3.19	12.00	0.45	30.1	0.5

nitrogen declined by 69% and 98% there were increases in ammonium nitrogen by factors of 18.7 and of 1.2. However, in the soil where nitrogen as nitrate only fell by 50% there was also a fall of 38% in nitrogen as the ammonium ion. This suggests that some nitrogen was lost as gas. These results indicate that the action of bacteria in compacted soils shifts from oxidation to reduction, thus reducing the amount of nitrate (NO_3) nitrogen available to plants. This is almost certainly due to the low oxygen levels, as nitrate will not accumulate in the absence of molecular oxygen (Kemper, Stewart and Porter, 1971).

Although some bacteria are famous for their ability to survive in hot springs at near 100°C, soil bacteria generally have an optimum temperature range of 25–30°C (FitzPatrick, 1971). Compaction of snow by snowmobiles and consequent lower soil temperatures have been shown to delay the resumption of soil bacterial activity by over a week in the spring following compaction (Wanek, 1974).

The special effect of animal and human faeces on nutrient and bacteria content of fresh water is discussed in Chapter 26.

I have only found one paper that deals with soil fungi in trampled situations. The numbers of soil fungi show small increases in compacted soils (Hubbell and Gardner, 1948) (Table 13.1). There are also records of reduced populations of mycorrhizal fungi after fires, and species composition may change for several years (Cole and Dalle-Molle, 1982). Fungi are important in soils for their symbiotic association with the roots of higher plants (mycorrhiza) as well as being agents in the breakdown of organic materials.

13.2 Fungi

Earthworms are members of the subclass Oligochaeta, five families of which are normally considered as terrestrial earthworms (Edwards and Lofty, 1977). They have long been known to be associated with soil fertility – Darwin (1881) made a careful study of the habits of

13.3 Earthworms

Table 13.2 Transformation of nitrogen between ammonium (NH_4^+) and nitrate (NO_3^-) ions, in compacted soils (parts per million, dry weight) (from Whisler, Engle and Baughman, 1965)

| | Sandy loam (1) | | Clay loam | | Sandy loam (2) | |
	NH_4^+	NO_3^-	NH_4^+	NO_3^-	NH_4^+	NO_3^-
Uncompacted	1.3	34.7	1.6	51.4	24.9	23.6
Compacted (1.6 or 1.7 g cm^{-3})	0.8	17.6	26.9	15.8	30.2	0.4

earthworms in England and calculated that their casts may amount to a surface layer of between 2 and 5.5 cm thick over a period of 10 years.

As earthworms burrow through the soil, mineral and organic material passes through their guts. Some of the organic material is decomposed and the remainder is formed into a homogeneous blend with the mineral part of the soil. The resulting mixture is rich in urea and it may also contain calcium carbonate, providing a suitable habitat for the rapid proliferation of micro-organisms (FitzPatrick, 1971). Members of the Lumbricidae family (Fig. 13.1) live in the litter and upper layers of the soil, feeding by seizing dead fragments of plant from the surface and dragging them into their passages which also run through the mineral horizons of the soil (Yur'eva, Matveva and Trapido, 1976). Another ecological group of earthworms, including *Eisinia rosea*, *Allolobophora caliginosa* and *Octolasium lacteum*, tend to occur in deeper layers of the soil, inhabiting mostly the mineral horizons, and feeding on the soil humus. Both groups create channels by which rainwater may percolate into the soil.

The birch woods near Moscow, Russia (Fig. 13.2) had been exposed to 'active recreational influence' for more than 15 years when they were studied by Yur'eva, Matveva and Trapido (1976). They are visited by 300–3000 people ha^{-1} month^{-1} from June to October. The undergrowth survives in some places, especially near the trees. This provided 'minimum damaged' zones where soil fauna that had not been subject to pressure could be recorded, acting as a control for the counts of organisms in the impacted soils. The bulk density of the soil

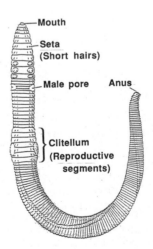

Fig. 13.1 A generalized diagram of a lumbricid earthworm (Hendreck, 1938).

was 0.91 g cm^{-3} in these zones. A zone of medium damage was adjacent to these thickets, several metres from path and forest roads; here the soil bulk density was 1.14 g cm^{-3}. The third, most damaged, zone was almost bare of plants and litter, and the soil bulk density was 1.47 g cm^{-3}. Yur'eva, Matveva and Trapido (1976) found seven species of earthworms in the zones of minimum and medium damage, three of which were surface-living *Lumbricus* species. In the minimum damaged area the greatest mean numbers were of *Eisinia rosea* (198 m^{-2}, 52%) and *Allolobophora caliginosa* (130 m^{-2}, 34%) (Table 13.3). In the maximum damage zone the worm numbers had been reduced to 32% of those in the control zone and the worm biomass was 50% of the biomass of the worms in the control area. However, the mean weight of each earthworm rose from 0.3 g in the soils of the minimum damage zone to 0.47 g in the maximum damage zones. The worms that survived in the trampled zones were therefore larger than those in the minimum damage zone and mainly (93%) of the deeper soil species *Allolobophora caliginosa*. The *Lumbricus* species were largely absent. Cluzeau *et al.* (1992) reported that *L. rubellus* and *L. friendi*

Fig. 13.2 The author in the birch (*Betula* sp.) woods south of Moscow, Russia, at a site similar to that studied by Yur'eva, Matveva and Trapido (1976) (photograph D.E. Douglas).

Table 13.3 Species composition, average numbers and biomass (g m^{-2} of earthworms (*Lumbricidae*) in birch woods exposed to recreation influence (recorded in May 1974; modified from Yur'eva, Matveva and Trapido, 1976).

Species	Damage zone		
	Minimum	Medium	Maximum
Numbers of			
Eisenia rosea	198.0	19.5	7.0
Allolobophora caliginosa			
f. *typica*	130.0	176.0	115.0
Dendrobaena octaedra	–	0.5	–
Octolasium lacteum	3.0	3.5	–
Lumbricus rubellus	44.0	18.0	2.0
L. castaneus	2.0	2.0	–
L. terrestris	7.0	7.0	–
Number of individuals m^{-2}	384.0 ± 41.6	226.0 ± 52.4	124.0 ± 63.5
Number of species	6	7	3
Biomass (g m^{-2})	115.4	124.3	58.0
Mean weight of each worm (g)	0.3	0.55	0.47

survived in cattle-trampled soils in Britanny, France, but their numbers were greatly reduced and *L. castaneus* was eliminated. Again, the larger species or individuals were the survivors in the compacted soils.

A study of earthworms in 'an old Scots Pine Stand' (*Pinus sylvestris*) near Uppsala, Sweden, also showed very similar results. *Allolobophora caliginosa* was the most numerous worm in the impacted area and the individuals were all in the class of longest length – over 5 cm (Ingelog, Olsson and Bodvarsson, 1977). The numbers and total biomass were similar to those in the southern English chalk grassland soils studied by Chappell *et al.* (1971) and, again, were reduced by recreation impact. Piearce (1984), working near Lancaster, England, also found that, in spring, the *Lumbricus* species were the most vulnerable to trampling, and that *Allolobophora longa* was the least affected by the conditions in trampled soils (Fig. 13.3). Interestingly, he found that there was no significant difference between total numbers of worms in path soils and adjacent areas during the previous autumn, when the path had received only very light use. However, there were at this time about twice as many *Allolobophora longa* individuals in the path soils compared to adjacent areas. All the species were found at greater depths under the path than in the adjacent areas. Edmond (1962), in New Zealand, found that the numbers of worms in soil compacted by sheep treading fell from 711 m^{-2} to 544 m^{-2} and that the species

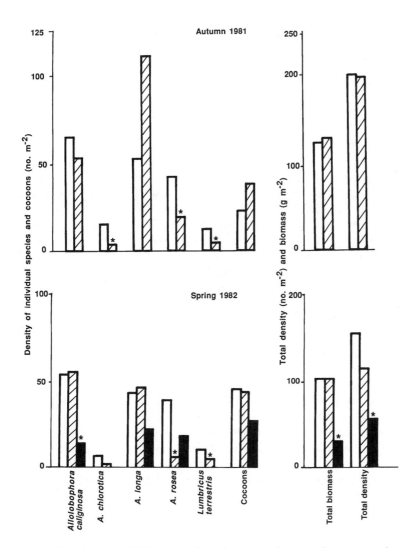

Fig. 13.3 Abundance and biomass of earthworms along and near a path at Barker House Farm, near Lancaster, England. Open bars, lightly compacted; hatched bars, moderately compacted; and solid bars, heavily compacted soils. Starred samples are significantly different from lightly trampled samples. (From Piearce, 1984.)

were *Allolobophora caliginosa* and *Lumbricus rubellus*, the species found in the birch woods near Moscow.

In general, recreation impact reduces the numbers and biomass of earthworms and only the deeper-living, larger individuals are able to survive, perhaps having more strength to push through the denser soils and living below the most compacted layers. Mixing of the litter with the surface layers by *Lumbricus* species ceases as they do not appear to survive trampling, and channels for the percolation of water into the surface layer of the soil will no longer be created by these worms in compacted areas.

13.4 Other soil animals

ARTHROPODS IN GENERAL

The numbers of soil arthropods vary greatly, in part according to their size. The springtails, (Collembola) (Fig. 13.4) may occur at densities of up to 43 000 m^{-2}, while the mites (Acari) may exceed 120 000 m^{-2} (Salt *et al.*, 1948). The larger animals, such as woodlice (Isopoda), millipedes (Diplopoda) and spiders (Areneida), which in general live on the soil or in the litter layer (although some spiders do construct burrows in the soil), have maximum populations of 500 m^{-2}, 200 m^{-2} and 300 m^{-2}, respectively, in the soils of the recreation sites considered in this discussion.

The effects of recreation on the soil arthropod fauna are quite different from damage to the vegetation. While some animals are undoubtedly wounded or killed by trampling, they could readily be replaced by others from adjacent areas (Little, 1974). However, their habitat is changed, above ground by vegetation and litter removal, and below ground by soil compaction. A few species react positively to the new situation, and those that can do this are usually able to jump or run away from approaching walkers or vehicles (Little, 1974).

The sampling methods used to estimate the numbers of individuals and species may have a considerable effect on the result. Pitfall traps sample only the mobile species that may be passing through the area, while extraction from soil cores will sample only the less mobile species or life stages. Some insects spend their larval stages in the litter or soil and the adult imago may be free-living and not appear as a member of the soil fauna.

A linear, 5 cm wide series of samples was taken from the soil between trees in a birch wood near Moscow, Russia (Yur'eva, Matveva and Trapido, 1976) (Fig. 13.2). The transects were chosen so that they crossed paths where 'maximum damage' had occurred (Fig. 13.5). The numbers of animals were in inverse proportion to the degree of damage to the habitat. In the maximum damage zone the total numbers of microarthropods in a $5 \times 5 \times 5$ cm sample varied from 0 to 10 (0–4000 m^{-2}), in the medium damage zone the numbers ranged from 20 to 200 (8000–40 000 m^{-2}) and in the minimum damage zone from 120 to 600 or more (48 000–240 000 m^{-2}). The microarthropod distributions were similar to the distribution of litter on the woodland floor.

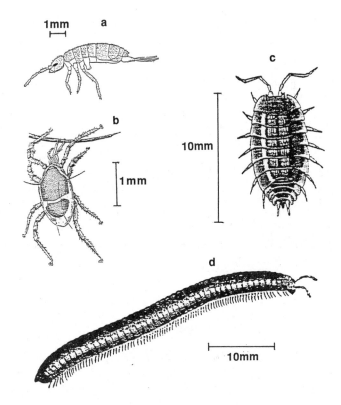

Fig. 13.4 Some examples of soil animals: (a) springtail (Collembola); (b) mite (Acarina); (c) woodlouse (Isopoda) and (d) millipede (Diplopoda) (from (a) Boradale *et al.* (1963); (b) Kuhnelt and Walker (1976); (c) Parker and Haswell (1963); and (d) Ainsworth-Davis (1903)). Notice the different scales.

SPRINGTAILS (COLLEMBOLA)

The springtails are small insects about 0.5–2 mm long. Many feed on decaying organic matter in the soils while others are also phytophagus, eating moss protonema, fungal spores and pollen. They occur in large numbers in soil and are generally reduced by the effects of trampling. Little (1974) found a mean of 75 600 individuals m^{-2} to a depth of between 9 and 18 cm in the untrampled areas of a coastal dune 'valley' in The Netherlands, falling to 5200 individuals m^{-2} in an experimental path which had been trampled 1300 times in 10 weeks. Similarly, there were 7000 m^{-2} in the top 5 cm of the minimal wear areas of the chalk grassland soil studied by Chappell *et al.* (1971), and these were reduced to 1200 m^{-2} in the trampled zone. In contrast, the numbers of springtails were found to be higher in the centre than at

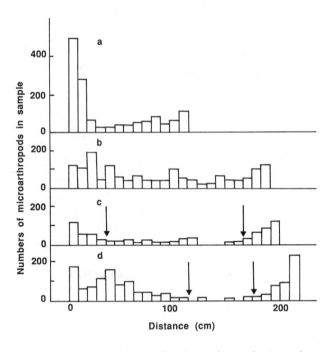

Fig. 13.5 Variation in numbers of microarthropods in a linear series of samples in birch (*Betula* sp.) woods near Moscow, Russia: (a) from a trunk to midway between trees; (b) from tree to tree; (c) and (d) adjacent samples between two trees across a 'badly worn-down' path in degenerated birch wood; limits of path shown by arrows. (After Yur'eva, Matveva and Trapido, 1976.)

the edges of two paths on a sandy heathland podzol at Kinver Edge studied by Newton and Pugh-Thomas (1979) (Fig. 13.6). These authors suggested that the springtails were not only more resistant to trampling than the mites (Acari) found in their soils, but that they may in fact be favoured by moderate trampling. The painstaking separation of their samples of springtails into separate species enabled Ingelog, Olsson and Bodvarsson (1977) to show that of the 16 species found in the soils of their Swedish woodland that were subject to only slight stress, seven did not occur in an area subject to extreme stress, while two 'new species' were found only in the areas of extreme stress (stress here means use for military exercises and by people walking). This shift in species composition is parallel to that observed in plants, and suggests that perhaps a few individuals (98 out of 2782 m^{-2}) found that the path situation had some advantages (Fig. 13.7).

In parallel with the earthworms, the deeper-living soil springtails appear to be the ones that survive in trampled areas. Yur'eva, Matveva and Trapido (1976) found that the deep-soil group made up 51% of

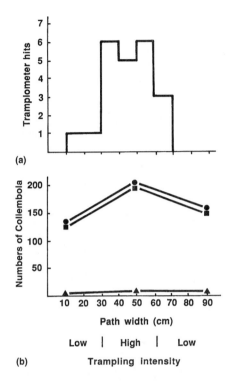

(a)

(b)

Fig. 13.6 Tramplometer readings indicating relative use (a) and the numbers of Springtails (Collembola) in the soil (b) of a transect across a path on a sandy heathland podzol at Kinver Edge, south Staffordshire. ●, total Collembola; ■, Arthropleona; ▲, Symphleona. (From Newton and Pugh-Thomas, 1979.)

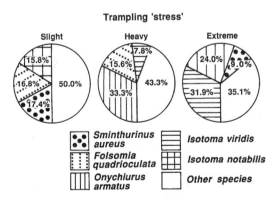

Fig. 13.7 Percentage distribution of springtails (Collembola) according to species, from soils of an old Scots pine stand at Uppsala, Sweden which had been subject to long-term use (from Ingelog, Olsson and Bodvarsson, 1977).

the total springtail population in the undisturbed areas near birch trees, but that this proportion rose to 77% in the most trampled zone. In contrast, those species living in the layers designated surface litter, lower litter and litter–soil were drastically reduced; the total numbers declined from 22 000 m^{-2} to 1100 m^{-2} and no new species were recorded in this zone.

In summary, springtails are generally reduced in numbers in soils of trampled areas but there is a suggestion that lower levels of wear might create conditions that lead to increases in the numbers of some species.

MITES (ACARI)

The mites are also small soil animals about 0.1–1 mm long, belonging to the class Arachnida. Some feed on decomposing plant remains and fungi while others are predators. The numbers of soil mites are about twice those of the springtails, and together they are the dominant soil arthropods, usually occurring in far greater numbers than any other groups. There were 23 200 m^{-2} to a depth of 5 cm in the minimal wear zone in chalk grassland (Chappell *et al.* 1971) and 40 600 m^{-2} in the top 5 cm of the minimum damage zone of the birch woods studied by Yur'eva, Matveva and Trapido (1976). Both of these studies showed a reduction in the number of mites in the most worn areas; to 22 200 m^{-2} and 1000 m^{-2}, respectively, for the short grasses and birch woods mentioned above. As in the case of the springtails, not all species are affected in the same way by the changes resulting from trampling. The larger species with heavier external skeletons, which are usually pigmented, were reduced from 28% to 16% of the mites present in the Russian birch wood samples. These are the litter feeders, and this result parallels the reduction in litter species of springtails. The smaller mite species, inhabitants of the minute soil pores, comprised a similar percentage of the total mite fauna in both the trampled and untrampled sites. The unspecialized species were almost eliminated, but the percentage of deep-soil species rose from 0.5 to 12%, although they are represented here by only 120 individuals m^{-2}. Ingelog, Olsson and Bodvarsson (1977) sorted their mites more precisely according to size, and demonstrated a fall in the numbers of mites over 700 µm in the soils of the extreme stress areas, from 14 to 1%, and a rise of those smaller than 500 µm, from 23 to 65% of the total mites present (Table 13.4).

In general, the numbers of mites are drastically reduced in the soils of the trampled areas used for recreation; the species that live in the surface litter and the larger species are reduced more than those that live in the small soil pores or deeper down the soil profile.

Table 13.4 Number of oribatid mites m^{-2} divided into size categories. Mean value from 27 core samples with 95% confidence interval (in brackets, the percentage share of the total population) (Ingelog, Olsson and Bodvarsson, 1977)

Size category	Degree of trampling 'stress'		
	Slight stress	Heavy stress	Extreme stress
I (700–1200 μm)	2120 ± 870 (14%)	10 ± 20 (< 1%)	20 ± 30 (1%)
II (500–700 μm)	9510 ± 1880 (63%)	1340 ± 620 (20%)	950 ± 580 (34%)
III (150–500 μm)	3510 ± 1080 (23%)	5220 ± 3880 (80%)	1820 ± 630 (65%)
Total	15140 ± 2660 (100%)	6570 ± 3950 (100%)	2790 ± 1080 (100%)

CRANEFLIES (TIPULIDAE)

Cranefly adults are normally free-living, while the larvae live in the soil and litter, feeding on the roots of living plants. On occasion they are important burrowers and channelers, increasing the porosity of the soil. The larvae declined in numbers in trampled chalk grassland soil but there was a fivefold increase in the number of free-living adults recovered from the soils (Chappell *et al.*, 1971). However, Leney (1974) found an increase in the numbers of craneflies in the soils of six of the seven picnic sites she investigated, in a range of different vegetation types (i.e. in young and old sand dunes, bog, heathland, loch margin and pine woods, but not in grasslands). It is also perhaps worth noting that Duffey (1975), in his litter-bag experiment (Fig. 10.5), found a non-significant increase in the numbers of cranefly larvae in compacted litter. He suggested that these higher numbers may have been favoured by the soil–litter mixture formed as a result of treading.

Bayfield (1979a) recorded a decrease in number of the larvae of *Molophilus ater* (Diptera) in trampled peat along a footpath at Inverpolly National Nature Reserve, Scotland. This author followed his field observations by a series of experiments to discover how far the reduction in numbers on the path had resulted from death by crushing and how far from habitat modification or other factors. He first tested whether the peat from paths was a suitable habitat when not trampled, by transferring cores of peat from the path to holes in the soil of the adjacent undisturbed vegetation, in March 1973, before egg-laying commenced. These were removed in the following

September and found to have as many adult flies as untrampled ground. This indicates that the peat was still a suitable habitat when set in a vegetated area and disturbances had ceased. There were, however, fewer flies in the peat of a larger area of bare ground created in March but not subsequently trampled. In a laboratory experiment mated female flies were presented with a choice of bare or vegetated cores in which to lay their eggs. Observations showed that they were only on bare cores for 17% of the time, while they were on the vegetated cores for 60% of the time. An additional trampling experiment on cores demonstrated that increasing numbers of impacts reduced the mean number of imagines (new adults) subsequently emerging from the cores (Fig. 13.8), and that there were fewer larvae surviving in the top 2.5 cm of the core than at depths of 2.5–4 cm (Table 13.5). In conclusion, Bayfield (1979a) suggested that 'The reduced numbers of *M. ater* on the footpaths may be partly explained by physical crushing, but fewer eggs may have been laid on the path than in the adjacent less open ground, or it may be that survival was poorer in the disturbed ground.' Thus this insect is reduced in numbers in part because of direct trampling damage and consequent deaths, and in part because of the altered habitat conditions (type 3 and type 2 disturbance, respectively; Chapter 17.)

OTHER 'MINORITY' GROUPS

The woodlice (Isopoda) (Fig. 13.4c) were rather more sensitive than spiders in Duffey's (1975) experiment on bags of plant litter, 'even though they are primarily feeders on dead plant material, and other organic material associated with it'. Numbers fell from 412.5 m^{-2} to 8 m^{-2} in the bags that received 10 treads per month for 12 months. However, the numbers of woodlice caught in pitfall traps in trampled

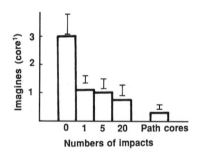

Fig. 13.8 Effects of increasing numbers of impacts by a trampling machine on the number of imagines of the cranefly (*Molophilus ater*) in peat cores taken from the Inverpolly Nature Reserve, Scotland (from Bayfield, 1979a).

Table 13.5 Effects of trampling on the numbers of slices of peat containing larvae of *Molophilus ater*. The slices were taken from the surface, middle and bottom of the cores, and data are numbers of cores out of 25. Probability values are based on Cochran's (1954) linear regression chi-square test (Bayfield, 1979a)

Number of impacts	Slice position			Total cores containing larvae	Mean number of larvae/core ± SE
	Surface	Middle	Bottom		
None	10	18	16	21	4.8 ± 0.8
1	11	15	17	19	4.1 ± 0.7
5	8	12	15	20	3.3 ± 0.6
20	5	7	12	13	1.8 ± 0.4
χ^2	8.40	9.43	2.18	–	– –
P	0.07	0.002	0.14	–	– –

areas of the coastal dunes at Nord-Finistere, France, were similar to the numbers caught by this method in vegetated parts of the dune system (Camberlein, 1976). This suggests that even though they may not be permanently on the open ground, these mobile animals are capable of moving through the vegetation as freely as they move in the other areas.

The aphids (Homoptera) have been shown to increase in numbers in the soils of trampled areas of chalk grassland (Chappell *et al.*, 1971), and two of the six species recorded increased in the trampled sand dunes at Nord-Finistere (Camberlein, 1976). However, Chappell *et al.* (1971) commented that 94% of their aphids were extracted from one core and the large aggregation may be misleading. Aphids were reduced from a mean of 43.2 m^{-2} to 7 m^{-2} in Duffey's (1975) plant litter bags.

Snails and slugs (molluscs) showed various responses to trampling in different areas. The molluscs *Ena obscura* and *Cochlicella acuta* were captured in greater numbers in Camberlein's (1976) pitfall traps placed in the trampled zones than in any of the other vegetation types. In contrast, the total number of snails and shells were reduced in the trampled zones of chalk grasslands (Chappell *et al.*, 1971), and snails together with slugs in trampled litter (Duffey, 1975). However, in the trampled chalk soils there was an increase in the species typical of grasslands, while the decrease was in the woodland species. Leney (1974) also recorded an increase in molluscs in her trampled old dune areas in Scotland, although they decreased in nearly all the other habitats she studied. The snails found in trampled areas are generally xerophytic (dry habitat) elements of the fauna, and this may explain their occurrence in the drier and warmer trampled areas.

Of the remaining records of the various groups of soil animals, the majority occur in lower numbers in recreation areas. The exceptions are the beetles (Coleoptera), which were up to twice as numerous in the trampled soils of five of the picnic sites studied by Leney (1974). However, there were small reductions at the other two picnic sites that she studied, and Little (1974) recorded a fall in numbers on his experimental sand dune paths in The Netherlands. In Finland disturbance associated with suburban development and consequent 'higher vegetational diversity' has been shown to increase carabid beetle numbers and species richness, while the original forest species declined or could not disperse to the remaining forest fragments (Halme and Niemelä, 1993).

Spiders found in grassland litter are mostly small and fragile (1–3 mm) (Duffey, 1975). 'They hunt in the spaces between leaf and stem fragments, or construct tiny webs which would be easily destroyed by disturbances.' The numbers of individual spiders and of spider species recorded in the three trampling treatments in Duffey's (1975) 10-month grass litter experiment were: in the controls, 275 individuals and 25 species; after five treads month^{-1}, 74 individuals and 12 species; and after 10 treads month^{-1}, 42 individuals and only 6 species. Most species were greatly affected by trampling, but three showed little response.

Most of the vertebrates that live in the soil do so for only a part of their life cycle, or utilize burrows as lairs or resting places. However, the moles (e.g. European mole, *Talpa europaea*; Mediterranean mole, *Talpa caeca*; North American mole, *Scalopus aquaticus*; and to a lesser extent the marsupial mole, *Notoryctes typhlops*) spend most of their time beneath the ground. There do not appear to be any records of their presence or absence in recreation areas. However, it has been suggested that they do not burrow through hard ground (Schaerffenberg, 1941). The author has observed that the piles of earth excavated from tunnels may occur alongside but not on footpaths in Sussex, England.

13.5 Summary   The general decrease in numbers of individuals and species of soil organisms in recreation areas appears to be the consequence of changes in habitat conditions, as well as direct injury and death. The removal of the surface layers of litter and soil profile (Figs 10.1, 10.7, and 15.10) clearly destroy the habitat for the animals that live in those layers, and even compaction of litter by light trampling was shown to have a major effect on the animal numbers. Compaction, and the consequent hardness of the deeper layers of soil, is also associated with a decline in the numbers of individual species of the deeper-living organisms. This is probably due to the physical difficulty of burrowing in the dense soil, but may also be associated with changes in soil moisture. Human trampling, with its associated lateral forces, was also shown to cause more injury to animals in the surface layer than simple vertical impact (Bayfield, 1979a). The soil animals that did survive trampling appear to fall into two groups. Generally, smaller members of the community survived in the surface layers, while the larger members were found deeper in the soil, and here the smaller members were absent. This suggests that survival in the surface layers depends on being able to 'shelter' in and move through the smaller pores that remain in trampled soils. At depth the requirement may be for greater physical strength to move through the compacted soils. Changes in microclimate, and particularly soil aeration, are also a causal factor in the community changes, especially in the case of the bacteria, but the larger organisms would also be affected by climatic change.

The removal of vegetation cover also has a major effect on the more mobile surface-living groups. The balance between the effects of direct injury and of habitat change is uncertain, and probably varies for the different species, as does the ability to recolonize affected areas from adjacent undisturbed soils.

14 Plant roots, soils and growth

Plants use the strength of the soil for their support, soil water for photosynthesis and transpiration, and soil nutrients for growth. The narrow-leaved (monocotyledonous) plants penetrate the soil with a fibrous root system in which there are many roots of similar size and form (Fig. 14.1b). The broad-leaved (dicotyledonous) plants commence their life with a single tap root which branches and forms a dendritic system below the surface of the earth (Fig. 14.1c). In both groups, the water and nutrients in solution are taken up by the root hairs which occur at the tips of new roots (Fig. 14.1a). As the root ages, the hairs are shed, so that new growth is constantly required for water uptake, unless the plant's metabolism is slowed down by excessively low or high temperatures or drought. The resulting matrix of roots is as much a part of the soil as are the other organisms that live there.

14.1 Plant roots   Plant roots vary in the depth to which they penetrate, according to their form and the soil conditions. For example, a mature cotton plant (a broad-leaved species) growing in good soil has the mass of its roots in the top 2 m (Trouse, 1971). The older primary and secondary roots occurred in the upper layers, but the new rootlets on which the root hairs grow were more or less evenly distributed down the profile. The plant was therefore drawing water and nutrients from all parts of the soil. Tree roots may extend further into the soil but the surface layers are still of considerable importance for water and nutrient uptake. In soils which are not waterlogged, the depth to which roots penetrate may depend upon the depth to which water percolates down the profile. This was illustrated for prairie plants at Hays, Kansas, after a series of moist years and a series of dry years (Albertson, 1937) (Fig. 14.2a,b). However, this dependence of depth of rooting on water supply holds only if the plant is producing sufficient carbohydrates to allow good root development. Trampling reduces the leaf area (Fig. 6.10); all carbohydrates produced may be utilized in restoring leaf structure, and plants may then become shallow rooted. This is particularly deleterious in climates with arid seasons, as many investigators

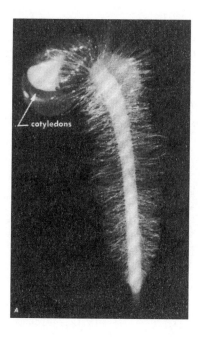

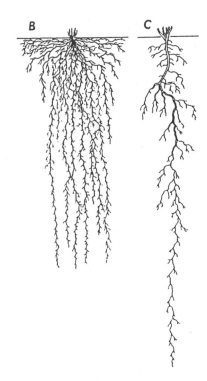

Fig. 14.1 The forms of various kinds of root: (a) root hairs, ×2; (b) root of a monocotyledonous plant; (c) root of a dicotyledonous plant (from Wier *et al.*, 1982).

have shown in campsites where watering is the primary management action that maintains the vegetation. A further point is that soil compaction also increases rainfall runoff (Chapter 15) and the wetted depth, and consequent rooting pattern, in a moderately trampled area may be changed from that of Fig. 14.2a to that shown in 14.2b.

In exceptional conditions, roots can extend many metres in a vertical direction – the author has observed tree roots up to 15 m deep inside caves near Rockhampton in tropical Queensland. The aerial roots of the rainforest strangling fig (*Ficus watkinsiana*) grow down the trunks of their host tree from the site where the seed has germinated, to the soil which may be as much as 50 m below the point of germination. In poor soil conditions, especially where the water table is near the soil surface, the mass of roots may occur in the upper layers of soil, making the plant particularly vulnerable to soil erosion and trees to windthrow.

The biomass of plant roots has a variable relationship with the biomass of the above-ground portion of the plants. The root/shoot ratio depends in part on the conditions in which the plant is growing,

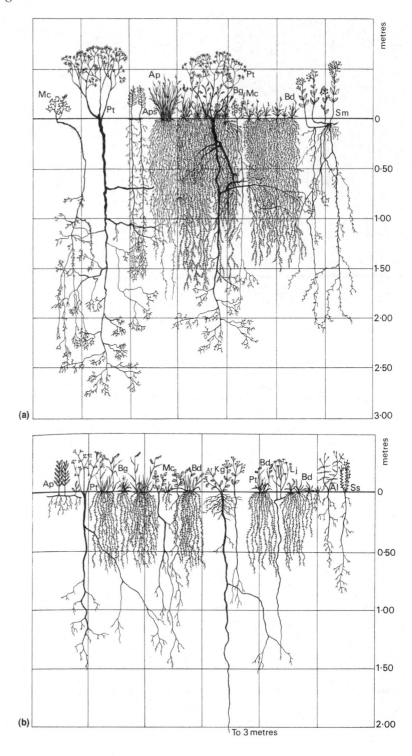

and is likely to be affected by trampling. However, an experiment in which 12 species of plants were subjected to simulated wear by a dropped weight showed an overall reduction in plant biomass, with roots and shoots being equally reduced (Sun, 1991). Wear by horses, often galloping, tended to reduce the proportion of shoot material at two sites on a southern English heathland (Liddle and Chitty, 1981). Where high tangential forces are exerted, such as in wear by vehicles, I would expect that the above-ground biomass might be markedly reduced in relation to the root system. Roots also function as soil stabilizing agents, discussed in Chapter 15.

The effect of recreation on the plant roots may operate indirectly, by compacting the soil and altering the conditions in which they grow, or directly, by erosion and subsequent mechanical damage to the root itself. Plant roots which are modified to store carbohydrates would appear to be especially vulnerable to soil changes resulting from trampling. Modified underground organs include corms (swollen stem bases), rhizomes (horizontal underground stems), bulbs (swollen leaf bases), tubers (swollen terminal portions of underground stems) and lignotubers (swollen, woody, partly underground stem bases with numerous cortical buds) (Fig. 14.3).

MECHANICAL EFFECTS

14.2 Roots in compacted soil

Meinecke (1928) was one of the first workers to discuss the ecological effect of recreation on the environment and I quote some parts of his paper here. First, a description of the roots of the giant redwood trees (*Sequoia sempervirens*):

> The mechanical support is effected by large and heavy roots while the water supply is furnished exclusively by the extreme tips of fine and much-branched feeding rootlets. It is transported through thicker roots to the bole. Redwood is characteristically shallow-rooted. The whole tree stands virtually on an inverted plate of heavy roots. Frequent graftings (connections) between these roots add to the mechanical strength of the plate without impairing its elasticity and give, which are necessary to take up the heavy

Fig. 14.2 Root systems of plants in a typical short grass prairie at Hays, Kansas. (a) After a run of years with average rainfall; (b) after a run of drought years. Al, *Allionia linearis*; Ap, *Aristida purpurea*; Aps, *Ambrosia psilostachya*; Bd, *Buchloe dactyloides*; Bg, *Bouteloua gracilis*; Kg, *Kuhnia glutinosa*; Lj, *Lygodesmia juncea*; Mc, *Malvastrum coccineum*; Pt, *Psoralia tenuiflora*; Sm, *Solidago mollis*; Ss, *Sideranthus spinulosus*. (From Russell, 1973, after Albertson, 1937).

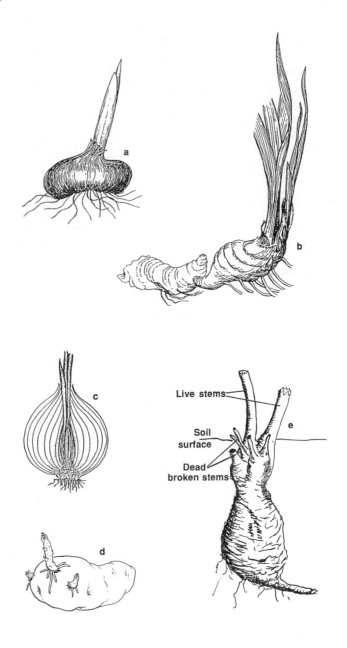

Fig. 14.3 Underground storage and perennation organs: (a) corm, swollen stem base; (b) rhizome, horizontal underground stem; (c) bulb, swollen leaf bases; (d) tuber, swollen underground terminal portion of stem; (e) lignotuber, lignified, swollen, partly underground stem base with numerous cortical buds (a–d, from Muller, 1979; e, by the author).

swinging of crown and bole in strong winds. The entire plate is surprisingly small for so large a tree. In the main, it hardly extends as far as the correspondingly heavy limbs of the crown, and it is rarely thicker than 4 to 5 feet (120 cm to 150 cm) ... From the main roots a great number of thinner roots from an inch or two (2.5 cm to 5 cm) down to pencil thickness reach far out through the soil, mostly within 1 to 3 feet (30 cm to 90 cm) from the surface of the soil. It is from these roots that the great mass of feeding roots and rootlets spring.

And about the growth of the roots of a redwood:

This tree with a diameter of about 12 feet (3 m) and an approximate height of 375 ft (112 m) stood close to the old country road which had been in use for at least forty years until it was abandoned for the new highway eight years ago ... Within 1 to 2 feet (30 cm to 60 cm) from the base of the tree, the soil was soft and mellow, full for several inches in depth, with a mass of fine feeding rootlets in perfect health. With the approach to the edge of the road, these feeders become more and more rare ... The rootlets were much flattened out and the majority of the fine feeders were black and dead.

Meinecke, in his elegant style, presented observations which have since been repeated by many workers. His conclusions about one particular site 'several dead and dying trees ... showed plainly enough the effect of concentrated and heavy traffic, and indicate unmistakably what will happen to the remaining trees' were clear enough to provoke much of the protection we can see in the national parks today (Fig. 14.4). Wooden boardwalks are now a common feature of intensely used areas in dunes (Carlson and Godfrey, 1989), on boggy areas such as taiga woodlands where they have been constructed for off-road vehicles (Slaughter et al., 1990), and even in subtropical mangroves at Moreton Bay, Queensland. All of these allow more or less natural growth to occur and avoid soil compaction and consequent damage to plant roots.

Root and root hair penetration of the soil is inhibited by compaction; Russell (1973) gives a figure of 1.8 g cm^{-3} as the maximum soil bulk density in which plant roots may survive. The negative relationship between the yield of corn (Zea mays) and the soil bulk density shown by Raghaven et al. (1978) (Fig. 14.5) suggests that in this clay soil growth would take place at higher densities. However, the extended portion of the relationship is likely to be curvilinear, and would reach zero yield at an earlier stage than extrapolating from the straight lines would suggest. From the results of his work, Lull (1959) suggested that 'Sunflower (Helianthus annuus) roots, for instance, cannot penetrate soils that have compacted to densities ranging from 1.75 g cm^{-3}

Fig. 14.4   Raised wooden walkways on the Cedar Trail at the Glacier National Park, Montana (photographs by M.J. Liddle).

for sands to 1.46–1.63 g cm^{-3} for clays'. Generally, root growth tends to be restricted in fine-textured soils when the bulk density gets much above 1.4 g cm^{-3}, and in coarser-textured soils above 1.6 g cm^{-3}.

Compaction also limits the volume of soil explored by plant roots. In an elegant series of experiments, Engelaar (1994) demonstrated the differential abilities of great plantain (*Plantago major*) and common sorrel (*Rumex acetosa*) to explore the soil in which they were growing (Fig 14.7). The more rapid spread of plantain roots in the compacted soils and the greater amount in the lower levels at the end of the experiment indicated its superior ability to survive in these conditions.

The information summarized in Chapter 11 indicates that the density of soil in most compacted recreation areas is not sufficient to prevent root extension by mechanical resistance, although associated changes in other soil characteristics may inhibit growth. For example, the dry weight of bluebell bulbs was reduced by over 50% by trampling every 7 days starting early in May (Blackman and Rutter, 1950) (Fig. 6.15), and the underground growing points of the rhizome of the sand sedge (*Carex arenaria*) were killed by driving over them, but the plant flourished in the tyre tracks once driving had ceased (Fig. 6.17). Although the mechanism by which the bulb dry weight was reduced is unclear, the results from the sand sedge suggest that the rhizomes were affected more by the mechanical forces of the compaction process than the altered soil conditions. The lignotubers of creeping lantana (*Lantana montevidensis*), a native of South America, growing beside a path in dry sandstone country at Cania Gorge, Queensland, had a mean dry

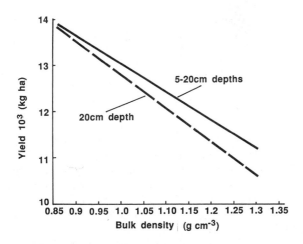

Fig. 14.5 The relationship between bulk density of a Rosalie clay soil in Quebec and yield of corn (maize). (Note that the *y* axis starts at 10×10^3 kg ha^{-1} and not at zero.) (From Raghavan *et al.*, 1978.)

Control   Compacted

Plantago major ssp. major

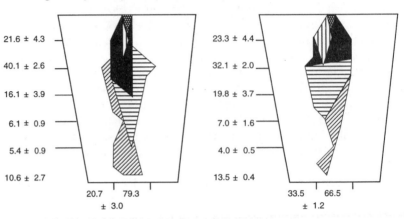

Rumex acetosa

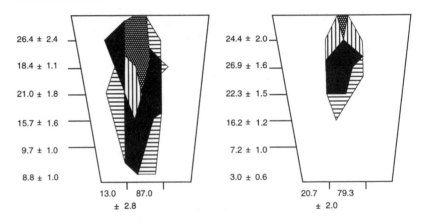

Fig. 14.6 Outlines of the root systems of great plantain (*Plantago major* ssp. *major*) and common sorrel (*Rumex acetosa*) in a loosely packed and a densely packed soil at different sample times. Percentages of horizontal and vertical distributions of measured root length at the end of the experiment are also indicated. Small glass tubes penetrated the soil and allowed observation of the presence of roots as they grew into the soil. The outlines connect the positions of the tubes where roots were observed at each observation. Once contact had been made it was continued in subsequent observations so the shapes in this figure represent an increasing volume of root. Cross-hatched, day 14; horizontal hatch, day 17; black, day 20; vertical hatch, day 23. (From Engelaar *et al.*, 1993.)

weight of 4.24 g while those on the path had a mean weight of 3.35 g, but the difference was not statistically significant and a larger sample could have influenced this result. However, there was a marked reduction in mean stem length, from 169 mm to 18.8 mm on the path, but the mean root length was almost the same in both situations, and the total biomass showed a similar pattern. This again suggests that the direct mechanical forces of trampling have a much greater effect than altered soil conditions, although the state of the soils can be very important.

PORE SIZES

Two mechanical factors appear to influence the penetration of roots into compacted soils. One is the pore size in relation to root diameter, and the other is the amount of pressure required to extend the pore size once it has been penetrated by the root tip. An experiment showed that there was a distinct difference in the behaviour of seminal (main) roots (diameter 0.3–0.45 mm) and lateral roots (0.32–0.37 mm; Russell, 1973) of barley (*Hordeum vulgare*) when grown in a matrix of glass beads (ballotini) of various dimensions. These glass beads have predetermined pore sizes and can be held at various pressures (Goss and Drew, 1972). All other requirements for growth were supplied in solution. The seminal roots were able to elongate while the pressure was such that they could enlarge the pores. When the pressure was increased so that the seminal roots could not move the glass beads and enlarge the pores, elongation completely stopped and the root apices ceased to function. This process normally happens when a root meets a stone and the growth of that particular root cannot be diverted. Lateral roots may then extend their growth and bypass the obstruction. When Goss and Drew (1972) set a pore size of 0.15 mm and there was no pressure applied externally to the system, normal growth was possible, as the seminal roots could enlarge the pores (Fig. 14.7a). When the pore size was held at 0.15 mm and pressure was applied at 0.5 bar (0.05 MPa) to the glass bead matrix, the seminal roots were inhibited, but laterals could still develop (Fig. 14.7b). When pore size was reduced to 0.06 mm, root growth was very restricted (Fig. 14.7c). Root hairs (diameter for wheat 0.008–0.012 mm) would, of course, be unrestricted in these conditions, so that plants may still survive, but with little root growth.

In an experiment with rice (*Oryza sativa*) it was found that coarse sands had more rigid structures than fine sands, so, in spite of their smaller pore sizes, roots could penetrate fine sands and enlarge the pores but not the coarse sands unless the pore radii were above 0.06 mm (Kar and Ghildyal, 1975). Roots can exert penetration pressures

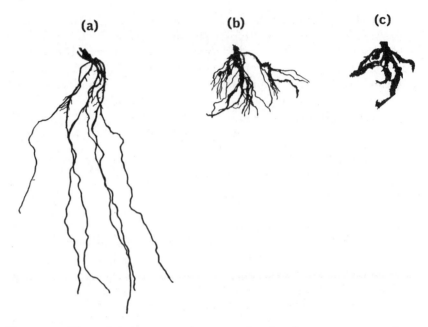

(a) **(b)** **(c)**

Fig. 14.7 Effect of the pore size between glass beads (an experimental 'soil') and applied pressure on the growth of barley roots.

Root	Minimum pore diameter (mm)	Excess pressure (bar)
a (control)	0.15	0
b	0.15	0.5
c	0.06	0.5

Note the reduction in seminal root length at 0.5 bar (0.05 MPa) (b and c) and impedance to lateral roots in finer pores (c). (From Goss and Drew, 1972.)

in excess of $1 \, kg \, cm^{-2}$ over the small area of their tips, and this can allow them to penetrate some compacted soils. The force a seedling can exert in a vertical direction depends to some extent on the compaction of the soil beneath the seed, as it must react against something to produce an upward force. A certain level of compaction can therefore aid seedling growth and development. Corn (*Zea mays*) can exert an emergence force of about 70 g when growing in soil of a bulk density of $1.4 \, g \, cm^{-3}$, but when the soil density was raised to $1.7 \, g \, cm^{-3}$ the emergence force increased to about $120 \, g \, cm^{-3}$ (Prihar and Aggarwal, 1975). This is sufficient to penetrate moderately crusted soil, especially if it is moist, but not enough to break through the soils when they are dry. However, as Russell (1973) remarked, 'the crust (produced by rainfall in this case) often prevents the shoots of germinating seeds penetrating it, even if it is only a few millimetres thick'. In addition to the strength of the hypocotyl, the presence or absence of cotyledons may affect the ability of a seedling to emerge from compacted soil. As pointed out by Silcock (1980), dicotyledonous seedlings with epigeal germination (where the two cotyledons are raised above the soil surface) have to push the large seed leaves through the soil, so the volume of soil that has to be displaced is greater than that displaced by hypogeal dicotyledonous and monocotyledonous species. The emergence of clover (*Trifolium repens*) was differentially reduced compared to that of rye grass (*Lolium perenne*), guinea grass (*Panicum maximum*) and elastic grass (*Eragrostis tenuifolia*) when they were all planted at 2 cm depth and the soil compacted by simulated trampling, in this case 30 drops of a weight equivalent to a human footfall (Mr A. Sharp, personal communication). It is clear that compaction of soils by recreation activities is likely to inhibit growth of many plant species by mechanical forces, especially if the soil has a high clay content.

SOIL PENETRABILITY

Field measurements of species distribution in relation to soil bulk density or soil penetrability clearly show that some species occur preferentially in more compacted soils. However, the causality of the relationships have not often been investigated. For example, Horikawa and Miyawaki (1954) recorded the cover of species growing, and the soil hardness, in a series of tracks at Hiroshima, Odawara and Tokyo (between 34.5° and 35.5°N on the island of Honshu, Japan), and produced a figure showing this relationship to illustrate their discussion of habitat segregation (Fig. 14.8).

However, the relationship between the various parameters and plant growth is not clear. A survey of plants growing in sand dunes at Aberffraw, North Wales, showed that plants occurred in a different sequence when placed in order of increasing bulk density of soils in which they were growing compared to the sequence of increasing penetration index (Liddle and Greig-Smith, 1975b) (Fig. 14.9). It is notable that the moss *Tortula ruraliformis* occurred in the soils with the highest bulk density but lowest penetration index. This may be because it occurred mainly on the drier soils (9% soil water by volume). Other deeper rooting species which tended to follow a similar pattern included glaucus sedge (*Carex flacca*) and white clover (*Trifolium repens*) and these occurred in the wetter areas (18–24% soil water by volume). There are apparently complex reasons for the different occurrences.

One point that does seem clear is that root penetration is greater in soils that have lower soil strengths as measured by penetrometers.

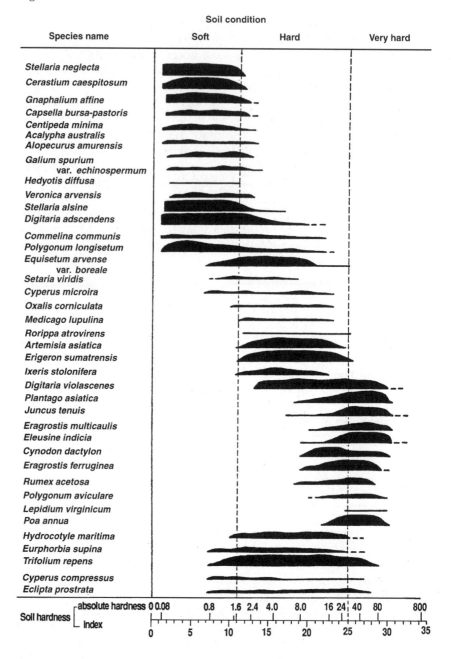

Fig. 14.8 The range of soil hardness in which various species were recorded growing on pathways at Hiroshima, Odawara and Tokyo. The shaded areas indicate estimated cover of each species. (From Horikawa and Miyawaki, 1954.)

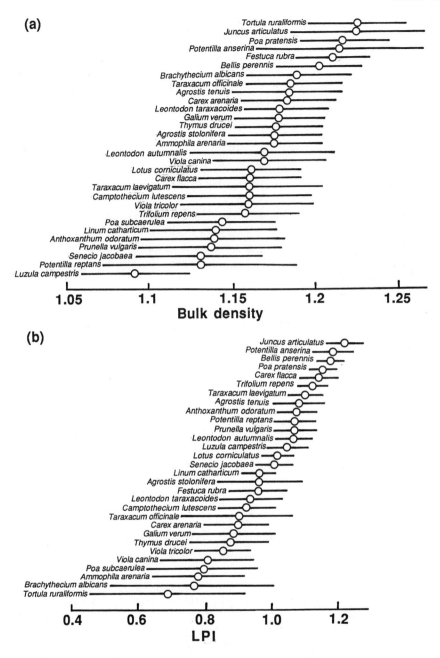

Fig. 14.9 The relationships between soil bulk density (a), soil penetration index (b) and the common plant species in the Aberffraw dune system, Anglesey, North Wales. ○, Mean and two standard errors. Note that the relative sequence of species is different for the two measures. (After Liddle and Greig-Smith, 1975b.)

Taylor, Roberson and Parker (1966) measured the percentage of cotton tap roots penetrating through cores of four different soil types and found that there was little penetration in soils over a strength of 20 bars (2 MPa) (Fig. 14.10), although limiting resistances have been found to vary widely between 8 and 50 bars (0.8 and 5 MPa), depending on species (Cannell, 1977).

14.3 Soil water and plant growth

The effect of compaction on soil water has been discussed in Chapter 11. Germination, seedling establishment and survival of plants in dry periods may be favoured by compaction in relatively dry and loose soils.

Red fescue (*Festuca rubra*) plants grown on compacted sandy soils remained alive longer after water was no longer supplied to the pots, than did the control plants grown in uncompacted soil (Liddle and Greig-Smith, 1975a) (Fig. 14.11). There was clearly more water available to plants and more lost from the pots of compacted sandy soil. A similar result was obtained by Yang and De Jong (1971) when they measured the amount of water taken up by wheat from soils that were already fairly dry. However, it should be noted that at higher soil

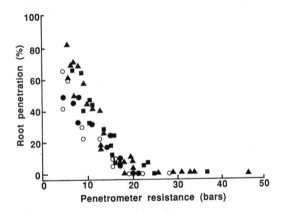

Fig. 14.10 Relationship between root penetration and penetration resistance in four contrasting soils (from Russell, 1977). The proportions of soil components were as follows:

	Sand, 750μ	Silt, 2-50μ	Clay <2μ
■	44	37	19
▲	73	20	7
●	80	11	9
○	83	8	9

water content wheat used in the experiment was able to transpire more water from the uncompacted than the compacted soils. Presumably compaction inhibited the water conductance of these clay and loam soils.

In summary it would appear that soil compaction as a result of trampling might limit the maximum growth of plants through limitation of water supply but that, at least in certain soils, their survival may be enhanced during dry seasons.

The effects of soil compaction on nutrients may operate through changes in mass flow of the soil water, root contact with the source of the ion, or diffusion across short distances (Parish, 1971). Soil

14.4 Soil nutrients and plant growth

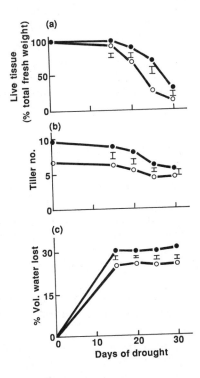

Fig. 14.11 The effect of soil compaction on the survival of plants in drought conditions. (a) Change in percentage of live fresh weight of total plant fresh weight plotted against days of drought, measured after the recovery period, (b) number of tillers per plant plotted against days of drought, measured after the recovery period, and (c) percentage volume of water lost from pots plotted against days of drought. ●, Compacted soils; ○, uncompacted soils; I, two standard errors. (From Liddle and Greig-Smith, 1975a.)

compaction will directly influence mass flow and diffusion and, by altering root growth, may also change the ability of roots to 'forage' through the soil matrix and make contact with ion sources. As the most important mechanism for uptake of the less mobile ions such as phosphorus and potassium appears to be mass flow (Parish, 1971), and this is clearly restricted by compaction, the supply of these ions is the most likely to be restricted.

However, it must be noted that compaction will increase the mass of soil in the root depletion zone so that uptake by root contact may be increased (Cornish, So and McWilliam, 1984). This is perhaps the situation in most recreation areas.

The effect of compaction may be to reduce the elongation of roots and increase their mean diameter (Lindburg and Petterson, 1985). These authors suggest that since nitrogen and phosphorus are mainly absorbed by lateral roots but calcium is taken up by seminal roots, the uptake of calcium may be differentially reduced.

The effects of soil compaction on nutrient uptake are complex; however, all experiments have shown restricted growth of roots in compacted soils with or without adequate nutrient supply, so it would seem likely that mechanical impedance is the primary factor limiting growth in most conditions.

14.5 Soil air and plant growth

The restriction of the number of air-filled pores is especially obvious in compacted wet or waterlogged soils. Under these conditions an anaerobic, usually black, layer develops and roots of plants not specially adapted to these conditions are unable to grow. This may lead to a layer below, for example, a grass turf, which is easily sheared and the lateral forces of walking or running may be sufficient to strip off the surface layers, leaving bare soil.

As with other elements of compaction, the relationship of oxygen deficiency with other aspects of soil change is not yet fully understood. However, one experiment showed that at a bulk density of $1.5 \, \text{g cm}^{-3}$ in a clay loam soil, oxygen concentrations below 10% inhibited root elongation (Tackett and Pearson, 1964) (Fig. 14.12). In this experiment root growth in soil bulk densities of more than $1.6 \, \text{g cm}^{-3}$ was uniformly inhibited at oxygen concentrations up to 20%.

The relative importance of deficient aeration, mechanical impedance and moisture stress on root elongation in sandy loam was indicated by Eavis (1972) (Fig. 14.13). Moisture stress appears to have only a secondary effect at high matrix potentials.

14.6 Summary

Plants require support, water and nutrients from the soil. The depth of a plant's roots depends in part on soil oxygen and the physical

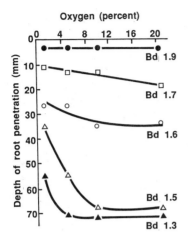

Fig. 14.12 The effect of bulk density (Bd) and oxygen content on the depth of cotton root penetration into a compacted subsoil (from Tackett and Pearson, 1964).

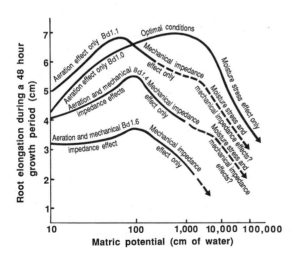

Fig. 14.13 Summary of the role of mechanical impedance, aeration and moisture stress on pea seedling root elongation in a sandy loam held at different matrix (water) potentials and bulk densities (from Eavis, 1972).

state of the soil. The effect of wear may compact the soil and limit root penetration or disturb the soil and mechanically damage the roots.

The yield of plants may be related directly to soil bulk density, and generally little growth will take place in sandy soils over $1.75 \, \text{g cm}^{-3}$ and in clay soils over $1.5 \, \text{g cm}^{-3}$. Recorded soil densities from recreation areas rarely, if ever, reach these levels, so direct mechanical damage appears to be more important.

The nature and size of the soil pores have a direct effect on root development – if they are reduced by compaction to less than 0.06 mm, root growth becomes very restricted. The other factor is the rigidity of the soil matrix – if the roots can exert sufficient force to widen the pores, then growth is possible. Crust penetration by dicotyledonous seedlings may be especially restricted – if they have epigeal germination, they have to displace a greater area of soil than the single cotyledon of monocotyledonous species. The use of a soil penetrometer may more accurately simulate plant root or underground shoot development and give a more realistic assessment of soil conditions from the plant's 'point of view' than measures of bulk density. Plants appear to respond differently to the soil characteristics measured by these two methods.

Soil compaction may increase the amount of water available to plants in coarse soils, although uptake was limited in clay and loam soils. Uptake of soil nutrients may be limited when compaction limits the mass flow or diffusion of water and dissolved nutrients. Restrictions on root growth will also limit the volume of soil through which the roots may forage for nutrients. Availability of phosphorus and potassium is most likely to be limited by compaction, but uptake by root contact could be increased if the soil mass is increased while still not restricting root growth, the situation found in most recreation areas. The uptake of calcium by seminal roots will be restricted if their growth is restricted.

The restriction of soil air by the reduction of the pore volume may be especially important when the soils are partially or fully waterlogged. An anaerobic layer beneath the soil surface is likely to restrict almost all root growth below the surface soil.

The effects of soil compaction on plant growth are variable, depending upon the type of soil and the degree of compaction. It would appear that most soils in recreation areas are capable of supporting plant growth, although where anaerobism or surface crusting occurs they can almost prevent survival of many species.

Soil erosion 15

Erosion is a natural process. It has occurred continuously since the first rocks solidified and rain fell on the earth. As the volcanic rocks are pushed upwards by the movements of the earth's crust, they are gradually worn down by climatic and mechanical forces. The detached particles are eventually deposited where great beds of material form the sedimentary rocks, which may in turn be uplifted. The erosive effects of recreation are minute when considered against the background of natural processes, but they are of considerable concern as their aesthetic and management consequences impinge upon both the visitors to, and the managers of, recreation areas. Recreation activities produce these effects by changing local conditions so that the natural erosion processes are accelerated and, in local terms, considerably increase the volume of soils eroded. In order to understand these effects, it is first necessary to understand the processes of natural erosion, and the factors that limit its rate.

15.1 Erosion processes

Erosion takes place when some force is applied to the soil, usually by water or air, but sometimes as an interaction between changed fluidity of soils and gravity. The most common forms of erosion are rain splash, overland water flow, rill and gully erosion, subsurface flow (which is important in the leaching of nutrients in solution; Morgan, 1979) and, in exposed areas, wind erosion. Much of the following introduction is based on Morgan (1979).

Rainsplash erosion takes place when a raindrop falls on an exposed, usually sloping soil surface. The downslope component of the raindrop's momentum is all transferred to the soil particles, the remainder being reflected in the splash. The transfer of momentum to the soil particles will, in part, compact the soil, but it will also detach some particles from the soil aggregates, launching them into the air. The consolidation effect may be seen in the formation of a surface crust which may limit further erosion. The detaching effect operates primarily on medium and coarse particles as the clay platelets (Fig. 11.11) are held together by chemical bonding forces. The detached particles are then 'available' for transport by water flow.

Overland water flow occurs on sloping ground during rain storms when soil moisture storage capacity is saturated, or in heavy rain (more properly called intense rain as all rain is nearly the same weight) when the soil infiltration rate is exceeded. Overland flow commonly occurs as a mass of anastomosing, braided water-courses with no pronounced channels (Morgan, 1979). The amount of soil carried by overland flow depends upon the velocity of the water, the size of the soil particles and the amount of energy required to entrain the particle in the flowing water (Fig. 15.1). Thus a soil particle of 0.01 mm requires a flow of 60 cm sec^{-1} to detach it, but it is not deposited until the flow velocity falls below 0.1 cm sec^{-1} (Morgan, 1979).

Rill erosion occurs where overland flow first becomes channelled. Rills are transient features but because of their greater erosive power and concentrated flow, rill erosion may account for the bulk of the sediment removal from a hillside. In some conditions, 80% of the sediment may be transported in rills (Mutchler and Young, 1975).

Gullies are relatively permanent, steep-sided water courses which experience ephemeral flows during rain storms (Morgan, 1979). They are almost always associated with accelerated erosion and therefore with landscape instability. Erosion occurs at the head of the gully where the head wall is undercut and further down the gully as the 'stream banks' erode.

Subsurface flow takes place throughout the soil profile as water percolates down through the water table. On sloping ground this may become channelled into tunnels or subsurface pipes, which may collapse and initiate gully formation. Subsurface flow is, as mentioned above, normally important in leaching mineral nutrients from the soil.

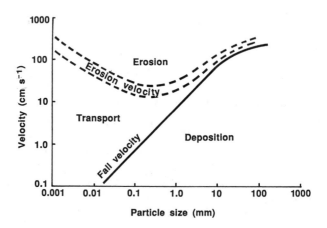

Fig. 15.1 Critical water velocities for erosion, transport and deposition as a function of particle size (from Morgan, 1979, after Hjulstrom, 1935).

Wind erosion occurs when the air flow is sufficiently strong either to move the soil particles over an exposed surface by rolling them along (surface creep), or to provide sufficient energy for the particles to move in a series of jumps (saltation). Fine particles, usually less than 2.2 mm in diameter, may be carried in suspension when they can rise high in the air and travel long distances.

Mass flow is well known in the form of dramatic landslides, which may block roads or, more harmfully, engulf campsites, villages or towns. However, small flows are more common in some environments, and they form the dominant erosion process (Morgan, 1979). They generally occur when the soil becomes fluid through the absorption of an unusual quantity of water, or when a natural barrier, such as a rock outcrop, becomes fractured by larger earth movements.

Erodibility of soils depends upon the soil texture, aggregate stability, shear strength, infiltration capacity, and organic and chemical content. All factors which may be changed by recreation use. The interaction with soil texture is demonstrated in Fig. 15.1, where it is evident that large particles require more energy to move them, and it has already been mentioned that fine particles are cohesive. Thus soils with high silt contents, generally between 40 and 60% (Fig. 11.9), are the most erodible (Richter and Negendank, 1977). The presence of cohesive aggregates where clay is combined with organic matter, especially where there is a high content of base minerals, also inhibits the effects of rain splash and overland flow. The shear strength is a measure of the cohesiveness of the soil. The infiltration capacity, the maximum sustained rate at which the soil can absorb water, has a direct relationship with the amount of water available for overland flow, until the soil reaches its saturation point. Organic and chemical constituents of soils have a strong effect on aggregate stability, and soils with less than 2% organic matter can be considered erodible (Evans, 1980). Other energy absorbing features, such as stoniness and surface roughness, will also affect the erodibility of soils.

The features of the environment which inhibit soil erosion and keep the rate below that of soil formation are essential for the biotic processes of the earth's mantle. The most obvious of these is level ground, where the movement of water is so reduced that it cannot maintain the sediments in suspension. However, this occurs on only a relatively small proportion of the earth's surface. On sloping ground, the primary natural erosion inhibitor is a cover of vegetation, which in one case reduced the erosion rate by 100 times, from $4.63\,kg\,m^{-2}$ on bare ground to $0.04\,kg\,m^{-2}$ from ground with a dense cover of grass (*Digitaria* sp.) (Hudson, 1971). Vegetation intercepts the raindrops so that their kinetic energy is dissipated by the plants rather

15.2 Erodibility of soils

than imparted to the soil. The effectiveness of plant cover in reducing erosion depends upon the height and continuity of the canopy, the density of the ground cover and the root density. Drops of water falling from the canopy gain 90% of the terminal velocity after 7 m, so the presence of low-growing vegetation is of vital importance. The relationship between the amount of rain falling per rain-day and the mean overland flow per rain-day, which is taken as a measure of potential erosion, was shown for three vegetation types by Kirkby (1980) (Fig. 15.2). It is clear that forest is the most efficient erosion inhibitor, followed by grass cover, although plant litter would perform the same function, and that scrub causes a lesser but significant reduction in overland flow. Plant roots will also inhibit erosion, as they tend to bind the soil and also maintain its porosity. Where there are marked seasonal variations in vegetation cover, the timing of the rainfall will have a marked effect on the consequent erosion (Kirkby, 1980).

Plant cover and associated dead litter also inhibit overland flow by breaking up the water courses and slowing down the rate of flow. Roots penetrate and may loosen the soil, thus increasing the infiltration rate, and if exposed on the surface, will also decrease the rate of overland flow. Plants also inhibit wind erosion by breaking the surface flow of air and providing pockets of almost still air where the entrained small particles may be deposited, and thus create dunes in coastal areas.

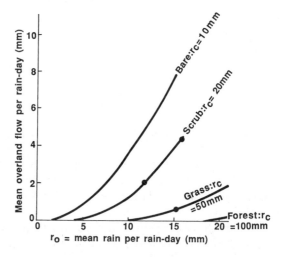

Fig. 15.2   The relationship between estimated overland flow and rainfall intensity at three different vegetation conditions. The overland flow gives an indication of degree of erosion. Rainfall intensity r = mean rain per day; vegetation cover, rc = storage capacity for each day's rainfall. (Note: scrub in this case refers to woody plants probably less than 2 m high.) (From Kirkby, 1980.)

The reduction in cover and biomass of vegetation (Chapter 3) and litter (Chapter 10) in impacted recreation areas clearly exposes the surface of the earth to the impact of raindrops, and increases the potential for erosion. When the cover of vegetation falls below 70%, significant erosion can be expected (Fournier, 1972; Elwell and Stocking, 1976). This suggests that it might be more practical to set the level for comparison of vulnerability of vegetation types at 70% rather than the 50% which the author adopted in 1975 (Liddle, 1975b) (Fig. 3.9). However, Quinn, Morgan and Smith (1980) and Kuss and Morgan (1984) suggested that erosion starts before the destruction of vegetation cover has commenced, so the level of cover set for comparisons must remain an arbitrary figure. The infiltration rates of impacted soils in recreation areas have also been shown to decrease to 3% of the rate in untrampled soil (Cole, 1982) (Chapter 10). This, combined with the loss of vegetation, can only increase the potential erosion hazard.

15.3 Measurement of erosion

There are essentially two approaches to the measurement of erosion. One is to collect the eroding material and, knowing the size of the area from which it has come, calculate the weight of soil lost per unit area. The second involves measurement of the space left behind from which the soil has been eroded.

The first method can be carried out by use of 'Gerlach troughs'. The design and method of use is described in Morgan (1979), and Kuss (1983) provides an example of their use. Measurement of the space left by eroding material is usually carried out on transects across paths, trails or tracks. This involves the placing of two fixed points on either side of the path between which the horizontal girder or cord is suspended (Fig. 15.3a). The distance between the horizontal 'base' line and the ground is then measured at 5 cm or 10 cm intervals, and the profile of the transect may then be drawn (Fig. 15.3b). After a period of time, a year is ideal, the measurements are repeated and the area of the profile that has been eroded can be calculated. Replicate transects can be made along the path and the mean volume of soil lost m^{-2} of path may be calculated.

15.4 Water and wind erosion and recreation impacts

Erosion in areas used for recreation is largely the consequence of changes to the habitat which facilitate rain splash and the amount of surface water flow. The higher the speed of the water flow and the greater the volume, the more sediment can be removed. Paths, trails and tracks provide a ready-made route for gully erosion, and the location of the path may have a major effect on water flow. A steep path will obviously allow a faster rate of water flow than one with a shallow

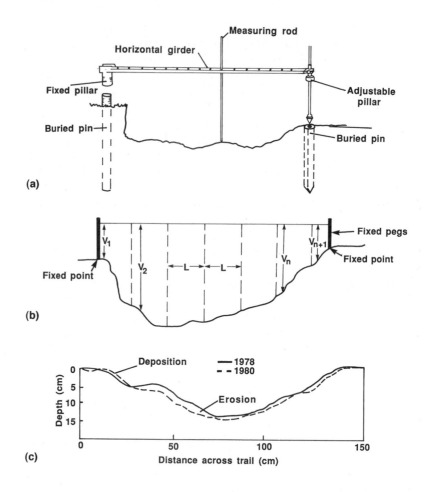

Fig. 15.3 The method of measuring erosion across a path transect (a, b) and the results of such a measurement (c). The cross sectional area:

$$A = \frac{V_1 + 2V_2 \ldots + 2V_n + V_{n+1}}{2} \times L$$

V_1 to $V_n + 1$ = the vertical distance measurements starting at V_1 the first fixed point, and ending at V_{n+1}, the last vertical measurement taken. L = the interval on the horizontal taut line between measurements. (c) The cross-sectional profiles for a path transect taken in 1978 and 1980. There was a net loss (erosion) of 110 cm^2 of material and a small amount of deposition. This, if evenly distributed, equals a loss of 7333 cm^3 m^{-2}. (a, From Streeter, 1975; b and c, from Cole, 1983a.)

slope, and if it is aligned down the slope, then an ideal channel is provided for running water (Fig. 15.4a) (Root and Knapik, 1972), while a trail that is more nearly parallel to the contours will not provide such a suitable channel (Fig. 15.4b). However, where such a trail cuts into the hillside it may intercept subsurface water flow if, for example, the soil is a compacted till (Fig. 15.4c) (Helgath, 1975).

Steepness of the trail was the most important physical factor related to trail condition in the survey by Bratton, Hickler and Graves (1979) in the Great Smokey Mountain National Park. Steepness was positively correlated to the percentage of water erosion, percentage of ruts, percentage of artificially exposed rock and percentage of exposed roots

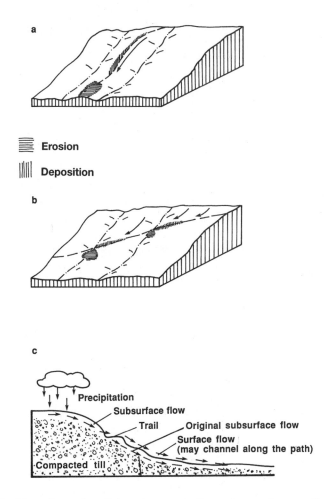

Fig. 15.4 The effect of trail location on trail erosion and water flow (a and b, from Root and Knapik, 1972; c, from Helgath, 1975).

(Table 15.1). The angle of the trail (the difference between the line of the trail and the aspect of the surrounding terrain) had significant negative correlations with all measures of erosion, except bank erosion, which was positive, and the percentage of bare rocks, which was not significant. Here too, trails that were constructed perpendicular to the slope of the surrounding terrain showed the least erosion. The strong positive correlations between the different types of erosion and elevation were attributed by the authors partially to different forest types, but they considered that they may also be a consequence of greater rainfall on the high ridges.

Steepness of trails was also considered to be a major factor in the erosion of tundra trails over permafrost once the increased depth of summer thawing had been initiated (Rickard and Slaughter, 1973). Here again the interaction of the environment with humans is a two-way process, and as the paths become wetter walkers try to avoid the worst parts and the paths become wider and a greater area is exposed to erosive forces (see also Figs 3.6 and 15.10).

In sand environments erosion is largely through wind action on denuded dunes. For example, where a pathway is created in line with the major wind direction it will be rapidly eroded. In coastal situations, this will usually be at right angle to the shore line. Most dune paths are routes between car parks and beaches, and active management (reinforcing or redirection of the path) is usually required if they are not to erode to the level of the water table and leave a passage for flooding at high 'spring' or 'king' tides.

The effects of off-road vehicles (ORVs) on desert soils can be totally destructive and, in extreme areas where there are thin soils on sloping ground overlying a compacted or impermeable layer that prevents free drainage of infiltrating rainwater, the soils may saturate and a slurry of soil and water flows downhill (Hinkley, Iverson and Hallet, 1983).

Table 15.1   Product moment correlations between environmental factors and trail condition. This table shows the relationship of a series of environmental variables to erosion variables. Note that there may be positive correlations to one type of erosion in some cases and negative in another (Bratton, Hickler and Graves, 1979)

Environmental factors	Trail condition factors							
	Mud	Rut	Rock	Roots	Water	Bank	Erosion rating	Computer erosion rating
Slope of the trail	0.016	0.181*	0.183*	0.292*	0.160*	0.036*	0.209*	0.213
Trail angle with slope	−0.052*	−0.082*	−0.027	−0.116*	−0.056*	0.092*	−0.105*	−0.125
Elevation	0.026	0.082*	0.294*	0.300*	0.190*	0.013	0.278*	−0.110*

*Significant at $P > 0.05$.

A conspicuous scar or channel is left and a lake of material is deposited on the valley floor. The same authors presented a diagram which clearly summarizes the effects of off-road vehicles on the surface water behaviour of sloping ground (Fig. 15.5). Two illustrated papers graphically emphasize the role of ORVs in damaging the surface crust of desert soils; the satellite photographs of dust plumes extending for many kilometres from ORV use sites provide all the evidence required to realize the increase in wind and water erosion that is consequent on this recreation activity (Wilshire, 1989; Stebbins, 1990).

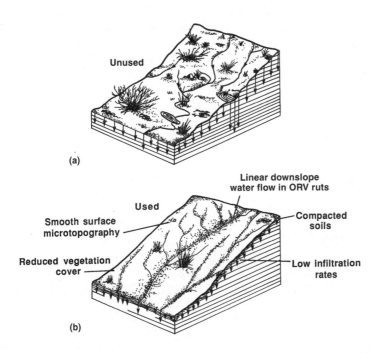

Fig. 15.5 Schematic diagrams illustrating changes in surface water behaviour resulting from off-road vehicle use in arid environments. (a) A typical unused surface characterized by high infiltration capacities, considerable microtopographic roughness (leading to indirect runoff paths and intermittent ponding), and abundant surface stabilizers. (b) A typical vehicle-used surface characterized by disrupted, compacted soil (leading to lowered infiltration capacities), considerable microtopographic smoothness and tread ruts oriented up and down the slope (leading to direct runoff paths and increased sediment transport capacity), and few surface stabilizers. The net effect of these changes is almost inevitably accelerated erosion (Hinkley, Iverson and Hallet, 1983).

15.5 Patterns of erosion in recreation areas

Apart from a general increase in susceptibility to erosion, the pattern of use will have a strong influence on the degree of observable soil movement in a recreation area. Grassy, relatively level areas that are used for picnicking, or other widespread activities, may well be undergoing considerable sheet erosion, which is intensified around centres of interest (Fig. 15.6). But the evidence for this is mainly to be found in the ditches leading water away from the site, where the silt has been deposited. Using an alternative approach, Settergren and Cole (1970) estimated, by a comparison of exposed roots and the relative amount of soil over the roots in the control area, that 'two or three inches (5–7.5 cm) of soil had been removed by sheet erosion' from the recreation areas in the Missouri Ozarks.

Widespread small-scale erosion was recorded in the Cairngorms, Scotland, by using such indicators as loose pebbles, partial or complete burial of plants, soil dislodged beside boot prints, rill erosion, exposed pale undersides of boulders where earth had been removed, absence of lichens, roots exposed beside boot prints and dislodged 'cobbles' (Watson, 1985). Subsequent work has shown that paths in this area have continued to erode and 'parts of a path with more than one

Fig. 15.6 Picnicking groups admiring the views from Box Hill, Surrey, Easter 1971. Sheet erosion is likely to occur on the grassy slope leading to rill and gully erosion at the bottom. Local intense wear around the viewpoint and along the path has exposed the chalk beneath the thin soil where it was undergoing more extreme erosion. (Photograph by M.J. Liddle.)

track were commonly ... focal points for erosion' (Lance, Baugh and Love, 1989). Paths with up to seven tracks were recorded.

Rill or gully erosion can often be observed where use becomes concentrated into linear features such as paths, trails, and tracks, in gateways or around centres of interest. This initially takes the form of removal of litter and then of the surface layers of the soil.

In an experiment carried out on a woodland trail in the Hubbard Brook Experimental Forest, New Hampshire, Kuss (1983, 1986) found that the erosion yield from his trail study plots increased according to the intensity of use (Fig. 15.7a). He also recorded two phases of erosion, one during the wear process, and a second phase after treatment had finished (Fig. 15.7b). This suggests that not only had trampling detached soil particles, particularly organic material from the mor soils, but that the soil surface was now exposed to continuous increased erosion.

More massive movements of the soil can occur when it becomes unstable. Watson (1985) found that on gradients of more than 15°

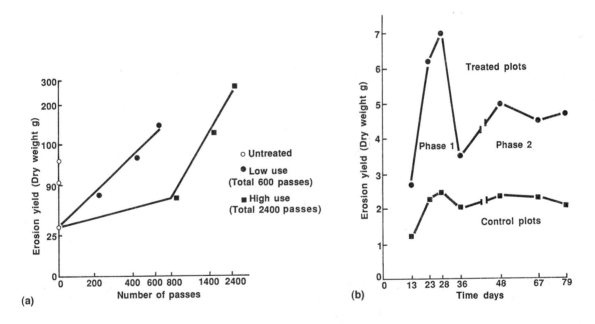

Fig. 15.7 The erosion yield from experimental paths in the Hubbard Brook Experimental Forest, New Hampshire. (a) The erosion yield from untreated plots and a sequential series from those subject to wear (yield extrapolated to kg per mile of path). (b) Erosion yield from trampled and untrampled plots over a period of 39 days when wear was being imposed (phase I), and a post-impact period of 41 days (phase II) (yield in grams per 3.05 m of path). (From Kuss, 1983, 1986.)

his feet slipped increasingly in disturbed areas, but not when walking on nearby undisturbed hillsides. The rate of slipping increased logarithmically with the rise in gradient, until nearly 80% of his foot placements slipped down the slope when it reached 29°. He recalls, 'not only were the disturbed slopes unstable, but it was obvious ... that slipped footsteps caused still further loosening of the soil and vegetation'.

VEGETATION AND EROSION

An extensive study was made of the condition of trails in the Great Smokey Mountains National Park in the Appalachian Mountains by Bratton, Hickler and Graves (1979). They compared various features such as total path width, width of bare ground, depth of erosion and the percentage of the trail's length on which water, mud, exposed rock and roots were present, as well as bank erosion. These factors were also integrated as a 'computer erosion rating', with the highest number being the most affected. All types of erosion were more prevalent in 'spruce fir' forest than other closed-canopy types (Table 15.2). Exposed roots were four times as common, and bare rock was seven times as common as in the 'oak and pine' forests. The degree of erosion was also associated with the successional stage of the vegetation. The most damaged were the virgin or undisturbed forests (erosion rating 3.2), while the early successional types also had a lot of damage (erosion rating 2.8), except that root exposure and bank erosion were nearly absent (Table 15.3). The major understorey types that were most damaged were the herbaceous communities, which had a mean erosion rating of 2.61, while the shrub/sapling and ericaceous types had lower erosion ratings of 2.31 and 2.32.

SOIL TYPE AND EROSION

The width of a pathway depends, among other features, on the vulnerability of the vegetation and soil to damage. In a survey of the Pennine Way, Bayfield and Lloyd (1971) recorded, at approximately every 50 m, the width and condition of the hill path that wound through a range of vegetation types for 435 km and, using a double weighting for bare ground, calculated an index of extent for each type (Fig. 15.8). With the exception of the *Empetrum* bog, the organic soils have the highest index, and are therefore the most vulnerable to damage and erosion. Where a peaty soil closely overlies bedrock, such as on Mount Algonquin in the Adirondack Mountains, New York State, a linear path can cut through the peat and the underlying humus

Table 15.2 The relationship of major forest type to soil erosion, maintained trails only (Bratton, Hickler and Graves 1979)

Forest type	N	Total width (cm)	Tread width (cm)	Tread (%)	Trail depth (cm)	Water (%)	Mud (%)	Rut (%)	Rock (%)	Roots (%)	Bank erosion (%)	Computer erosion rating
Early successional	47	117.0	73.2	63	7.8	14.6	6.6	22.5	16.1	3.5	1.8	3.3
Spruce fir	302	125.1	98.1	78	8.2	16.2	10.9	16.7	39.9	17.3	8.1	3.2
Mesic types	575	152.1	97.8	64	3.9	6.7	5.0	8.4	13.7	9.7	5.7	2.9
Northern hardwood	180	129.4	95.0	73	4.8	10.9	5.7	6.5	10.7	12.0	4.2	2.7
Xeric types (oak and pine forest)	731	132.4	73.0	55	3.3	5.1	3.0	5.7	5.9	3.9	6.4	2.3
Total	1835	136.6	87.4	64	4.6	8.2	8.9	8.9	14.6	8.7	6.1	2.7

Table 15.3 The relationship of successional state to erosion, maintained and unmaintained trails

Stage	N	Total width (cm)	Tread width (cm)	Trail depth (cm)	Mud (%)	Rut (%)	Rock (%)	Roots (%)	Bank erosion (%)	Computer erosion rating
Virgin	304	143.3	107.1	6.7	11.8	12.5	21.4	16.1	7.5	3.2
Mature	485	148.5	101.9	4.2	4.5	9.3	18.4	12.5	5.7	2.7
Successional	1693	176.0	124.6	3.8	5.4	7.3	10.3	4.1	6.6	2.6
Early successional	164	219.4	133.0	5.4	9.3	10.4	13.7	2.5	1.9	2.8

until the bedrock is exposed (Fig. 15.9) (Ketchledge *et al.*, 1985). The exposed edge of the vegetation mat and the soil may then be subject to wind erosion, further increasing the denuded area (N.A. Richards, personal communication).

In many cases trails occur over deep soils which have a greater potential for the development of deep erosion channels. The development of one such erosion gully in the Alberta and British Columbia Rocky Mountains was illustrated by Root and Knapik (1972) (Fig. 15.10). Here the soils were unconsolidated, glacial or alluvial sediments overlain by a thin turf layer. The alluvia consist of particles of silt to fine sand, and the glacial till is mainly clay, sand and boulders. The authors surveyed approximately 50 km of trail, and found that 35% of the trail on alluvial soil was damaged, 32% on glacial till, while damage was apparent on only 6% of that portion of the trail that passed over colluvium, a broken, course and angular bedrock material. There was no apparent damage where the trail passed across naturally exposed bedrock, although surface algae and lichens would have been destroyed.

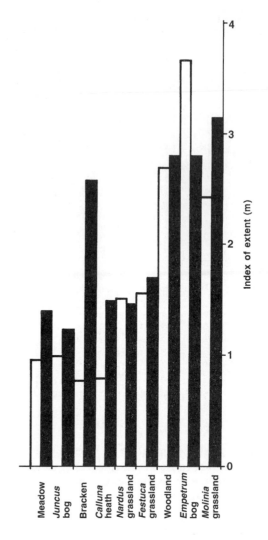

Fig. 15.8   The mean index of extent of the Pennine Way path where it passed through various vegetation types. Open bars, mineral soil; solid bars, organic soil. (From Bayfield and Lloyd, 1971.)

Two contrasting desert soils, one with an A horizon of gravelly or sandy permeable loam, the other with a gravel pavement often forming a gravel mulch, were tested for sediment 'production' under an artificial rainfall at the rate of 3.4 cm h^{-1} for 30 min (Eckert *et al.*, 1979). The loamy soil, which had the highest infiltration rates, produced only 9 kg ha^{-1} of sediment, whereas the more impermeable soil produced 244 kg of sediment ha^{-1} (Table 15.4). When these soils were subjected

(a)

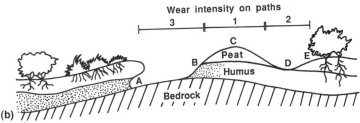

(b)

Fig. 15.9 Path erosion where bedrock has been exposed on Mt. Algonquin, New York State. (a) Exposed bedrock path. (b) Soil profile. 3, Highest; 2, medium; and 1, lowest wear intensity. A, Undercut trail edge; B, trail edge with exposed humus; C, exposed peat; D, humus accumulation in hollow; E, edge of less damaged shrubs or krumholz. (a, Photograph by M.J. Liddle; b, from Ketchledge *et al.*, 1985.)

to the impact of motorcycles (50 passes) and four-wheel drive trucks (20 passes) there was a marked rise in sediment production, with both soils yielding most after the truck treatments, and less after the motorcycle treatments, although the differences on the permeable soil were not statistically significant (Table 15.4). The amount of water held by the soil and the level of the water table can be very important factors in its vulnerability to trampling damage. When the water content is high, the soil can pass from a solid to a liquid condition under pressure (thixotropy). This can be seen on water-filled beach

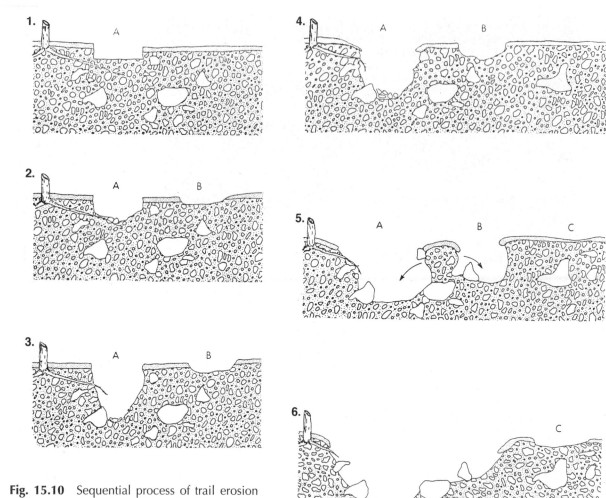

Fig. 15.10 Sequential process of trail erosion and the creation of parallel trails (Root and Knapik, 1972).

Table 15.4 Yield of sediment (kg ha^{-1}) of desert soils in southern Nevada after 3.4 cm h^{-1} of simulated rainfall for 30 min (from Eckert *et al.* (1979)

Treatment	Soil type	
	Permeable	Impermeable
Control	9[a]	244[a]
Motorcycle (50 passes)	36[a]	509[b]
Truck (20 passes)	45[a]	857[c]

Sediment means followed by different letters are significantly different at the 0.05 level of probability as determined by Duncan's multiple range test.

sand where repeated pressure alters the structure so that the continuous phase ceases to be the sand grains in contact with each other, and becomes the water. At this point the sand and water mixture becomes liquid.

Erosion as a consequence of recreation activities is difficult to separate from natural changes in habitats, where the substrate is highly mobile. Yet there is little doubt that walkers and vehicles have a marked effect on the soils in, for example, sand dunes. Apart from the fact that vegetation may be damaged and the natural erosion processes accelerated, considerable quantities of sand may be moved downslope, especially by vehicles. One visitor was estimated to move an average of 269 280 g of sand 1 cm downslope on an overnight visit to the beaches beside the Colorado River in the Grand Canyon (Valentine and Dolan, 1979). Godfrey (1975) estimated that as much as 148 000 cm^3 m^{-2} of sand (this equals a depth of 14.8 cm of surface sand m^{-2}) may be displaced by a four-wheel-drive vehicle travelling across the face of a dune. Compared to the effects of trail erosion in alpine areas, which may take hundreds of years to be obscured, the aesthetic effects of driving on sand dunes that have no surface vegetation are relatively transient. Seely and Hamilton (1978) recorded that the wind had erased all but one of the vehicle tracks on the day they were made where they crossed the upslope part of their sand dune impact experiment in the Namib Desert, South Africa.

Erosion in tundra tracks which have been used by vehicles is rather different from the cases considered above, as it is more the consequence of changed surface conditions permitting a deeper melting of the permafrost layer in the soil, than of exposure of the surface layers to erosive forces. Permafrost landscapes are particularly susceptible to degradation when they consist of frozen silty materials with a high water content, which settle and have low adhesiveness when thawed (Rickard and Slaughter, 1973). These soils occur in most situations north of the Arctic Circle, and discontinuously over most of Alaska, where off-road-vehicle use for recreation is quite common. Rickard and Slaughter (1973) reported that on trails in central Alaska, where the vegetation had been removed 3 years previously, the surface of the permafrost layer had penetrated from a pre-impact depth of 36 cm to 189 cm and their transect figure shows a surface erosion depth of 135 cm.

There have been some comparative experiments in which the effects of different types of use on the amount of bare ground created have been compared. This has a direct effect on the first phase of the erosion process, namely exposure to rainfall energy. Different types of boots generally have similar effects on trail soils, especially on dry surfaces

15.6 The effect of different types of use on erosion rates

(Kuss and Jenkins, 1985). However, these workers did find that a greasy-smeary mineral soil surface was most damaged by boot soles that had a 'mini-check' tread, while soft friable to moist mineral surfaces were most affected by lug soles. There was also a noticeable difference between the effect of shod walkers and those with bare feet, which were much less damaging when walking on beach grass (Nickerson and Thibodeau, 1983) (Fig. 3.17).

When compared to the effect of a walker, horses create between 1.7 and 4.4 times more bare ground and a trail-bike between 1 and 16.6 times more bare ground (Fig. 3.22). In general, the greater the slope the greater the increase in damage caused by these two modes of transport, and this is especially true of the trail-bike in the grassland habitat. The pleasure of utilizing the mechanical power and the satisfaction gained in exercising the skill involved in reaching the top are hard to resist when faced with a steep or muddy slope. The light vehicle is also quite destructive, creating between 6.3 and 9.1 times more bare ground than a walker (Table 15.5). Although horses may appear to do more damage than vehicles because of the undoubted disturbance to the soil surface, they apparently cause less damage to the vegetation cover, at least in these habitats. The first phase of erosion appears to be accelerated more by vehicles than by animal transport.

Table 15.5 Relative amount of bare ground created by different recreation uses, compared to the amount created by a walker. Based on comparisons at 50% bare ground. All slopes were at 15°

Walker	Trail-bike
All habitats	Grassland[a]
(Comparative standard)	1.0 level ground
1	16.6 slope uphill
	4.4 slope downhill
Horses	Forest[a]
	4.0 level ground
Grassland[a]	2.3 sloping ground[c]
1.7 level ground	
4.0 slope uphill	**Light (760 kg) vehicle**
2.0 slope downhill	Level sand dune grassland[b]
Forest[a]	6.3 winter
4.0 level ground	9.1 summer
4.4 sloping ground[c]	

[a] Data of Weaver and Dale (1978).
[b] Data of Liddle (1973b).
[c] Comparisons made at 75% bare ground as graph not distinct at 50%.
All habitats standardized. See Figs. 3.19–3.22 for comparisons of habitats and use.

In an extensive experiment, four-tracked vehicles of different weights were driven over the tundra in the Mackenzie Delta region of arctic Canada (Bellamy, Radforth and Radforth, 1971). Utilizing these data, Liddle (1973b) showed that there was a positive correlation between the vehicle weight and the amount of terrain 'damage', a measure that involved vegetation cover as well as 'destruction of the secondary terrain structure' and rut formation. Intuitive knowledge supported by these data is, I think, sufficient to suggest that, in general, the heavier the vehicle, the greater the erosion initiated. Trail-bikes are an exception as they may create more erosion per unit weight, and soft, wide-tyred vehicles may cause less erosion (cf. Bellamy, Radforth and Radforth, 1971).

The only comparative data the author has located on the effect of different uses specifically on the later phases of erosion are those of Eckert *et al.* (1979). They demonstrated that, taking the number of passages into account, a 750 kg pickup truck caused between 3.1 and 4.3 times more erosion of desert soils than was caused by a 159 kg trail-bike (calculated from data in Table 15.4).

Given that increased runoff is one of the causes of the later phases of erosion, and that there is a link between the penetration resistance and infiltration rate, the vertical hardness profiles of a walking track and a car track after 256 passages showed clearly that the car created a harder layer, especially in the upper 2 cm, and therefore caused a greater potential for erosion (Fig. 15.11). The bulk density measures also support this conclusion (Fig. 11.1). However, the condition of the surface layers will also depend upon the amount of lateral force applied by the vehicle (Fig. 11.7) and there may be wide variations in the effects of different vehicles, or even the same vehicle with different drivers.

Weaver and Dale (1978) published data on the increases of trail depth caused by hikers, motorcycles and horses, which they attributed to both soil compaction and erosion (Fig. 15.12). They came to four clearly stated conclusions:

(i)   trail depth increased with use up to at least 1000 passages;
(ii) trail depth tends to be greater on slopes than on level sites;
(iii) trail depth tends to be greater in a stone free meadow soil than a stony forest soil, at least for hikers and cycles; and
(iv) trail depths were greatest under horse use and least under hiker use at all sites.

Weaver and Dale's (1978) results showed that although motorcycles generally created more bare ground than horses, they caused rather less erosion and they certainly caused less soil compaction in the same experiment.

It seems likely that the second phase of erosion is also strongly linked with the ground pressure exerted by the different means of transport (Table 2.1). At least it is clear that in the habitats used by Weaver and Dale (1978), after bare ground had been created, walkers caused the least erosion, followed by trail-bikes and then by horses, which do the most damage. The erosive effect of four-wheel-drive vehicles probably depends upon their weight and ground pressure, but initially they would appear to cause more damage than four walkers (the maximum number of passengers normally carried).

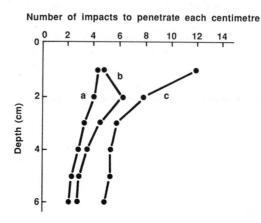

Fig. 15.11 Penetration resistance of the top 6 cm of sand dune soils at Aberffraw, Anglesey, North Wales. (a) Uncompacted; (b) after 1024 passages by walkers; (c) after 256 passages in a car. All points at the same depths are statistically different. (Modified from Liddle and Greig-Smith, 1975a.)

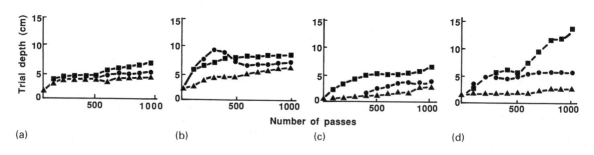

Fig. 15.12 The effect of the number of passes of hikers (▲), motorcycles (●) and horses (■) on trail depth in grassland and forest sites. (a) Flat grassland; (b) sloping grassland; (c) flat forest floor; (d) sloping forest floor. The effect of walkers was statistically different from that of horses by 1000 passes at all sites. (From Weaver and Dale, 1978.)

Digging may also be a part of recreation activities in searching for 'buried treasure' with metal detectors, as a water deflection trench around a tent in rainy weather or as a pleasure in itself on beaches and beside streams. In most cases this is a minor effect and it does not seem to have been recorded by quantitative techniques.

It is clear that erosion has only been considered a problem worthy of mensuration in certain habitats, particularly alpine, tundra, sand dune, woodlands, deserts and chalk grassland. The greatest volumes lost per year appear to be where trails have been created on permafrost tundra (Table 15.6). Alpine trails are perhaps the next most vulnerable, especially where they are subject to high levels of use by large numbers of visitors. The volume of sand moved by vehicles may be very high in sand dune areas, but as the sand is often mobile the loss in vegetated areas of the dunes is probably more important. This is especially true of vegetated fore dunes in coastal systems which form a barrier against the sea. Removal of the beach grasses (*Ammophila breviligulata*), marram grass (*A. arenaria*) or spinifex (*Spinifex hirsutus*) may initiate a serious 'blow-out' where wind erosion can make a considerable breach in the sea defences. Gilbertson (1983) suggested that ORV (off-road vehicle) use might have accelerated dune advance in the Coorong Dune and Lake Complex, South Australia, but he pointed out that the dune movement is also associated with earlier clearance and grazing which occurred around the old homesteads, and that the ORVs may just have taken advantage of an existing, previously abandoned, track system. The same author gave an interesting presentation of the interaction between the users and the environment, showing how the natural human desire for isolation and adventure may lead to inhibition of natural dune stabilization processes, a reduction in relative relief and accelerated erosion (Fig. 15.13).

The author knows of no measurements of the volume of sand lost by recreation-induced blow-outs but I would expect it to be much greater than the volumes recorded in Table 15.6. The volume of chalk soil eroded on Box Hill is at an intermediate level, but the aesthetic problems are very great as the site has many thousands of visitors each year.

Erosion in desert habitats where surface disturbance may endure for several hundreds of years is very low in terms of the volume displaced per year. This is primarily because of the very low rainfall that occurs in these areas. The magnitude of erosion following or accompanying recreation activities appears to depend upon an interaction between the level of impact, the softness or vulnerability of the substrate and the magnitude of the erosive forces of rainfall and subsequent water flow or wind strength. As discussed below, the volume of soil displaced

15.7 Comparison of erosion in various habitats

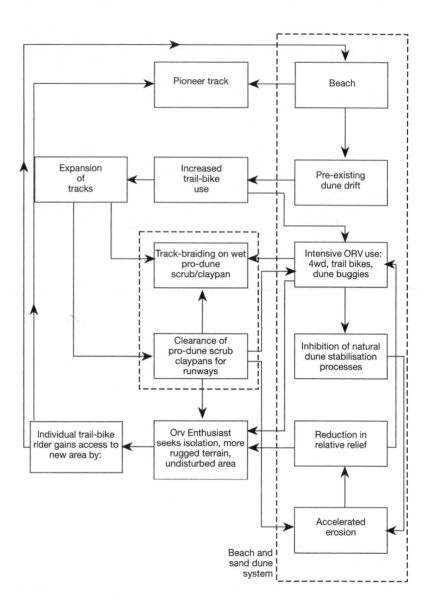

Fig. 15.13 A model of the interaction between off-road vehicle (ORV) users, and between ORV users and geomorphic change in the lower Coorong Dune and Lake Complex, South Australia. Note that the possible outcomes from the individual bike-rider gaining access to a new area (bottom left box) are a pioneer track, expansion of tracks or riding on the beach. (From Gilbertson, 1983.)

Table 15.6 Comparative volumes of soil eroded from various habitats

Habitat	Local feature	Cross-section area (cm²)	Estimated volume (cm³ m⁻² yr⁻¹)	Source of original data
Alpine 1200–1800 m	Trails	-110 $+97$	-3666 $+3233$	Cole (1983a)
Alpine Adirondack 900 m (1970)	Trail	$-$	-25000	Ketchledge and Leonard (1970)
Box Hill, England Chalk grassland	Path	$-$	-7058	(8½ months' measurement extrapolated to 1 yr) Streeter (1975)
Arctic Alaska	Bulldozer trail on permafrost areas	$-$	$-450\,000$	Rickard and Slaughter (1973)
	Hand-cleared trail avoiding permafrost	$-$	$250\,000$	
Cape Cod sand dune	4WD 100 passages New routes			Godfrey (1975)
	cross dune	$-$	$-148\,000$ after	
	parallel to contours	$-$	$-89\,000$ treatment	
	diagonal to contours	$-$	$-62\,000$ on 1 day[a]	Eckert et al. (1979)
Nevada Desert[b]	Motorcycle 50 passages	Mean two soil types	-31	
	4WD truck 20 passages	Assumed BD 1.3	-59	

[a] Volume cm³m⁻².
[b] After only 30 min of rain at 3.4 cm hr⁻¹. This occurs about once every 4 years.

from the processes consequent on the recreation activity should not be taken as the only indicator of environmental or aesthetic damage.

15.8 Experimental erosion

The most convenient way to establish quantitative relationships between the amount of use for recreation and changes in the environment has been by experiment. The study of erosion has also been pursued by this means.

Soil loss was shown to be related in a linear fashion to the number of tramples in an experiment on planted turfs with 45% soil moisture, placed in a series of increasing steepness (at angles 5° to 30°) (Quinn, Morgan and Smith, 1980) (Fig. 15.14a). The relationship between slope angle and soil loss was curvilinear, with rapidly increasing soil losses at the steeper angles to 20° (Fig. 15.14b).

At the 30° gradient the soil loss was less than at 20° and the authors suggested that this may be due to a 'flattening out of the shear force component of trampling during the transition from walking to climbing'. While it is not realistic to compare quantitative amounts of soil

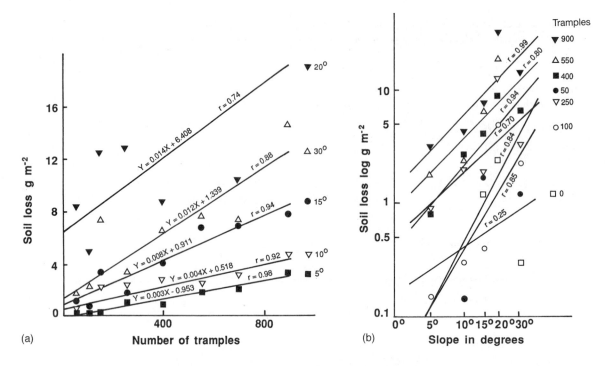

Fig. 15.14 The relationship (a) between soil loss and number of tramples at different slope angles and (b) between soil loss and slope angle at different numbers of tramples. (From Quinn, Morgan and Smith, 1980.)

erosion in this experiment with natural situations, the general trend of increasing erosion with increase in slope has been found in the field (Bratton, Hickler and Graves, 1979) (Table 15.1). Perhaps the most important point to note from this experiment is that 'patches of bare ground did not appear until after 250 tramples, by which time much of the grass had become smeared with a layer of soil as the grass blades were forced into the soil mass and the soil was displaced downslope by the shearing action of the toe' (Quinn, Morgan and Smith, 1980), and, most importantly, erosion was already well under way, especially at the steeper angle where, for example, over 8 g m⁻² of soil was lost from the plots at a 20° slope. The authors also pointed out that 'under the wet conditions of the second experiment most wear results from soil deformation and smearing, and is thus related to shearing forces associated with the action of the toe in walking rather than compaction forces associated with impact of the heel'. They also stated that 'the loosening and breakdown of the soil occurred in the early stages of trampling whilst vegetation cover is still dense, and before there is any visual evidence of vegetation wear, suggesting that the protective effect of vegetation is less than generally expected. By the time vegetation wear is noticed a critical period in which accelerated erosion is initiated has already passed, and the land is already adjusting to the new condition of greater trampling and increased soil loss'.

The two points, that shearing action is more important than compaction and that erosion starts before vegetation cover is lost, have important implications for our understanding of the trampling process and need further verification, especially under dry conditions on relatively level ground which normally pertain in many recreation areas.

Trampling experiments, in which the effects of different footwear on erosion were compared, were carried out in various habitats. In general, there appeared to be little difference between the different soil types when hiking in boots, but trampling in tennis shoes caused less erosion than trampling in boots (Saunders, Howerd and Stanley-Saunders, 1980; Kuss, 1983).

The natural and artificial factors that contribute to various degrees of erosion are indicated in Fig. 15.15.

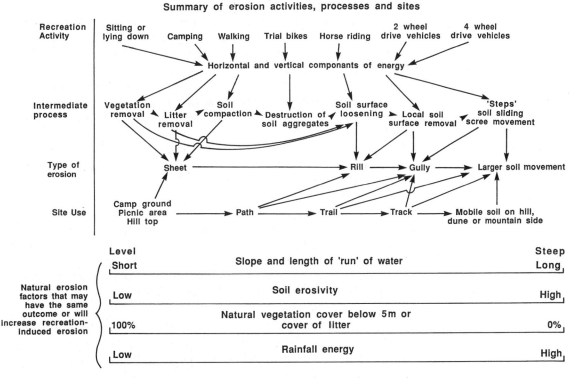

Fig. 15.15 Summary of erosion activities, processes and sites, showing the relationship between recreation activities, erosion factors and erosion processes.

15.9 Mathematical analysis and modelling of erosion

The ultimate aim of much recreation research is to establish quantitative data on which predictions of the nature and extent of environmental change, as a result of a known amount of use, can be based. Scientists in the USA have utilized a formal relationship between various environmental factors and soil loss, known as the universal soil loss equation (USLE), to predict erosion successfully under a wide range of conditions. This is defined by Mitchell and Baubenzer (1980) as:

$$A = (0.224)RKLSCP$$

where:

A = the soil loss in $kg\ m^{-2}\ sec^{-1}$
R = the rainfall erosivity factor
K = the soil erodibility factor
L = the slope length factor
S = the slope gradient factor
C = the cropping management factor (vegetation cover % in recreation areas)
P = the erosion control practice factor (not usually applied in recreation areas).

Rainfall erosivity depends essentially on the amount of kinetic energy in a rainfall event over a particular period of time. Soil erodibility is a factor that depends upon soil properties that affect infiltration rate, permeability, total water capacity, dispersion, splash abrasion and forces required for transporting soil particles. The general magnitude of K for various texture classes is given in Table 15.7. A detailed discussion of the other factors is given in Mitchell and Baubenzer (1980).

Kuss and Morgan (1980) suggested that the USLE may be applied to recreation areas for the calculation of the potential for sheet or rill erosion, a phenomenon which may be more important than previously considered if the suggestions of Quinn, Morgan and Smith (1980) about erosion occurring before damage to the vegetation has been observed are correct. Kuss and Morgan (1980) suggested that a predictive equation could be used as an estimator of physical carrying capacity. They propose that P, the erosion control factor, is generally 1 for recreation areas and may be eliminated from the equation. The C factors for forested areas have been calculated (Patric and Brink, 1977).

In order to solve the equation for C, the minimum vegetation cover that will maintain the soil loss at or below the level that is acceptable, the equation is rearranged as follows:

$$C = \frac{T}{RKLS}$$

Table 15.7 Indications of the general magnitude of the soil erodibility factor, K^a (Mitchell and Baubenzer, 1980)

Texture class	Organic matter content		
	< 0.5% K	2% K	4% K
Sand	0.05	0.03	0.02
Fine sand	0.16	0.14	0.10
Very fine sand	0.42	0.36	0.28
Loamy sand	0.12	0.10	0.08
Loamy fine sand	0.24	0.20	0.16
Loamy very fine sand	0.44	0.38	0.30
Sandy loam	0.27	0.24	0.19
Fine sandy loam	0.35	0.30	0.24
Very fine sandy loam	0.47	0.41	0.33
Loam	0.38	0.34	0.29
Silt loam	0.48	0.42	0.33
Silt	0.60	0.52	0.42
Sandy clay loam	0.27	0.25	0.21
Clay loam	0.28	0.25	0.21
Silty clay loam	0.37	0.32	0.26
Sandy clay	0.14	0.13	0.12
Silty clay	0.25	0.23	0.19
Clay		0.13–0.29	

[a]The values shown are estimated averages of broad ranges of specific soil values. When a texture is near the borderline of two texture classes, use the average of the two K values. For specific soils, use of Fig. 11.9 or Soil Conservation Service K-value tables will provide much greater accuracy.

where T is the soil loss tolerance factor, which represents the maximum rate of soil erosion that will permit the 'productivity of the land to be sustained economically and indefinitely'. T is substituted for A in the USLE; P is omitted at this stage. Kuss and Morgan (1980) give examples of solutions to their equations for three picnic sites in the Patapsco Valley State Park, Maryland (Table 15.8).

They demonstrated that the physical carrying capacity varied in the different areas they studied, with site one being the most sensitive and site three the least sensitive. The cover factor (last column in Table 15.8) is converted to a percentage vegetation cover required to prevent soil loss from exceeding the soil loss tolerance (Table 15.9), by use of tables published by Wischmeier and Smith (1978). The cover factor is dependent on the soil loss tolerance factor (T) which is an arbitrarily decided acceptable erosion loss in kg ha^{-1} yr^{-1}. Thus very little loss of vegetation ground cover may occur at site one (5–14%) before erosion losses become unacceptable, whereas at site three the ground cover in open areas may be as low as 59% before the acceptable soil loss is exceeded.

Kuss and Morgan (1984) calculated the soil losses that they predicted would occur at various levels of vegetation cover for a range of situations under specified vegetation types. An example is shown in Fig. 15.16. It is evident that erosion is occurring even at 100% ground cover on all except the most resistant soils; however, given an acceptable erosion loss (T value) of 826–1652 kg ha^{-1} yr^{-1}, the acceptable loss of vegetation cover ranges from nearly 100% (soil loss index of 10.5 or less) to 10% (soil loss index of 230 or less) at 826 kg ha^{-1} yr^{-1}.

Kuss and Morgan (1986) further developed the application of their theoretical base by designating ground cover values as 'low' where 80% or more ground cover is required to keep the annual soil loss below the designated maximum, 60–80% cover as moderately resistant, and below 60% as having a high resistance or high recreation carrying capacity. They applied these figures to seven sites in the Presidential Management Unit of the White Mountain National Forest, New Hampshire (Table 15.10). As this table indicates, grassy (graminoid) ground cover has a higher resistance than woodland or forested herbaceous cover. In a more extended survey they mapped the areal distribution of the carrying capacity (or erosion resistance) in the same park (Fig. 15.17). In this case the high carrying capacity areas ranged from soils on foothills, outwash plains and terraces to those on lower side slopes and flood plains (Kuss and Morgan, 1986). They suggested that such a map could then be used in conjunction with other ecological, aesthetic and social information as a basis for broad planning decisions. These would be followed by small-scale site-specific investigations for detailed planning.

Table 15.8 Soil index and calculated C factor values for three picnic sites in the Maryland Piedmont (Kuss and Morgan, 1980)

Site	Soil series	R	K[a]	Lt (ft)	St (%)	LS	Soil loss index (RKLS)	T	Calculated C factor (T/RKLS)
1	Manor	175	0.43	300	12	3.13	235.53	2	0.009
2	Montalto	175	0.28	200	14	3.25	159.25	3	0.019
3	Glenelg	175	0.43	300	4	0.62	46.66	2	0.043

[a] Because the soil types for the three picnic sites are moderately eroded, the K factors are for the B horizon.
The L and S factors for the three picnic sites were determined by field observation. The remainder of the data in the table is from Stephens (1978).

Table 15.9 Percentage ground cover needed to maintain the calculated *C* factor on three Maryland picnic sites, given three vegetative cover conditions (calculated from Wischmeier and Smith, 1978)

Vegetative cover condition	Site	Percentage ground cover needed to Maintain calculated C factor
No appreciable canopy:	1	86
grass ground cover,	2	76
herbaceous ground cover	3	59
Appreciable brush:	1	95
50% canopy cover,	2	91
herbaceous ground cover	3	79
Trees but no appreciable brush:	1	95
50% canopy cover,	2	91
herbaceous ground cover	3	80

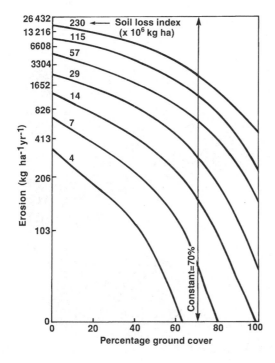

Fig. 15.16 The relationship between erosion yield and ground cover (*C* factor) with a range of values from 0 to 100% and *RKLS* values of 230, 115, 57, 29, 14, 7 and 4 (× 10⁶ kg ha). The line for 70% cover is indicated. This example was calculated on the basis of an undisturbed forest with 70% canopy, soil reconsolidation factor 0.45, depression 0.03, 50% fine roots and litter 5 cm deep, factor 0.07. (From Kuss and Morgan, 1984.)

Table 15.10 Percentage of cover required to maintain soil productivity in the face of recreation use of seven potential primitive campsites (50% canopy cover, 4 m drop height) (Morgan and Kuss, 1986)

Site	% Cover graminoid	Carrying capacity	% Cover herbaceous	Carrying capacity
Salmon, very fine sandy loam	48	High	68	Moderate
Hermon, very stony fine sandy loam association	40	High	56	High
Marlowe-Peru, very stony fine sandy loam association	45	High	66	Moderate
Canaan-Redstowe, very rocky gravelly fine sandy loam association	34	High	50	High
Peru, very stony fine sandy loam association	65	Moderate	84	Low
Adams, loamy sand	12	High	16	High
Nicholville, silt loam	53	High	74	Moderate

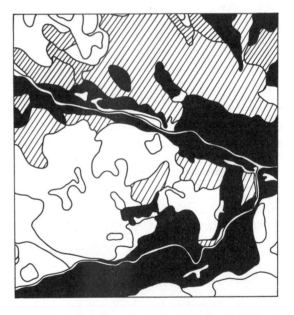

Fig. 15.17 Map of a 20.4 km² area of the Presidential Planning Unit of the White Mountain National Forest, New Hampshire, showing the distribution of soil mapping units and carrying capacity classes. Clear areas are the most vulnerable, hatched areas intermediate and the black area is the most resistant. Sides of the square are approximately 4.5 km. (From Kuss and Morgan, 1986.)

Morgan and Kuss (1986) also suggested that their scheme may be used as a basis for monitoring by calculating a 'coefficient of condition':

$$\frac{\%\ \text{cover observed}}{\begin{array}{c}\%\ \text{cover calculated (as a minimum}\\ \text{to maintain good condition)}\end{array}} = 1\ \text{coefficient of condition}$$

The cover loss factor requires a series of measurements to assess change over time and to utilize this as a prediction of the effects of a given amount of use. This may then be used as a basis for appropriate management action, such as closure or increasing permitted use levels.

A criticism of the USLE by Morgan (1985) is that it does not take account of the impact of recreation on species composition (Chapter 4) or changes in soil properties that affect erodibility, for example the changes in soil aggregates (section 11.2). A new generation of models has been developed in which soil erosion is viewed as a two-phase process of detachment and transport (Morgan, 1985).

Morgan (1985) constructed a model based on this principle and a flow chart of the model is shown in Fig. 15.18. The main advantage appears to be the fine tuning that is possible with all of the factors as they change in response to recreation impacts, and the consequent changes in the dynamic relationships between the factors. One obvious omission at this stage is wind erosion, which is not considered in any of the models, and in some situations this could be an important factor (N.A. Richards, personal communication, 1985).

The model was 'run' using data from various sources, but mainly the work of Weaver, Dale and Hartley (1979) where the impact of hikers, horses and motorcycles was compared. The results of the simu-

Table 15.11 Results of simulated soil erosion rates and changes in soil depth based on the experiments of Weaver, Dale and Hartley (1979) (Morgan, 1985)

	Hikers	Horses	Motorcycles
Estimated soil loss from field observations (kg m^{-2} yr^{-1})	0.219	0.624	0.386
Predicted soil loss in second Year (kg m^{-2} yr^{-1})	0.106	1.057	1.340
Year when topsoil is completely removed	9	6	6
Year when total soil is completely removed	17	13	13

See also Figs 15.12 and 15.18.

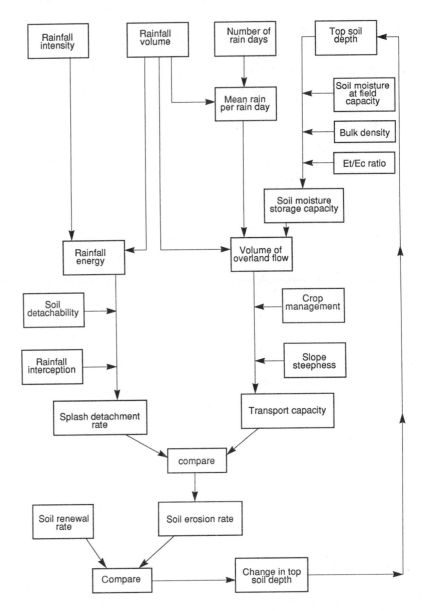

Fig. 15.18 Flow chart of a soil erosion model developed by R.P.C. Morgan which was used to calculate the values given in Table 15.11 (Morgan, 1985).

lation are given in Table 15.11. As Morgan (1985) commented, the model simulations agree with field observations that horses and motor-cycles are more destructive than hikers; however, the model also suggests that motorcycles are more destructive than horses, whereas the reverse is the case. Also, the model assumes that there is no deposition on the pathways but in practice deposition is often recorded (Chapter 16), and the author points out the need for validation of the models. However, the model very clearly illustrates the magnitude of the destabilizing processes set in train by recreation activities: the end points of complete removal of topsoil, or of the total soil profile, are frighteningly near, only 17 years for hikers and less for the other activities.

A rather simpler model was developed by Garland (1990) in which rainfall, as the annual sum of all rain falling on the wettest day of each month (MDR index), path slope and soil type were combined to produce a total score, ranging from 3 to 13 (Tables 15.12, 15.13). This is then interpreted as four erosion risk classes (Table 15.14). Garland (1990) then interpreted the erosion risk classes as follows:

- Class 1: low-risk areas, where extensive path maintenance is not required.
- Class 2: medium-risk areas, where a few sites may need mainte-nance.
- Class 3: high-risk areas, where several sites must be regularly main-tained.
- Class 4: very high-risk areas, where most path sites will require regular and frequent maintenance.

This scheme has the advantage of relative simplicity but obviously lacks the accuracy of the USLE approach.

Overall the models do have a predictive value, and the ideas should also be followed further as they increase our understanding of the very complex process of soil erosion.

15.10 Summary

Recreation-induced erosion of soils is a small effect when compared to natural forces but has considerable aesthetic importance in recre-ation areas. Natural erosion may be sheet erosion by overland water flow, rill or gully erosion, subsurface water flow, wind erosion or mass flow. Erodibility of soils depends upon the soil texture, aggregate stability shear strength, infiltration capacity and chemical content, all factors that may be changed by recreation use. Vegetation cover is a primary factor in reducing erosion.

Removal of vegetation directly increased erosion, and the degree of damage may be affected by vegetation type. In one study, virgin forests were most vulnerable, early successional types next and shrub/sapling

Table 15.12   Scoring of erosion risk parameters (Garland, 1990)

Score	MDR index (mm)	Lithology	Topographic slope (degrees)
1	< 300	Basalt/alluvium	< 6
2	300–400	Sediments/dolerite	6–10
3	401–500	Sediments/dolerite	11–15
4	501–600	Sediments/dolerite	16–20
5	> 600	Sediments/dolerite	21–25
6			> 25

Lithology scores 1 or 2 only.

Table 15.13   Parameter scores and erosion risk classes

Possible combined MDR and lithology scores	Slope scores					
	1	2	3	4	5	6
2 Class 1	3	4	5	6	7	8
3	4	5	6	7	8	9
4 Class 2	5	6	7	8	9	10
5	6	7	8	9	10	11
6	7	8	9	10	11	12
7	8	9	10	11	12	13

Class 3   Class 4

Table 15.14   Main characteristics of land in each erosion risk class

Class	Total score	Main characteristics
1 Low risk	< 4	Sediments or dolerite only, sloping at < 5°; MDR < 300 mm
2 Medium risk	4–6	Any rock type; MDR < 600 mm *and* slopes < 20°
3 High risk	7–12	Any rock type; MDR probably > 400 mm and/or slopes probably > 10°
4 Very high risk	> 12	Any rock type; MDR > 500 mm

types and ericaceous types least vulnerable. The soil type will affect the width of paths, as will the amount of use. Organic soils are the most vulnerable and deep soils have a higher potential for deep erosion gullies. Soils with high infiltration rates are likely to produce less sediment in the drainage water than those with low infiltration rates. On sandy and loess soils walking or driving may directly cause the downhill movement of the substrata, and tundra soils erode when their surface conditions are changed to allow a deeper melting of the permafrost layers.

Different types of use have different impacts on vegetation and soils. Horses created more bare ground than trail-bikes and over 16 times more than walkers. More damage is also caused on steeper slopes. In general, heavier vehicles cause more erosion than lighter vehicles, and the amount of erosion caused by vehicles also depends upon the way they are driven.

Paths provide a route for surface flow and are consequently vulnerable to rill and gully erosion. They may also intercept overland and subsurface flow and in this way increase the volume of water flowing down the path. Steepness of the path will also affect the rate of water flow and hence the energy available to pick up particles of soil and carry them away. In sand dune areas recreation-induced erosion is usually by wind after vegetation has been destroyed.

Erosion may be measured by collecting the material eroded or by measuring the space left behind. In both cases it is usually expressed as volume or weight of material removed from a given area of land. The time that erosion scars may endure is very variable, being less than a day on sand dunes and hundreds of years in deserts and alpine areas.

Experiments have shown that by the time vegetation wear becomes visible, erosion is well under way and the land is already adjusting to conditions of greater soil loss. The same experiments also showed that a shearing action could be more important than compaction as an erosive force.

The universal soil loss equation has been applied to recreation studies and used to predict vulnerability of certain areas. The high-capacity areas were foothills, outwash plains, terraces and flood plains. A new generation of models has been developed in which the processes of detachment and transport are considered separately.

The models do have a predictive value and increase our understanding of the very complex process of soil erosion.

16 Soil deposition and recovery

Erosion of soils and the underlying rocks is only one aspect of the movements of the solid materials of the earth's crust. The material removed from one place must eventually be deposited somewhere else. A part of the soil that is eroded as a suspension in moving water will eventually be deposited in estuaries and on the sea floor. But the remainder may come to a temporary rest much nearer the source and, indeed, many path transects show deposition as well as erosion (Figs 15.3c, 15.4).

16.1 Distance of particle movement

In a survey of major trails and roads in the Guadalupe Mountain National Park, Texas, Fish, Brothers and Lewis (1981) found that deposition had occurred at 47% of the measurement points while erosion was recorded at the remaining 53% of the points. The same survey revealed that the cross-sectional area of deposition over a period of 8 months was 6043 cm^2, whereas erosion was recorded as only 5559 cm^2. There was a slightly greater cross-sectional area (484 cm^2) of deposition than of erosion. These data indicate that in this case the paths were themselves acting as sediment traps as well as sites of erosion. Wind-borne particles of grit were deposited up to 100 m from their source when blown on to the icy snow in the Cairngorm Mountains, Scotland (Watson, 1985), but wind-blown grit normally lay in a 1 m band on the windward side of vegetation clumps from which it was eroded, and up to 5 m to the lee of big patches of grit. The same author also recorded the wind erosion of soil organic material at this site. However, human-induced erosion deposited material in less scattered positions than the natural erosion described above. Wind-blown grit was moved from flat terraces to pile up on vegetation below the steps, from whence water erosion was observed to remove some material during heavy rain.

Deposition occurred in 10 out of 30 ski grounds on Mt. Teine, Hokkaido Island, Japan (Tsuyuzaki, 1990) and in four out of six of the adjacent forest areas. In the forest the mean accumulation was +3.7 cm and +1.5 cm in summer and winter, respectively, while the erosion rates were −0.4 cm and −1.1 cm, respectively. In the ski areas the deposition was +0.4 cm in summer and +0.5 cm in winter, and the losses were −5.3 cm and −0.7 cm, respectively. Apparently deposition was much less than erosion in the ski grounds and greater in the forests.

Except in the gullies of the Cairngorms, buried vegetation and top soil lay underneath loose stones and grit 'often so deep that plants had died and the value of the top soil for vegetation recolonisation had been lost' (Watson, 1985). During 1968 extensive areas of erosion appeared in the Cairngorms and at one site deposition covered an area of 2.8 ha (Bayfield, 1974). This stimulated extensive reseeding of the area and an experiment to examine the reaction of the native vegetation to burial.

16.2 Effect of deposition on vegetation

Observation showed that plant cover was reduced in proportion to the depth of sediment, with nothing surviving under 7 cm or more of sediment (Fig. 16.1) (Bayfield, 1974). Two types of natural vegetation were covered with sediment to depths of 1.25 cm, 2.5 cm, 5 cm and 10 cm. Records were made at yearly intervals for the next 2 years. The effect on the plant cover depended partly upon its form. Mosses and lichens tended to be buried by the sediment depths greater than 1.25 cm. Most species were largely covered by 2.5 cm or more of sediment. Recovery from burial was, however, fairly rapid when sediment depths were 2.5 cm or less, but 5 cm of sediment allowed regeneration of only 6% cover of vegetation on the *Rhacomitrium* heath after 2 years of growth (Fig. 16.2a). The more robust species showed etiolation growth from under the sediment, with stiff sedge (*Carex bigelowii*) increasing its stem length by up to 4 cm under the deeper sediment (Fig. 16.2c). Bayfield (1974) commented that the environment under a deposition of sediment is greatly modified, being wetter at the base and drier and more exposed on the surface than the normal growing conditions experienced by the plant. Both of these factors, as well as loss of light, will influence the rate and nature of recolonization, which may be extremely slow.

The vegetation growing on soils deposited from eroded ski slopes near Kazan, Tatarstan, in Gorky Park and Lebyazhe Forestry Preserve was quite different from that growing on eroded and undisturbed areas (Baiderin, 1978). Plants with long roots and perennating buds, in this case tillering nodes, below the soil surface (geophytes) were dominant on the deposited soils, whereas 'thick bushed' plants (hemicryptophytes) and annuals were present on the washout areas. The species particularly mentioned on deposited soils were vernal sedge (*Carex caryophyllea*) and awnless brome (*Zerna inermis*) and on washout areas feather grass (*Stipa capillata*) and other annuals.

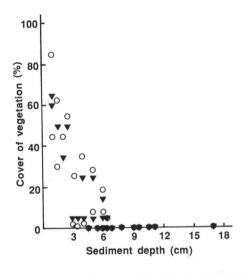

Fig. 16.1 The effect of sediment depth on plant cover of *Empetrum–Rhacomitrium* heath. ○, visual estimates and ▼, point quadrat estimates of cover. (From Bayfield, 1974.)

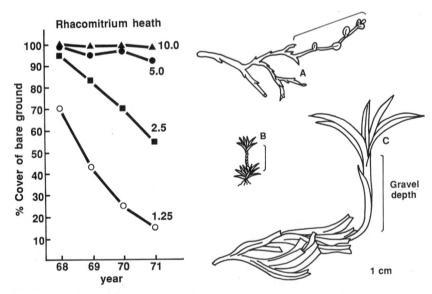

Fig. 16.2 (a) Changes in the cover of bare ground following burial and growth of three species through erosion gravel. Gravel cover: ○, 1.25 cm, ■, 2.5 cm, ●, 5.0 cm, ▲, 10 cm. (b) A, *Vaccinium myrtillus*; B, *Polytrichum juniperinum*; C, *Carex bigelowii*. Parts bracketed are etoliation growth. Scale bars = 1 cm. (From Bayfield, 1974.)

Deposited material is by no means stable and may itself be subject to subsequent erosion. The 10 cm thick layer used in Bayfield's (1974) burial experiment reduced in thickness between 5 and 25% during 3 years of observation. Like erosion, deposition is a natural process which may be greatly accelerated by use of the land for recreation, and the increase may have considerable effects on the local communities of organisms.

Just as the regeneration of vegetation is very variable in time, and the degree to which it may return to its original condition, so too is the regeneration of soils. The formation of topsoil in desert conditions where the surface has been eroded takes from 300 to 500 years per 2.5 cm, and, of course, revegetation depends upon the condition of the soil (Webb *et al.*, 1978).

16.3 Regeneration of soils

Because of the time factor involved and the extreme visual scarring from ORV use, reformation of desert soils appears to have been studied more than regeneration of soils in any other habitat.

Two sites have received particular attention, military camps and manoeuvre areas from the Second World War (40 years after their closure) and desert ghost towns between 51 and 75 years after they had been abandoned. The soils in 'medium' tank tracks were, on average, 50% harder than those under adjacent vegetation, while the camp roads had a 285% higher penetration reading than the adjacent areas and only 36% of the attempted penetration measurements reached below 5 cm (Prose, 1985) (Fig. 16.3). Excavation of a tank track showed that the subsurface sand lenses had been deformed into a concave figuration that extends laterally at least 50 cm from the track edges. Interestingly bulk density measurements for the top 10 cm of soil only showed variations of −5 to 12% compared to adjacent areas. Clearly, in this situation penetrability is a much more sensitive measure and more relevant to the vegetation response (Chapter 8).

The studies of soil recovery in abandoned towns, also in the Mojave Desert, also demonstrate long recovery times (Webb and Wilshire, 1980; Webb, Steiger and Wilshire, 1986). In the second study a linear model based on three points – the quality of the abandoned road, the quality of a currently used road (considered to be the old road's original state) and the quality of an undamaged area – was used to forecast the overall time to return to the original condition. Calculations based on penetration depth, bulk density and peak shear strengths all gave estimates of 75–120 years (Table 16.1).

However, Webb and Wilshire (1980) suggested that bulk density is better modelled on an exponential decay curve (Fig. 16.4), which gives an estimated recovery time of 680 years. Since this is based on only

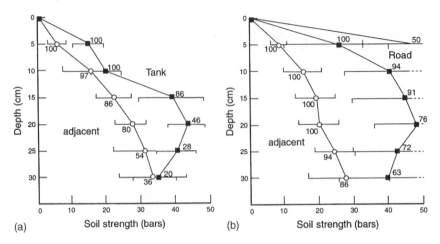

Fig. 16.3 Recording penetrometer values in 40-year abandoned military manoeuvre areas, Mojave Desert, California. (a) In the track of a 'medium' tank; and (b) a camp road, and their adjacent uncompacted areas. The numbers beside each point indicate the percentage readings to reach that particular depth. (From Prose, 1985.)

Table 16.1 Estimated recovery times, determined from three indices, for compacted soils in five Mojave Desert ghost towns (Webb, Steiger and Wilshire, 1986)

| Site | Estimated recovery time (years) based on | | |
	Penetration depth	Bulk density	Peak shear strength
Gold Valley	110	120	90
Skidoo townsite	80	90	–
Skidoo road	140	100	–
Greenwater townsite	100	–	–
Greenwater road	110	–	–
Furnace	100	–	–
Harrisburg	< 75	–	–

four points it is also subject to adjustment, but intuitively the curve would appear to be more reliable. In any case recovery of these Mojave Desert soils appears to be a process that exceeds the normal human life span and may take up to 10 generations.

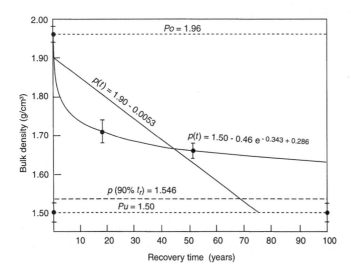

Fig. 16.4 Recovery functions for bulk densities at the Wahmonie Town site, Mojave Desert, California $p(t)$ recovery time (dependent variable); Po, dependent variable at time = 0 (active road state); Pu, dependent variable at time = infinite (undisturbed state), the intersect for recovery estimates; tr, time required for $p(t)$ to equal Pu (complete recovery time); 90% tr = time required for $p(t)$ to equal $Pu + 0.1 (Po - Pu)$ (90% complete recovery time). The straight-line estimate gives 75 years, and the curvilinear estimate gives 680 years. Both extrapolated beyond recovery data points. (From Webb and Wilshire, 1980.)

16.4 Summary

Deposition of eroded material often occurs near the eroding source and may cause more aesthetic disturbance to the site. Paths themselves may act as sediment traps or the material may be deposited on adjacent vegetation. The survival of plant cover depends on the depth of burial and only the more robust species may survive. The environment of buried plants is wetter than the surrounding vegetation at the base of deposited material and drier on the surface. Deposition may have a considerable effect on the local plant and animal communities.

The regeneration of soils has been relatively little studied, mainly because of the exceedingly long time period required for the development of mature soils. Desert soils are perhaps extreme examples, but in these habitats estimates of regeneration times range from 75 to 680 years. Given the higher rate of biotic activity, mesic or tropical soils may not take quite so long to regenerate.

17 A classification of disturbance to animals

17.1 Definitions of disturbance

The interactions between humans and other animals have been investigated in many contexts. The effects of recreation activities on wildlife form a relatively small part of this literature. Disturbances such as those associated with exploration for oil or direct exploitation often take the same form as recreation activities and I have drawn freely on this type of literature in the following chapters. They are not intended to be comprehensive reviews but to point to possible interpretations or predictions that may be useful in suggesting management solutions or research directions. However, the application of any of the ideas recorded here should be preceded by careful studies of all available information, including especially any local knowledge of the animals concerned. As Geist (1978) stated: 'the literature on behaviour of different species and populations within a species invariably reveals unforeseen surprises'. In particular, he referred to the variations of behaviour between different forms of American mountain sheep reported in Trefethan (1975).

The definitions of the term 'disturbance' vary considerably between authors and many are not suitable for the recreation context. So, following the approaches by Hammitt and Cole (1987), Wall and Wright (1977) and many others, I will redefine the interactions between human recreation and wildlife in terms of three types of disturbance.

The Oxford English Dictionary definitions gives some indication of my classification:

- Disturbance type 1: 'Interruption of tranquillity'.
- Disturbance type 2: 'Interference with rights or property'.
- Disturbance type 3: 'Molestation'.

Although these terms originated in relation to people, they form a useful starting point for understanding the following scheme.

DISTURBANCE TYPE 1

I define this, the most obvious form of recreation disturbance, as occurring in situations where the animal is aware of the physical presence of the recreationalist. The animal sees, hears, smells or otherwise perceives the human but there is no contact and it may or may not alter its behaviour. The consequences of this type of disturbance may be positive or negative for the animal.

DISTURBANCE TYPE 2

This is probably a much more common form of disturbance where the habitat is changed in some way by pathway creation, camping, the presence of food or, more drastically, clearing native vegetation. This may involve building rest areas, facilities, visitor centres or even the development of large tourist complexes (such as Yepoon in Queensland, which covers a large area with buildings, surfaced roads, golf courses and the like). Provision of food in winter for wild animals or the adaptation to town living (see Geist, 1971b) also fits this category. The consequences of this kind of disturbance may also be positive or negative for the animal.

DISTURBANCE TYPE 3

This extreme kind of disturbance involves those actions in which there is a direct and damaging contact with the animal. This form is primarily hunting and fishing, although I include under this heading such things as treading on smaller animals, collision with vehicles and other accidental contact which is not strictly 'meddling' as there is no intention to make such contact, but the results of the contact are similar to the consequences of hunting.

The classification of various types of recreation activity according to the three disturbance types is given in Table 17.1.

AGREEMENT WITH OTHER TERMINOLOGIES

I am avoiding the graphic term 'impact' (*The Oxford English Dictionary*: striking, collision or strong effect) commonly found in the recreation literature, as it is more accurately type 3 disturbance in my classification. The sequence from type 1 to type 3 reflects increasing amounts of physical interaction (Fig. 17.1) and can readily be fitted to the continuum of human–wildlife interaction proposed by Duffus and Dearden (1990).

Wall and Wright (1977) had similar concepts to the three types described: disturbance equals type 1, alteration of habitat equals type 2, and killing equals type 3. Hammitt and Cole (1987) first divided

Table 17.1 The subjective classification and ranking of the disturbance level of various recreation activities

Activity	Type of disturbance[a]		
	1	2	3
Walking	2	1	
Walking with dogs	4	1	
Horse riding	3	2	
Trail-bike	5	2	?
Bird watching	1	1	
Animal photography	1	1	
4 × 4 (off-road vehicle)	5	3	?
Presence of toilets, etc.	4	4	?
Car park frequently used	4	4	?
Large development	5	5	0/5
Presence of roads	2/3	2/3	
Canoeing (excluding landing and camping, etc.)	2	0	
Sailing (excluding landing and camping, etc.)	3	0	
Launching boats	3	2/3	
Camping			
in wilderness	2	2	?
in prepared sites	4	4	
Hang gliding			
launching area	3	2/3	?
in air	3	0	
Orienteering	2/3	2/3	
Skiing cross-country	2	1	
Piste – (downhill skiing) including			
preparation of slopes	3	4	?
Skiing village development	5	5	?
Hunting with rifle on foot	2	1	0/5[b]
Hunting with a shot gun	2	1	0/5[b]
Hunting with horses and hounds	4	2/3	0/5[b]
Hunting with dogs only	5	2	0/5[b]
Fishing from bank	2	2	0/5[b]
Fishing from boat	2	0	0/5[b]
Fishing by wading	2	2	0/5[b]

1 = Low effect.
5 = High effect.
Some element of the animals' response has been considered in these rankings.
[a] See text for definitions of type 1, 2 and 3 disturbance.
[b] 0 if hunting is unsuccessful and 5 if animals are killed. Injury has intermediate ranking.

disturbance into direct impacts and indirect impacts (Fig. 17.2). Direct impacts in this case included harassment, which they later discussed under the heading 'Animal disturbance and harassment', and defined as 'Events which cause excitement and/or stress, disturbance of essential activities, severe exertion, displacement and sometimes death

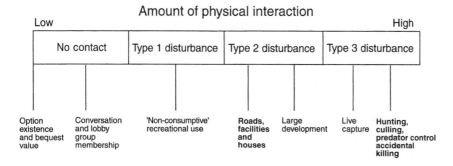

Fig. 17.1   Classification of disturbance to animals as type 1, 2 or 3 according to the degree of physical interaction between the source of the disturbance and the animals.

(Ream, 1979)'. Hammitt and Cole (1987) went on to say that 'although intentional harassment of wildlife does occur, the major impact is caused by recreationalists who unknowingly and innocently produce stressful situations in wildlife'. This would seem to fit disturbance type 1 fairly well. The other 'direct impact' of Hammitt and Cole was 'Harvest', which included recreational hunting and trapping. This is clearly disturbance type 3. Indirect impacts were defined as habitat modification and accord with disturbance type 2.

One merit of Hammitt and Cole's scheme is that their grouping into direct and indirect impacts accord well with Niven's classification of an animal's environment (Niven, 1987). The direct impacts are all the result of processes in the animal's 'direct environment' and humans would be classed as a malentity or predators except where they are observing the animal but not hunting it. This interaction cannot be classified in Niven's scheme as there is no term incorporating the well-being or recreation we feel from such an interaction (Niven, personal communication). The indirect impacts of Hammitt and Cole (1987) fit well with the concept of first- or higher-order modifiers of Niven's scheme.

The three types of disturbance in the scheme proposed here also fit with Niven's scheme fairly well. In type 1 the human is a malentity or unclassifiable, in type 2 the humans are first- or higher-order modifiers, and in type three the humans are predators or malentities.

The alpha, beta, gamma scheme (Liddle, 1984) (Fig. 8.3) is more relevant to the study of specific interactions than to the classification of disturbance types. The alpha stage is involved in all three types of disturbance but the beta and gamma stages are only relevant to type 3 disturbance.

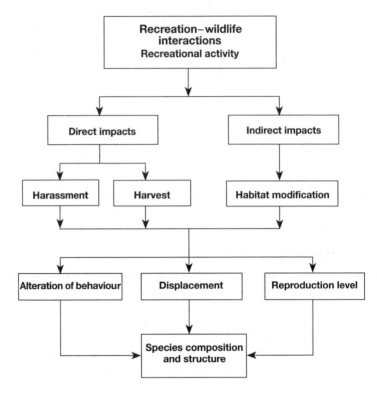

Fig. 17.2 Major impacts of recreation–wildlife interactions (from Hammitt and Cole, 1987).

17.2 Patterns of disturbance in space

THE ISLAND OR CORRIDOR OF DISTURBANCE

From the foregoing discussion we can understand that a walker or someone using a vehicle may be perceived by various senses from considerable distances. The animals' perception of disturbance may be visual or through sound or vibration, through smell, heat or electrical senses or by touch. An animal with good vision may be able to see a human 4 or 5 km away in good conditions, and, on an open heath-land for example, human speech may be heard at a distance of up to 300 m and a motorcycle up to 2 km. An animal with a good sense of smell may detect humans a kilometre or more distant by this means.

A useful way of conceptualizing these interactions is to imagine a stationary human and around that person a series of concentric circles, each indicating the range of a particular stimulus (Fig. 17.3a). The concentric circles may be modified by wind (Fig. 17.3b), obstructions (Fig. 17.3c) or other causes. The actual diameter of the circle will, of course, depend upon the sensitivity of the animal we are considering.

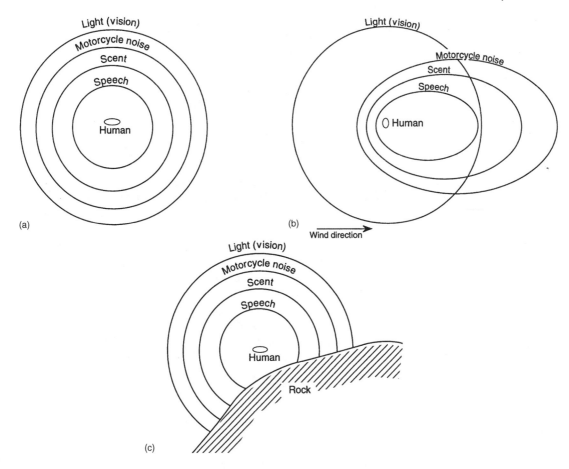

Fig. 17.3 Schematic representation of the hemisphere of disturbance at the point where it intersects with flat ground: (a) in still conditions with no visual obstruction; (b) hemisphere affected by wind; (c) hemisphere obstructed by physical barrier.

If the human moves through the environment, its aura of disturbance becomes elongated into a corridor which, in the absence of distortion or obstructions, is parallel to the path taken through that environment. It is then theoretically possible to draw up a map of the range of disturbance a human will create along a particular path, based on the sensitivity of a particular animal and particular weather conditions.

Liddle (1987) made a spatial analysis of the areas of open woodland between paths in Toohey Forest (500 ha) in which Griffith University is placed (Fig. 17.4). The disturbance distance for the forest animals is unknown but to allow a distance of 200 m from paths

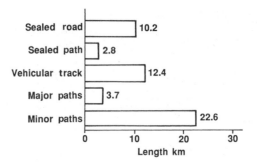

Fig. 17.4 The total length of different categories of roads, tracks and paths in urban Toohey Forest (*c.* 500 ha) (Liddle, 1987).

requires an area of over 10 ha. Less than 20% of the forest consists of blocks of over 10 ha; the mean size was 4.1 ha. Given the linear nature of most blocks it is unlikely that any disturbance-free sites 200 m from the pathways are within the forest (Liddle, 1987).

In conclusion, it is possible to think of the disturbance environment as a series of overlying maps, each showing the area of disturbance from one particular source under specified weather conditions and for one specific animal's range of sensitivities.

17.3 Responses to disturbance

RESPONSES TO TYPE 1 DISTURBANCE

It is clear that the world perceived by different animals differs according to each species and the development of their receptor systems. It is therefore important that we do not judge the level of disturbance created by a particular recreation activity by our human receptor systems or a mechanized substitute such as a decibel meter. The only effective way of evaluating a particular disturbance is by observing the animal's behaviour. However, this has its own problems, as the observed animal may be disturbed but not respond, or respond positively either because it has judged that the disturbance is not harmful or because it is habituated to that particular disturbance and may even regard it as beneficial.

A model for the flight or fight reaction of vertebrates, based on control theory, was presented by Archer (1976) (Fig. 17.5). In this case the disturbance is observed (1) and compared (2) with previous experience (3). If the situation is not disturbing the animal, then no further action takes place; if it is judged to be beneficial, the animal may move towards the disturber. On the other hand, if the situation is either unknown or has previously been found to be deleterious, decision process 1 is set in train. This determines whether the animal will attack or be frightened. If it is frightened then decision process 2 takes place and the animal decides whether to escape or freeze. Whatever the response, the discomfort caused by the disturbance is eased by driving off the disturber (4), escape (5) or cutting off disturbance and/or maintaining observation (6). If the disturbance is unknown to the animal, as may happen when a new area is opened to recreation or if the animal is young and inexperienced, its reactions may be quite inappropriate and can lead to injury or death of either the animal or the human intruder. However, if the latter consequence eventuates, the response may be considered appropriate for the animal concerned (a crocodile for instance), if it does not trigger a reprisal response from other humans.

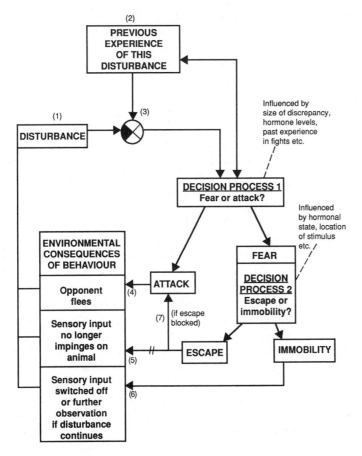

Fig. 17.5 A simplified version of a control theory model of aggression and fear in vertebrates (modified from Archer, 1976).

Some known responses to disturbance include the raised heartbeat experienced by bighorn sheep in 73% of responses in the disturbance experiments of MacArthur, Geist and Johnston (1982a). A change in the rate of heartbeat occurred either preceding or in the absence of any observable activity by the sheep, although this only amounted to 15.6% of the mean duration of the heartbeat responses. A series of nicely graded motor responses by the crested tern (*Sterna bergii*) to increasing loudness of a recording of aircraft noise was used by Brown (1990). He identified four responses:

1. scanning behaviour involving head turning;
2. alert behaviour involving movement on the nest;
3. startle/avoidance behaviour, an incomplete intention movement to fly up or escape; and
4. escape behaviour.

He recorded the percentage of the colony that responded in each of the above ways when exposed to 30 sec of recorded approaching and leaving aircraft noise at various levels of loudness (Fig. 17.6). This work shows that even when an animal does respond with an observable movement, it would not be clear to an inexpert observer that disturbance had taken place.

The flight reaction may lead to the exposure of young to predators (Brown, 1990) or to a loss of vital feeding time for ungulates in marginal winter conditions when the animal's survival is precarious (Geist, 1971a). In all cases some energy is used by the animal and resource use is restricted. A technique for estimating the loss of feeding resources due to disturbance and consequent reduction in feeding time was tested by Gill *et al.* (1996).

Aggressive responses towards humans are often noted and whole books have been written on animals dangerous to people. The consequences of a snake biting a human are often that the snake is killed, either in revenge or so that it might be identified and the correct anti-venom administered. Bears have often attacked humans in the USA and many bears have been killed or transported into more remote areas. Given this type of reaction, it is probably no coincidence that many large mammals have become extinct during the past 100 000 years as the world population of humans has increased exponentially.

The ultimate negative response may be considered to be that which leads to a reduction or extinction of the animal population, and this is all too often the consequence of recreation activities. More detailed responses to type 1 disturbance are considered in the chapters on the various groups of animals.

It should also be noted that different species of wildlife have different tolerances for interactions with humans. Even within a species, tolerance level for interaction will vary according to the time of year,

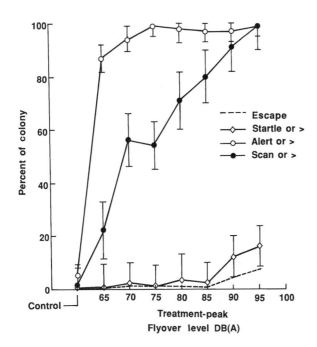

Fig. 17.6 Mean proportions of the crested tern (*Sterna bergii*) colonies exhibiting different behavioural responses to aircraft noise stimuli. Scanning, head turning; alert, neck extended to a few steps on the spot; startle, incomplete intention movement to fly up or escape; escape, flying up for short or long time. (From Brown, 1990.)

breeding season, animal's age, habitat type, and the individual animal's experience with recreationalists (Hammitt and Cole, 1987). In general, the larger the animal, the greater range of its receptor systems and therefore the greater the effect of type 1 disturbance.

RESPONSES TO TYPE 2 DISTURBANCE

These are harder to determine as in many cases the wildlife disturbed has not even been seen by the persons causing the disturbance. The concepts of 'increasers', 'invaders' and 'decreasers' proposed by Dyksterhuis (1957) for plants may be useful in considering responses to type 2 disturbance where the habitat is changed (Chapter 4).

There are many recorded increases in wildlife populations as a result of habitat change: the increase in the numbers of the mouse, *Microtus montanus*, in the camping area of the Yosemite Valley (Garton, Bowen and Foin, 1977), the many birds that are to be found feeding at picnic and campsites (e.g. the common sparrow, *Passer domesticus*) and, of course, several species of rats, cockroaches and, more recently, foxes found around buildings in many parts of the world. These increases are probably caused in the main by an inadvertent increase in the food supply, although shelter and protection from predators may also contribute.

The invading species are not always easy to distinguish from the increasers unless the area had been studied before the type 2 disturbance occurred. Clear examples are the introduced sparrows and foxes, and probably rats, in parks around Brisbane, Queensland. The reasons for these invasions are likely to be similar to the increases above. The beneficial aspects of this type of disturbance are likely to feed back to the animals' responses to type 1 sensory disturbance, so that instead of flight or fight reaction, the animals respond positively, either by not moving or by going towards the disturber.

The examples of decreasers are many and often regarded as the usual consequence of type 2 disturbance. The clearance of mangroves in many parts of the tropics and subtropics in order to build hotel complexes or residential areas obviously eliminates the whole mangrove community of crabs, fish, birds and invertebrates. The use of mountain areas for campsites has reduced the habitat available to bighorn sheep in the Rocky Mountains and the spread of non-native Americans has drastically reduced the habitat of the grizzly bear (see Fig. 21.2). Many soil animals (described in Chapter 13) are also decreasers.

Another approach to the effect of structures, in this case a road, was proposed by van der Zande, van der Ter Keurs and van der Weijden (1980). According to these authors the type 1 and type 2 effects cannot be separated but the presence of the road and its traffic are '**primary activities**' and the activities that are enhanced by the presence of the road are '**secondary activities**', including the '**settlement**' of plants. Figure 17.7 shows the primary ecological effects they attribute to the road–traffic complex.

RESPONSES TO TYPE 3 DISTURBANCE

By definition, the consequences of this type of disturbance are deleterious to wildlife, leading to injury or death. Animals injured by vehicles or inaccurate shooting are clearly at a disadvantage and may not survive. It has often been reported that hunted animals are much more timid and easily disturbed. If type 1 disturbance occurs to previously hunted animals in winter when food is in short supply, the effects may become quite harmful (Hammitt and Cole, 1987) as the flight or fight reaction is the only possible response. Hunted animals are also less available for the human spectator who has no intention of hurting

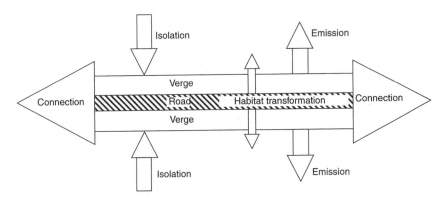

Fig. 17.7 Schematic representation of the primary ecological effects of the road–traffic complex: habitat transformation, connection (dispersal corridors), isolation and emission of matter and energy (from van der Zande, van der Ter Keurs and van der Weijden, 1980).

them. It should be noted that the group that tends to be most affected by type 3 disturbance are larger mammals, while smaller animals are more influenced by type 2 disturbance.

17.4 Summary

A scheme in which disturbance to animals by recreation activities is classified into three types is proposed. Type 1 – interruption of tranquillity – involves those actions where the animal is not in direct contact with the disturber and the land use is not changed. Type 2 is a change to the habitat and type 3 is where the animal is wounded or killed.

The distribution of disturbance in space may be considered as a series of concentric hemispheres around a stationary visitor. Each hemisphere represents the range of sensitivity of a particular animal to a particular stimulus and these vary between individuals and species as well as with the strength of the stimulus. Some, such as scent or sound, may be distorted by air movement, and all can be altered by obstructions. When the visitor starts to move through the environment the hemisphere becomes a corridor of disturbance through the landscape, with its edges parallel to the visitor's direction of travel.

A model for the sequence of vertebrate responses to disturbance based on control theory applied to the flight or fright reactions is discussed. Some examples of responses to the three types of disturbance are presented and the difficulty of knowing the extent of metabolic changes and their consequence for the animals' survival is discussed and illustrated with examples.

18 Invertebrates

The invertebrates make up by far the greatest part of the animal kingdom and include all those animals that do not have a backbone. They range in size from the giant squid (*Architeuthis* spp.) which may be 16 m long, to the small, unicellular protozoa which may be found in almost any drop of natural water. Their complexity also ranges from animals with highly specialized organ systems, such as the eyes and nervous system of the octopus (e.g. *Eledone cirrhosa*) and the arthropods (insects, spiders and crustaceans), to the simple eye spot on the unicellular *Euglena*.

Some groups of invertebrates, such as butterflies and other insects (Insecta), earthworms (Polychaeta) and octopus and shellfish (molluscs), are very familiar to everyone but others, such as the sea mats (bryozoans), rotofers or the comb jellies (Ctenophora), are less well known and one or two specialists are the only people acquainted with many of the invertebrate groups.

The senses with which invertebrates may become aware of humans include sight (insects, some molluscs and some polychaetes); chemical detection (some insects and many other phyla); detection of sound or other vibrations (annelids, (including earthworms and lugworms), leeches and arthropods) and touch (almost all phyla).

Since we are generally only familiar with one or very few members of each taxon, this chapter is presented on the basis of two habitat groups, the terrestrial, above-soil invertebrates and the marine invertebrates, including those of rocky shores, sandy or muddy shores and coral reefs. Soil invertebrates were discussed in Chapter 13. It has not been possible to separate the types of disturbance for most of these groups from information given in the various papers. This may be because 'only habitat modification (type 2 disturbance) is really important for invertebrates' (E. Duffey, personal communication).

18.1 Terrestrial above-ground invertebrates

DISTURBANCE TYPES 1 AND 2

The sensitivity and response of an experimental subject to the presence of an observer has long been debated in the life sciences, and now in physics, so there is little doubt that type 1 disturbance to invertebrates does occur. A shadow cast by a watcher on a resting butterfly, causing it to fly, is a good example. However, there do not seem to be records of this level of interaction in the recreation context, but type 2 disturbance is frequently documented.

The rest of this section is ordered according to increasing intensity of disturbance. The reduction of vegetation parameters by trampling is intuitively obvious and has been subject of many investigations (Chapters 3–8). There is a direct relationship between arthropod numbers and vegetation structure or volume, and this has been demonstrated by van der Ploeg and Wingerden (1974) and Luckenbach and Bury (1983). Even diffuse human trampling was shown to reduce drastically the numbers and diversity of grasshoppers and crickets (Orthoptera) and possibly of other groups, in mountain pastures of the French Alps (Vosin, 1986; Table 18.1). This reduction in all parameters was shown to occur even before there were any visible changes in the vegetation. This suggests that very sensitive vegetation measurements are required to detect the diffuse effects of recreation before the fauna are affected either directly or indirectly.

A rare study of the effects of recreation on phenotypic variation of animals by Emetz (1983, 1984a,b,c, 1985a,b,c, 1986 (only 1985c in English)) concerned the ground beetle *Pterostichus oblongopunctatus*. The author examined the variation in pitting (fossae) of the elytra (cover wings) between populations in a part of an oak grove near Voronezh, Russia, that was subject to recreation pressure (mainly walking), and in a relatively unused area. Normally each elytron has five or six pits; rarely four, seven or eight; very rarely three or 9–12 (Fig. 18.1) (Emetz, 1984c, 1985c). During the period of study (1974–82), use of the impacted area increased and, at the same time, the proportion of less-pitted forms increased at the expense of the more-pitted morphs. This change did not occur in the less-used area. Apparently, this was not a direct effect of recreation impact, but was an indirect effect, through a reduction in thickness of the ground litter

Table 18.1 Parameters of the Orthopteran fauna of mountain pasture in the French Alps subject to different degrees of utilization (Vosin, 1986)

	None or very few pedestrians		Diffuse human trampling		Pastures with intense human trampling
	No or very little grazing	Grazed	No or very little grazing	Intense grazing	
Linear index of mean abundances	16.8	12.2	3.1	1.1	2.9
Mean number of abundant species	6.1	4.9	4.3	2.1	3.1
Shannon–Weiner diversity index	4.7	3.5	3.7	1.4	2.7
Mean of 'banalization' indices	1.5	1.2	1.1	0.8	0.8

and consequent reduction and instability of the moisture content during the period of larval development (Emetz, 1985c). The new phenotypic structure of the group was preceded by a sharp and sudden increase in the number of morphs and the proportion of rare morphs in 1977 which then stabilized in 1979–80 (Table 18.2). There was an associated drop in numbers of individuals (Emetz, 1983), although the mean number of mature eggs per female beetle increased (Emetz, 1985a) and they were spatially aggregated in the impacted area while their distribution remained random in the less-used area (Emetz, 1986).

The imagos moved further in the impacted area and the migration was directed into 'the depths of the forest' (Emetz, 1984b). The asymmetry of the imagos (different number of pits on left and right elytras) increased in the impacted area (Emetz, 1984a) and the author concluded that the destabilizing selection acted to increase the speed of microevolution in the insect population and their adaptation to the

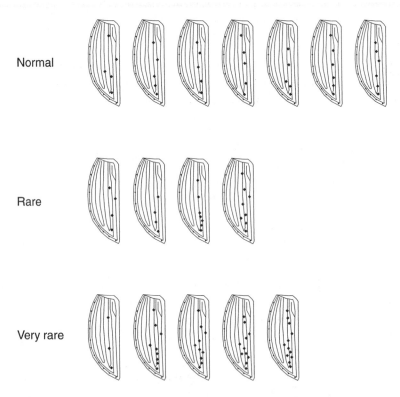

Fig. 18.1 The location of the pits in the surface of the elytra of the ground beetle (*Pterostichus oblongopunctatus*). The animals with three and four pits increased as a result of recreation inpact, while the numbers of the more pitted ones decreased. (Modified from Emetz, 1984c.)

anthropogenic change in the environment (Emetz, 1985c). However, Duffey (personal communication) has suggested that it is difficult to believe that these changes are the consequences of trampling; some other factor may be involved.

Measurements by Luckenbach and Bury (1983) of the nocturnal tracks left in the sand by beetles such as the large, sand-dwelling scarab, *Pseudocotalpa andrewsi*, showed remarkably lower numbers in areas that were driven over by off-road vehicles compared to control areas (Fig. 18.2). They noted that ORVs can break up the desert surface and destroy some desert beetle habitats, such as pockets of accumulated vegetative material or crusted deposits. The entire ground-dwelling beetle population of an area may congregate in an area of 2–3 m² on the surface for breeding, and are particularly vulnerable to ORVs at this time.

A study of the secondary succession of arthropods and plants after vehicle use, in the Sonoran Desert, led Johnson *et al.* (1983) to state: 'restoration of numbers of arthropods on the disturbed area is dependent on the total plant cover on the plot, apparently regardless of the composition of the plant species involved'. However, they also presented data demonstrating correlations between certain arthropod groups and particular plant groups (Table 18.3). In their table, comparison between associations with annual and perennial grasses and herbs shows that there are 12 significant relationships with annual

Table 18.2 Variability of *Pterostichus oblongopunctatus* groups on the basis of the number of pits on the elytra in the recreational (A) and less-visited (B) areas during 1974–1982 (environs of Voronezh) (Modified from Emetz, 1985c). Note the increase in number of morphs and proportion of rare morphs when recreation impact increased in 1977

	A			B			Criterion of identity of groups (A), (B) and I[a]	Parameters of affinity (r) between groups A and B
Year	No. of adults in sample	Mean number of morphs	Proportion of rare morphs	No. of adults in sample	Mean number of morphs	Proportion of rare morphs		
1974	591	3.507	0.300	563	3.504	0.299	4.6*	0.998
1975	584	3.361	0.328	638	3.221	0.356	4.9*	0.998
1976	796	3.699	0.260	723	3.642	0.272	9.1*	0.997
1977	987	4.919	0.508	655	3.734	0.253	114.2	0.962
1978	689	4.228	0.396	793	3.787	0.369	109.9	0.962
1979	751	4.002	0.333	867	3.899	0.350	160.9	0.946
1980	593	3.502	0.300	631	3.622	0.276	151.6	0.938
1981	637	3.828	0.362	694	3.809	0.365	67.8	0.973
1982	596	3.911	0.348	544	3.691	0.385	179.6	0.917

[a] I is the identity criterion used to check the statistical difference between populations A and B.
The asterisks show that the samples do not differ (the affinity parameter, *r*, does not differ significantly from 1).

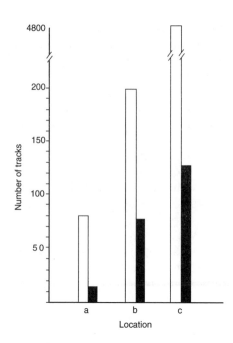

Fig. 18.2 Comparison of the numbers of beetle (*Pseudocotalpa andrewsi*) tracks recorded on 0.5 × 100 m sand sweeps: (a) and (b) desert psammophylic scrub; (c) desert microphyll woodland. Control (□) and ORV-impacted areas (■). (Modified from Luckenbach and Bury, 1983.)

plants and only five with perennials. Also, if the Lepidoptera and Thysanoptera are excluded, there are 16 associations with plants in the disturbed area and only nine in the control areas, suggesting perhaps that choices become more restricted between species in an environment which has been modified by disturbance.

More drastic changes in land use have a major effect on invertebrates, as with all flora and fauna. In heavily populated West Germany, the density of paved roads in 1990 was 3.6 km km^{-2} (Mader, Schell and Kornacker, 1990) and is increasing. These authors found that while grassy field tracks have no significant effect on arthropod movement, paved and gravel field tracks and railway tracks stimulate longitudinal movements and reduce the rate of crossing. Marked carabid beetles moved freely between pitfall traps set out in a wheat field, whereas they only crossed a gravel road nine times compared to 59 longitudinal movements (ratio 6.6 : 1) (Fig. 18.3). The corresponding ratios at a paved field track and a railway track were 4.4 : 1 and 5.8 : 1, respectively. These roads may present a serious problem for the dispersal of flightless ground-dwelling species. The average distance that these species can move may also be reduced by a network of barriers (Fig. 18.4) and their energy exhausted before they can reach another suitable habitat.

The increasing isolation of natural dune habitats on the southern Mediterranean coast of Spain (Costa del Sol) caused by destruction of intermediate locations is also isolating invertebrate species (Haesler, 1989). These locations are being developed as tourist resorts and the dunes are becoming urbanized areas. The species of arthropods that build nests, such as wild bees, digger wasps, spider wasps and other solitary wasps, all members of the Hymenoptera, are liable to be affected. Haesler (1989) was apparently unable to find a single member of the wild bee species *Osmia rutila* along a 200 km stretch of coast between Malaga and Almeria, where there were 12 previously recorded locations. 'In large areas where the food plants are still present and where the remaining dune fragments do offer sufficient nesting places', *O. balearica* is also absent.

Johnson *et al.* (1983) also found that although the total numbers of individuals were markedly reduced by clearance of the vegetation for construction, there was a predominance of herbivores in all sites,

Table 18.3 Arthropod–plant correlations, on a seasonal basis (Johnson *et al.*, 1983)

Arthropod group	Plot type	Ah	Ph	Ag	Pg	TSC	Tc
Acarina	E	–	–	–	0.69**	–	0.60*
	C	–	–	–	–	–	–
Araneida	E	0.78**	–	–	–	0.95**	0.86**
	C	–	–	–	–	0.52*	–
Lygaeidae	E	0.94**	–	–	–	–	0.83**
	C	0.67**	–	0.94**	–	–	0.73**
Cicadellidae	E	0.96**	–	–	–	0.63*	0.87**
	C	0.63*	–	0.91**	–	–	0.80**
Coccoidea	E	0.90**	–	–	–	0.77**	0.95**
	C	–	–	–	–	–	–
Chalcidoidea	E	–	0.74**	–	–	–	–
	C	–	–	–	–	0.58*	0.53*
Formicidae	E	–	0.74**	–	–	0.80**	–
	C	–	–	–	–	–	–
Lepidoptera	E	–	0.58*	–	–	–	–
	C	0.82**	–	0.95**	–	–	0.70**
Thysanoptera	E	–	0.64*	–	–	0.59*	–
	C	0.79**	–	0.95**	–	–	0.71**

Dashes indicate no correlation. E, experimental (disturbed) plots; C, control plots; Ah, annual herbs; Ph, perennial herbs; Ag, annual grasses; Pg, perennial grasses; TSC, trees, shrubs and cacti; Tc, total cover. Significance levels: * = 0.05, ** = 0.01.

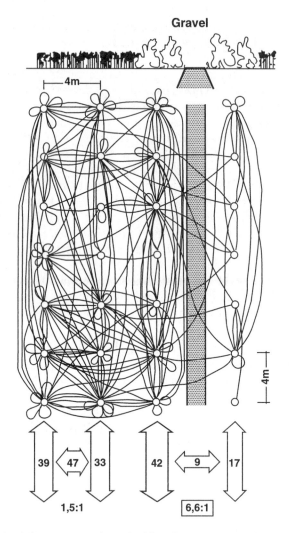

Fig. 18.3 Mobility pattern of carabid beetles at a gravel field track. Each line represents a movement of a marked animal between the traps. (From Mader, Schell and Kornacker, 1990).

with higher numbers of mixed feeders and non-herbivorous species in the experimentally disturbed area (Fig. 18.5).

The grasshopper assemblages in a recreation area in Natal, South Africa, were also shown to be very sensitive to the planting of exotic conifers (*Cupressus arizonica*, *Pinus elliotti* and *P. roxburghii*) (Samways and Moore, 1991). The pines reduced species and numbers of grasshoppers, while *Cypress* patches acted as within-system diversity generators.

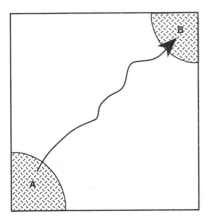
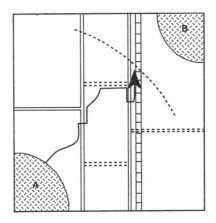

Fig. 18.4 Possible effects of roads and railway tracks on the dispersal of animals. Frequent movements parallel to roads may decrease the effective dispersal range. Energy resources may be exhausted before reaching a suitable habitat. (From Mader, Schell and Kornacker, 1990.)

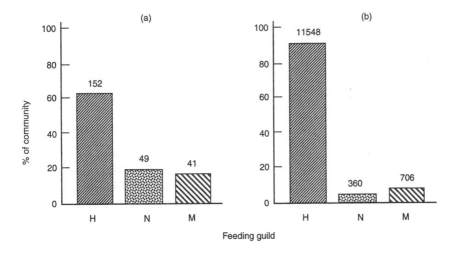

Fig. 18.5 Relative proportions of three feeding guilds expressed as percentage of total number of arthropods collected. (a) Experimental and (b) control arthropod communities after construction. Actual numbers collected are given above the bars. The dominant orders Acarina and Thysanoptera are included. H, herbivores; N, non-herbivores; M, mixed-feeders. (From Johnson *et al.*, 1983.)

Although largely unrecorded, there is no doubt that increasing land use and development for recreation is having a drastic local effect on many terrestrial invertebrate species and possibly driving some of them to extinction.

TYPE 3 DISTURBANCE

In spite of their interest and our direct and indirect utilization of invertebrates, the general view is that they are nuisances at best, and pests to be exterminated at worst. The increasing utilization of marginal land for recreation and other purposes has brought humans into increasing conflict with invertebrates. However, one study by Munguira and Thomas (1992) reported that accidental kills by vehicles of road-verge populations of butterfly and burnet populations only accounted for 0.6–1.9% of the adults of closed populations and 7% of those from open populations. 'These mortalities were insignificant compared to those caused by natural factors.' In other cases, areas that were left unused have been avoided precisely because they were the habitat of some insects that either gave painful bites or spread diseases. Interestingly, Russian researchers have shown that people will tolerate up to 10 midge bites in 15 minutes but would not use an area where they would receive over 40 bites in 15 minutes (Fedovora, 1985).

The literature more often deals with the 'problems' created by arthropods and their control, usually by killing in one way or another, than their preservation or survival, although some popular groups, such as butterflies, are exceptions. In a paper entitled 'Arthropod problems in recreation areas' Newson (1977) reviewed this conflict in the USA. He considered especially mosquitos (eastern equine encephalitis, malaria, Californian encephalitis), blackflies (bite allergies), stable flies (large numbers), biting midges (bites) and ticks (Rocky Mountain spotted fever, Colorado tick fever, relapsing fever), and concluded that entomological considerations should be an integral part of planning new resort facilities to avoid undesirable site selections. I suspect that the effectiveness of pesticides is such that it has been, or will be, a rare decision to abandon a project on these grounds alone, although Merritt and Newson (1978) reported that 'mosquitos have delayed and complicated the development and operation of recreational facilities wherever they have been present in large numbers'. They also quote a case where metathion fogging also killed the predators of pine needle scale (*Chionaspis pinifoliae*), producing an outbreak of these insects on lodgepole and jeffrey pines in the area. They conclude that, 'Man must accept a certain degree of discomfort and show a greater willingness to tolerate rather than dominate, the rigours of nature.'

ALL DISTURBANCE TYPES

Sea-shores of all kinds attract large numbers of visitors. They are one of the best examples of an ecotone and it is therefore not surprising that they have also attracted many ecologists, a few of whom have looked at recreation impacts. The soft estuarine substrates provide a home for many invertebrates who, in turn, provide food for large flocks of waders and sea birds, as well as bait for fishermen who spend considerable effort in their collection. The effects of bait collection of estuarine lugworm (*Arenicola marina*) from the shores of northern Europe was studied in three locations. Heiligenberg (1987) showed that most of the major invertebrate species were severely reduced after hand digging for lugworms in the Dutch Wadden Sea area (Fig. 18.6). Some species, such as the bivalves *Macoma baltica* and *Scoloplos armiger*, were highly mobile and able to migrate back into the disturbed area. This would have initially led to a reduction of these species in the surrounding area until juvenile recruitment restored the population level. This experiment was started on the 17 March, settlement of juveniles continued until August. If the experimental digging had been done after that date, populations might have been much slower to recover. Even in ecology it is 'an ill wind that turns none to good' and the crustacean *Gammarus* increased in the dug areas that had been colonized by the green alga *Ulva lactuca*, which was able to adhere to the irregular surface. About half the lugworm population was removed by hand digging and the population remained low until the end of September, when measurements ceased. Similar results were found by Cryer, Whittle and Williams (1987) who carried their measurements through to the following January. Their control and treated populations continued to decrease to about 25% of the August numbers. Jackson and James (1979) suggested that increased bait digging was responsible for the decline of the local cockle fishery at Blakeney Point, Norfolk, UK, by the end of the 1960s. In Budle Bay, Northumberland, UK, the black anoxic layers approximately 10 cm below the surface contain high levels of bio-available lead and cadmium, and these heavy metals increased in the surface layers as a result of anglers digging for lugworms (Howell, 1985).

In general, the practice of digging lugworms for bait reduces the invertebrate fauna of estuarine shores, especially when commercial machinery is used. Depending upon the proportion of the populations taken, digging lugworms may lead to long-term declines in invertebrate numbers. There is also concern for the effects of the presence of diggers on feeding shore birds (Heiligenberg, 1987).

Sandy beaches also have considerable populations of invertebrates and there is a conflict of interest between the natural denizens and the owners of off-road vehicles who use the foreshore as a thoroughfare

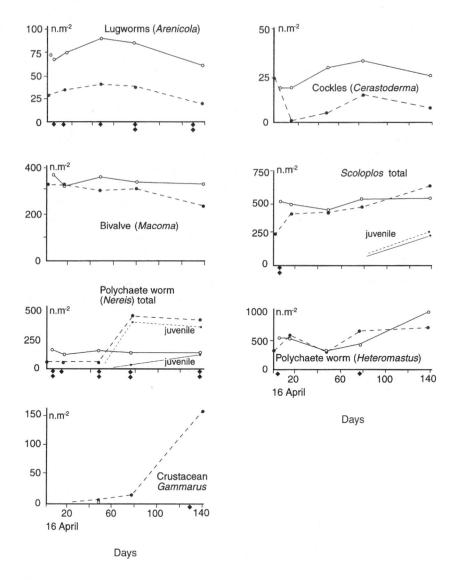

Fig. 18.6 Densities of benthic animals within and outside the hand-digging site. ●, Densities in the area dug over; ○, the control area. The horizontal axis represents days after digging. Significance of differences: ◆ = $P < 0.05$; ◆◆ = $P < 0.01$. (From Heiligenberg, 1987.)

to reach desirable camping, surfing or fishing locations. The ghost crab (*Ocypode quadrata*) on the Atlantic beaches of North Carolina and Virginia is a nocturnal animal living in burrows as shallow as 5 cm and present in large populations, up to 10 000 km^{-1} of beach (Wolcott and Wolcott, 1984). Experiments involving driving over the beach at night suggest that up to 98% of the population could be killed by 100 passes of an ORV. Although Wolcott and Wolcott (1984) did not find any marked reduction on the North Carolina beaches, densities were reduced to zero near the ORV crossover point on Assateague Island, Virginia. Densities were as high as 32 per hectare on less-used sites (Steiner and Leatherman, 1981) (Fig. 18.7). The mean densities on Assateague were 10 on unused 1000 m^2 (0.1 ha) plots, 19 per 1000 m^2 on pedestrian-impacted beaches, 1 per 1000 m^2 on lightly ORV-used beaches and 0.3 per 1000 m^2 on a heavily ORV-impacted beach. Pedestrians appeared to have no harmful effects on ghost crabs. Instead the crabs may have been capitalizing on the food scraps left by bathers. ORVs could be crushing or burying the crabs inside their burrows, interrupting the reproductive cycle or making substrates drier and unsuitable for the crabs (Steiner and Leatherman, 1981). Some substrates suitable for clams (*Mya arenaria*) also pack hard as a result of ORV impact, and prevent the extension of the clams' syphons to the surface as well as crushing their soft shell bodies. Both effects will result in the death of the clams (Godfrey, Leatherman and Buckley, 1978). In general, driving on the foreshore at night appears to do the most harm to ghost crabs, and Wolcott and Wolcott (1984) suggested that this activity should be banned, thus not only protecting the crabs but also the nesting turtle population.

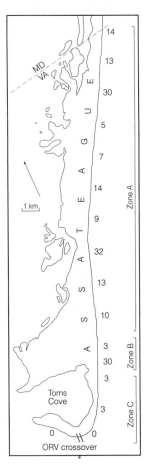

Fig. 18.7   Densities of ghost crab (*Ocypode quadrata*) per 0.1 ha for the night census of 19 June 1978 (21.00–01.00 hours), Chincoteague National Wildlife Refuge beach (from Steiner and Leatherman, 1981).

18.3 Rocky and boulder shores

ALL DISTURBANCE TYPES

As might be expected from the exposed nature of rocky shores, fauna in this habitat seem to be much more robust, resisting the effects of trampling, although some species are attractive and may be removed by visitors. However, the density and diversity were found to be higher at less-trampled sites on a rocky shore near the Natural Bridges State Park, California (Beauchamp and Gowing, 1982). The densities of mussels (*Mytilus californianus*) and barnacles (*Balanus* spp. and *Chthamalus* spp.) and the diversity of algae were unaffected, although the brown alga, *Pelvetiopsis limitata,* was absent from the most trampled site, and small bivalves were found in lower densities. Parallel results were found in Point Nepean National Park, Victoria, southeastern Australia (Povey and Keough, 1991). They noted that individual footsteps damaged some animals but the percentage crushed was very low. Some molluscs were dislodged but the survival of two gastropod species, *Bembicium nanum* and *Austrocochlea constricta,* was not affected because they quickly righted themselves. One limpet, *Cellana tramosercia,* was not damaged by being kicked or stepped on. However, the dominant alga, *Hormosira banksii,* was easily damaged. The mussels *Mytilus californianus* and *Lottia gigantea* were found to decline at nine sites on the California coastline in Los Angeles county, where visitors removed them for food or bait (Ghazanshahi, Huchel and Devinny, 1983). Starfish, *Pisaster ochraceus,* were also collected from this site and the seastar, *Stichaster australis,* was collected in New Zealand (Paine, 1971). One species of limpet, *Collisilla conus,* did show a considerable increase at the Los Angeles sites, possibly because it is not attractive and seeks sheltered depressions where it does not get damaged so easily. Also, two of its competitors, the barnacle, *Balanus glandula,* and limpet, *Collisilla digitalis,* were reduced, *B. glandula* because it does not retreat to depressions and grows on rock surfaces used for walking and *C. digitalis* because it is attractive to collectors (Fig. 18.8). Ghazanshahi, Huchel and Devinny (1983) argued that because competition for space is common in undisturbed situations, trampling reduces the dominant algal species. This favours the rarer animals and thus increases animal diversity (Fig. 18.9). Since people do not walk on the steep sides of the rocks or in the area between rocks, the overall diversity of the community is increased, an effect also found by Beauchamp and Gowing (1982) in their less-trampled sites. This also accords well with the effects recorded for light trampling on plants. A South African study in which experimental trampling was employed found that although there was some damage to algae and dead barnacles were dislodged there were no significant long-term effects (Bally and Griffiths, 1989). These authors also recorded that over 85% of the visitors were barefooted

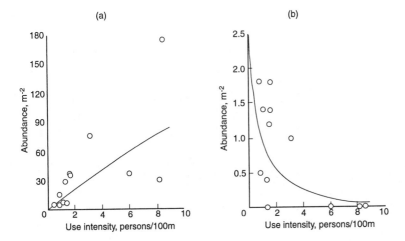

Fig. 18.8 Average abundance of (a) *Collisilla conus*, (b) *C. digitalis* and (c) average cover of *Balanus glandula* at different use intensities. The correlation coefficients were 0.70, 0.62 and 0.90, respectively. (From Ghazanshahi, Huchel and Devinny, 1983.)

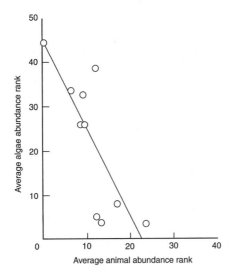

Fig. 18.9 Average algal abundance rank v. average animal abundance rank at each site. Linear regression for the remaining sites produces the statistically significant relationship (algal abundance rank = 46 − 2.0 × animal abundance rank), with a correlation coefficient of 0.80. The regression line is shown. (From Ghazanshahi, Huchel and Devinny, 1983.)

and therefore took extreme care in order to avoid injury to themselves (cf. Chapter 3).

Boulder shores used by visitors are subject to disturbance by having smaller boulders overturned either out of curiosity or, in one case, searching for crabs in ecdysis to use for bait (Cryer, Whittle and Williams, 1987). Correctly replacing the boulders has some beneficial effects as crabs were more likely to be found under these than the inverted ones. Boulder turning damaged most of the attached species within a few weeks and was most harmful to sponges (Zedler, 1978, cited in Ghazanshahi, Huchel and Devinny, 1983) at Cabrillo National Monument, San Diego, California.

18.4 Invertebrates of coral reefs

The primary invertebrates of reefs are the corals themselves. The hard coral colonies consist of many polyps with tentacles which are normally extended to collect food from the surrounding water. Surrounding and beneath these is a hard calcium carbonate base which is built up from carbon fixed by the symbiotic algae that live within the polyps.

This carbonate forms the fantastic variety of structures of the various species of coral, and when the polyps have died it becomes compacted into the limestone that underlies and supports the reef. While corals are alive, or the dead coral maintains its complex structure, the surfaces and lacunae provide a sheltered environment for all the associated animals of the coral reef community. Each cay of the Great Barrier Reef has a variety of coral structures. A transect from Heron Island would first pass over the reef flat, an area of sandy pools with increasing cover of coral as the outer edge is approached (Fig. 18.10a). Exposed parts of the reef may have an area of detached pieces of dead coral 'the boulder field' and then beyond the deeper gutter is the consolidated coral 'the reef crest'. All of this may be traversed on foot at low tide and tourists are taken for reef walks on a daily basis (Fig. 18.10b).

ALL TYPES OF DISTURBANCE

Type 1 disturbance can be observed when a hand is held very close to the polyps and they withdraw their tentacles, but recreation-associated studies have focused on the physical effects of trampling (Woodland and Hooper, 1977; Kay and Liddle, 1984; Hawkins and Roberts, 1992). Pollution and the large sediment-laden flows of fresh water from the rivers that pass through cleared agriculture land change the waters of coral reefs and have given rise to concern about their long-term survival. Local pollution from tourist resorts has also been examined and found to limit invertebrates in the locality of the outfalls.

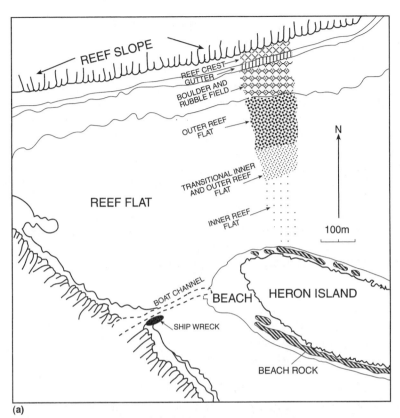

Fig. 18.10   (a) A section of Heron Island reef flat showing the area to the north of the cay which is used by the P&O tourist resort for guided reef walks. The stippling shows the extent of the area and the zones within it. (b) A guided reef walk from the Heron Island Resort. (From Kay and Liddle, 1984.)

The effects of trampling vary greatly for the different coral forms (Fig. 18.11). Some, such as plate and foliaceous types, are very vulnerable and the whole colony is very easily detached and broken, while only the surface living tissue of the massive and encrusting forms is damaged and the skeleton remains intact (Fig. 18.12). The various branching forms which reach their more luxuriant growth on the outer face of the reef or, to some extent, on the reef flat, are the ones that show the most obvious damage from trampling. Whole colonies, or more often branches both alive and dead, become detached. These may lodge amongst the coral matrix and in this position continue to live and even become fused with attached colonies. The pieces that fall to the sandy interstitial areas or into the sandy pools are not able to survive the effects of sand smothering the polyps (Kay and Liddle, 1984). Sand stirred up by walkers may also be a source of stress to living corals (Niel, 1990). Experiments showed that the fragments of some forms such as *Acropora millepora* had high survival rates, while relatively fewer fragments of *Pocillopora damicornis* survived (Fig. 18.13).

The patterns of regrowth rates were similar, with 8 cm fragments of *A. millepora* showing as much as 5 cm growth in 10 months (Fig. 18.14). The fragments were also liable to be moved around by wave action at high tides, the larger ones being more likely to be washed into pools or over the edge of the reef crest, but becoming lodged in the more open structure of the reef flat corals (Fig. 18.15).

From all of these experiments the authors identified three types of survival strategy followed by reef corals subject to human trampling (Kay and Liddle, 1984):

- **Resistant**: where the coral had high resistance to damage but low survival and recovery rates (e.g. *Porites lutea* and *Acropora palifera*).
- **Resilient**: where the resistance to damage was low but survival and recovery rates were high (e.g. *Acropora millepora*).
- **Recruitment (or ruderal)**: where the other properties were low or intermediate but there was a high rate of colonization from planktonic spat (e.g. *Pocillopora damicornis*).

The broader effect of trampling on the reef flat can be categorized as a reduction in the cover of live coral. Woodland and Hooper (1977) organized 18 traverses of an area of the reef flat 4 m wide × 12.5 m long by four people who picked up the detached coral as they walked. This treatment reduced the cover of live coral from 41 to 8% and a total of 607 kg of living pieces were removed. Some of this might have survived, but only a small proportion. There is no doubt that human trampling may be a powerful agent of damage on the reef flats but

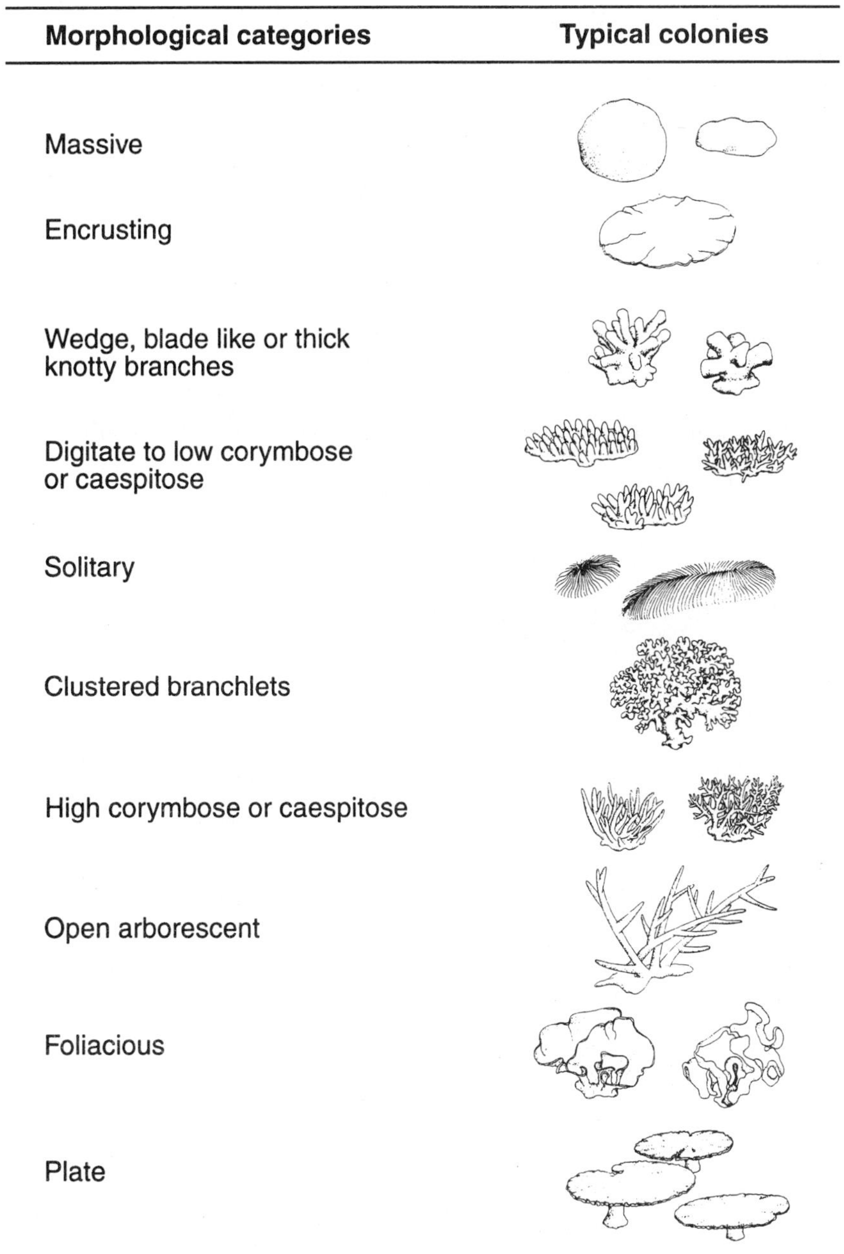

Morphological categories	Typical colonies
Massive	
Encrusting	
Wedge, blade like or thick knotty branches	
Digitate to low corymbose or caespitose	
Solitary	
Clustered branchlets	
High corymbose or caespitose	
Open arborescent	
Foliacious	
Plate	

Fig. 18.11 Generalized coral morphologies found on intertidal reef flats. Their relative vulnerability increases downwards from the top of the figure. (From Kay and Liddle, 1989.)

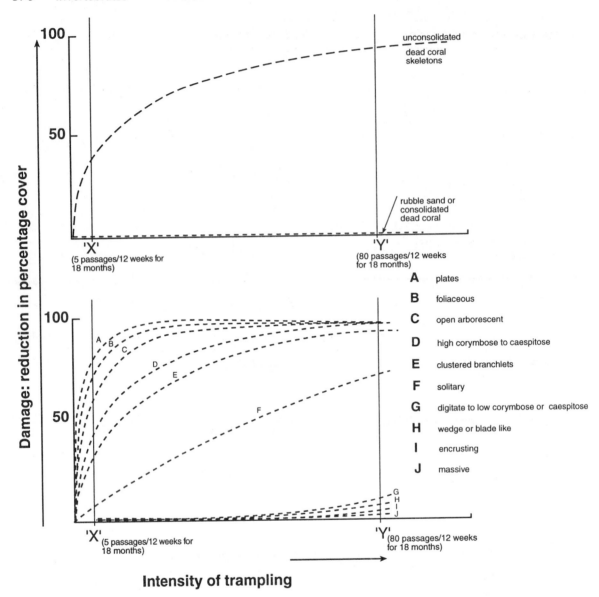

Fig. 18.12 The relationship between damage, measured as the percentage reduction of percentage cover, and intensity of trampling for different substrate categories and for different morphological categories of coral. A, Plates; B, foliaceous; C, open arborescent; D, high corymbose to caespitose; E, clustered branchlets; F, solitary; G, digitate to low corymbose or caespitose; H, wedge or blade-like; I, encrusting; J, massive. (From Kay and Liddle, 1989.)

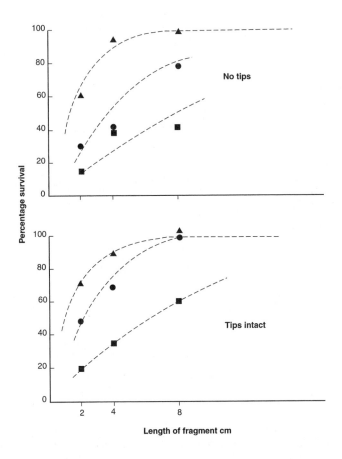

Fig. 18.13 The percentage survival of coral fragments 3 months after detachment from the parent colony: ▲, *Acropora millepora*; ●, *A. palifera*; ■, *Pocillopora damicornis* (from Kay and Liddle, 1987).

the reef crest is a much denser and more consolidated substrate. Trampling experiments on both areas showed an increase in the cover of dead material and a corresponding decrease in cover of live corals (Fig. 18.16). The crest is subject to the forces of the breaking waves and serves as a barrier which protects the more vulnerable reef flat so it is perhaps not surprising that it appears to be less damaged by human trampling. However, the reduction in cover of live coral should be noted although there was no obvious increase in rubble or visible trenches in the structure of the substrate as there was on the reef flat.

Scuba diving also has a considerable effect on the corals growing in deeper waters of the Red Sea. Hawkins and Roberts (1992, 1993) measured the amount of broken and abraided coral and fragments of

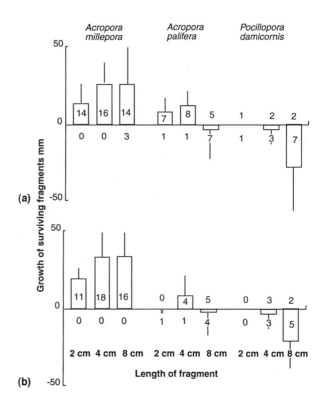

Fig. 18.14 The growth of surviving fragments over a period of 10 months. (a) Fragments from which the tips had been removed, (b) intact fragments. The numbers of fragments which gained length are given above the x axis and the numbers of fragments which lost length are given below the x axis for each treatment. Bars are standard errors. (From Kay and Liddle, 1987.)

coral as well as reattached broken pieces and the number of 'part dead' colonies. All measures of damage were significantly greater in dived than undived areas. They also noted that 'of the 171 broken hard coral colonies recorded ... 168 were branching, 1 foliaceous, 1 plate and 1 massive' and that branching corals received a greater amount of damage relative to their abundance than any other form. The growth forms evidently have a similar relationship to vulnerability on both the reef flat of Heron Island and in the deeper areas of the Red Sea.

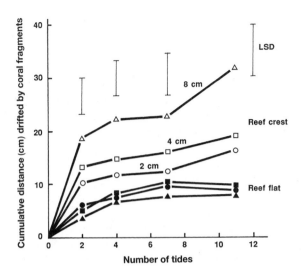

Fig. 18.15 Cumulative distance drifted by coral fragments on the reef crest (hollow symbols) and the reef flat (solid symbols): △▲, 8 cm fragments; □■, 4 cm fragments; ○●, 2 cm fragments (from Kay and Liddle, 1989).

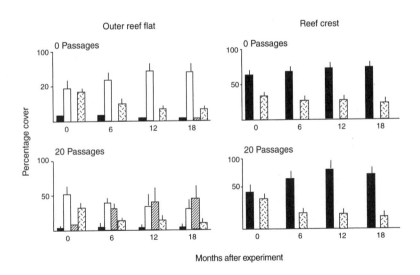

Fig. 18.16 Percentage cover of consolidated dead coral (solid bar), unconsolidated dead coral (open bar), rubble (diagonally lined bar) and live coral (dotted bar), on the reef flat and the reef crest. The controls and 20 passage treatments are shown from an 18-month trampling experiment. Bars are standard errors. (From Kay and Liddle, 1984.)

18.5 Summary   In this chapter terrestrial arthropods, invertebrates of soft and sandy shores, rocky and boulder shores and coral reefs are considered separately.

In general, terrestrial arthropods are reduced in number and diversity and, in at least one case, their distribution became more clumped as a result of recreation impacts. Orthoptera numbers and diversity were reduced by trampling before any changes could be detected in the vegetation. Roads have been shown to limit the movement of ground-living species, but cars were responsible for very small numbers of butterfly deaths. There was also evidence of a change in the numbers of different phenotypes of ground beetles in a recreation-impacted area.

Digging for bait, especially lugworms, was shown to affect the inhabitants of muddy shores, while ORV compaction was the main effect of recreation on sand biota. Where machinery was used to collect bait, the impact was greatest and there was also disturbance to feeding birds. The sand animal most at risk appears to be the crabs, many of which are crushed in their burrows or unable to return to the surface through the compacted sand.

The fauna of rocky shores appears to be much more robust, although molluscs may be crushed when the intensity of use is high. Light use may increase animal diversity in a parallel way to that occurring in plants. Boulder shores are vulnerable to the curiosity of tourists who turn the small rocks to discover the animal living underneath, thus exposing them to desiccation and death.

Coral reefs have strong crests where little damage occurs from trampling but the sheltered reef flat areas are more vulnerable. Not only may the living corals be easily broken by trampling, but the matrix of dead material may be destroyed, thus reducing the habitat available to invertebrates and small fish. Corals vary in their vulnerability, massive forms being quite robust while the foliose and branchy forms are very vulnerable. Of the branching forms, some are moderately robust and slow to recover while others are fragile but regrow quite quickly. Three strategy designations are suggested for trampled corals: resistant, resilient and recruitment (or ruderal). Scuba diving may also damage corals, especially where use is intensive.

Reptiles 19

The four groups discussed here are the crocodiles (Crocodilia), sea turtles and tortoises (Chelonia), lizards (Squamata) and snakes (Ophida). With the exception of the turtles and sea snakes, the other groups are mainly terrestrial in tropical and subtropical regions, with a few extending to temperate zones. They may be found in almost any habitat, from desert to rainforest.

Reptiles are egg-laying animals but many lizards and snakes retain the eggs in the oviduct until the young are ready, or nearly ready, to hatch. Alternatively, the eggs may be placed in an underground nest chamber or in one constructed of piles of rotting vegetation. The female crocodile may guard the nest and young hatchlings, but more generally, reptiles leave the young to survive on their own. The animals are often solitary, although some may stay together as a mating pair (e.g. brown snakes, *Pseudonaja* spp.) or in groups (Galapagos iguanas, *Amblyrhynchus*). Others, such as the red-sided garter snakes, may overwinter in groups or form groups for mating purposes.

19.1 Group characteristics

The order Crocodilia, the crocodiles, have apparently been little changed for 190 million years. They include *Crocodilus*, the most widespread genus, occurring in Central America, Africa, Asia, Malaya and the East Indies, and northern Australia; *Alligator* from North America and China; its relative, *Caimen*, from Central and South America; the Indian ghavial (*Gavialis*) and *Tomistoma* of the East Indies (Young, 1962). All species are amphibious, living in fresh water or the sea or both. They are tropical or subtropical animals. Crocodiles are carnivorous amphibians, living on fish, birds and mammals, and they are large enough when mature to take humans easily. Many species will hold their prey under water until it drowns. They lay their eggs in piles of rotting vegetation or warm sand and the young may receive some parental care.

19.2 Crocodiles and alligators

TYPE 1 DISTURBANCE

These reptiles generally receive a bad press and while almost any attack on a human is recorded, the effect of humans on the crocodilian is rarely noted. A partial exception was the record of the locations of 'alligator–visitor incidents' in the Everglades National Park from 1972 to 1980, which indicated that the majority occurred in areas where visitors had unrestricted access (Jacobsen and Kushlan, 1986) (Fig. 19.1). Air boats are another cause of disturbance to alligators in the Florida Everglades.

One major recreation impact on crocodiles is tourist disturbance of females guarding their nests and the consequent loss of eggs to other predators. Nile crocodiles nest on the banks of the Victorian Isle in the National Park below Mercheson Falls, Uganda. Tourists visit the nests by launch and Cott (1969) recorded that of the 36 nests in visited areas 30 (83%) were destroyed by predators, whereas in unvisited areas with 72 nests only 23 (32%) were predated. The predators included olive baboons, spotted hyena, white-tailed mongoose, marabo stalk, palmnut vulture and especially monitor lizards. Unfortunately, a part of the tourist experience was to watch the lizards and hyenas at work. Nests of the Mississippi alligator (*Alligator mississippiensis*) in northern Florida which had been opened by investigators were also twice as liable to predation as undisturbed nests (Deitz and Hines, 1980). These authors recorded that the females changed their behaviour after the nests had been visited, in general their attendance

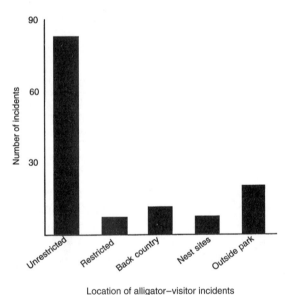

Fig. 19.1 Number and location of alligator–visitor incidents in the Everglades National Park from 1972 to 1982. Areas are classified on the basis of degree of access to the visiting public. (From Jacobsen and Kushlan, 1986.)

at the nests and defensive behaviour was reduced. Here the main predators were raccoons (*Procyon lotor*) while otters (*Lutra canadensis*) were also suspected. The Paraguayan caiman (*Caiman yacare*) in Brazil also showed a similar negative habituation: 'it is our impression that females are so sensitive to being disturbed at the nest that even one contact [with humans] may be enough to change guarding behaviour' (Crawshaw and Schaller, 1980). They continue, 'several females ceased to protect nests after one visit by us ... it did make the eggs more vulnerable to coati (*Nasua nasua*) and other diurnal predators'.

Habituation also occurs in the Florida alligators and they showed aggressive behaviour to visitors, often because they had come to expect food from the tourists (Jacobsen and Kushlan, 1986). On 34 occasions between 1972 and 1982 the aggressive animals were captured and transported to creeks up to 50 miles from the site of the original encounter. Some animals did return to their original location but on only one occasion was the animal still aggressive. Onadeko (1992), in his study undertaken to characterize the level of management necessary to ensure safe relationships between alligators and the public in the Brazos Bend State Park, Texas, concluded that 'the numbers of "large" alligators in the park is too big for a park intended for high intensity human use'. He suggested that removal of aggressive or habituated alligators would be preferable to culling the large alligators, which 'might result in negative ecological chains of reaction'.

TYPE 2 DISTURBANCE

The familiar story of reduced habitat area being the main cause of reduction in populations is true of these reptiles, and Ogden (1978) estimated that in Florida a population of between 1000 and 2000 American crocodiles (*Crocodylus acutus*) was reduced to between 100 and 400 by changing land use.

TYPE 3 DISTURBANCE

Hunting (either for their skins or to remove them from areas where they are considered a danger to people) and road accidents appear to be the main causes of premature death in the American crocodile in Florida. Ogden (1978) recorded five shootings and four hits by cars.

Most turtles are aquatic and herbivorous, the females of marine species coming ashore on specific sandy beaches to dig their 'nests' in which the eggs are buried and from which the hatchlings emerge and find   **19.3 Turtles** their way to the ocean. The best-known sea turtles are the green or soup turtles (*Chelonia*), the hawksbill (*Eremochelys*) and the logger-head (*Caretta*) (Parker and Haswell, 1963).

TYPE 1 DISTURBANCE

Although turtles may be caught at sea in fishermen's nets and along the coasts in shark nets (Table 27.5) or injured by boat propellers (Arianoutsou, 1988), they are most vulnerable to disturbance from recreation when the females come ashore to lay their eggs in the sand, or as very young hatchlings making their way to the sea.

The nesting activity of the loggerhead turtle (*Caretta caretta*) was found to be inversely related to the intensity of disturbance on the beaches of the Greek islands of Zákynthos (Arianoutsou, 1988). This author considered that sun umbrellas, sea paddles and various sunbathing paraphernalia minimized the area available for nesting, and that compaction of the sand by trampling tourists made it difficult for the turtles to dig their nests. Stancyk and Ross (1978) also considered that there is a strong possibility that human interference is causing green turtles (*Chelonia mydas*) to nest on less-disturbed beaches of Ascension Island, turtles approaching the shore to nest at night may be affected by lights or the presence of people. There is no doubt that turtles digging their nest holes and laying their eggs are a consider-able tourist attraction. The author has seen parties of up to 30 people surrounding an egg-laying turtle on Heron Island on the Great Barrier Reef. Although it seemed unperturbed, it must have been aware of their presence and their torchlight. Lights have also been shown to disrupt the orientation of hatchling loggerhead turtles (*Caretta caretta*) making their way to the sea after leaving their nests (Witherington and Bjorndal, 1991). White light was most disruptive and low-pressure sodium-emitting long waves (red) was least harmful.

Vehicles driving on beaches may also damage the nests and the eggs they contain, or prematurely stimulate the hatchlings to emerge when the temperatures are too high for survival (Arianoutsou, 1988). A further stress is placed on the hatchlings making their way from the nest down to the surf by the uneven topography of beaches after they have been trampled or rutted by vehicles (Hosier, Kockhar and Thayer, 1981). Young loggerhead turtles were found to travel at 2.75 cm sec^{-1} over undisturbed beaches, but this was reduced to 1.64 cm sec^{-1} when the sand had been disturbed by a Honda tricycle, and to 1.25 cm sec^{-1} when the sand had been trampled. They were also inclined to travel along the ruts that were parallel to the shore for distances of up to 20 m and were often inverted as they attempted to negotiate the barrier caused by a rut. These factors obviously increased stress and energy use and may make them more subject to predation at this time (Hosier, Kockhar and Thayer, 1981).

TYPE 2 DISTURBANCE

Several authors have cited development or habitat destruction as the cause of declining numbers of sea turtles nesting at particular sites (Shoop, Ruckdeschel and Thompson, 1985; Cypher *et al.*, 1986; Arianoutsou, 1988). In particular, Shoop, Ruckdeschel and Thompson (1985) suggested that reasons for present nesting distributions of loggerhead turtles in the south-eastern United States include human-induced changes in beached areas. They make the point that as distribution changes, beaches which are little used at present may become important nesting areas in the future – a consideration that is clearly important in long-term planning decisions.

An additional factor that may influence the distribution of green turtles (*Chelonia mydas*) is the damage that is caused to the seagrass beds on which they feed. Williams (1988) found that recovery of the beds may take years, and that in two bays on St. John, US Virgin Islands, they are being destroyed by the anchors of pleasure craft at the rate of 1.8% of their area per year.

TYPE 3 DISTURBANCE

The only deaths recorded as a direct result of recreation activities appear to be the two or three per season found on the island of Zákynthos, which had injuries attributed to collisions with fast-moving craft (Arianoutsou, 1988). However, large numbers are killed by shark nets which are permanently placed offshore from popular tourist beaches such as the Gold Coast, Queensland, Australia and the coasts of Natal, South Africa (cf. Table 27.5).

19.4 A note on tortoises

These animals are very attractive to tourists when seen in the wild, but they are also very vulnerable. An interesting experiment has been carried out on Curieuse Island in the Seychelles where a colony of the giant tortoises (*Geochelone gigantea*) has been re-established for tourists. Eight hundred of the animals were transported from Aldabra Atol where there was a population of approximately 150 000 animals (Stoddart *et al.*, 1982). It was considered that the advantages of having a new colony far outweighed the slight disturbance to the original population, an example of one of the positive effects that tourism and recreation can have. Negative impacts of recreation were reported on the desert tortoise (*Gopherus agassizi*) in the Mojave Desert, California (Bury and Marlow, 1973). Here animals had been killed both accidentally by vehicles and deliberately by vandals. They also reported that the animals' burrows, on which they depend for survival when conditions are too dry or too cold on the desert floor, had been destroyed by ORVs and trail-bikes. The status of this tortoise is also threatened by increasing numbers of ravens (*Corvus corax*) who eat the young reptiles. The ravens depend on the increasing use of the desert by humans and the accompanying increase in road kills and rubbish tips which provide extra food (Edgar, 1990). Additionally the remaining adult tortoises are suffering from a stress disease. The interactions are complex but recreation activities are a major part of the threat to the survival of the desert tortoise.

19.5 Notes on lizards

Like tortoises, there have been relatively few studies of the impact of recreation on lizards and there is not sufficient quantitative information to enable separation of the different types of impact.

One study examined the numbers of four species of lizard in areas heavily used, moderately used and unused by off-road vehicles in the Mojave Desert (Busack and Bury, 1974). They found more animals and greater biomass in the unused areas (Table 19.1), although one species, *Callisaurus draconoides*, only occurred in the moderately used area and was absent from unused and heavily used areas. Perhaps it is an increaser species according to the theory of Dyksterhuis (1957). An experimental approach was adopted by Vollmer *et al.* (1976) who generally found a decrease in the numbers of the lizard *Uta stansburiana* as a result of use by ORVs, although there were more lizards in the test areas than in the control areas at the time of the spring recording. However, the heavily impacted areas censused by Volmer *et al.* (1976) appear much less impacted in the photographs than the areas utilized by Busack and Bury (1974). Lizards may also suffer by having their burrows collapsed by ORVs, so they may then lose their refuges and be exposed to climatic extremes. Edington and Edington (1986) and Harris (1973) also reported that land iguanas (*Conolophus subcristatus*) on the Gallapagos Island of South Plaza had become so accustomed to tourists feeding them that their territorial mating system had broken down and breeding ceased. However, when tourists were banned from feeding the animals the breeding activity resumed. The opening of areas to ORVs may also result in thieves removing loose rocks (bushrock) for use in gardens, and Schlesinger and Shine (1994) have shown that there is a tendency for humans and velvet geckos (*Oedura lesueurii*) to require the same size of rocks. The removal of the rock may well reduce the population sizes of the lizard and other large reptiles.

The Mojave fringe toad lizard (*Uma scoparia*) was subjected in the laboratory to various levels of noise recorded from a dune buggy run at high power. It was found that the animals' hearing was impaired by this treatment (Fig. 19.2). It was thought that they would take 4 weeks to recover, a greater time than the weekly rest period some areas receive between weekend use (Brattstrom and Bondello, 1983).

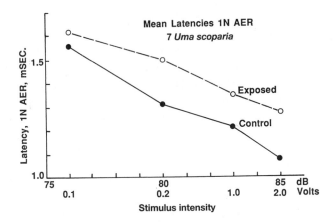

Fig. 19.2 Effects of click intensity on the mean latency in milliseconds of the first negative (1N) average evoked response (AER), for the control and exposed runs of seven *Uma scoparia* at four sound intensities (dB). Five hundred responses averaged for each subject at each intensity. Note that the latency is increased (or hearing reduced) in animals exposed to the loud noise. (From Brattstrom and Bondello, 1983.)

Table 19.1 Numbers and total biomass (g ha^{-1}) of lizards in three areas of off-road usage (modified from Busack and Bury, 1974)

Species	Heavy Use		Moderate use		Unused	
	Number	Total biomass	Number	Total biomass	Number	Total biomass
Callisaurus draconoides	0	0	3	20.8	0	0
Crotaphytus wislienii	0	0	0	0	3	135
Uta stansburiana	2	6.4	2	8.3	7	24.9
Cremidophorus tigris	0	0	10	259	14	350

19.6 Notes on snakes   The literature has many references to the effect of snakes on humans (Edington and Edington, 1986) but apparently few on the effect of recreation on snakes themselves. However, it is clear that areas around habitation, especially tourist resorts or camps, are likely to be cleared of poisonous snakes, although the harmless carpet snake (*Morelia spilota*), which may reach a length of 3 m, is often regarded as an attraction in camps in Australian National Parks. As the Edingtons remarked, rats attracted to tourist sites by inadequate garbage disposal may in turn attract snakes and thus create 'a problem' for the tourists.

In most parts of the world when snakes enter areas of human habitation they are inevitably killed. This means that any area that is developed for tourism is likely to reduce, if not eliminate, the snake population, and therefore the impact of tourism on snakes is closely linked to development and land use. The exception to this killing may occur in some National Parks. The effects of other kinds of disturbance do not seem to have been documented.

Snakes are killed on roads either accidentally or deliberately. The greatest death rate on the two-lane highway that crosses the Pa-hay-okee wetlands of the Everglades National Park, USA, occurred when the water levels were declining and the snakes were migrating to follow the edge of the water. Unfortunately, this coincided with the greatest tourist traffic densities (Fig. 19.3) and resulted in the recorded deaths of 712 animals in the 2 years from February 1987 to January 1989 (Bernardino and Dalrymple, 1992). These authors recommended the construction of drift fences and underpasses to provide alternative crossing places for the snakes.

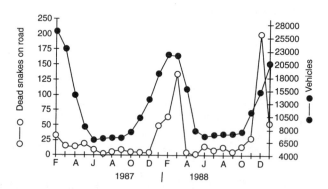

Fig. 19.3 Monthly number of snakes found dead (O) on the Main Park Road in the Pa-hay-okee wetlands and number of vehicles recorded (●) at Flamingo between February 1987 and January 1989 (from Bernardino and Dalrymple, 1992).

Table 19.2 Reptiles observed by 'road-cruising' during the summer of 1985 in northern Alabama; the number dead on the road (Dor), followed by the number alive on the road (Aor) (modified from Dodd, Enge and Stuart, 1989)

Species	Dor	Aor
Snakes	112	14
Lizards	8	9
Turtles	119	41
Totals	239	64
Number per km	0.0126	0.0034

A survey of reptiles on roads in northern Alabama found one dead animal every 79 km, including a total of 112 dead snakes in 19 041 km (Table 19.2) (Dodd, Enge and Stuart, 1989), far fewer than in the Everglades, but in both cases the proportion of the populations killed in this way is unknown. A number of authors have claimed that populations are being affected by road kills (Dodd, Enge and Stuart, 1989).

There is a considerable fascination with reptiles, largely derived from the fact that a number of them are a threat to humans. This leads to a curiosity and a desire to view them but also to the reaction that they must be killed or removed from recreation areas where people congregate.

Crocodiles and alligators are frequently disturbed and nest guarding interrupted, often leading to predation of the eggs. Habituated or aggressive alligators are transported to remote areas in the Florida Everglades. Hunting and road accidents are the main cause of death of American crocodiles.

Many turtles are among the most threatened reptiles and the main causes of population reduction appear to be changing land use at their nesting sites or drainage of freshwater habitats, and hunting. They are most vulnerable as hatchlings making their way from their nests to the sea. It is at this time that recreation can have a major impact, either by disturbing the surface of the sand or the presence of disorienting white lights on the shore. Both effects delay the small turtles from reaching the sea and thereby prolong their exposure to predators. Spectators may also disturb the females when they come ashore to lay their eggs.

Some tortoises are very rare and, in the case of the Mojave Desert tortoise, recreational use of ORVs presents a real threat to their

19.7 Summary

survival. Some desert lizards are in the same position and the Galapagos iguana is a constant attraction to visiting tourists.

The facts that many snakes are venomous and present an archetypal symbol of the kind of knowledge that has been largely rejected by our Western society, have led to deliberate killing of many species, especially near populated areas or campsites. The other major cause of death linked with recreation occurs when they are run over by tourist vehicles, and at the Pa-hay-okee wetlands, Florida, the peak migration occurs coincidentally with the peak tourist season.

Birds 20

Birds depend more on their eyes than their other senses; they are perhaps more visual than any other animal. They 'undoubtedly discriminate colours, apparently on a trichromatic basis, although many birds are rather insensitive to the blue end of the spectrum' (Young, 1962). Some species have especially large eyes, usually associated with nocturnal habits. Hearing is acute and some birds are able to discriminate subtle changes in other bird calls. Birds are also more sensitive to low frequency vibrations, such as distant gun fire, than are humans. Birds' ability to localize sound is high and nocturnal hunting birds are also able to find their prey by sound, using, in some cases, an asymmetrical arrangement of ear cavities (e.g. owls, *Strix*) or external ears (e.g. eared owls, *Asio*). The other senses are not well developed except in a few special cases (Young, 1962).

This chapter is divided into five groups of birds on the basis of their ecology, which largely coincides with different habitats and recreation activities. Game birds are not discussed as a separate group as there is an extensive literature relating to hunting of birds for pleasure.

20.1 Characteristics of the phylum

GROUP 1: PERCHING BIRDS

This group comprises solely the perching birds (order Passeriformes), which encompasses about half of all known species of birds (Young, 1962) including, for example, thrushes (*Turdus* spp.), starlings (*Sturnus vulgaris*), sparrows (*Passer domesticus*), American robins (*Turdus migratorius*), and the crow family (Corvidae).

GROUP 2: WATER BIRDS

The birds considered here include the ducks, geese, swans, (the Anseriformes), the loons or divers (Gaviiformes) and the grebes (Podicipediformes).

20.2 Description of bird groups

GROUP 3: SEA BIRDS

Here are included the gulls and terns (Laridae), the auks (Alcidae) (both members of the order Charadriiformes) and the pelicans (the Pelecanidae).

GROUP 4: EAGLES, HAWKS AND OWLS

This group includes the carnivorous birds of prey (Falconiformes) and the owls (Strigiformes).

GROUP 5: WADERS (SHORE BIRDS AND MEADOW BIRDS)

This group includes the 'long-legged' birds, many of whom feed by wading at the water's edge. The birds considered in this section include the waders (order Charadriiformes) and herons (family Ardeidae).

As with other animals the study of birds has its own methods and terminology. Research into the effect of disturbance on birds has a limited history, with the main focus being on either the distance at which a bird perceives the human disturbance or on the nesting success of the birds. This is often reported as the number of young birds reared per breeding pair of disturbed adults compared to the success rate of birds in undisturbed areas. However, many workers have measured the flight or flushing distance, which is the distance between an approaching human and the bird when it takes flight (Cooke, 1980). Van der Zande, van der Ter Keurs and van der Weijden (1980) used the term 'disturbance distance' as the distance at which maximum nest density occurs in relation to a fixed source of disturbance, in this case a road, and they referred to 'disturbance intensity' as the total population density loss suffered over the whole disturbance distance (see Yalden and Yalden, 1990) (Fig. 20.1).

20.3 Group 1: perching birds

The perching birds are generally rather small and may be found in almost any terrestrial habitat, but are most common in shrubby or wooded areas. A number of species have readily adapted to urban conditions.

The perching birds generally construct nests in which their eggs are laid, hatched and the young fed until they can fly. Nests are mostly above ground and usually camouflaged to some degree.

This group includes bud-, seed- and fruit-eating birds, although there are also some insectivorous and even some carnivorous species (shrikes, e.g. *Lanius*). Some members, such as swallows (*Hirundo*), are migratory, wintering in or near the tropics and nesting in temperate areas in summer (Young, 1962).

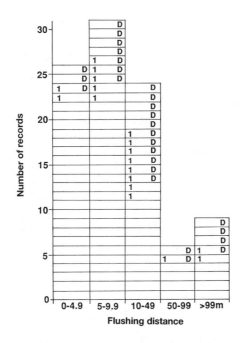

Fig. 20.1 Distances at which incubating golden plover flushed, grouped into irregular distance categories. Records when dogs were present (D) and first visits (1) are identified (n = 96 records). (From Yalden and Yalden, 1990.)

TYPE 1 DISTURBANCE

The effect of walkers on birds has mainly been recorded in three ways:

- the nearest distance to which the birds can be approached before they take flight;
- the density of birds seen in various areas subject to different levels of use; and
- reproductive success in terms of nesting and young birds produced.

The effect of a pedestrian on perching birds depends on the 'shyness' or timidity of the birds and this, in turn, may be affected by habituation (either familiarity or fear), and its reinforcement (van der Zande, 1984). Where groups of birds are concerned, they will generally all fly when the shyest individual or species flies away (Gendbien and Morzer-Bruijns, 1970, in van der Zande, 1984).

Cooke (1980), who compared the fly-away distance of rural and suburban birds, used a careful technique of walking slowly towards the birds after he was sure they knew he was present, and recording the distance between himself and the bird when it took flight. He found that in 13 out of 14 comparisons they could be approached more closely in a suburban area (7 of the 14 were statistically significant) (Table 20.1), which suggests that they had become positively habituated to humans. The smaller species (e.g. bluetit) also permitted a closer approach than the larger ones (e.g. rook). As with other animals, conditioning may be negative and Cooke (1980) commented that persecution may account for the long distances of nearly 60 m at which rooks (*Corvus frugilegus*) took flight from a walker. The smaller species recorded in this study were usually found in hedges and trees and they may be habituated to the harmless presence of people as they were partially hidden in this habitat and not seen by the visitors (Cooke, 1980).

Several studies have recorded the presence or absence of birds in areas used for outdoor recreation. One of the problems in interpretating these results is that most sites used extensively for recreation also have modified vegetation structure or even buildings to provide visitor facilities, combining the effects of type 1 and type 2 disturbance.

Table 20.1 Numbers of observations, tolerance distances and body length for species recorded in rural and suburban areas (modified from Cooke, 1980)

Species	Body length (cm)	Rural area, tolerance distance, mean ± SE (m)	Suburban area, tolerance distance, mean ± SE (m)
Rook (*Corvus frugilegus*)	46	59.9 ± 2.6	
Jackdaw (*C. monedula*)	33	18.1 ± 0.5	16.0
Blackbird (*Turdus merula*)	25	20.9 ± 2.5	11.4 ± 0.3***
Song thrush (*T. philomelos*)	23	15.8 ± 0.9	6.0 ± 0.5***
Starling (*Sturnus vulgaris*)	21.5	17.3 ± 0.7	11.8 ± 0.4***
Skylark (*Alauda arvensis*)	18	14.9 ± 1.0	
Pied wagtail (*Motacilla alba*)	18	8.6 ± 0.9	8.3
Yellowhammer (*Emberiza citrinella*)	16.5	12.4 ± 0.7	
Greenfinch (*Carduelis chloris*)	14.5	19.0 ± 1.5	9.7 ± 1.3***
Dunnock (*Prunella modularis*)	14.5	9.2 ± 0.5	6.1 ± 0.4***
House sparrow (*Passer domesticus*)	14.5	12.4 ± 0.7	8.2 ± 0.1***
Robin (*Erithacus rubecula*)	14	7.0 ± 0.6	5.1 ± 0.5*
Great tit (*Parus major*)	14	5.9 ± 0.6	5.6 ± 0.9
Whitethroat (*Sylvia communis*)	14	11.7 ± 1.2	9.0
Goldfinch (*Carduelis chloris*)	12	11.5	7.2 ± 0.8
Blue tit (*Parus caeruleus*)	11.5	4.8 ± 0.3	5.2 ± 0.4
Chaffinch (*Fringilla coelebs*)	15	8.7 ± 0.9	7.1 ± 0.4

Significantly different from mean in rural area: *, $P < 0.05$; ***, $P < 0.001$.

Van der Zande and van der Vos (1984), using sophisticated analytical techniques, found that an increase in recreation use by walkers, cyclists and picnickers of a lake shore in The Netherlands had a negative effect on the number of birds nesting in adjacent groves and hedges. They analysed data on the 12 most abundant species of Passeriformes. While there were no statistically significant results for the individual species, the fact that 11 of the 12 species had negative differences was considered important. In another study of recreation woodlands adjacent to urban residential areas, van der Zande *et al.* (1984) also concentrated on breeding bird species. Of the 13 abundant species recorded, they found significant negative correlations between bird densities and recreation intensities for eight species. They were also able to rank the species according to their susceptibility and to break down their information according to habitat variables (Fig. 20.2). In general there was a stronger recreation effect in deciduous than in coniferous plantations, probably because visibility is lower in coniferous plantations. In this case the larger species did not appear to be more timid than the smaller ones, the wood pigeon (*Columba palumbus*) being among the least responsive. This may, however, be due to the fact that these birds (along with the other two least-timid species) nest in the treetops and are thus further from disturbance at ground level. It is significant that these authors found the strongest correlations between bird densities and recreation intensity on weekdays rather than at weekends (the time of maximum levels of use). This suggests that sustained use was having a greater effect than the intermittent peaks of use. They concluded their paper with an ingenious application of the idea of lethal dose for 50% of organisms (LD_{50}). Areas were mapped where the recreation intensity was above that at which 50% of the individuals of selected species had disappeared. Data on two susceptible species, the garden warbler (*Sylvia borin*) and the turtle dove (*Streptopelia turtur*), and on two moderately susceptible species, the chiffchaff (*Phylloscopus collybita*) and the wood pigeon, respectively, were combined to produce the maps shown in Fig. 20.3. The LD_{50} concept is a convenient one for management, although the 50% level may clearly be altered to, say, 70% or higher for the rarer species.

Disturbed areas around car parks on the Cairngorms were found to have more pied wagtails (*Motacilla alba*), crows (*Corvus corone*) and rooks (*Corvus frugilegus*), as well as gulls (Laridae) and snow buntings (*Plectrophenax nivalis*), than other areas, as these birds fed frequently on waste human food (Watson, 1976). In this case the influx and increase of these scavenging bird species has occurred on ground adjacent to two national nature reserves on fairly natural arctic–alpine habitats, and may be considered as deleterious for the integrity of this environment.

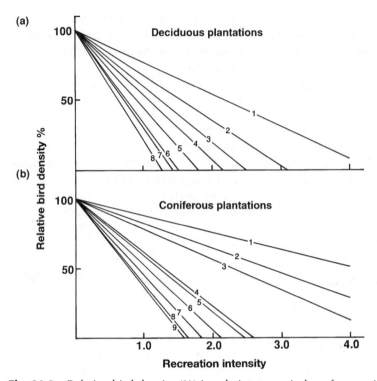

Fig. 20.2 Relative bird density (%) in relation to an index of recreation intensity. The species are: (a) 1, chaffinch (*Fringilla coelebs*); 2, blackcap (*Sylvia atricapilla*); 3, wood pigeon (*Columba palumbus*); 4, chiffchaff (*Phylloscopus collybita*); 5, song thrush (*Turdus philomelos*); 6, willow warbler (*Phylloscopus trochilus*); 7, garden warbler (*Sylvia borin*); 8, turtle dove (*Streptopelia turtur*); and in (b) 1, chaffinch (*Fringilla coelebs*); 2, robin (*Erithacus rubecula*); 3, wood pigeon (*Columba palumbus*); 4, wren (*Troglodytes troglodytes*); 5, chiffchaff (*Phylloscopus collybita*); 6, turtle dove (*Streptopelia turtur*); 7, songthrush (*Turdus philomelos*); 8, willow warbler (*Phylloscopus trochilus*); and 9, garden warbler (*Sylvia borin*). (From van der Zande *et al.*, 1984.)

TYPE 2 DISTURBANCE

There are many studies of the effect of urbanization on birds, but few relating to recreation facilities. The bird species on campgrounds in riparian zones along the Logan and Blacksmith Fork rivers in northern Utah were recorded by Blakesley and Reese (1988). They found that seven species were positively associated and seven species negatively associated with campgrounds (Table 20.2). Five of the seven species positively associated with campgrounds nest in trees and, with one

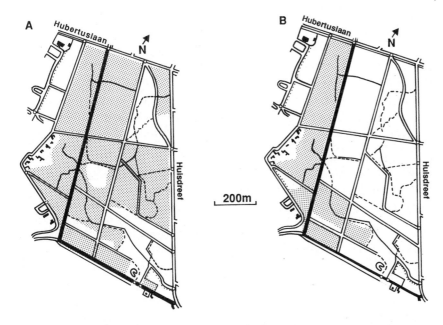

Fig. 20.3 Maps of Ulvenhout, The Netherlands, showing hatched areas in which 50% of the individuals of certain species will disappear (LD$_{50}$). The area contains both deciduous and coniferous woodlands. (a) Two very susceptible species, turtle dove and garden warbler, and (b) two moderately susceptible species, chiffchaff and wood pigeon. (From van der Zande *et al.*, 1984.)

exception, all the species avoiding campground sites nest on the ground, in shrubs or in small trees. The location of the nesting sites was also a factor in the distribution of species recorded by van der Zande (1984).

Feeding habits and human activity may also influence distribution. Blakesley and Reese (1988) also commented that of the six ground feeders in their list, two of the three associated with campgrounds, the American robin (*Turdus migratorius*) and grey catbird (*Dumetella carolinensis*), are attracted to food sources created by humans. The more wary ground-feeding species, such as the fox sparrow (*Passerella iliaca*), may avoid human activity.

The presence in campgrounds of species which were associated with vegetation was attributed to the patches or stretch of shrubs and/or trees along the stream banks and between individual campsites. An alternative adaptation to the presence of humans in campgrounds was noted in the Yosemite Valley where white-headed woodpeckers (*Dendrocopos albelarvatus*) were only seen to forage in the heavily populated campgrounds between dawn and 7.00 a.m. when the

Table 20.2 Bird species associated with campgrounds in Utah (Blakesley and Reese, 1988)

Species positively associated with campgrounds	Species negatively associated with campgrounds
Swainson's thrush (*Catharus ustulatus*)	Broad-tailed hummingbird (*Selasphorus platycercus*)
American robin (*Turdus migratorius*)	Dusky flycatcher (*Empidonax oberholseri*)
Grey catbird (*Dumetella carolinensis*)	Willow flycatcher (*E. traillii*)
Warbling vireo (*Vireo gilvus*)	Black-capped chickadee (*Parus atricapillus*)
Yellow warbler (*Dendroica petechia*)	Lazuli bunting (*Passerina ameona*)
MacGillivray's warbler (*Oporonis tolmiei*)	Song sparrow (*Melospiza melodia*)
Black-headed grosbeak (*Pheucticus melanocephalus*)	Fox sparrow (*Passerella iliaca*)

campers became active. The authors of a study of the high-altitude (2590 m) lodgepole pine (*Pinus murrayana*) forest campground of Tuolumne Meadows in the Yosemite National Park also discussed the importance of the feeding habits, or guild (a group of birds with similar habits), on the birds' reaction to visitors (Garton, Bowen and Foin, 1977). In their analysis of the birds present in the campground and adjacent forest they also took account of the effect of type 2 habitat change by censusing the birds in June before many visitors were present and in August when there was over 90% occupancy of the campsites (Table 20.3). The statistically significant changes occurred among the ground-feeding omnivores. The brown-headed cowbird (*Molothrus ater*), Brewer's blackbird (*Euphagus cyanocephalus*) and American robin (all forest-edge or meadow species commonly associated with humans) were all more abundant near the campsites, while the Oregon junco (*Junco oreganus*), which is a forest species, preferred dense lower foliage and ground vegetation and was less abundant in the campsites.

Campsite avifauna was also generally found to decrease when the site in the Coconino National Forest, Arizona, was in use (Aitchison, 1977). In one year (1973), apart from removal of trees and cut plant material from the site for management purposes, campers destroyed 30% of the Steller's jay (*Cyanocitta stelleri*) nests and 20% of the American robin nests by removing branches for firewood and making room for tents. In 1973 and 1974 the species numbers dropped from 12 to 8 and 16 to 13, respectively. During 1974 the species numbers changed from 10 to 12, perhaps due to an unusual influx of birds as the number on the unused control site went from 7 to 17 at the same

Table 20.3 Activity indices of birds along transects in Yosemite National Park during two time periods (modified from Garton, Bowen and Foin, 1977)

Species	Diet	Location	Campground		Non-campground		Significance of differences between transects	
			I (type 1 disturbance)	II (type 2 disturbance)	I (type 1 disturbance)	II (type 2 disturbance)	I (type 1 disturbance)	II (type 2 disturbance)
Audubon warbler (*Dendroica auduboni*)	Insectivore	Foliage	6	38	8	9	NS	NS
Nashville warbler (*Vermivora ruficapilla*)	Insectivore	Foliage	16	36	16		NS	NS
Mountain chickadee (*Parus gambeli*)	Insectivore	Foliage	26	192	42	47	NS	NS
Brown creeper (*Certhia familiaris*)	Insectivore	Trunk	13				NS	NS
Western wood peewee (*Contopus sordidulus*)	Insectivore	Air	6			9		
Oregon junco (*Junco oreganus*)	Omnivore	Ground	82	35	218	332	$P < 0.01$	$P < 0.01$
Brown headed cowbird (*Molothrus ater*)	Omnivore	Ground	80	81			$P < 0.01$	$P < 0.01$
Brewer's blackbird (*Euphagus cyanocephalus*)	Omnivore	Ground	142	463	24		$P < 0.01$	$P < 0.01$
American robin (*Turdus migratorius*)	Omnivore	Ground	184	48	24	21	$P < 0.05$	$P < 0.05$
Varied thrush (*Ixoreus naevius*)	Omnivore	Ground	16	9	8			
Steller jay (*Cyanocitta stelleri*)	Omnivore	Foliage	39	9	8		NS	NS
Pine siskin (*Spinus pinus*)	Granivore	Foliage	18					
Purple finch (*Carpodacus purpureus*)	Frugivore	Foliage	16					
Pine grosbeak (*Pinicola enucleator*)	Frugivore	Foliage	29	10		11		
Evening grosbeak (*Hesperiphona vespertina*)	Frugivore	Foliage	8		8			

Period I, June 1974; Period II, August 1974. Activity index is expressed as the mean number of birds/2.0 km².
[a] Wilcoxon rank sum test.

time. There is little doubt that the use of campgrounds does cause a change in the numbers of birds present. The type 2 disturbance had a positive effect for 10 species whereas the presence of people (type 1 disturbance) only had a positive effect on four species (Table 20.4). The overall effect at Coconino was positive with a corresponding increase in species diversity for the campsite area. Garton, Bowen and Foin (1977) commented that the habitat changes caused by trampling, log and small tree cutting and burning, which increased the foliage volume diversity in the campground, would cause a corresponding increase in bird species diversity according to the theories of McArthur and McArthur (1961).

Garton, Bowen and Foin (1977) also reported that many rarer species are known to occur in the area but were not seen in their studies of the Tuolumne Meadow area of Yosemite National Park, because humans are likely to have a negative effect on such birds even when there are very few people around.

There has been rather more research on the effects of development and urbanization. In general, the numbers of bird species decrease while the numbers of individuals of some species increase with the increasing development (De Graaf and Thomas, 1976). Omnivores or species that nest on or in buildings are favoured, while those that nest on or near the ground are eliminated. Wooden or ornate stone structures provide more nest sites, while modern 'slabsided' buildings have no places for nesting or even perching. However, a study of birds in

Table 20.4 Impact of the valley campground in Yosemite National Park on members of feeding guilds (modified from Garton, Bowen and Foin, 1977)

		Number of species for which the impact was positive, negative or neutral								
		Indirect impact (type 2 disturbance)			Direct impact (type 1 disturbance)			Overall impact		
Diet	Location	Positive	Negative	Neutral	Positive	Negative	Neutral	Positive	Negative	Neutral
Insectivore	Foliage	2		1	3			3		
Insectivore	Trunk		1				1		1	
Insectivore	Air	1				1				1
Omnivore	Ground	3*	1* + 1		1*	2*	2	2* + 1	1* + 1	
Omnivore	Foliage	1					1	1		
Granivore	Foliage	1					1	1		
Frugivore	Foliage	2	1			1	2	2	1	
Total		10	4	1	4	5	7	10	4	1
Total for significant differences		3	1	–	1	2	1	2	1	

*Differences significant at 5% level or higher using Wilcoxon rank sum test

Tucson, Arizona, showed a massive 26-fold increase in numbers and biomass of mostly omnivorous ground-feeding species, when compared to the surrounding desert (Emlen, 1974) (Fig. 20.4). Forest fragmentation and the associated increase in forest edge is also associated with development for recreation and may have a major impact on forest birds (Robinson, 1990).

In contrast to the studies discussed above, Tilghman (1987) investigated winter bird communities in urban woodlands and found that 19 species were positively correlated (11 significantly) and 11 species were negatively correlated (zero significantly) with woodland size (Table 20.5). Four habitat variables, the size of woodland, the density of adjacent buildings, the amount of edge and the distance to the nearest body of water accounted for 65% of the variation in numbers. Her list contained one duck, one hawk and two game birds as well as 26 perching birds. The study of the birds on two islands, one developed and one untouched, near Charleston, Carolina (Chamberlin, 1982) did not yield sufficient information to be analysed statistically, but the placing of each species according to its feeding guild (Table 20.6), as well as nesting sites if breeding, is one approach that is likely to yield results in future work on recreation impacts. There are, of course, some grassland-feeding or nesting passerines which are favoured by the human activities maintaining this habitat.

An approach using response guilds was applied to the bird and mammal populations in a disturbed and an undisturbed watershed in Pennsylvania (Croonquist and Brooks, 1991). Their guilds depended on five broad characteristics, each divided into five ranked categories (Table 20.7). Their test census showed that the percentage of birds with high response (guild scores), especially species with specific habitat requirements or neotropical migrants, declined as the intensity of disturbance increased. Disturbance included residential and

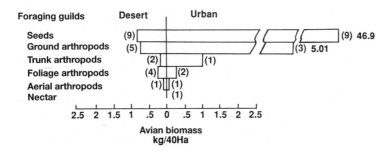

Fig. 20.4 Structure of desert and urban bird communities at Tucson, Arizona, compared by foraging guild biomass (kg/40 ha). The number of species in each guild in each habitat is shown in brackets (from Emlen, 1974).

agricultural development. Species with low guild scores were present in higher percentages in disturbed areas.

Road building, and particularly alteration of roadside habitat, can have a major effect on passerine species. Arnold and Weeldenburg

Table 20.5 Abundance indices for birds wintering in urban woodlands of different sizes (Tilghman, 1987)

Species		Woodland size class (ha)[a]				Simple correlation coefficient[b]
Common Name	Scientific Name	1–5	6–17	18–42	43–69	(r)
Birds observed six times or more in two winters						
Mallard	*Anas platyrhynchos*	0.45	0.07	0.01	0	−0.13
Red-tailed hawk	*Buteo jamaicensis*	0	0.003	0.004	0.02	0.49**
Ring-necked pheasant	*Phasianus colchicus*	0.002	0.006	0.005	0	0.06
Ruffed grouse	*Bonasa umbellus*	0.005	0.003	0.01	0.04	0.36**
Mourning dove	*Zenaida macroura*	0.13	0.22	0.06	0	−0.07
Downy woodpecker	*Picoides pubescens*	0.22	0.22	0.29	0.50	0.35**
Hairy woodpecker	*Picoides villosus*	0.10	0.12	0.18	0.24	0.36**
Blue jay	*Cyanocitta cristata*	0.60	0.40	0.67	0.40	0.05
American crow	*Corvus brachyrhynchos*	0.19	0.27	0.57	0.55	0.11
Black-capped chickadee	*Parus atricapillus*	1.22	1.00	1.24	2.20	0.32*
Tufted titmouse	*Parus bicolor*	0.58	0.55	0.54	0.81	0.12
Red-breasted nuthatch	*Sitta canadensis*	0.27	0.36	0.43	0.66	0.42**
White-breasted nuthatch	*Sitta carolinensis*	0.006	0.002	0.04	0.12	0.44**
Brown creeper	*Certhia americana*	0.06	0.04	0.07	0.38	0.51**
Winter wren	*Troglodytes troglodytes*	0	0.002	0.003	0.02	0.44**
Golden-crowned kinglet	*Regulus satrapa*	0.06	0.05	0.04	0.21	0.20
Northern mockingbird	*Mimus polyglottos*	0.09	0.04	0.04	0.03	−0.20
European starling	*Sturnus vulgaris*	0.37	0.28	0.38	0.05	−0.14
Northern cardinal	*Cardinalis cardinalis*	0.21	0.09	0.10	0.11	−0.11
American tree sparrow	*Spizella arborea*	0.004	0.01	0.04	0.02	0.13
Song sparrow	*Melospiza melodia*	0.04	0.02	0.02	0.005	−0.14
White throated sparrow	*Zonotrichia albicollis*	0	0.006	0.01	0.01	0.15
Dark-eyed junco	*Junco hyemalis*	0.28	0.16	0.23	0.26	−0.02
House finch	*Carpodacus mexicanus*	0.26	0.18	0.30	0.03	−0.16
White-winged crossbill	*Loxia leucoptera*	0	0.02	0.02	0.11	0.29*
Common redpoll	*Carduelis flammea*	0.09	0.22	0.09	0.62	0.15
Pine siskin	*Carduelis pinus*	0	0	0.10	0.20	0.41**
American goldfinch	*Carduelis tristis*	0.33	0.22	0.20	0.17	−0.14
Evening grosbeak	*Coccothraustes vespertinus*	0.32	0.03	0.09	0	−0.10
House sparrow	*Passer domesticus*	0.22	0.30	0.05	0.02	−0.25*

[a] The average number of individuals of a given species detected per observation point for all woodlands in a particular size class.
[b] Regression of abundance indices for a particular species on the size of woodland. Significance of relationship: *, $P < 0.05$, **, $P < 0.01$.

(1990) demonstrated a clear relationship between roadside vegetation structure and particular bird species in western Australia. The volume of traffic along roads in The Netherlands appeared to have a quantitative effect on the density of nesting birds (Reijnen *et al.*, 1995). There was a reduction in numbers of breeding birds up to a distance of 2800 m from roads with 60 000 cars per day travelling at an average speed of 120 km h^{-1}. A road with only 10 000 cars per day affected birds up to 1500 m distant for the most sensitive species. The least

Table 20.6 Feeding guild distribution on two South Carolina Barrier islands (Chamberlin, 1982)

	Transect	Ground–seed	Foliage–seed	Foliage–nectar	Ground–insect	Air–insect	Timber drilling	Timber searching	Foliage–insect	Misc.
Caper's Island	1	12	1	1	7	3	2	1	18	6
Wildlife	2	6	2		1	1	1	1	15	4
Sanctuary	3	7	2		4	3	2	1	13	3
Kiawah Island	1	5	2	1	2	3	4	–	17	5
Resort, intensely	2	10	1	1	6	2	3	–	21	6
developed	3	9	1	1	4	3	3	–	21	5

Table 20.7 Explanations of response guilds for bird and mammal communities (Brooks and Croonquist, 1990)

Response guilds	Scores	Response guilds	Scores
Wetland dependency		Habitat specificity	
Obligate species (> 99% in wetlands)	5	Alpha species – stenotypic, specialist	5
Faculative wet (usually in or near wetlands)	3	Gamma species – landscape dependent	3
Faculative (wetlands not essential)	1	Beta species – generalist, edge	1
Faculative dry (occasional or no use)	0		
Upland (> 99% in uplands)	0		
Trophic level		Seasonality (birds only)	
Carnivore, specialist (restricted diet)	5	Neotropical migrant	5
Carnivore, generalist	4	Short-distance migrant	4
Herbivore, specialist (e.g. nuts, nectar)	3	Year-round resident	3
Herbivore, generalist	2	Non-breeding-season resident only	2
Omnivore (plants or animals)	1	Migratory transient	1
		Occasional	0
Species status			
Endangered, endemic, of concern	5		
Commercial, recreational value	3		
Other native species	1		
Exotic	0		

sensitive species were only affected up to 40 m distant. These figures relate mainly to passerine species and to situations where there was an average of 70% woodland along the road. Noise appears to be the main disturbance factor, with visibility relatively unimportant. These figures for the presence of nests suggest that the disturbance corridor for the bird's awareness may be much wider than 2800 m (see also section 20.7 and Figs 20.18, 20.19).

However, it should be mentioned that the verges of roads with low densities of traffic, if managed correctly, may provide valuable nesting sites in agricultural areas (Warner, 1992).

TYPE 3 DISTURBANCE

In some parts of the world passerines are hunted for food or trapped for use as cage birds. However, most problems are caused by children collecting eggs or nests. Well-meaning observers or photographers may attract such 'recreational predators' by their activities or the presence of hides. The author well remembers the shock of entering a carefully constructed hide to photograph the long-tailed tit (*Aegithalos caudatus*) and their hanging nest to find that the nest had been broken open and the eggs had been taken. Children had been attracted to the nest by the presence of the hide. Most of these kinds of predation take place in areas that are visited by rural workers or at forest edges; Oniki (1976) found that 3% of nests had been disturbed by humans near the Amazon town of Manaus, Brazil.

20.4 Group 2: water birds

All the birds in this group are able to swim and generally nest on or near fresh water. They include ducks, geese and swans as well as lesser known groups such as greebs and loons. Many species flock on the water before migration but some need areas of shore on which to graze. Some will nest and feed in urban parks and streams. (Sea birds are discussed in the next section.)

Water birds generally have breeding territories to which they migrate in the summer months. Breeding pairs select a territory and build a nest in the shoreline vegetation, on the shore or sometimes in trees. They often form flocks at the end of the breeding season and migrate to warmer areas for the winter. Some species remain by or near the sea, while others use freshwater lakes and streams. Most species are herbivores, feeding on shore, in reed beds or diving to the bottom to feed on water plants or algae. The carnivores are usually diving birds, such as loons or grebes, eating fish and other small animals.

TYPE 1 DISTURBANCE

Some distances at which passerines fly from a walker were given in Table 20.1. The responses differ between species, flock sizes and to different stimuli at different seasons, under different conditions of food supply, and with different kinds of preconditioning. Several authors have considered the effects of different stimuli with respect to various species of water bird (Table 20.8). It can be seen that power boats are the most disruptive, presumably because of their speed and consequent wash, as well as noise, although the loon (*Gavia immer*) in the Boundary Waters canoe area, was able to habituate to more intensive recreational use, including motorboat access, admittedly on quite large lakes (Titus and van Druff, 1981). In this case some of the habituated birds even defended their nests, staying put and pecking at humans who came within reach, a behaviour also noted in habituated great crested grebes (Keller, 1989).

Activities that caused decreasing levels of disturbance to a mixture of sea and water birds were people jogging (100% of occurrences), men working near the shore, walkers and, finally, worm diggers, which caused zero disturbance (Burger, 1981a). The species included Canada geese (*Branta canadensis*), mallards (*Anas platyrhynchos*), black duck (*A. rubripes*), American wigeon (*A. americana*), greater scaup (*Aythya marila*) and coot (*Fulica americana*). Burger attributed the degree of disturbance to the rate of movement of the person. She noted that the birds were also unaffected by people on horses and suggested that birds are only aware of the horse and not of the person riding it. Game or fly fishing on reservoirs, either from the shore or boat (often both), had a marked effect on the number of wintering wildfowl, including wigeon (*Anas penelope*), teal (*A. crecca*), pochard (*Aythya ferina*) and mallard (*Anas platyrhynchos*). They were all driven from their preferred feeding or roosting sites, and some birds departed from the Llandegfedd Reservoir, South Wales, prematurely when visitation suddenly increased at the start of the fishing season (Bell and Austin, 1985) (Fig. 20.5).

In this case sailing occurred throughout the winter but had little effect as birds preferred the shallows and sailing was restricted to the deeper central area. However, the birds were forced to move to this deeper water when angling from the shore commenced. The combination of the two activities resulted in the birds being driven from the reservoir. At Grafham Water, England, there was a rapid redistribution of water fowl when the permitted fishing season ended in the autumn (Cooke, 1974). Mallard, teal, pochard, tufted duck (*Aythya fuligula*) and coot (*Fulica atra*) had previously been restricted to the unfished, reserved part of the water, where sailing was also prohibited. This occurred in spite of the fact that there was no direct interference of birds by fishermen.

The flight distance of water birds also increases with the size of the flock, up to a certain maximum (Madsen, 1985) (Fig. 20.6). It may alter with the seasons, especially if they are hunted in the autumn, as was the fate of the Brent geese studied by Owens (1977) (Fig. 20.7). Burger (1981a) noted that water birds disturbed when on the water tended to land on adjacent water.

As might be expected, flight distances vary according to the amount of cover available to the birds. Pierce, Spray and Stuart (1993) found that in *Paspalum vaginatum* grass mats, the flight distances of water birds on Lake Sougkla Thailand, were shorter than in *Scirpus litoralis* beds (41 m and 45 m, respectively). The flight distances in open water were greater again (59 m). They also found that the flight distances from a motorboat (42 m) were slightly greater than from a boat driven by a pole (39 m).

Table 20.8 Flight (or flushing) distance of water birds in response to different stimuli

| | | Flight distance (m) | | | |
| | | Winter flocks | | | |
Stimulus	From nest	Floating	Feeding on land	Species and flock size	Reference
Walkers on shore		100/200		Goldeneyes, 60	Hume (1972)
Sailing boat		350/400		Goldeneyes, 38	Hume (1972)
Powerboat		550/700		Goldeneyes, 28	Hume (1972)
Powerboat towed on road trailer		350		Goldeneyes, large group	Hume (1972)
Sailing boat		450		Tufted duck and Pochard, 300	Batten (1977)
Sailing boat		350		Tufted duck and Pochard, 300	Batten (1977)
Sailing boat		350		Tufted duck and Pochard, 300	Batten (1977)
Sailing boat		230		Tufted duck and Pochard, 200	Batten (1977)
Sailing boat		275		Tufted duck and Pochard, 100	Batten (1977)
Rowing boat		200		Tufted duck and Pochard, 90	Batten (1977)
Walkers			50/200	Brent geese, 6–400	Owens (1977)
Car on road: autumn			500	Pinkfooted geese, 400–600	Madsen (1985)
Car on road: spring			300/400	Pinkfooted geese, 400–600	Madsen (1985)
Rowing boat, undisturbed lake	50–100			Great crested grebe, 1	Keller (1989)
Rowing boat, road-disturbed lake	0–20			Great crested grebe, 1	Keller (1989)
Rowing boat, very disturbed lake	0–10			Great crested grebe, 1	Keller (1989)
Canoe, undisturbed lake (no motorboats)	'Far' (100s)			Loon, 1 (great northern diver)	Titus and van Druff (1981)
Canoe, more disturbed lake (including motorboats)	'Near' (0–2)			Loon, 1	Titus and van Druff (1981)

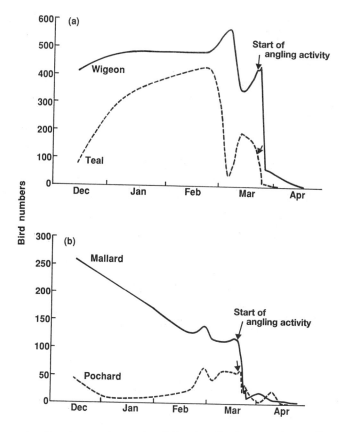

Fig. 20.5 Decline in numbers of wildfowl with the start in angling activity in March on Llandegfedd Reservoir, South Wales. Species shown are wigeon (*Anas penelope*), teal (*A. crecca*), pochard (*Aythya ferina*) and mallard (*Anas platyrhynchos*). (From Bell and Austin, 1985.)

The consequences of the disturbance are much harder to determine. One frequently used measure of the effects of disturbance has been nesting success. Loons (*Gavia immer*) were found to hatch more eggs and raise more 2-week-old juveniles per egg laid and per breeding pair when nesting in lakes where motor canoes were not allowed than in lakes where motors were permitted (Titus and van Druff, 1981) (Table 20.9). This was in spite of the fact that the birds in frequently used lakes had become habituated to humans (Table 20.8). However, some of this disturbance resulted from the conflict between loons and humans for the use of small islands for nesting or camping. McIntyre (1986) noted that the numbers of loons were increasing throughout North America, e.g. an increase from 200 adults in New Hampshire in 1975 to 300 in 1985, after a decline from an estimated total of

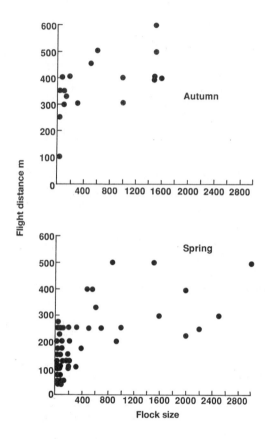

Fig. 20.6 Flight distances of flocks of pinkfooted geese (*Anser brachyrhynchus*) approached by a car in autumn (September–November) and in spring (February–May), in West Jutland, Denmark (from Madsen, 1985).

800 in 1900. So there are indications that habituation could be leading to greater survival. A similar effect for habituated great crested grebes (*Podiceps cristatus*) was found in three small lakes in Switzerland (Keller, 1989). Here the birds in two lakes used for recreation had 53% and 75% clutch losses, compared with 83% loss in the unused lake. In the used lakes, 76% and 54% of the pairs were successful in rearing one or more young, and 76% of the pairs were successful in the lake that was almost unused, where the birds were shy and had much greater flight distances (Table 20.8).

An additional behavioural feature of grebes is that when they leave their nests unattended they often protect their eggs by pulling nest material over them. However, those birds that were flushed from their nests at flight distances of 6 m or less covered their eggs significantly

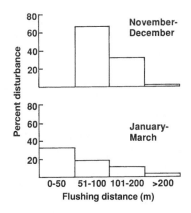

Fig. 20.7 The distances at which people on the ground put Brent geese (*Branta bernicla*) to flight in early and late winter on Norfolk saltmarshes. The flight distance decreased in late winter as birds became more habituated to people (from Owens, 1977.)

less frequently (Keller, 1989). In contrast, the dabchick (*Podiceps ruficollis*) was found to cover its eggs only when leaving its nest as a result of disturbance, but the flight distance was not specified (Broekhuysen, 1973). The breeding success of birds is well known to be endangered by people visiting their nests, whether the purpose of the visit was recreation or study. Over 22% of the eggs of Canada goose (*Branta canadensis*) (288 eggs) were lost between 1965 and 1969 in a study area near Hudson Bay, Canada (MacInnes and Misra, 1972). These authors concluded that 55% of the losses could be directly attributed to human-assisted predation. The predators that human researchers attracted to the nest by their well-intentioned study visits included jaegers (*Stercorarius* spp.), gulls (*Larus argentatus* and *L. thayeri*) and arctic foxes (*Alopex lagopus*). A study of mallard nesting in Moravia, Czechoslovakia, also found that the birds that had been flushed from their nests by the scientists were subsequently more easily frightened by passing anglers, and 57% abandoned their eggs and, in three cases, even newly hatched young (Balát, 1969). In this location habituation with a negative response had occurred.

Studies in the Archipelago of Torku, Finland, showed that disturbance from boats could adversely affect velvet scoter (*Melanitta fusca*) ducklings (Mikola *et al.*, 1994). They were subject to 3.5 times as many gull attacks and diverted valuable feeding time to swimming so that they could escape from the boats. Broods that were disturbed more frequently tended to be smaller. Similar effects were found for disturbed eider (*Somateria mollissima*) ducklings in Scotland (Keller, 1991). Here they were also exposed to fishermen and other people

walking along the shore, some with dogs, and these caused more disturbance than the water-based activities of windsurfers and rowing boats.
Keller (1991) also suggested that motorboats are likely to have a much
greater effect. Canoes that were paddled or towed by a walking man
were thought by Hulbert (1990) to have minimal effect on adult ruddy
shelduck (*Tadorna ferruginea*) in the Royal Chitwan National Park of
Nepal. However, model powerboats on Herdsman Lake, Perth,
Western Australia, did not permanently displace the water birds,
although they sheltered or moved away while the boats were present
(Bamford, Davies and van Delft, 1990).

The noise of supersonic airliners taking off from Kennedy Airport,
New York (120–30 dB(A), see Fig. 17.6) was always found to flush a
mixture of water birds, sea birds and waders whenever the aircraft
passed overhead, whereas other passenger subsonic aircraft almost
never caused disturbance, except adjacent to the runways (Burger,
1981a). However, Owens (1977) found that Brent geese (*Branta
bernicla*) did not habituate to small, slow-flying recreational aircraft,
possibly because their shape resembled that of a hawk. Greater snow
geese (*Chen caerulescens atlantica*) were also found to be sensitive to
low-flying private aircraft at the Montmagny bird sanctuary, Québec
(Bélanger and Bédard, 1989). Low-flying aircraft, at less than 500 m,
caused about half of all disturbances, which occurred at rates of 1.46
h^{-1} and 1.02 h^{-1} in the fall and spring, respectively. The entire flock
took wing in 20% of the cases, and they did not resume feeding for
2–6 minutes. When disturbance frequency exceeded 2 h^{-1} half the flock

Table 20.9 Success in hatching and brood rearing of common loons on lakes where motors were allowed and on lakes where they were not allowed during the 1975 and 1976 nesting seasons, Boundary Waters canoe area study area (Titus and van Druff, 1981)

Year	Type of lake use	Number of breeding pairs	Number of eggs	Numbers of batches	Number of juveniles 2 weeks of age	Mean batch per egg	Mean batch per breeding pair	Mean number of juveniles 2 weeks old per egg egg laid	Mean number of juveniles 2 weeks old per breeding pair
1975	Motorboats	27	47	17	14	0.36	0.63	0.30	0.52
	No motorboats	37	72	33	26	0.46	0.89	0.36	0.70
1976	Motorboats	17	32	11	9	0.34[a]	0.65[b]	0.28[b]	0.53
	No motorboats	31	51	27	22	0.53[a]	0.87[b]	0.43[b]	0.70
Both years	Motorboats	44	79	28	23	0.35[a]	0.64[b]	0.29[b]	0.52
	No motorboats	68	123	60	48	0.48[a]	0.88[b]	0.39[b]	0.71

[a] Significant difference at $P < 0.05$.
[b] Difference at only $P < 0.10$.

did not return to feed the next day. The authors recommended strict regulation of low-level aircraft flights.

TYPE 2 DISTURBANCE

Clearly, any alteration to local water levels or vegetation removal for recreation purposes in nesting areas could have an effect on breeding water birds. However, the major changes in feeding or breeding areas affect water birds to a greater degree, and structures built for recreation purposes are not themselves likely to have other than very local consequences.

TYPE 3 DISTURBANCE

Hunting water birds, especially during their migrations, is a common sport. The numbers shot have been the subject of considerable debate and bag limits for some or all species have been imposed in many countries. There is also no doubt that previously hunted birds are more timid and they only become habituated to humans, even when not being hunted, as shortage of food or nesting sites drive them closer (Owens, 1977) (Fig. 20.8).

20.5 Group 3: sea birds

Sea birds spend most of their life at sea or on the sea shore. However, important exceptions are fairly frequent, such as the herring gull (*Larus argentatus*) which has been spreading inland in parts of Europe, and other species which may be found at large bodies of inland water (e.g. pelicans (*Pelecanus conspicillatus*) on Lake Eyre, an ephemeral salt lake in Australia).

Practically all sea birds nest in colonies, some of which are very large. Species such as gulls (Laridae), terns (Sternidae) and gannets (Sulidae) utilize beaches and flatter areas, while auks (Alcidae), fulmars (*Fulmarus glacialis*) and kittiwakes (*Rissa tridactyla*) utilize ledges and crevices in steep or vertical cliffs. Emperor penguins (*Aptenodytes forsteri*) breed on the Antarctic ice and brown pelicans (*Pelecanus occidentalis californicus*) may nest in mangrove trees (*Rhizophora mangle*) in areas where they have been constantly hunted (Anderson and Keith, 1980). Some, such as the shearwaters (*Puffinus* spp.) nest in colonial burrows on island cliff tops and short distances inland.

During their non-breeding season sea birds may wander extensively, a classic case being the wandering albatross (*Diomedea exulans*) which may traverse the southern oceans between 30°S and 60°S on its own, only mating and nesting every 2 years. However, most sea birds will

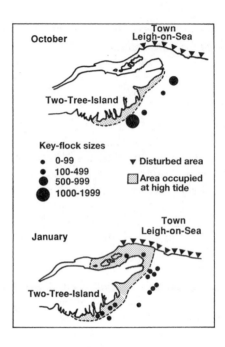

Fig. 20.8 The distribution of Brent geese at Leigh Marsh before and after habituation to disturbance by people on the shore at Leigh-on-Sea. As food supplies reduced the birds moved nearer the town. (From Owens, 1977.)

remain in smaller or even large flocks in the non-breeding season. They all feed on fish which are usually caught at sea, although some groups, such as gulls, adapt readily to feeding on human refuse. They nest in colonies, and parental care is usually shared between parents. Conspecific predation on chicks may occur.

TYPE 1 DISTURBANCE

The massed array of colonial seabird nests, the natural beauty of many sites chosen by the birds, plus the excitement of wheeling flight and cacophony of distress calls from disturbed birds have long attracted sightseers and those interested in the natural environment. Seabird colonies have been a major victim of the conservation 'backlash' (Anderson and Keith, 1980) in which the desire to see animals, fostered by conservation interests, has led to increasing disturbance and consequent predation of eggs and young. Predators are mostly other birds that are less timid than the inhabitants of the colony. Some species, such as crows (*Corvus carone*), great black-backed gulls (*Larus marinus*), herring gulls (*L. argentatus*) and dolphin gulls (*Leucophaeus scoresbii*), have learned to utilize the opportunities created by human visitors, following them around and feeding on the exposed seabird eggs and young, and then leaving the area when visitors are not present (Anderson and Keith, 1980; Hand, 1980; Randall and Randall, 1981). Conspecific predation may also occur when disturbed chicks wander into adjacent nest territories or when the parents have lost their own eggs (Hand, 1980).

Flight distances of sea birds from their nests are similar to perching passerine flight distances (Table 20.10). Roseate terns (*Sterna dougallii*) may fly from their nests when humans are as much as 50 m distant, whereas noddy terns (*Anous minutus*) need to be quite frightened before they will depart (Hulsman, 1984). The birds' sensitivity is higher at the start of the breeding cycle and the birds are more likely to stay on their nests in late incubation or chick-rearing stages (Nisbet and Drury, 1972; Hulsman, 1984). For example, Anderson and Keith (1980) noted that in early incubation some brown pelican nests were abandoned after disturbance, although this may also occur if food supplies are restricted. The breeding success of brown pelicans was very strongly reduced (by at least half and in one case to zero) by the presence of visitors. The direct causes of death of small pelican chicks included trampling by adults or larger chicks as they were frightened off the nest; hypothermia; and larger, more mobile, young pelicans often became impaled on cactus (*Opuntia*) when they fled from visitors, and were then unable to free themselves. Exposed seabird eggs and young chicks may also die from hypothermia while the parents are away (Hunt, 1972).

When nesting herring gulls were disturbed regularly by picnickers on an island off the coast of Maine, their hatching success was only 22%, compared to 49% in undisturbed colonies (Hunt, 1972). The cause of death was attributed to predation and hypothermia. Over 96% of the world population of Hermans gull (*Larus heermanni*) nests on one island on the Gulf of California: a vulnerable species. When subcolonies were not subject to human disturbance they produced 17.7 young to every 100 adults, when moderately disturbed 13.6 young survived, but heavy disturbance depressed production to only 4.5 young. Clearly, disturbance is a significant threat to this species of gull. Mortality in this case was the result of intense territoriality of neighbouring gulls who destroyed both eggs and young when they or the adults are displaced by visitors (Anderson and Keith, 1980). Similar conspecific predation as a result of disturbance was observed in western gulls (*L. occidentalis*) colonies in the same area (Hand, 1980). Mortality of chicks of the glaucous-winged gull (*L. glaucescens*) on Colville Island, Washington, mainly occurred when they scattered into neighbouring territories in the first 7 days after hatching, and was much

Table 20.10 Flight distance of sea birds

Species	Distance	Reference
King shags (*Phalacrocorax albiventer*)	5 m	Kury and Gochfeld (1975)
Western gull (*Larus occidentalis*)	17 m	Hand (1980)
Roseate terns (*Sterna dougallii*)	50 m	Hulsman (1984)
Noddy terns (*Anous minutus*)	Few m	Hulsman (1984)
Least terns (*Sterna albifrons*)	Walker 20 m (?) Vehicle 10 m (4% flew) Vehicle 5 m (54% flew) Vehicle 1 m (94% flew)	Blodget (1978)

greater in plots disturbed two or three times each day than in undisturbed control plots (Gillett, Hayward and Stout, 1975) (Fig. 20.9). As in most other cases, the intensity of disturbance has a proportional effect on the mortality of eggs and small young of the western gull. Weekly disturbance only caused 8% loss, compared to between 18% and 28% losses when they were disturbed three times a day (Robert and Ralph, 1975) (Fig. 20.10).

However, the effects of disturbance on reproduction are not always negative. For example, the survival of young western gulls once they had hatched, was greatest on the most disturbed plot, very slightly less on the undisturbed plot but depressed by nearly 20% on the plot disturbed only once a week (Table 20.11). Similar beneficial effects of disturbance have been recorded in other species, including the black guillemot (*Cepphus grylle*) on islands in the estuary of the St. Lawrence River, Canada. Chicks in disturbed colonies were significantly heavier (mean, 412 g, compared to 370 g for chicks from lightly disturbed nests) (Cairns, 1980). The author considered that 'those birds which succeeded in hatching their eggs in the face of daily disturbance may have been more attentive or more experienced as parents than the average successful nester in the lightly disturbed area'.

Many gull species are numerous and often seen in large flocks in populated areas. We often do not consider that recreation or other activities could affect them, yet as the following study shows, they are also a part of the vulnerable living world around us. In this case the gulls were subject to disturbance when away from their nests, either as non-breeders or as feeding adults (Burger, 1981b). Burger and Galli (1980) studied the responses of five members of the genus *Larus* to

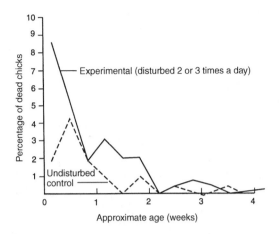

Fig. 20.9 The approximate age at death for glaucous-winged gull (*Larus glaucescens*) chicks in experimental and control areas. In each graph mortality is plotted as percentage of total (live and dead) chicks for that group. (From Gillett, Hayward and Stout, 1975.)

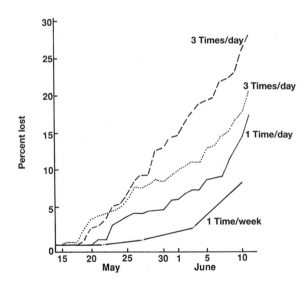

Fig. 20.10 Cumulative percentage of eggs and chicks which died as a result of different levels of disturbance in western gull (*Larus occidentalis*) colonies. This was mostly during the incubation period for these gulls on the Farallon Islands near San Francisco, California. (From Robert and Ralph, 1975.)

aircraft, motor vehicles, people walking, jogging, with dogs or with children, fishermen and clam diggers in two bays on the coast of New Jersey, USA. They classed the responses to disturbance as:

- birds remained where they were;
- birds flew but returned;
- birds flew away.

The responses were complex and could be explained as an interaction between flock size and disturbance. When there were more disturbances, a higher proportion of gulls responded by flying – evident as disturbance increased from April to May (Fig. 20.11). When more gulls were present a higher proportion also responded by flying in Delaware Bay (Fig. 20.12a) but when there were few disturbances and small flock sizes more gulls responded by flying in Raritan Bay (Fig. 20.12b). They explained the fact that more birds flew when more birds were present as the consequence of non-resident and therefore non-habituated birds joining the flocks. They also found that the proportion of gulls that flew away was inversely related to the number of aircraft and directly related to the number of disturbances involving dogs.

Dunnet (1977) found that a small, twin-engined propeller-driven aeroplane flying over birds on the Buchan coast, Scotland at 100 m altitude, did not disturb birds in a mixed colony. A large Sikorsky S61 helicopter also had no effect on nesting birds at an altitude of 150 m, although some non-nesting kittiwakes (*Rissa tridactyla*) did fly off.

A more general effect of disturbance has been the abandonment or reduction in size of colonies of common terns (*Sterna hirundo*), least turns (*S. albifrons*), black skimmer (*Rynchops niger*) and herring gulls (*Larus argentatus*). Differences in human disturbance between New

Table 20.11 Percentage of young and eggs of the western gull (*Larus occidentalis*) surviving on all study plots on south-east Farallon Island, 43 km west of San Francisco, California. Calculations are based either on the total number of eggs known to have been laid and/or on counts made on 11 June (Robert and Ralph, 1975)

Disturbance schedule	Eggs laid	Eggs or young 11 June	% surviving of eggs laid (phase I)	Young 16 July	% surviving of eggs laid	% surviving of no. on 11 June (phase II)
Three times a day	239	174	72.8	82	34.3	47.1
Once a day	316	262	82.9	94	29.7	35.9
Once a week	334	315	94.3	88	26.3	27.9
None		244		110		45.1

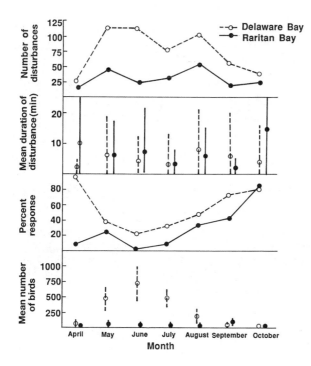

Fig. 20.11 Seasonal pattern of number of disturbances, duration of disturbances (mean ± one standard deviation), percentage of gulls that remained (no response), and mean number of birds present on Delaware and Raritan Bays, New Jersey (mean ± one standard deviation). (From Burger and Galli, 1980.)

Jersey, where unrestricted recreation is allowed on 75% of all ocean frontage, and Virginia, where only 15% of the beaches are totally unprotected, have a marked effect on the locations of nesting sea birds (Erwin, 1980). In New Jersey the vast majority of sea birds have retreated to nest on dredged deposition material or natural marsh islands, whereas in Virginia they nest in their optimal habitat on natural barrier island beaches. Dewey and Nellis (1980) also suggested that there may have been selection for a preference for abnormal cliff nesting amongst rosiate terns (*Sterna dougallii*) and sooty terns (*S. fuscata*) as a result of human disturbance in the Virgin Islands.

TYPE 2 DISTURBANCE

There seem to have been very few investigations into the effects of changed habitats on sea birds but, of course, the presence of herring

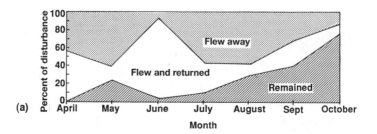

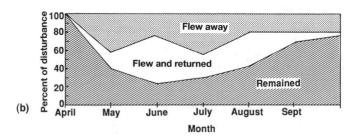

Fig. 20.12   Response of gulls by month (a) in Delaware Bay and (b) in Raritan Bay (from Burger and Galli, 1980).

and other gulls feeding on rubbish dumps is well known. Herring gulls seem to adapt readily to human presence and the author recollects seeing a small nesting colony on the roof of the Hoover factory, in Merthyr Tydfil, Wales in the early 1960s. Silver gulls (*Larus novae-hollandiae*) are also common around coastal picnic and barbecue sites in southern Queensland and northern New South Wales. In southern Australia they are a major problem at garbage dumps (D. Jones, personal communication). Clearly, if facilities or paths are developed within disturbance distance of a seabird colony, there would be a serious impact – Gaddy and Kohlsaat (1987) found that breeding success of least terns was 0.39 fledglings per nest on an undeveloped South Carolina barrier island and 0.09 and 0 for two colonies on islands that had been developed for recreation.

TYPE 3 DISTURBANCE

Although seabird colonies were vital sources of food and fertilizer (guano) in the past, and heroic tales are still told about the exploits of the Hebridians collecting eggs from precipitous cliffs, there are few, if any, instances of present-day hunting for recreation. However, the decline of breeding Australian gannets (*Morus serrator*) on Cat Island,

Victoria, has been attributed primarily to human predation and vandalism; little penguins (*Eudyptula minor*) were also taken for use as bait (Harris and Norman, 1981). This kind of predation may well be inflicted on other sea birds in other locations, especially near large human populations.

Members of this group may be found in almost any terrestrial habitat, from rainforest to desert and high mountains, but they do not usually venture very far out to sea.

20.6 Group 4: eagles, hawks and owls

They are universally carnivorous, usually catching their prey live and, when rearing chicks, taking it back to their young. Their nests are often in an elevated position in the branches of a mature tree or, in the case of owls, in some secluded tree-hole or roof cavity (e.g. the barn owl, *Tyto alba*).

TYPE 1 DISTURBANCE

Like other top carnivores, these birds are intelligent, learn readily and are very individual in their responses to human disturbance. For example, the flight distances from their nests of one population of bald eagles (*Haliaeetus leucocephalus*) in north-central Minnesota during the incubation and rearing season varied between 57 m and 991 m (Fraser, Frenzel and Mathison, 1985) (Fig. 20.13). The effect of habituation in this population was negative and the flight distances increased with repeated disturbances. Stalmaster and Newman (1978) previously recorded a mean distance of 196 m in a bald eagle population in Washington, and they discounted any flights that occurred at distances over 500 m because they felt the flight could not be attributed with certainty to the intrusion. Fraser, Frenzel and Mathison (1985) claimed that bald eagles' long flight distances were genuine responses to disturbance as the birds showed typical flight behaviour. This is in accordance with the observations of Grubb and King (1991) who found that eagles showed alert behaviour at distances over 1000 m. They did, however, also record greater than 1000 m flight responses to aircraft, gunshots and, on two occasions, people, one in a canoe. This means that the disturbance corridor is over 2 km wide in this case. Pedestrians were the strongest disturbance factor, with aquatic, vehicles, other noise and aircraft the least disturbing (Grubb and King, 1991) (Fig. 20.14). In general, eagles that were perching were more easily flushed than those on nests. In contrast, the population of bald eagles on Amchitka Island, Alaska were not easily displaced from their nests but were very aggressive, attacking humans and even helicopters at considerable distances from their nests (Sherrod, White and Williamson, 1976). An attack on a researcher was also recorded on Kodiak National Wildlife Refuge (Grubb, 1976). An Alaskan bald eagle population on the Kenia Peninsula had its nesting success reduced from 88% to 23% during incubation and rearing between May and September, as a consequence of disturbance from recreation, including boating, canoeing and camping (Bangs *et al.*, 1982). However, Grier (1969) reported that climbing to nests to band nestlings had no effect on productivity in north-western Ontario, and human activity had no effect on nesting success in Cheppewa National Forest, Minnesota (Mathisen, 1968).

Fraser, Frenzel and Mathison (1985) commented that variations occur between populations because of geographic differences in human persecution and that different management plans are required for different populations and even different birds, because of their individual variation in response. They added that although these birds are legally protected while nesting, when they are intolerant of disturbance, by the Bald Eagle Act in the USA, their staging and foraging areas are not protected. That bald eagles are susceptible to disturbance when overwintering along the Nooksack River in north-west Washington was demonstrated by Stalmaster and Newman (1978). They found that

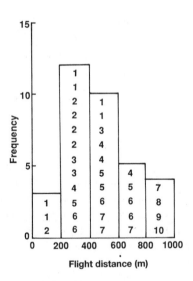

Fig. 20.13 Frequency distribution of 34 bald eagle flush distances, Chippewa National Forest ISA, 1977. The numbers inside the columns are test numbers, where 1 represents the first test at a particular nest, 2 represents the second test, etc. (From Fraser, Frenzel and Mathison, 1985.)

eagle distribution was negatively correlated with the level of human activity, with only 8% of the birds occurring in areas of high activity and 57% and 35% in areas of moderate and low activity, respectively. Anthony and Isaacs (1989) recommended that human activities within 800 m of bald eagle nests should be restricted from 1 January to 8 August. Stalmaster and Newman (1978) found that flight distances from roosting trees were greater for adults than for young birds and that this difference was greatest when the disturber approached by wading along the river channel and least when the birds were approached under a heavy vegetation canopy (Fig. 20.15). They also found that birds disturbed at their feeding sites did not return to the same feeding area for several hours, and that, compared to other eagle activities, feeding birds were most sensitive to human disturbance. However,

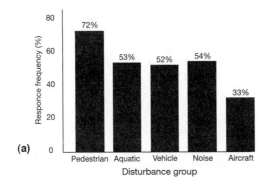

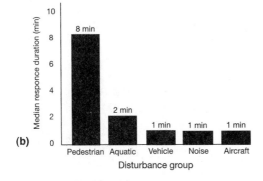

Fig. 20.14 (a) Breeding bald eagle response frequency and (b) median duration of response for five disturbance groups recorded in Arizona, 1983–85 (*n* pedestrian, 401 aquatic, 261 vehicle, 392 other noise, 2849 aircraft). (Frequency = number of responses divided by number of disturbances × 100%.) (From Grubb and King, 1991.)

Bangs *et al.* (1982) also considered them to be intolerant of human activity while raising young. Stalmaster and Newman (1978) recommended activity restriction zones for bald eagle wintering grounds. This may be very important as Craig, Mitchell and Mitchell (1988) suggested that 'increased human disturbance with its consequent higher energy demands could have a negative impact on winter survivorship [of bald eagles] in this population'. This applies especially to juveniles who feed inefficiently and they may have difficulty meeting energy requirements.

Disturbance to other raptors has understandably been less studied than the bald eagle in the USA. Birds of prey were found to be affected by recreationalists in the Duivelsberg Woods near Nijmegen in The Netherlands (Saris, 1976). In particular, buzzards (*Buteo buteo*), sparrowhawks (*Accipiter nisus*), hobbys (*Falco subbuteo*), kestrels (*F. tinnunculus*), tawny owls (*Strix aluco*) and long-eared owls (*Asio otus*) were found to nest only in areas not open to the public. The breeding success of golden eagles (*Aquila chrysaetos*) and peregrine falcons (*Falco peregrinus*) has been reduced by human disturbance in the Cairngorms, Scotland (Watson, 1976). Tourists have even frightened golden eagles off their eggs sufficiently long for the embryos to die

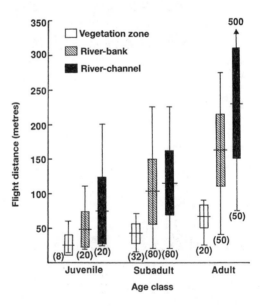

Fig. 20.15 Comparison of the responses (flight distances) of juvenile, subadult and adult bald eagles to vegetation zone, river bank and river channel simulated disturbance types. Mean ± standard deviation, range and sample sizes (in parentheses) are presented. (From Stalmaster and Newman, 1978.)

from the cold. Golden eagles were also studied in New Mexico and 'human disturbance accounted for 85% of all known nest losses' over a 6-year period (Bocker and Ray, 1971). The recovery of the red kite (*Milvus milvus*) population in Wales after the drastic reductions due to pesticide-related egg losses was also inhibited by human disturbance (Newton, Davis and Moss, 1981). Osprey (*Pandion haliaetus*) nesting was disturbed on the Choptank River, Maryland, by sharply increasing numbers of boats with outboard motors and activities relating to launching boats (Reese, 1972). Ospreys have also abandoned their eggs after bird watchers had built an observation blind near the nest and, in another case, the observer actually climbed to the nest which was immediately abandoned (Dunstan, 1973). It is also worth noting that bald eagles do not appear easily disturbed by fixed-wing aircraft, even when they are only 20 m from the eagle's nest (Fraser, Frenzel and Mathison, 1985). Red-tailed hawks (*Buteo jamaicensis*) apparently habituate to low-level (35–40 m) helicopter overflights in east-central Colorado, only 1 in 12 flushing from their nests, compared to 9 out of 17 birds which left their nests in an area that had not previously been exposed to helicopters (Anderson, Rongstad and Mytton, 1989).

In general, it is clear that disturbance can affect many species of raptor and while habituation may allow humans to encroach on their territories without harming the birds or their young, they may also habituate by becoming more sensitive to the presence of humans.

TYPE 2 DISTURBANCE

Changing land use often has a major effect on raptor populations, and as long ago as 1890 Schick pointed out that osprey populations of Seven Mile Beach, New Jersey, declined from 100 pairs to 25 pairs in a 6-year period. He attributed this decline to the destruction of woods on the beaches to make room for summer cottages (Schick, 1890). Similar development of shorelines has affected bald eagles of the Chippawa National Forest, Minnesota, nearly 100 years later. Here the birds nest between 10 and 400 m of the uninhabited shoreline, but they are as much as 1200 m inland when the shoreline has been developed, and up to 4800 m inland when houses have been placed in clusters by the shore. At random localities, the nests were all within 100 m of the shore where it was undeveloped (Fraser, Frenzel and Mathison, 1985). Ospreys were also shown to produce fewer young in areas within 1500 m of human disturbance in west-central Idaho (van Daele and van Daele, 1982).

A very similar effect of development on the distribution of bald eagles was found in the northern end of Chesapeake Bay, Maryland (Buehler *et al.*, 1991). Only 4.9% of 1117 locations occurred in developed areas and eagle use was inversely related to building density. A similar effect on wintering eagles was noted by Paruk (1987) by the Mississippi River, except that river segments that included dams consistently had more eagles than those that lacked dams, as the pools below the dams provided food for the eagles. The Spanish imperial eagle (*Aquila adalberti*) has also been shown to avoid nesting in areas affected by human disturbance (González, Bustamante and Hiraldo, 1992). Fire has also destroyed many of the old trees suitable for bald eagle nest sites on the Kenai Peninsula, Alaska, and what appears otherwise to be a suitable habitat no longer supports nesting bald eagles (Bangs *et al.*, 1982). Some land use changes have favoured raptors – Newton, Davis and Moss (1981) attributed some of the recovery of red kites in Wales to changes in agricultural practice where removal of hedgerows created a more open landscape in which they could hunt for their prey. However, an increasing human population worldwide usually means that land is developed and this inevitably leads to 'raptor population decline' (Craighead and Mindell, 1981).

TYPE 3 DISTURBANCE

Raptors have long been the victims of hunting, both as desired trophies (eggs included) and as supposed pests who prey on lambs and other domestic animals. They are now protected by law in most Western countries but in extreme cases, such as the short-eared owl (*Asio flammeus*) in Belgium, 'hunting remains the main cause of mortality' (van Gompel, 1979). The breeding success of golden eagles in Scotland from 1949 to 1980 depended inversely upon the degree of persecution from game keepers maintaining grouse moors for hunting (Watson, Payne and Rae, 1989). In Minnesota, humans were also a significant factor in the death of adult ospreys (by shooting) (Dunstan, 1968). In North America, Cooper's hawks (*Accipiter cooperii*) also suffered increased mortality after they had become accustomed to people when they were handled for study purposes when young. The handled individuals were more likely to die from human predation, especially shooting, than individuals that had not been handled (Snyder and Snyder, 1974).

While there is an increased awareness of the value of raptors as part of the world's wildlife, it is evident that many still suffer from human predation, including some activities which, although illegal, are considered by their perpetrators as recreational.

Waders are long-legged birds, generally found (as their name implies) wading at the edges of fresh or salt water. Many species of waders are migratory, having certain breeding ranges where they nest and rear their young (usually near water), and other, often warmer, wintering ranges. Many species are found in small groups or, especially at migration times, quite large flocks.

20.7 Group 5: waders

TYPE 1 DISTURBANCE

Disturbance to birds in this group has been studied while they were nesting and at other periods of their life cycles.

The effect of walkers, sometimes with dogs, on breeding golden plovers (*Pluvialis apricaria*) in the Peak District National Park, England, was generally deleterious (Yalden and Yalden, 1990). They found that pre-breeding birds could be disturbed from 'settling into' their territory by walkers, and in one case a male was driven into two adjacent birds' territories by hikers, with the consequent disturbance to all three groups (Fig. 20.16). These authors also found that once the birds were incubating their eggs their flushing distances were very variable, but considerably greater when dogs were present; however, they were much lower than those recorded for nesting bald headed eagles (Fig. 20.13). Yalden and Yalden suggested that the variation may be due to different characteristics of the birds, some being 'sitters' and others 'flyers' . They also noted that the birds were more agitated and took longer to return to their nests when people were present on the moors. Lapwings (*Vanellus vanellus*) also tended to flush at about the same distance when disturbed by humans and much less when disturbed by other birds except raptors (Iversen 1986) (Table 20.12).

Yalden and Yalden (1990) also recorded the amount of time golden plovers spent in anxiety behaviour (alarming, running, flying) compared with the proportion of the day when people were present (Fig. 20.17). They calculated that, on average, the energy expended by this activity may require an extra 58 minutes of foraging time,

Table 20.12 Flushing distance of lapwing (*Vanellus vanellus*) from their nests to various disturbances (modified from Iversen, 1986)

Source of disturbance	Flushing distance (m)
Crows	30–60
Gulls	15–20
Birds of prey	80–100
Humans	70–80

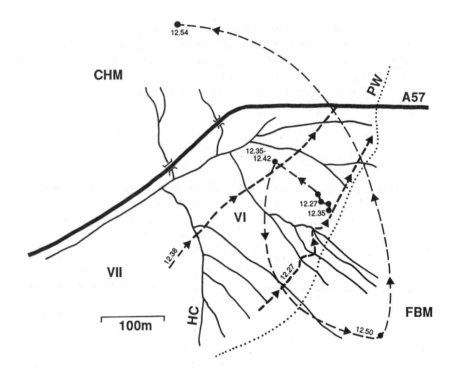

Fig. 20.16 Disturbance to a male golden plover at Snake Summit on Sunday 27 April 1986. Times (BST) given beside tracks of golden plover (– – – →) and people (——→). CHM, Cold Harbour Moor; FBM, Featherbed Moss; HC, Holden Clough; PW, Pennine Way; VI, area VI; VII, area VII. (From Yalden and Yalden, 1990.)

which is spent on pastures away from the moorland, reducing the efficiency of guarding their chicks. Once the chicks were old enough to move away from the nests, the adult and young moved away from walkers and this led to fights with other family groups and further unnecessary expenditure of energy (Fig. 20.18). These authors concluded that the principal mechanism producing a low breeding population was avoidance behaviour when golden plovers are selecting their territories at the start of the breeding season. Common sandpipers (*Actitis hypoleucos*) were disturbed by anglers and other visitors to the Ladybower Reservoir in the Peak District National Park, England (Yalden, 1992). They took flight about 29% more than they would if undisturbed; as with the golden plover, this also led them to encroach on neighbours' territories, with the consequent extra fighting. Their normal flight distance from a person walking had a mean of only 27 m, but if they had chicks they give alarm signals at a mean of 75 m. Single anglers may be 25 m apart along the popular shores, the disturbance level is very high and those parts of the lake edge are not available for sandpiper territories. The breeding population is therefore reduced by disturbance at this location.

Other waders that have been shown to have their activity curtailed by human disturbance are the curlew (*Numenius arquata*) on dune land in The Netherlands (van der Zande, 1984), oystercatcher (*Haematopus ostralegus*), redshank (*Tringa totanus*) and kentish plover (*Charadrius alexandrinus*) (de Roos, 1981). However, not all species show this reduction and Watson (1979, 1988) has shown that both dotterel (*Charadrius morinellus*) and snowbunting (*Plectrophenax nivalis*) were unaffected, or even increased in the case of snowbuntings, with increasing numbers of visitors. In contrast to de Roos' (1981) study of the effects of walkers, the distribution of oystercatcher nests was unaffected by up to 54 000 cars per day on a nearby road (van der Zande, van der Ter Keurs and van der Weijden, 1980). Roads did appear to affect the nesting density of two other species, black-tailed godwit (*Limosa limosa*) and lapwing (*Vanellus vanellus*), which gradually increased up to a distance of 2 km from the road (Fig. 20.19). They confirmed this distance with a second study published in the same paper. They also calculated disturbance distances and relative disturbance intensity relationships with the number of cars on adjacent roads for these species (Fig. 20.20). Disturbance to the terrain itself by all terrain vehicles was also said to reduce nesting success of piping plovers (*Charadrius melodus*) on lake shores in north Dakota (Prindiville-Gains and Ryan, 1988).

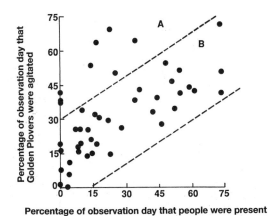

Fig. 20.17 The proportion of each observation day (5-min recording periods) in which golden plovers displayed anxiety behaviour (alarming, running, flying) compared with the proportion of that day when people were present in the territory (from Yalden and Yalden, 1990).

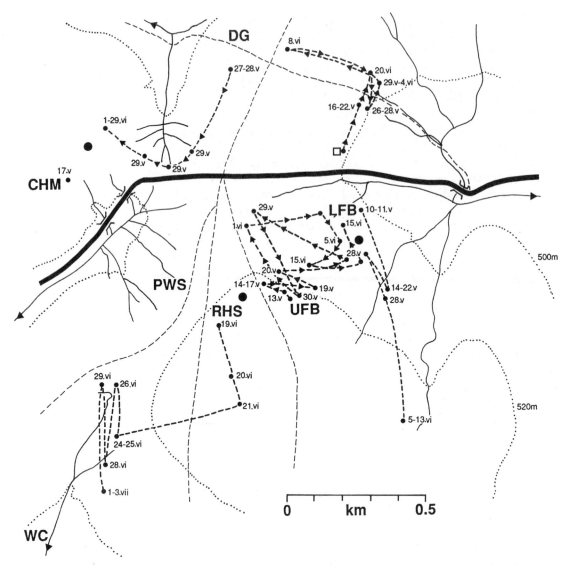

Fig. 20.18 Areas used by five broods of golden plover chicks under observation in 1988, and tracks (with dates) of four broods that moved away from their nest site, apparently in response to excessive human disturbance. ●, Sites where fighting was recorded; CHM, Cold Harbour Moor; PWS, Pennine Way South; RHS, not explained in original paper. LFB, Lower Featherbed Moss; UFB, Upper Featherbed Moss; DG, Doctor's Gate Path; WC, Withins Clough. The UFB birds did not move far from their natal area, but did take over the natal area of LFB birds, and constrained both the LFB and RHS broods. (From Yalden and Yalden, 1990.)

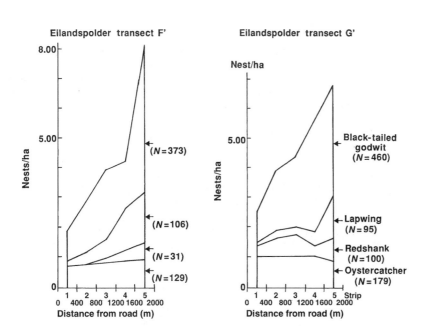

Fig. 20.19 Nest density of lapwing, black-tailed godwit, oystercatcher and redshank as a function of the distance from the road in Eilandspolder, transects F and G. Densities in 1968, 1970 and 1971 combined. (From van der Zande, van der Ter Keurs and van der Weijden, 1980.)

Herons and egrets are colony-nesting birds and activities such as logging have reduced the number of active nests in a great blue heron (*Ardea herodias*) colony from an average of 107 to 36, with other nests left vacant (Werschkul, McMahon and Leitschuk, 1976). In an experiment involving overflying a colony of mixed black-crowned night heron (*Nycticorax nycticorax*) and great egret (*Casmerodius albus*) by a single-engine propeller aircraft, the birds did not appear to be affected by sound levels up to 92 dBA (Grubb, 1978). However, visits to a black-crowned night heron colony in the estuary of the St. Lawrence River caused nest abandonment, inhibited laying and increased egg predation by crows (*Corvus brachyrhynchos*) (Tremblay and Ellison, 1979). There was also an associated nestling mortality. These birds were particularly sensitive to disturbance just before and during laying. Disturbance to waders while feeding with ducks and gulls on the shores of Jamaica Bay Refuge, New York, generally drove them from the area, whereas the gulls, terns and ducks returned to the beaches when people had passed (Burger, 1981a). This is in contrast to the behaviour of the species observed by (Blodget, 1978), which included short-billed dowicher (*Limnodromus griseus*), stilt sandpiper (*Micropalama*

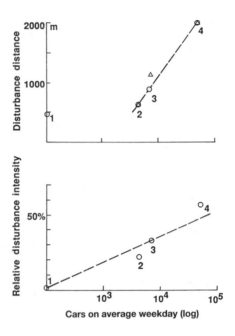

Fig. 20.20 Disturbance distances of lapwing (○) and black-tailed godwit (△) in the four study plots, and relative disturbance intensities of the lapwing, both plotted against traffic volume (on linear and semi-logarithmic scales) (from van der Zande, van der Ter Keurs and van der Weijden, 1980).

himantopus), godwits (*Limosa* spp.), black-bellied plover (*Squatarola squatarola*), piping plover (*Charadrius melodus*), semi-palmated plover (*C. semipalmatus*) and sandpipers (*Calidris* spp.). These birds, when disturbed by off-road vehicles on the upper shore of North Beach, Orleans and Chatham, Massachusetts, during high-tide periods, were very reluctant to fly and 'would often run as a flock rather than flush' and when they did fly there was rapid realighting.

Burger (1981a) explained the timidity of the waders she observed as a consequence of their lack of habituation to disturbance as they were migrating through the area, whereas the gulls and ducks were residents. However, the Massachusetts birds were also migratory (Blodget, 1978). Staging areas are important for migratory bird populations as they feed intensively to build up fat needed for long-distance flights which may be thousands of kilometres in length (Pfister, Harrington and Lavine, 1992). Some populations of Nearctic migrants have declined dramatically since the 1970s (Howe, Geissler and Harrington, 1989). Loss or degradation of habitat by human disturbance in staging areas may be one factor involved here. Censuses of shore birds at Plymouth Beach, Carolina, showed that there was a decline in some species over the 18 years from 1973 to 1989 (Pfister, Harrington and Lavine, 1992). Disturbance by vehicles, pedestrians and sometimes dogs caused the nesting front-beach species to move to the back beach. There was a clear negative relationship between the number of vehicles on the beach and abundance of short-billed dowicher and sanderling (*Calidris alba*). Recreational disturbance could have a crucial influence on the migrating birds, especially if they do not have alternative resting areas adjacent to their feeding sites (Pfister, Harrington and Lavine, 1992). Burger (1981a) also made a strong recommendation that 'human activity should be restricted to a distance from waders' nesting areas'.

Blodget (1978) found that compaction of the shore by 419 passages of a vehicle inhibited the feeding of deep-probing feeders and surface-feeding sandpipers, but that plovers fed more frequently in the treated sections of the plots. There was no associated change in the fauna of the treated areas. It would appear that substrate compaction does have some effect on feeding shore birds but direct disturbance is likely to have a much greater impact.

TYPE 2 AND 3 DISTURBANCE

I have found no records of these effects on waders but, clearly, shore-line development diminishes their habitat, and some species such as snipe (*Gallinago* spp.) are well-known prey of hunters.

20.8 Summary

There can be no doubt about the high value of birds as a recreation resource. The literature discussed above indicates both the widespread interest in certain groups of birds, and the positive or negative effects of that attention. Like most of the studies discussed in this book, with the exception of campground studies (Aitchison, 1977; Garton, Bowen and Foin, 1977), the impact of recreation is usually combined with the impact of other human activities and development. However, recreation alone is clearly responsible for some changes in bird activity and success, as well as changes to bird communities (Fig. 20.21).

Vehicles on roads can also cause considerable disturbance to birds. One estimate put the average number of traffic victims in The Netherlands in 1973 at 653 000 birds, in addition to 159 000 animals (van der Zande, van der Ter Keurs and van der Weijden, 1980). The proportion of these attributable to recreation activities is unknown. Lynch and Johnson (1974) concluded that 'where extinction or immigration is reasonably well documented, human influences are most

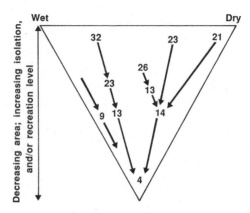

Fig. 20.21 Increasing similarity of the bird communities of different types of heathland (arranged along the top line according to a wet–dry gradient) with decreasing number of species. Numbers within the figure indicate the number of species in each heathland and type. (from Opdam and Helmrich, 1984).

often strongly implicated'. There is no doubt that people watching birds for recreation can also cause considerable disturbance and need careful control, especially where breeding colonies or raptors are the subject of attention.

Bears 21

The family Ursidae, in the order Carnivora, includes the brown or grizzly bear, *Ursus arctos* (subspecies include the grizzly bear, *U. arctos horribilis,* and the Alaskan brown bear or Kodiac bear, *U. arctos middendorffi*), the smaller American black bear, *U. americanus,* and the polar bear, *Thalarctos maritimus.* Other members of the family occur throughout the world except for the African and Australian regions.

Bears are now confined to wild, often mountainous country; in general, the brown or black bear inhabits wilderness areas more or less untouched by humans. The polar bear spends about 6 months of the year on sea ice, hunting seals, and the rest on land. One population enters the town of Churchill in Hudson Bay quite often as this is within part of its normal annual range (Davids, 1978).

Bears are active animals and can move fast when they want to, polar bears have been paced at 55 km h^{-1} (Davids, 1978). Grizzly bears usually have a home range or territory which may be as much as 422 km^2 but can be as small as 25 km^2 (Schneider, 1977). In the Rocky Mountains it was found that the daily average distance wandered by males over a 6-year period was 3.8 km in spring, 2.8 km in summer and 2.2 km in the autumn. Over the whole period males covered an average of 3.2 km daily and females 2.9 km (Aune and Stivens, 1983). During the winter they use a den, which is a cave or similar sheltered place, for their partial hibernation, for giving birth to their young and for the first few weeks of the cubs' life (Fig. 21.1).

Grizzly bears have an amazing sense of smell, their hearing is excellent but their vision is weak (Schneider, 1977). Polar bears also have a strongly developed sense of smell (Davids, 1978).

21.1 Characteristics

TYPE 1 DISTURBANCE

21.2 Disturbance

There are many dramatic stories about encounters with bears, especially grizzly bears in North America and, as is well known, there have been quite a number of maulings and deaths of humans. The same species in western Europe, known as the brown bear, does not seem to have the same reputation, perhaps because these are only relict

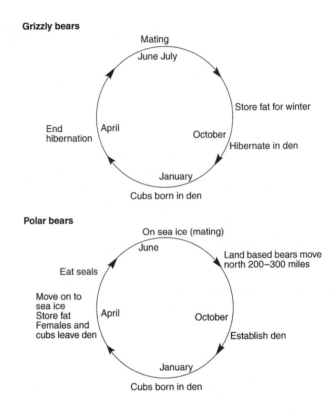

Fig. 21.1   Seasonal cycle of activities of grizzly and polar bears.

populations in remote areas. Norway had about 200 individuals in 1982 (Kolstad *et al.*, 1986), Sweden an unknown number and Finland about 175 individuals in 1971 (Pulliainen, 1972), there are small numbers in northern Greece and Italy, and only 5 or 6 in the French and Spanish Pyrenees (Duffey, personal communication; see also Duffey, 1982), with rather more further East. There is no doubt that a walker may disturb a bear but the outcome of an encounter with such a large and powerful animal is uncertain. Some individual bears have in the past become habituated to humans and, especially before the National Parks in the USA decided to make garbage and humans' foodstuffs unavailable to bears, looked to them as a source of food. Black bears in Yosemite National Park apparently obtained as much as half their requirements from humans in the valley (Graber and White, 1978). This caused many problems and created a situation where damage, injury and sometimes death of the human, followed by the death of the bear, was the result of an encounter (Edington and Edington, 1977; Hammitt and Cole, 1987). Herrero (1976) recorded 23 incidents between 1970 and 1973. However, since the change in management attitudes, removal of garbage and use of secure bins, the number of incidents and cost of damage has notably declined. The annual ranges of Yellowstone grizzly bears have increased since they were denied access to garbage dumps (Blanchard and Knight, 1991).

Bears are strong enough to stand up to, or attack, humans if they feel threatened or if their natural behaviour patterns have been altered by contact and habituation, but the real question is how do they respond when the people are just a passive part of their environment? Jonkel and Servheen (1977) categorized the aggressive behaviour of grizzlies as either **natural** aggression, including:

- females defending their young;
- a bear defending its food; and
- a bear defending itself when approached too closely;

or **unnatural**, when:

- bears move towards people or pursue them when people are seen from a distance;
- bears chasing people or horses when no natural aggressive explanation is available; and
- extreme boldness by bears in approaching people or repeated visits to a campsite or cabin.

They recommended that the natural aggression, which often does not result in severe physical injury to people, is dealt with by temporarily closing the area to people, but unnatural aggression may also require the killing or removal of the offending animal. In general, grizzly bears appear to be self-effacing and shy animals (Russell, 1968). For example, 'I saw several grizzlies on back country trails. Without exception they fled like spooked cottontails upon sensing my presence' (Schneider, 1977) or, 'Having hunted the grizzly for many years as a professional guide and in predator control work ... I am also aware to what lengths a grizzly will sometimes go to avoid a man even when sorely wounded' (Russell, 1967).

But caution is required. As the same author stated, 'A cardinal rule is never run from a bear' (Russell, 1968). He claimed that grizzlies will stop an attack if the human stands still, and pointed out that 'Nobody can outrun a bear' and that black bears can climb trees while a grizzly's reach may exceed 10 feet. He stated that they will stand up to identify the human as they have such poor eyesight and bears nearly always retreat upon identifying a human (Russell, 1968). In addition, contrary to an often repeated belief, black bears are not attracted to human menstrual odours (Rogers, Wilker and Scott, 1991).

Herrero (1980) also reported that even when people threw rocks or chased black bears at Jasper town garbage dump the bears never struck, bit or touched a person. The interaction with bears in North American National Parks has often been to the detriment of the bears, as those that are considered troublesome around campgrounds or visitor areas are either trapped or drugged and transported to remote areas, and if they reappear or have caused extensive injury or deaths to visitors they are sent to zoos or shot. For example, 22 grizzly bears were lost from the Yellowstone population in 1970 (Craighead and Craighead, 1971), and in the years 1973–94, 165 of the total of 209 bear deaths were 'man caused' (Knight and Blanchard, 1995). However, research suggests that the Yellowstone population is now increasing at 4.6% annually (Eberhardt, Blanchard and Knight, 1994).

The feeding activities of bears may also be reduced in the vicinity of human constructions (Tietje and Ruff, 1983). However, some authors maintain that the food bears obtained from garbage dumps before they were closed maintained a greater number and more healthy population (Chase, 1983). The whole question of bear control and feeding is still very controversial.

In general, the effect of winter recreation on hibernating bears has not received much attention. Goodrich and Berger (1994) studied the denning habits of black bears in the Sierra Nevada and the Sweetwater Mountains and found that they preferred north-east slopes – the same as those sought by skiers – as the insulating snow was deeper and stayed longer in these situations. They recorded that over three winters 21 bears abandoned dens in the Sierra Nevada and three in Sweetwater Mountains. In addition, the researchers disturbed 12 bears over this period. With one exception the bears moved to lower elevations and remained active. The authors emphasize that protecting denning areas from human disturbance would minimize cub abandonment and excessive winter activity.

The brown bear in Europe does not seem to attract a dramatic press coverage, and Elgmork even entitled one paper 'The cryptic brown bear populations of Norway' (Elgmork, 1987), this in spite of the fact that the populations of the brown bear in eastern Europe have increased over the past two decades (von Hell and Bevilaqua, 1988). Duffey (personal communication) commented that 'the survival of 70 or so bears in the Abruzzi National Park in Italy is remarkable. They are extremely secretive and seldom seen although thousands of people visit the park in summer'. He suggested that 'they may have evolved a form of behaviour which enables them to survive in spite of people'. In general, the response of humans to type 1 disturbance of bears has been better, if dramatically, documented than the response of the bears.

TYPE 2 DISTURBANCE

The effect of humans' pastoral, agricultural and urban development has been to reduce dramatically the areas occupied by bears. There are only a few countries in western Europe with relic populations, while in the past they were present throughout the area. The more recent reduction of the region in which the grizzly bears are living in North America was illustrated graphically by Schneider (1977) (Fig. 21.2). A closer linkage between recreation development and reduction of brown bear numbers is shown by the reduction in the number of sightings of bears in an area of south-central Norway coincident with the building of about 1200 new holiday cabins between 1960 and 1973 (Elgmork, 1983) (Fig. 21.3). Elgmork pointed out that the number of bear observations run counter to the augmentation of the chance of discovery, and that the results indicate a negative influence on bear occurrence near cabin concentrations due to human disturbance and changes in the environment. Another link with recreation was demonstrated by the reduction in the numbers of sightings of the brown bear in Norway as the total length of forest roads increased (Elgmork, 1978) (Fig. 21.4). This author concluded that the 'small remnant European brown bear populations are especially shy and do not readily adapt to human interference'. In his study area there was not a single incidence reported of a bear having visited cabins for garbage or other food.

The effect of the presence of roads in grizzly bear habitat centred on the Flathead River in British Columbia just north of the US border (McLellan and Shackleton, 1988) (Fig. 21.5). They found that the area within 100 metres of all classes of roads was used by bears significantly less than would be expected, and even the area up to 250 m was used less than the calculated expected figure (Fig. 21.6). Thus the mere presence of roads was sufficient to change the bears' habits. The authors concluded that it represented an effective loss of 8.7% of the bears' habitat and they considered it significant because these areas contained important sources of food.

TYPE 3 DISTURBANCE

Hunting outside of National Parks appears to be the primary cause of injury or death of bears. Several authors have reported that some bears that attacked people were later found to have been previously injured by hunters. There is still controversy over whether hunting should be permitted, especially for the grizzly with its drastically reduced North American population. However, Jonkel and Servheen (1977) implied that light hunting would keep them wary of humans

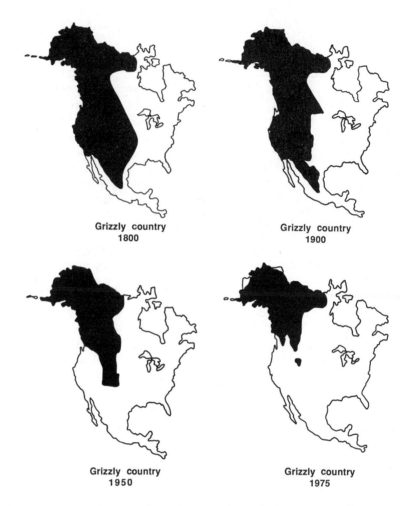

Fig. 21.2 Maps showing the reduction of grizzly bear country from 1800 to 1975. The reduction in habitat that is occupied is a result of both hunting and land use changes (from Schneider, 1977).

and thus reduce the possibility of attacks from the bears. Losing their fear of humans was said to be the first step in bears becoming unnaturally aggressive towards people and to a 'cultural deterioration' in the behaviour of a population (Jonkel and Servheen, 1977). In one study it was found that out of 13 bears legally killed by hunters, seven had been shot from roads (McLellan, 1989b), indicating how type 2 disturbance, road building in this case, can increase type 3 effects.

Deaths from road accidents seem to be rare, but four grizzlies were reported killed between 1970 and 1972 in Yellowstone National Park (Cole, 1974), and in 1967 three were killed in a collision with a vehicle

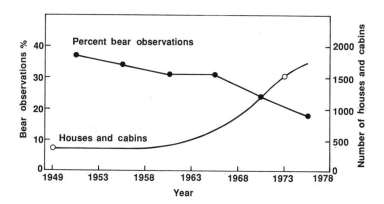

Fig. 21.3 Percentage of bear observations within 2 km of houses and cabins in relation to the total number of bear observations (5 year means) plotted against total numbers of houses and cabins. The graph indicates that bears are becoming less frequent in the neighbourhood of dwellings. (From Elgmork, 1983.)

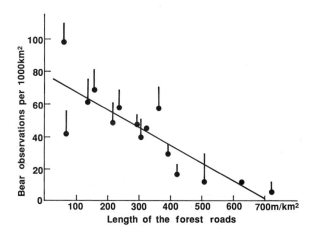

Fig. 21.4 Relationship between the length of forest roads per km² in central Norway and the number of bear observations. The vertical lines indicate doubtful reports. The reduction in numbers of animals as roads increased is clearly demonstrated. (From Elgmork, 1978.)

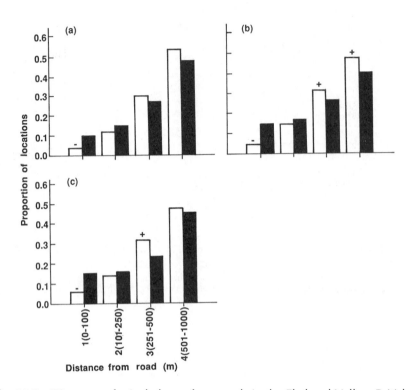

Fig. 21.5 Distances of grizzly bears from roads in the Flathead Valley, British Columbia. Observations are shown in clear columns and expected numbers in solid columns. Significant differences; between observations and predictions (+ or –) show modification of bear use of habitat. (a) Primary roads; (b) primary and secondary roads; (c) tertiary roads. Data from summer and autumn seasons from 1979 to 1985 (from McLellan and Shackleton, 1988).

in Abruzzi National Park, Italy (Zunino and Herrero, 1972). The Great Dismal Swamp Wildlife Refuge, in eastern Virginia and north-eastern North Carolina, has one of the few populations (of 56 black bears) on the Atlantic Plains still survives. Among various causes of death between 1984 and 1986, eight were killed by collisions with vehicles on major highways (Hellgren and Vaughan, 1989).

21.3 Summary   There is no doubt that changing land use, hunting and, very occasionally, conflicts with recreationalists have reduced the population of bears, especially grizzlies. They are sensitive animals and usually shy. The main problems have occurred where the bears have become habituated and lost their fear of people. This used to happen as a result of humans' food of various kinds being available to the bear

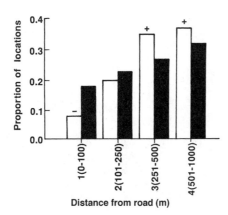

Fig. 21.6 Combined data for all types of road, showing distances of grizzly bears in the Flathead Valley, British Columbia. Clear columns are proportion of observations, solid column expected numbers. Significant differences between observed and expected shown as + or –. (From McLellan and Shackleton, 1988.)

(Fig. 21.7). It also happens as the result of the occasional meeting with walkers or others, and according to McCullough (1982) this can make the bears as dangerous as did the garbage dumps in the past. He suggested they should be 'negatively conditioned' to maintain their fear and thus prevent attacks on humans. He advocated a relationship between bears and people based on fear and respect in both populations, which he suggested will favour a more hopeful prospect of long-term coexistence in parks. (I have no problems with the idea of respect.) However, this would be in conflict with the decline in numbers in the Yellowstone National Park reported by Chase (1983). But Keating (1986) suggested that grizzlies have increased in numbers in the Glacier National Park partly because they have become habituated to people in certain districts. In this case to dehabituate them would also reduce their numbers so the park would be maintaining the population below its carrying capacity. However, the picture is not all bad since in south-eastern British Columbia they are slowly increasing in numbers (McLellan, 1989a) and possibly in Norway (Kolstad *et al.*, 1986), but Elgmork (1988) disputed this finding.

Bears are a memorable part of the wilderness experience for those who do meet them, and the public, especially in the USA and Canada, have a high awareness of bears. They are an important part of the value of wildlife, even though relatively few people actually see them since garbage removal was initiated. The bear is even used as a symbol, for example on the flag of California and in Russia. There is no doubt that the idea of wilderness would be the poorer if the bears were to become extinct in recreation areas.

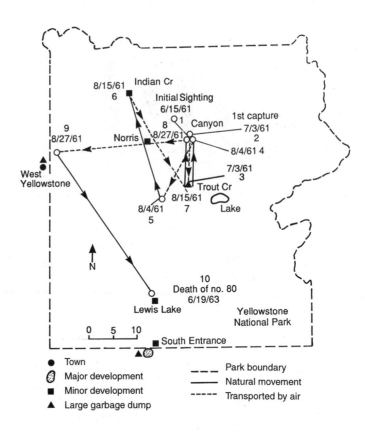

Fig. 21.7 Movements of one 'troublesome' grizzly bear, No. 80, from June 1961 to June 1963 in the Yellowstone National Park, Wyoming. The attraction to tourist locations and accompanying garbage dumps, which were present at this time, is clearly shown by its return to such sites after four aerial transportations. Numbers below dates indicate sequence of records. (From Craighead and Craighead, 1971).

Deer 22

The family Cervidae includes amongst others the red deer, *Cervus elephus*; the American elk, *Cervus canadensis*; the North American moose, *Alces alces*; the barren ground caribou or reindeer, *Rangifer tarandus*; the roe deer, *Capreolus capreolus*; the white-tailed deer, *Odocoileus virginiatus*; and the mule deer, *Odocoileus hemionus*.

22.1 Characteristics

The habitats utilized by deer are as varied as are the numbers of genera and species. Red deer in the UK are largely confined to the bare moorlands and mountains of Scotland, and occur in forests, their original habitat where not exterminated, throughout northern Europe (Saunders, 1937). The mule deer and white-tailed deer occur in the northern states of the USA, generally in or near mountainous areas with considerable tree cover, although they feed on open, often agricultural, land. The elk and caribou occur in the northern latitudes, and the caribou is particularly plentiful in Canada.

Red deer herds are said to choose as their midday resting place, a sheltered situation where 'two winds meet' because their hearing and sight are about equal to those of man but their sense of smell is infinitely keener and can 'never be deceived' (Saunders, 1937). However, Curtis (1974) commented that 'most animals rely on a highly developed sense of hearing to warn them of danger and some species such as moose have very poor eyesight'. Moose are also said to have a well-developed sense of smell and hearing (de Vos, 1958). Red deer can detect a human up-wind at just over a mile distant and are particularly shy of a human once seen and then lost to view. Disturbed deer travel up-wind, usually in single file with the stag going last. They soon learn the limits of areas where they are not in danger (sanctuaries) and make for them if alarmed. Stags who have been tamed become very dangerous during the rutting season (Saunders, 1937).

Deer are ruminants with four-chambered stomachs and the usual keen senses and swift running ability found in many large herbivores. The small, 60 cm high musk-deer (*Moschus*) has no horns, permanently growing canine teeth and lives alone in mountain forests of central Asia. The males of all the other species have antlers which are shed each year and grow progressively more branches (tines) as the animal grows older. In some species such as reindeer (*Rangifer tarandus*) the females also grow antlers. Deer are common in the

Holarctic region and South America but not in Africa. They live in herds with an elaborate social organization based on the supremacy of a leading male, which he maintains by a succession of fights with his rivals (Young, 1962).

Many species of deer have relatively small home ranges while some, such as the caribou, migrate over long distances. The North American white-tailed deer have an average home range of only 251 ha in summer, and in winter the area may vary according to the depth of snow from 949 ha in a snow-free winter to 88 ha in a winter of deep snow (Drolet, 1976). One group used an area as large as 1130 ha in winter (Progulske and Baskett, 1958) and a captive herd used only 60 ha (Ozoga, Verme and Bienz, 1982). In one case the summer and winter ranges were 20 km apart (Drolet, 1976); however, caribou may move over 700 km in one season (Fig. 22.1). Deer mostly move in small herds. The female white-tailed deer seek isolation at the time of parturition and remain alone with their fawn for about 4 weeks (Ozoga, Verme and Bienz, 1982), whereas caribou share a common area at this time (Calef, 1976). Red deer are gregarious but the sexes live apart except during the autumn mating season. The stags over-winter separately but mostly gather together, moving to higher ground in the spring. Towards May the hinds separate from the herds to drop their calves in deep cover, assembling again as soon as the calves can follow. The stags search for the hinds and fight for control of the

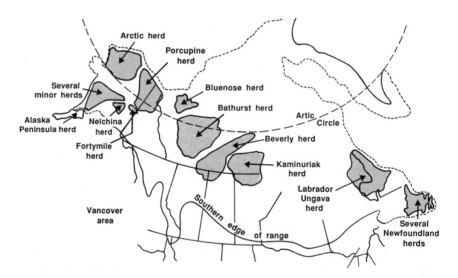

Fig. 22.1 North American range of the caribou and the territories travelled by the major herds. They may range over many habitats, including tundra, mountains, deciduous and evergreen forests, bogs, lakes and ponds, the edge of the sea and even the sea ice. (From Calef, 1976.)

groups of hinds in the autumn. The herds of hinds have defined territories (Saunders, 1937). They are crepuscular, feeding in the early morning and evening.

Although not affected by recreationists from urban areas, caribou are hunted by oil workers and 'new fashioned' hunters on snowmobiles. So the migratory habits of the barren-ground caribou are worth noting as they are increasingly coming in contact with human artefacts as a consequence of the opening up of northern Canada and Alaska. Twelve large herds or groupings are recognized in this area (Fig. 22.1) and in 1976 they numbered an estimated 1 200 000 animals (Calef, 1976). They winter at the northern edge of the great boreal forests, where they feed on lichen. The winter is intensely cold and has long periods of darkness. In early spring, usually late March, they start their migration to the calving grounds, which are often hundreds of miles away. The young are born in early June in bare areas which are still too cold for the insects which will plague them later in the summer. The calves are almost immediately mobile and are independent of their mothers by early August. The pregnant females reach the calving ground first, and by July they are joined by the bulls and the other females. At this time they form very large herds of 100 000 or more. These herds move up to 24 km (15 miles) a day and may be stampeded by insects or predators. Only about half the calves survive this period. The frosts of August herald the breakup of the herds, which fatten on the ample food supplies of August and September. The migration back to the wintering grounds starts gradually in early September and is in full movement after the first winter blizzard.

The daily activity patterns of deer may also be important if the deer are being managed as a recreation resource for viewers. There is a daily rhythm, with peaks of activity at sunrise and sunset reported for red deer (Georgii and Schröder, 1978), white-tailed deer (Kammermeyer and Marchington, 1977), moose (Geist, 1959) and roe deer (Espmark, 1974). There is also a seasonal change in the activity pattern, with more bursts of activity in summer than winter. The highest activity for white-tailed deer was in June and July (Weiner, 1975), as is the case in roe deer, which have a second peak in early autumn (Cederlund, 1981).

TYPE 1 DISTURBANCE   **22.2 Disturbance**

Walkers – orienteering

Disturbance of roe deer by orienteering in the Pamhule Forest, Denmark, was studied by Jeppesen (1987), following reports by Sennstam and Stålfelt (1976) that some roe deer and moose died after a large orienteering event in Sweden. There were 700 runners taking part in the Pamhule event and the movements of the deer were recorded by 14 observers and by the participants. When the deer were encountered they ran from the runners and since 120 observations were made in such a small area many must have been disturbed more than once (Fig. 22.2) and, if I interpret the figure correctly, 35 left the cover of the forest for the surrounding open agricultural land. The deer responded to orienteers by seeking shelter or leaving the area on first encounter with the runners, and this is reflected in the declining number of contacts after 11.00 a.m. (Fig. 22.3). In another study, at Kalo Forest, Jeppesen (1984) recorded that seven radio-marked roe deer sought cover in small, dense plantings or marshes, areas unlikely to be utilized by orienteers, within their home ranges and stayed there for 2 hours until the event was over.

The movements of two radio-collared red deer hinds were recorded during an orienteering event in Vejers Forest by Jeppesen (1987) (Fig. 22.4). In one case the hind left the forest on the first encounter with runners and ran 6 km across open land to a neighbouring plantation, returning to the home range 48 hours later. The second hind was disturbed not by the runners directly, but by other deer fleeing from

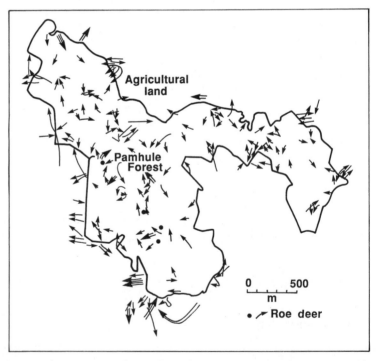

Fig. 22.2 Observations of roe deer during an orienteering event with 700 participants in Pamhule Forest, Denmark, 18 May 1980. Arrows indicate directions of flight. (From Jeppesen, 1984.)

the runners, it also ran to the neighbouring plantation but returned after 24 hours. All of these populations were hunted and this may account for the timidity of the animals. Von Petrak (1988) suggested that 'experienced lead animals as well as an intact social organisation [herd] are essential for adaptability to disturbances'.

The orienteering events took place in May, June and October and the additional energy cost of running may not have been very harmful to the animals. Had the events been held in January or February when roe deer are able to take in only the amount of energy they need for normal metabolic processes (Drozdz and Osiecki, 1973) the additional cost of running, about 20 times that of normal activities (Brodie, 1945, in Geist, 1975), could well have proved fatal. Various estimates of energy consumption suggest that deer may be in a marginal or even negative energy balance during winter (Moen, 1976; Parker, Robbins and Hanley, 1984). They adopt slow walking speeds, stay on level land and avoid deep snow. Harassment by skiers, dogs and snowmobiles clearly increases energy expenditure and may place the animals in jeopardy. Many authors suggest that deer should be undisturbed in their winter refuges (Moen, 1976; Parker, Robbins and Hanley, 1984; Chang and Xiao, 1988; Nixon, 1989). Jeppesen (1987) commented that 'hunting, orienteering, gathering mushrooms, tourism and other activities all constitute elements of disturbance in the forest environment, disturbances to which the animal can adapt to a certain extent'. I would add that Geist's work (e.g. Geist, 1971c) shows that without hunting their behaviour would be very different, although the populations might become too large to live comfortably with agricultural activities in a densely populated country such as Denmark.

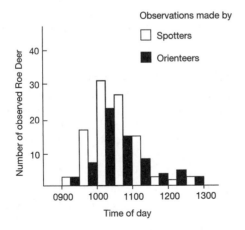

Fig. 22.3 Distribution of roe deer observations (made by spotters (white columns) and orienteers (black columns)) throughout 18 May 1980, during an orienteering event in Pamhule Forest, Denmark (from Jeppesen, 1984).

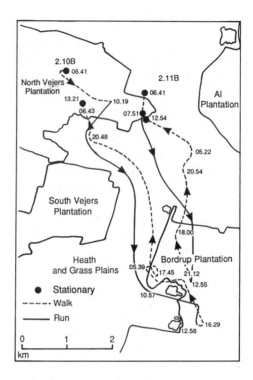

Fig. 22.4 Movements of two radio-marked female red deer for 31 hours during and after an orienteering event in North Vejers Plantation, Denmark, 25 October 1981 (from Jeppesen, 1987).

Elk are another species that appear to be rather wary of humans, but this conclusion may be based on the fact that the studies I have found have also concerned hunted populations. In the Medicine Bow National Forest, Wyoming, elk were observed to avoid people engaged in out-of-vehicle activities such as camping, picnicking and fishing. Aerial spotting showed that only 19 elk were within a quarter of a mile (400 m) of these activities, but 38 were between a quarter and half a mile (400–800 m) and 344 over half a mile (800 m) away (Ward, 1973). In contrast, I have observed a herd of 15 elk grazing within 10 m of housing in the Cascade Head area of Oregon where they were generally not hunted.

When disturbed from grazing, a cow elk tried to shield the calves and yearlings from, in this case, a human on a horse, and one even approached the disturber, presumably to distract a potential predator or drive them off (Altman, 1956) (Fig. 22.5). However, elk did not appear to react strongly to cross-country skiers in Elk Island National

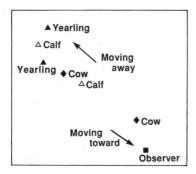

Fig. 22.5 Movements of elk when disturbed. Note one cow moving towards observer, shielding the remainder of the group who are moving away. (From Altman, 1956.)

Park, Alberta, where they just 'tended to move away from areas of heavily used trails during the ski season' (Ferguson and Keith, 1982). On the other hand moose, which have very acute senses of smell and hearing (Altman, 1958), showed a much stronger (statistically significant) avoidance of the heavily used trails during the January–March ski season. On occasions de Vos (1958) found that wading moose could be approached to within 30 m but they then reacted by moving slowly to the shore where, once concealed by brush, they broke into a run, indicating more alarm than one might expect.

Vehicles – snowmobiles

Two adjacent areas in eastern-central Minnesota and two in nearby north-west Wisconsin were chosen for two radio-tracking studies of the effects of snowmobiles on white-tailed deer (Dorrance, Savage and Huff, 1975 and Eckstein *et al.*, 1979, respectively). One of the two areas studied by Dorrance, Savage and Huff (1975) was also used by hunters, but no mention of hunting was made by Eckstein *et al.* (1979). Dorrance, Savage and Huff (1975) found that in the heavily used but unhunted area of St. Croix State Park, the number of deer observed per day from snowmobiles was reduced on Saturdays and Sundays and was negatively correlated with the number of snowmobiles entering the park (Fig. 22.6). However, the areas of the two-day home ranges did not differ significantly between weekdays and weekends. This suggests that while the deer took cover from the snowmobiles, they did not run any significant distance. In their second study area at Mille Lacs Wildlife Management Area, where hunting was allowed but snowmobile traffic was very light, they tracked five adults and one fawn in the first year and three fawns in the second year. In the second year they found that the home ranges became significantly larger during the snowmobile periods. This could have been the result of tracking only young animals because they were relatively unaccustomed to snowmobile traffic, or because hunting was permitted in this area. However, they did record that even in St. Croix State Park deer moved away from the immediate area of the snowmobile trails during daytime (Fig. 22.7). The resulting small reduction in home range could be significant in severe winters (Dorrance, Savage and Huff, 1975). Their main conclusion was that deer do become habituated to snowmobile traffic but those deer that are hunted annually would never become as habituated as those which have not been hunted for several years.

The results of Eckstein *et al.* (1979) from Chequamegon National Forest, Wisconsin, showed some small responses of deer to snowmobiles. When deer were within 61 m of the trail they moved away in 11 out of 21 encounters, but of the 22 encounters which occurred when they were over 61 m from the trail, deer moved away in only

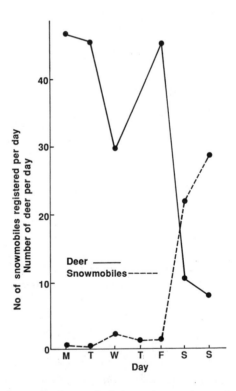

Fig. 22.6 Numbers of white-tailed deer observed per day, according to the day of the week, along a 10 km snowmobile trail in St. Croix State Park, Minnesota, in the winter of 1973 (from Dorrance, Savage and Huff, 1975).

5 instances, they stayed still in 9 instances, and in the remaining 8 out of 22 encounters deer moved closer to the trail. In this study deer also moved away much more during the day than at night. When skiers replaced snowmobiles, deer moved away in 8 of the 9 encounters, and they reacted much more to a person walking than a person on a snowmobile. In the authors' conclusion they recommend that snowmobile trails be routed away from areas where deer concentrate during the winter, but they felt that the environment of white-tailed deer is not very seriously changed by the advent of snowmobiles.

This is in contrast to the reactions of caribou and reindeer, who are severely disturbed by snowmobiles. Calef (1976) recorded that the Fortymile Caribou herd, (Fig. 22.1) completely stopped crossing the Taylor highway and reversed its migration after they had been hunted by humans on snowmobiles at this point. This author also commented that since the Eskimo hunters now work at other development-related activities in some areas, slow, careful hunting has given way to 'the destructive practice of chasing herds with snowmobiles so that the animals stampeded and ran often over long distances'. The responses of moose, which belong to the same genus as reindeer, to disturbance are very different. The calves and yearlings may be surprisingly tame (Geist, 1963) but the stimuli to which adult moose will take flight are variable. Geist (1963) recorded incidents where one old bull jumped at the sound of an axe chop approximately 500 m away, and where

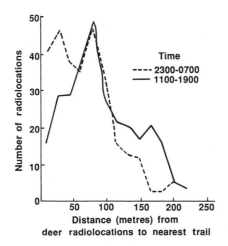

Fig. 22.7 Distance of white-tailed deer from the nearest trail at night and during the day in St. Croix State Park, Minnesota, in the winter of 1973. Notice deer proximity to trails at night when they are not used. (From Dorrance, Savage and Huff, 1975.)

a power saw was being used 'a bare 90 m' from the bull moose which was 'apparently not disturbed'. Importantly for recreation viewing it fed and rested within 270 m of an observation site used in winter and did not react to the observer's movements or voices.

The distance from the observer at which white-tailed deer become so disturbed that they moved away, was found to be between 17 and 54 m (Behrend and Lubeck, 1968). Mule deer are apparently more sensitive, moving from people on foot when they were 191 m distance and from snowmobiles at 133 m (Freddy, Bronaugh and Fowler, 1986). However, the threshold responses occurred at much greater distances, 334 m and 470 m from pedestrians and snowmobiles, respectively (Fig. 22.8a). In contrast, the duration of the response was much greater for people on foot than for snowmobiles after the deer had changed their activities or moved away from the disturbance (Fig. 22.8b). Reindeer are even more sensitive to snowmobiles, having a distance at which they showed alert behaviour of 640 m (Tayler, 1991). On this occasion the vehicle was driven directly towards the animals and the author observed smaller alert distances when the snowmobiles were driven past the animals. In this work the additional energy used by the animals was estimated to be less than 1% of the daily energy expended.

As long ago as 1972 Kopischke thought that all motorized vehicles should be forbidden in the winter refuges of the white-tailed deer in south-central Minnesota, but there are some advantages, as deer were observed to utilize the hardened snow of the snowmobile trails to move around their feeding areas in Somerset County, Maine (Richens and Lavigne, 1978).

With respect to vehicle noise, elk also appeared to avoid a major

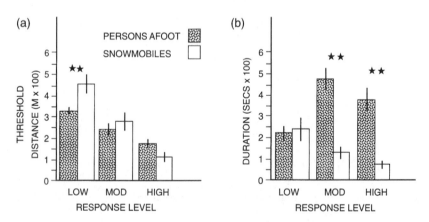

Fig. 22.8 (a) Threshold distances for initial responses; and (b) total durations of responses, by adult female mule deer. Both compared between people on foot and snowmobiles at each level of deer response. (From Freddy, Bronaugh and Fowler, 1986.)

highway which had noise levels up to 70 dB(A) from the trucks. They were observed only 14 times within 300 yds (270 m) of the highway, 21 times between 300 yds and a quarter of a mile (270–400 m), 17 times between a quarter and half a mile (400–800 m), and 55 times between a half and one mile (800–1600 m) (Ward *et al.*, 1973). In contrast, desert mule deer are reported to have habituated to low-flying aircraft in Arizona, where even flights lower than 50 m caused little visible response (Krausman, Leopold and Scarbrough, 1986).

TYPE 2 DISTURBANCE

The displacement of deer by recreation facilities and changing land use does not seem to have been recorded directly, although there are extensive areas of Europe and North America where this has occurred. The effect of the oil development at Prudhoe Bay has been noted. There the pregnant female caribou and females with newborn calves are generally intolerant of stressful surroundings and they avoid intensively developed areas through the summer months. There is also concern that they may have been displaced from the best calving grounds and that their summer movements between insect-free and richer forage areas may have been inhibited (Cameron, 1983) (Fig. 22.9). However, Geist (1971c) recorded that a population of mule deer living in Waterton Lakes National Park have settled in the small town: 'the deer feed and rest on the lawns, horn carefully planted trees and shrubs and use the houses as shelters during the gales that sweep the town so very often'.

Moose also seem to be able to adapt to the presence of humans, and in the Yellowstone National Park, 'where moose are in almost

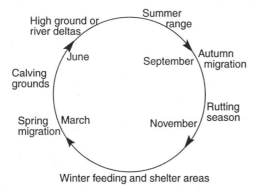

Fig. 22.9 The seasonal migration pattern of caribou. Roads or pipelines may hinder or prevent migration.

daily contact with human beings during the tourist season, animals soon learn to continue their feeding and other activities apparently undisturbed' (Denniston, 1956).

TYPE 3 DISTURBANCE

Hunting and drowning

There is considerable literature on deer hunting and I do not propose to review it here but it is important to note that hunting has an effect on the behaviour of the survivors, making them more timid and less visible to other people who may just wish to view and perhaps photograph the animals.

Antler deer, which are desired by hunters, were quite timid in areas where they were hunted (mean flight distance 52 m) and much tamer where there was no hunting (flight distance 27 m). There was also no noticeable difference in their movement according to the area and the way they were approached, with 95% of unantlered deer running away in the hunted areas and 80% in the unhunted areas. When they were beside a road and approached in a vehicle 89% ran and 11% walked away, but when they were on a lake shore and approached by a canoe only 62% ran and 38% walked away (Behrend and Lubeck, 1968). This has clear implications for the management of people wanting to view these animals.

Sage *et al.* (1983) suggested that the hunting of white-tailed deer bucks could only be compatible with viewing if strict control over the illegal hunting of females is achieved. However, they also recorded that viewers preferred seeing fawns and antlered males, which is not compatible with hunting the males. Increased wariness of hunted deer has also been reported by Behrend and Lubeck (1968), Grau and Grau (1980), and of moose by Altmann (1958). Major shifts in habitat use can occur as a result of hunting, and elk have been observed to move away from preferred winter grazing areas after hunting started, causing 'over-grazing of marginal sectors of potentially available grassland' (Morgantini and Hudson, 1980) (Fig. 22.10). This could, of course, also have a negative result on the critical winter energy budget of the elk. It has also been shown that white-tailed deer in Texas may change their home range as a result of disturbance by cowboys rounding up cattle and this, or similar forms of disturbance by hunters, could easily reduce their accessibility to non-destructive viewers (Hood and Inglis, 1974). Although deer may respond to such obvious disturbance as round-ups, an unsuccessful stealthy approach may cause more stress to the animals than a 'noisy tourist' approach (Denniston, 1956).

In two unusual incidents on California beaches elk were disturbed by visitors or researchers in situations where they were unfamiliar with

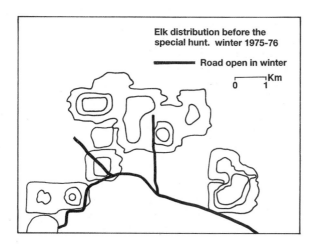

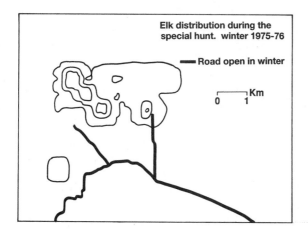

Fig. 22.10 Elk distribution before and during winter hunting at the Ya Ha Tinda Ranch, western Alberta. Note the absence of elk near the access roads. (From Morgantini and Hudson, 1980.)

the escape routes. On both occasions they took to swimming in the ocean, which was rough at the time, and they were drowned. The authors concluded that 'people encountering apparently distraught elk near large or rough bodies of water, when inland escape appears to be blocked, should avoid even slight disturbance of the animals' (Lieb and Mossman, 1974). Jeppesen (1984) recorded that the responses of roe deer to hunting battue where beaters deliberately disturbed animals and frightened them to run towards the 'sportsmen' with guns were similar to their responses to orienteers.

The migratory variations of the caribou may lead them into situations where recreational 'hunting' can occur, as for example in the case recorded at Coppermine (Calef, 1976). Part of the Bathurst herd (Fig. 22.1) wintered within sight of the village and 'young children on snowmobiles amused themselves by shooting Caribou with 0.22 rifles. The Tundra around this village was littered with dead animals by Spring'. This kind of vandalism is unfortunately still a factor that has to be considered in the conservation of the world's wildlife.

Vehicles

There is a considerable literature on the effects of roads, and particularly road accidents, on deer. While people travel on roads for many purposes other than recreation, there are notable occasions and places where road travelling is essentially for recreation, and I include some information for that reason. An analysis of sites where white-tailed deer collided with vehicles on two-lane roads in a mountainous area of Pennsylvania demonstrated that accidents occurred more frequently at some sites than others (Bashore, Tzilkowski and Bellis, 1985). Areas with least driver visibility along the road and areas with low, non-woody vegetation at the sides were most likely to be collision sites. Areas where there were residences, commercial buildings, other buildings, low visibility, speed limits, short distances to woodland and highway fencing all had a low probability of being the site of a collision between a deer and a vehicle. In addition, Bellis and Graves (1971) found that more collisions occurred in places where the highway lay in troughs with steep sides, than in places where the troughs ended, as well as in the flat areas noted by Bashore, Tzilkowski and Bellis (1985). Planted areas adjacent to the highway were utilized for grazing in the forested areas, but had little significance in the agricultural situation (Carbaugh et al., 1975). The use of salt as a highway de-icer may create pools of brackish water which attract moose and other animals to the roadside, thus increasing the chances of an accident (Fraser and Thomas, 1982).

Bashore, Tzilkowski and Bellis (1985) also noted that more deer kills occurred in spring and autumn than the other two seasons. They concluded that fencing could be used to reduce the number killed in the situations where collisions were most likely to occur. Studies of fencings along highways have suggested that they are not always effective in reducing road kills, but higher fences (2.7 m) work better than low ones (2.2 m) at keeping deer off the right of way (Feldhamer et al., 1986). The provision of underpasses has been effective for the migratory mule deer (Reed, 1981). Finally, it is interesting to note that in winter in Montana automobiles tended to 'select' young or old

white-tailed and mule deer in poor condition, compared with mountain lion or coyotes which killed mainly 2- or 3-year-old deer in good
condition (O'Gara and Harris, 1988) (Fig. 22.11).

22.3 Summary

Bearing in mind the limited range of studies considered here, it is
evident that deer may be strongly affected by recreating humans. They
are stressed by snowmobiles and humans skiing, particularly if they
are also hunted. It seems, again, that hunted animals should not also
be exposed to other well-intentioned, but under these conditions
stressful, recreational activities. The creation of roads or other structures in areas where deer have their ranges may also lead to their
death in collisions with automobiles, or reduce their access to important parts of their natural habitat.

Perhaps the greatest recognized value of deer is as prey for hunters,
and in many places populations are managed solely for this purpose

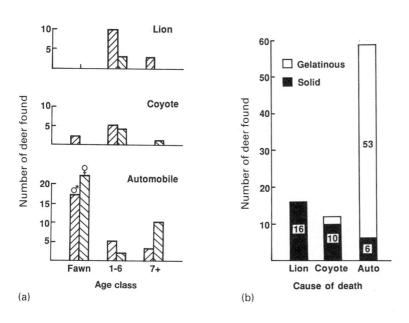

Fig. 22.11 Age structure and bone marrow consistency in mule deer and
white-tailed deer killed by mountain lions, coyotes and automobiles in western
Montana during the winters of 1969–81. (a) Age classes killed by different
'predators'. (b) Bone marrow consistency of animals killed by different 'predators'. Note that gelatinous marrow indicates that the deer were sick and in
poor condition and that automobiles selectively kill sick or weak animals.
(From O'Gara and Harris, 1988.)

or are sufficiently isolated to be unaffected by other human activities. However, they also have a great appeal to non-hunters and in some places they are managed solely or partly for the pleasure of visitors. This is perhaps an increasing activity.

Sheep, goats and chamois, with a note on gazelles 23

The Bovid family has over 100 genera, including among others the sheep (*Ovis*), goats (*Capra*), chamois (*Rupicapra pyrenaica*), cattle and yaks (*Bos*), bison or buffalo (*Bison*), gazelles (*Gazella*) and eland (*Tragelaphus*). They are mostly animals of open plains living in large herds, many in Africa and Eurasia with a few species in North America. A few genera, such as the goats and to a lesser extent the sheep, normally live in mountainous areas.

23.1 Bighorn sheep

The two species of mountain sheep discussed under this heading are the Rocky Mountain bighorn sheep (*Ovis canadensis*) and the dall sheep (*Ovis dalli*). At least one subspecies, *Ovis canadensis californiana*, has been classed as a rare animal in the United States. The first are confined to the Rocky Mountains of North America ranging from northern Mexico to about 55°N in Alberta and British Columbia (McCann, 1956). The dall sheep are a more northerly species.

Their normal habitat is steep, broken terrain, although they can move about by means of narrow ledges. Any particular flock has a limited range of movement. In the case of the flock in the Goss Venture Mountains, Wyoming, the range covers about 1300 km² (McCann, 1956). Selected parts of the areas are used for bedding and feeding and they also require access to water and salt licks. There are separate breeding territories held temporarily by the males, and special birth areas where the young are born. They occupy higher ground in the summer and move to lower areas in winter (Fig. 23.1). The bighorn flock in Fort Robinson State Park, Nebraska, preferred slopes of 61–80% and avoided slopes of less than 20%. They also preferred southern slopes and avoided open areas with 80–100% visibility (Fairbanks, Bailey and Cook, 1987) although Wakelyn (1987) stated that the 'ranges supporting greater numbers of Bighorn sheep had more high visibility habitat'. According to Fairbanks, Bailey and Cook (1987) they tended to select areas of low visibility near escape terrain.

The eyesight of bighorn sheep is very acute and they detect moving objects much more readily than stationary ones. They have a very wide angle of vision and may detect people even at a 90° angle from their line of vision (Holroyd, 1967).

TYPE 1 DISTURBANCE

In general, bighorn sheep tend to escape from disturbance by running uphill. The reaction to disturbance is therefore much greater if the disturber is above the sheep, threatening their escape route (Hicks and Elder, 1979).

Walkers

Cardiac monitoring telemetry was attached to a number of sheep and they were exposed to various forms of human disturbance (MacArthur,

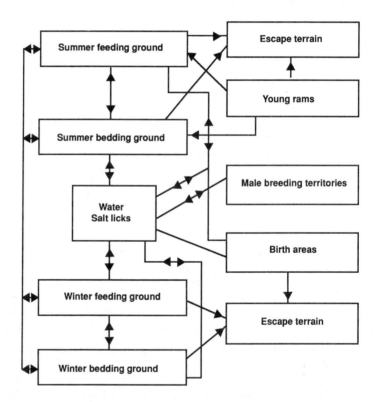

Fig. 23.1 Some movements of bighorn sheep. Barriers such as fences or snow walls beside swept roads in winter may cause a major problem for their survival (compiled from references quoted in the text).

Geist and Johnston, 1982a). The sheep were on their winter range in the Sheep River Wildlife Sanctuary, Alberta, and they had previously been hunted on their summer range. Within the sanctuary they were regularly exposed to human disturbance, at peak periods they may have encountered 25–30 vehicles per hour. The sheep were then disturbed experimentally by a person walking to within 25 m; 3-minute stops were made at 150, 100, 50 and 25 m. In the first trials the person approached from a parked vehicle on the road. Secondly, the approach was from over a ridge away from the road, and finally from the road, accompanied by a leashed dog. The resulting rise in the rate of heart beats was least for the unaccompanied approach from the road and highest when the disturber had a dog (Fig. 23.2). This shows clearly how important the direction of approach is in determining the reactions of the bighorn sheep. The reaction to the dog is associated with the fact that canids are the sheep's natural predators (MacArthur, Geist and Johnston, 1982b).

When the results of these experiments were analysed to take account of the sheep's activity at the time of disturbance, it was found that their heart beat rose more when they were standing alert or walking than when they were bedded or feeding (Fig. 23.3). It is of interest that 74% of heart rate responses preceded or occurred in the absence

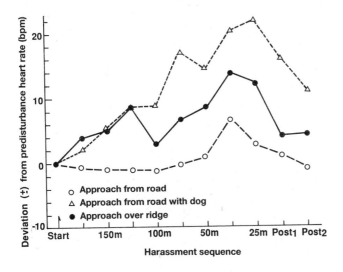

Fig. 23.2 Mean heart rate responses of mountain sheep to standardized approaches by a human. Distances (m) at which the person made 3-minute stops are indicated along the x axis. Post$_1$ = 2 minutes post-disturbance; Post$_2$ = 4 minutes post-disturbance. (From MacArthur, Geist and Johnston, 1982a.)

of any visible motor activity by the sheep. Although the total time of these heart rate increases was only 16% of the overall recorded disturbance time, it does show that the lack of behavioural response does not mean that the disturbance has no metabolic cost to the animal. The authors do not record the way the sheep perceived the disturber, but the description of the experiment suggests that it was visual. It is clear that hikers are likely to disturb the sheep if they come within 150 m, unless the hikers stay on a path that is familiar to the sheep and do not come between the sheep and their escape route. Dogs should, of course, be kept out of sheep habitats. The importance of visibility was clearly demonstrated in relation to Thomson's gazelles, where if humans remained within the shelter of the car, they did not run, but when a person stepped away from the vehicle the gazelles immediately fled (Walther, 1969).

The response of desert bighorn sheep (*Ovis canadensis nelsoni*) to hikers crossing the Baxter Pass at 12 000 ft (3660 m) in the Sierra Nevada was studied by Hicks and Elder (1979). They observed 10 bighorn–hiker interactions from a concealed position, on two of these occasions groups of sheep were surprised at close range. In all, three

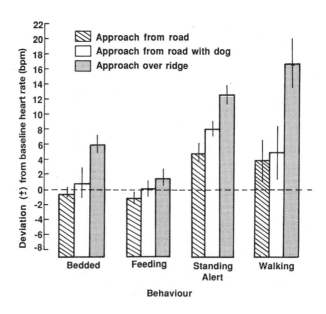

Fig. 23.3 Heart rates of mountain sheep in relation to behaviour during harassment trials. For each activity heart rate is expressed as a deviation (±) from mean base-line value recorded for the same activity when sheep were undisturbed. Vertical line, 95% confidence limits. (From MacArthur, Geist and Johnston, 1982a.)

groups did not run at all, four groups ran a short distance (about 100 yards (91 m)) and three groups ran a considerable distance. Altogether, 6 of the 10 bighorn groups left the pasture area after they had been disturbed, although the authors concluded that 'No permanent displacement was occurring'. They also recorded that the reaction of the ewe–lamb groups to people elsewhere in the summer range was generally of a 'mild nature'. Immediate retreat was observed only when encountered from below at 100 yards (91 m) or closer, and delayed retreat after an encounter at 100–150 yards (90–140 m). Greater distances caused no flight at all. When the humans appeared from above, the reactions were much more extreme. However, these populations were increasing in spite of the disturbance at the time the studies were made.

Aircraft

Price and Lent (1972) recorded the reactions of a population of dall sheep (*Ovis dalli*) to aircraft noise in the Atigun Canyon in the Brookes Range, over 68°N, in Alaska. They observed a group of 12 sheep start running uphill when a Jet Ranger helicopter was over half a mile distant. The wind was blowing from the helicopter towards the sheep. On another occasion a helicopter passed below a group of 36 sheep which only looked down at it. However, it returned later and flew about 100 ft (30 m) above them, 200 ft (60 m) to one side and the sheep all ran uphill. The desert bighorn population in the Grand Canyon National Park reacted to helicopters in much the same way (Stockwell, Bateman and Berger, 1991) when they were closer than about 450 m during winter. There was a reduction of feeding efficiency of 45%. During spring the sheep moved to lower elevations and were less liable to disturbance. In contrast, when a fixed-wing aircraft (Super Cub) flew within 50–100 yards (46–91 m) of 29 sheep in and near a mineral lick, 14 sheep walked out of the lick and climbed 800 ft (244 m) up the slope after the plane had passed over.

In general, Price and Lent concluded that reactions are governed by the distance of the aircraft or helicopter, the intensity and duration of noise generated by the aircraft, and position in relation to the sheep. Helicopters fly more slowly, are noisier and may fly lower than fixed-wing aircraft and so generally cause greater disturbance.

Habituation

The interaction between humans and bighorn sheep is not always negative, and close approaches are quite possible after a period of habituation as long as the animals have not been hunted or disturbed previously. The photographer Les Blacklock recorded how he gradually approached a herd of dall sheep over a period of a week, getting closer each day until they became habituated to his presence (Blacklock, 1977). There is also a record of a bighorn ram wandering through the streets of Palm Springs (Tevis, 1959). Habituation over a period of years has also been recorded for chamois (*Rupicapra pyrenaica ornata*) in the upper Val di Rose, Abruzzo National Park, Italy (Patterson, 1988). In 1977 the flight distance was 25 m, in 1981 19 m and in 1986 only 11 m, coincident with increasing use by visitors over the same period. Patterson (1988) concluded that 'continued "benign" exposure to people should encourage further habituation of the animals and so reduce the effects of visitors on them'.

TYPE 2 DISTURBANCE

The borderline between type 1 and type 2 disturbance is, of course, unclear, but there is little doubt that the increase of man's use of wild lands for recreation and retirement/holiday homes has reduced the size of some bighorn habitats over the past three decades. The reduction in the number of bighorn sheep was attributed to humans' increasing use of their habitat by Tevis (1959), Blong (1967), Wilson (1969), Hansen (1971), De Forge (1972) and others. Nelson (1966) listed recreational use, including picnicking, camping, exploring, hiking, rock hounds (amateur geologists), hunting and desert dwelling, as competitors for space with the bighorn sheep in New Mexico.

The displacement of herds from their normal ranges may result in their having to utilize areas that are inferior in some way and lead to a reduction in numbers. Woodard, Gulierrez and Rutterford (1974) recorded a loss of 85% of the spring lambs in the Sangre de Cristo Mountains, Colorado, in part due to pneumonia, and they stated that 'the ultimate cause [of death] is unknown but may be related to loss of historical winter range due [in this case] to a livestock operation'.

Mountain sheep (*Ovis canadensis mexicana*) in the Pusch Ridge Wilderness, near Tucson, Arizona, have also reduced their range, in this case from 206 km² to 46 km², as a result of human disturbance including two paved roads and a recreational tramway (Etchberger, Krausman and Mazaika, 1989). However, the back country recreationists using the higher altitudes were considered to be a greater cause of disturbance, but these authors do not give any population data. It is of interest that a part of this area was closed to access by hikers with dogs, but compliance with the regulation was low and the measure was regarded as ineffective (Smith and Krausman, 1988).

On the positive side, Leslie and Douglas (1980) reported that artificially created permanent water sources associated with water supply lines had allowed a population of desert bighorns to remain in the

River Mountain, Nevada, throughout the year, where they had a high degree of habituation to humans.

TYPE 3 DISTURBANCE

Hunting

Hunting, both legal and illegal, appears to be the main cause of type 3 disturbance to bighorn sheep. Hunting of any sort can have a serious effect on the accessibility of animals to wildlife viewers, through increasing the timidity of the animals and possibly redistributing the groups to the most remote part of their range. It can, of course, reduce the numbers of sheep and, if uncontrolled, drive populations to extinction. Three herds, one in and two near the Yellowstone National Park, Wyoming, were studied by Irby, Swenson and Stewart (1989). The horns of the males make them the desirable hunting or poaching trophy, and these authors found 'low survivorship of males 6–7 years of age and older, and 29 reported illegal kills versus 35 legal kills, over a seven year period' (Table 23.1). They suggested that 'poaching might have serious consequences for legal harvest, game viewing and possibly genetic quality of herds in the area', but concluded that the facts that sheep still tolerated close approaches and that there were small observed population changes over a period of 8 years indicated that the impacts were not serious. The idea that hunting the older rams, which have the longest horns and therefore make the best trophies, has no effect on the population was sharply dismissed by Geist (1971b). He commented that 'the worst combination of harassment was hunting combined with hiking, especially for wildlife photography. There is every reason to suspect that such a combination can be the most damaging and finally fatal'.

The concept that a wild animal should be 'wild' was also well stated

Table 23.1 Survival, by age class, for male bighorn sheep captured on the Cinnabar Mountain and Tom Miner (Upper Yellowstone Valley, Montana) winter ranges, 1980–85 (Irby, Swenson and Stewart, 1989)

Year class (beginning–ending)	Sample	Number surviving	% Survival
1–2	3	2	67
2–3	5	5	100
3–4	5	4	80
4–5	6	4	67
5–6	4	3	75
6-7	3	2	67

by Geist (1972): 'In particular it is demanded of mountain sheep that they stand on pinnacles and look down proudly on the passer-by. The behaviour favoured and demanded happens to be that of a thoroughly upset and frightened animal, with enough harassment it is not difficult to achieve'. Although our view 20 years on has changed, there is still a substantial body of opinion that rejects the tame behaviour of unhunted populations as being zoo-like or artificial.

Vehicles

Sheep/vehicle accidents have also occurred in many places, and Singer (1975) recorded that accidents were reduced in the Jasper National Park when warning signs were erected (but see section 25.4).

23.2 Bighorn sheep responses

The bighorn sheep have a complex set of ecological requirements which they meet in part by movement throughout their chosen range. They learn the pattern of these movements from the oldest sheep in the flock (Geist, 1971c) and although they can adapt to new conditions introduced by humans, the process usually results in the reduction of occupied habitat and hence the potential population size (Fig. 23.4).

There is no doubt from the many articles, papers and published photographs that the bighorn sheep are a significant part of the 'wildlife culture' of North America. They are also an important tourist attraction for people like myself from another country. Their survival and accessibility is therefore of both cultural (recreational) and economic interest.

23.3 Mountain goats

The goat discussed here is *Oreamnos americanus,* the Rocky Mountain goat. Their habitat is mountainous terrain, similar to that occupied by the bighorn sheep, and one particular herd in Mount Wardle Kootenay National Park occupied an area of 52 km^2 for their summer and winter ranges (Fig. 23.5). They occupy high ground, up to 9000 ft (2740 m), in summer and move to lower areas in winter.

They have very acute eyesight and can readily detect movement from 'some distance' (Holroyd, 1967). Hearing is not as highly developed as their eyesight, and according to Holroyd (1967) they react to human-created noises only when the noise is comparatively loud (see below). Their sense of smell is well developed, although Holroyd (1967) found that the Mount Wardle herd, which had not been hunted, would approach him upwind until they could confirm his presence visually.

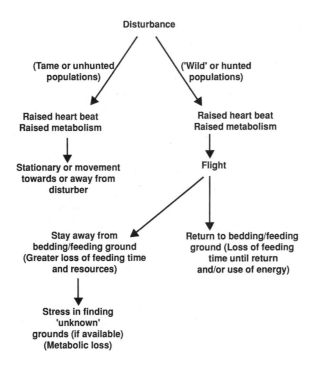

Fig. 23.4 Some consequences of disturbance to bighorn sheep.

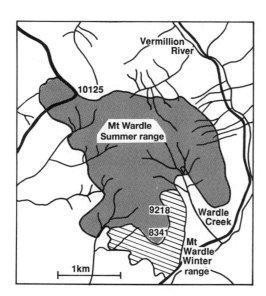

Fig. 23.5 Higher summer and lower and smaller winter ranges of the Mount Wardle herd of rocky mountain goats. Spot heights are given in feet. (From Holroyd, 1967.)

Their habits are similar to those of the bighorn sheep. In particular, Holroyd (1967) noted that their kidding area was very rough, with cliffs and ledges, and in late May or early June each nanny chose a high and inaccessible shelf, usually on a cliff face, on which they gave birth, remaining there for not more than 24 hours.

TYPE 1 DISTURBANCE

Walkers

The reactions of the Mount Wardle herd to disturbance varied according to season (Holroyd, 1967). At a low-elevation lick, an approach by a human walking to within a few feet was possible, but they were harder to approach at high elevations. The larger the group, the less likely it was that they would allow a person to approach them. When one animal fled, the rest followed, March and April were the best months to get near them. The nannies with new-born young were very defensive. Curiosity often caused the goats to approach rather than retreat from an intruder, and the fact that they were not hunted may have allowed them the freedom to react in this way (Holroyd, 1967).

Vehicles

Observations of mountain goats crossing the US Highway 2 in Glacier National Park from high grazing areas to the lower Walton salt lick, which had a visitor viewing area, showed that they responded to vehicle noise and would hesitate to cross the road to the salt lick when they heard a truck changing gear over 1 km away (Singer, 1978). They also ran along a trail when 400 m from the highway and ran off the lick, which could be seen from the parking lot by the road, in response to the sound of approaching trucks. Goats also reacted with fear to visual sightings of visitors or vehicles on or near the highway, especially when seen together (48 responses, compared to 38 responses to passing vehicles and 23 responses to visitors alone).

The whole behaviour pattern of the goats was altered by the presence of vehicles and visitors, and they made most of their road crossings at dawn or dusk and used the licks primarily at night, especially during the August holiday season, a pattern that was not followed by goats at the Little Dog Lick which was 21 km away and totally undisturbed by humans.

TYPE 2 DISTURBANCE

Roads

The presence of the road at Glacier National Park should be classed as a type 2 disturbance and Singer (1978) noted that 70% of the goats showed signs of being afraid of the highway pavement. This also occurred when no visitors or vehicles were present. The number of crossings to and from the salt lick are shown in Fig. 23.6. There were six observations of kids being separated from their nannies when they were disturbed crossing the highway (three times) or frightened by the pavement itself (three times), and in two cases separation was permanent, although one of the abandoned kids did become associated with another goat. Clearly, this example of type 2 disturbance could lead to the death of the kid.

TYPE 3 DISTURBANCE

Hunting and vehicles

Singer (1975) recorded that six goats were killed by hunters, two by poachers and two by collisions with cars, but he stated that the records are incomplete.

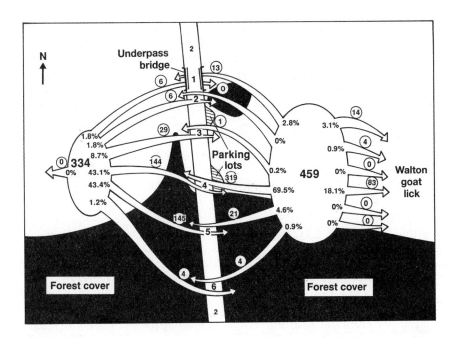

Fig. 23.6 Use of six different routes across US highway 2 by 334 mountain goats observed moving downhill to the Walton goat lick and 459 goats departing uphill from the lick. Total numbers and percentage of the total observations are both indicated for each route. (From Singer, 1978.)

23.4 A note on gazelles   Reactions of the Thompson's gazelle (*Gazella thomsoni*) to disturbance from humans and vehicles in the Serengeti National Park, Tanzania, showed that the distance from an approaching car at which they took flight varied according to sex, whether they were territorial or wandering and solitary or in a herd (Walther, 1969) (Fig. 23.7). In herds, the adult males have a lesser flight distance than adult females and subadult males. Territorial, and therefore mating males, have a lower flight distance than 'bachelors', but solitary wandering males have a greater flight distance than any other category. Solitary females with fawns have a relatively low flight distance.

These data show how complex the impact of recreation really is. It also suggests that even though the data on some groups of animals are relatively extensive, we probably have only the simplest picture for the interactions and effects of recreational disturbance.

23.5 Summary   Human-induced change has had a major effect on bighorn sheep populations, reducing their numbers in 100 years from approximately 2 000 000 to about 25 000 (Buechner, 1960). Today, bighorn sheep are found only in some of the most rough and remote areas left in North America (De Forge, 1976). De Forge suggested that competition from livestock, diseases introduced by domestic sheep and indiscriminate hunting are the major reasons for this decline. These factors, together with loss of habitat (to recreation?), cause stress which he defined as 'excessive stimulation of the endocrine system' (De Forge, 1976). The consequences of endocrine stimulation are summarized in Table 23.2 and they may all potentially lead to a reduction in the bighorn populations. However, there are many examples where bighorn sheep have been tamed (Wells and Wells, 1961) and sometimes, to the delight of tourists, come to live alongside humans, taking advantage of lawns and roadside planting for their forage

Table 23.2 Effects of levels of stress that produce excessive stimulation of the endocrine system in bighorn sheep (after De Forge, 1976)

Increased mortality	Inhibition of reproduction functions	Behavioural disturbances
Lowered resistance to: disease parasitism infection shock disease	Sexual maturation delayed or inhibited Delayed spermatogenesis Oestrous cycle prolonged Ovulation and implantation reduced Intrauterine mortality of fetus increases (possible resorption of embryos) Inadequate lactation	Perverted behaviour passed on to young Development of brain impaired

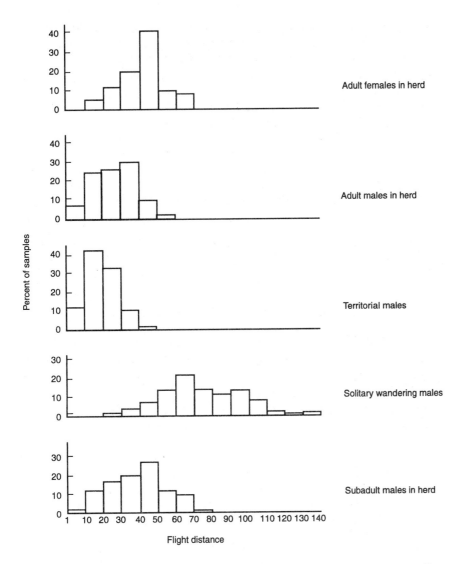

Fig. 23.7 Flight distances from an approaching car of Thompson's gazelles of different sexes and social condition (from Walther, 1969).

(Geist, 1971c). There is no reason why the interaction between bighorn sheep and people should not be advantageous to both species.

It would appear that mountain goats are affected by recreation activities in much the same way as bighorn sheep, and the conclusion and effects listed in Table 23.2 probably apply to them in the same way. Holroyd (1967) demonstrated that unhunted populations could readily be tamed. The mountain goats are rare mammals and can be seen by visitors at places such as the Walton mineral lick (the author observed a male right beside a path and a female and two kids within 20 yards in the Glacier National Park). They are clearly a recreation resource, although not so internationally famous as the bighorn sheep.

Small mammals 24

The previous chapters have dealt with conspicuous groups of mammals about which there was a considerable body of work documenting their response to various recreation activities. There remain many mammals about which there are a few and/or difficult-to-locate reports. Those describing impacts on smaller mammals are dealt with in this chapter.

The most detailed work on this group concerns the effects of campgrounds on mainly rodent and insectivore populations. The small mammals in and around campgrounds are the only wildlife that many people will see during their visit to a national park (Clevenger and Workman, 1975) so they are of central importance to the satisfaction gained from the recreation experience.

TYPE 1 AND 2 DISTURBANCE

24.1 Disturbance to small mammals

The effects of hikers, their choice of route and possible accompanying dogs has, as far as I can ascertain, only been investigated for marmots (*Marmota marmota*) (Mainini, Neuhaus and Ingold, 1993). This study was carried out at an alpine site where hunting was prohibited near Grindelwald, Switzerland. The authors approached 10 groups of marmots by walking along a marked trail, cross-country, walking across the marmot burrows, with dogs on short leads and with dogs on very long leads (free running). The resulting responses are very similar to those shown by mountain sheep (McArthur, Geist and Johnston, 1982a; Chapter 23). The free-running dogs caused the marmots' longest distance reaction, greatest number of warning calls, longest flight distance and longest times hidden in their burrows after retreating there on seeing the dogs (Fig. 24.1). The authors found that marmots were least disturbed by hikers who kept to trails and the animals were, to some degree, habituated to this form of disturbance. Hikers who left the trails were thought to break down this habituation and dogs and marmots clearly do not go together.

Campgrounds change the small mammals' habitat by destroying or replacing existing vegetation, by the availability of waste or dropped human food, by changing the distribution and presence of predators and the by presence of the campers themselves. It might be expected

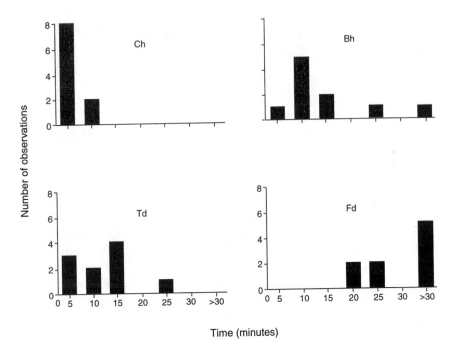

Fig. 24.1 Period of time during which the marmots stayed in the burrows for 'cross-country hikers' (Ch), 'burrow hikers' (Bh), 'trail hikers with dogs' (Td) and 'hikers with free-running dogs' (Fd). $n = 10$ each except for Fd ($n = 9$). (From Mainini, Neuhaus and Ingold, 1993.)

that the numbers of shyer native species that cannot avail themselves of the extra food would decline and that the opportunists would increase.

A survey of small mammals in the Yosemite National Park, California, found that of the three rodents that were present in sufficient numbers to give reliable results, deer mice (*Peromyscus maniculatus*) increased in numbers while the mountain mouse (*Microtus montanus*) decreased (Garton, Bowen and Foin, 1977). Squirrel (*Citellus beldingi*) numbers showed no significant relationship with the intensity of campground use, but populations were related to the level of soil moisture. The deer mouse has an omnivorous diet, can utilize human food waste and is known as an adaptable and widespread species; an increaser in the sense of Dyksterhuis (1957) (section 4.3). The mountain mouse has a specialized diet of certain types of 'herbage' and will avoid areas devoid of vegetation. Their density is governed by a complex set of factors, including social behaviour, limiting nutrients, genetic changes and predation (Garton, Bowen and Foin, 1977). This complex of factors suggests that their density responses to disturbance (decreasers) are equally complex. Campgrounds at two desert national parks in Utah were in the natural habitats of deer mice (*Peromyscus* spp.), Colorado chipmunks (*Eutamias quadrivittatus*), wood rats (*Neotoma* spp.), Ord's kangaroo rat (*Dipodomus ordii*) and antelope ground squirrels (*Ammospermophilus leucurus*) (Clevenger and Workman, 1975). Larger mammals included desert cottontails (*Sylvilagus audubonii*) and rock squirrels (*Spermophilus variegatus*). The Colorado chipmunks and deer mice were present in the campgrounds in significantly higher numbers than in adjacent areas, and there were more wood rats in the campgrounds at Squaw Flat and fewer in the camping area at Devil's Garden (Table 24.1).

The authors raised the possibility that the extra food available at the campgrounds may be preferred to the apple slices and seeds used in the traps to make the survey, and thus biased the results. They added that the ability of small mammals to adapt to and utilize the waste food in a manner that could affect population levels is not well understood. Wood rats and chipmunks, as well as deer mice, are opportunistic herbivores and granivores and this may account for the higher numbers trapped in the campgrounds. The lower numbers of wood rats in the Devil's Garden campground may be due to the lack of ground cover and higher number of campers at that site. The other species are more specialist feeders. The large number of predators, especially coyotes (*Canis latrans*) at both areas and Swainson's hawks (*Buteo swainsoni*) at Squaw Flat, may also have affected the population levels.

Small mammals and reptile numbers in the Algodones dunes in the California desert also showed a correlation with the vegetation cover (Luckenbach and Bury, 1983) (Figure 24.2). In this case the area had been used by off-road vehicles (ORVs) and the lower coverage of perennial vegetation was a result of destruction by these vehicles. Interestingly, the total number of rodents correlated well with the volume of vegetation, while the total number of lizards correlated with

Table 24.1 The numbers of small mammals trapped in the Squaw Flat (SF) and Devil's Garden (DG) campgrounds and adjacent areas of canyon lands in Arches National Parks, Utah (from data of Clevenger and Workman, 1975)

	Colorado chipmunk		Woodrats		Deer mice		Desert cottontail		Antelope ground squirrel		Ord's kangaroo rat	
	SF	DG	SF	DG	SF	DG	SF	DG	SF	DG	SF	DG
Campground	88	35	52	11	101	88	16	35	67	46	79	51
Adjacent area	25	19	3	24	88	58	18	26	67	43	87	55
Significance of difference	5.91	2.17	6.64	2.21	NS	2.51	NS	NS	NS	NS	NS	NS

its density and cover (Table 24.2). This suggests these two groups may be related to the plants in different ways. Luckenbach and Bury (1983) commented that alteration of bulk density and destruction of the surface crust of sand in areas of heavy ORV usage may prevent rodents from successfully constructing burrows and negate recolonization from less disturbed populations in adjacent areas. The relationship with vegetation may not be the causal one suggested for the wood rats in the Devil's Garden campground, but the reduction in numbers seems to be ultimately due to the impact of recreation.

Small mammals can lead to conflicts with campers, and in one case the same solution as that used for bears was adopted, namely capture and transportation. Rock hyraxes (*Procavia capensis*) became tame, aggressively solicited food from tourists and occasionally bit visitors to the Storms River Mouth rest camp in the Tsitsikamma Coastal National Park, Republic of South Africa (Crawford and Fairall, 1984). Twenty of the tamer hyraxes were captured and released about 70 km away. Two of the older males returned to the vicinity of their home range, having moved considerable distances across unknown terrain, an ability also demonstrated by grizzly bears (Fig. 21.7) and crocodiles (Chapter 19).

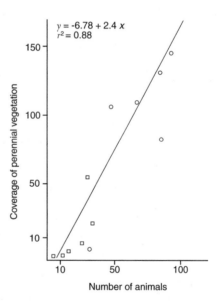

Fig. 24.2 Relationship between the number of animals (reptiles and mammals) and vegetation cover on the Algodones dunes, California. ○, Control plots; □, ORV-impacted plots. (From Luckenbach and Bury, 1983.)

Table 24.2 Comparison of vertebrate fauna sampled on the Algodones Dunes, Imperial County, California, and measures of perennial vegetation (Luckenbach and Bury, 1983)

| | Perennial vegetation | | |
	Density	Coverage	Volume
Total number of lizards	0.78*	0.88*	0.41
Total number of rodents	0.12	0.02	0.76*
Total number of individuals	0.73*	0.94*	0.63*
Total number of species	0.26	0.19	0.86*

*Correlation coefficient (r:$P < 0.05$).

The activity patterns of small mammals are also changed by human disturbance. A laboratory study of the activity cycles of captive elephant shrews (*Elephantulus myurus*) showed that individuals that had been exposed to high levels of human disturbance increased their nocturnal activity and reduced movements during the daylight hours (Woodall, Woodall and Bodero, 1989) (Fig. 24.3). This change in pattern would not be evident in the results of trapping experiments unless they were accompanied by close observation of the subject species.

More diffuse recreation activities may also affect small mammals. The birch (*Betula* spp.) woods near Moscow (Fig. 13.2) are subject to various levels of use in the summer months and Bykov (1985) noted that in 'strongly disturbed stands with grasses and herbs in the ground cover, non-forest species take a considerable part in the population of small mammals'. Also, the numbers were apparently reduced compared to relatively undisturbed forests with a ground cover of sedges (*Carex pilosa*). In a contrasting open habitat on the summit of Cairngorm Mountain, hares (*Lepus timidus*) were present in greater numbers on the disturbed high area most frequented by visitors, but this may have been an indirect effect as most of them were flushed from patches of fertilized re-seeded grass (Watson, 1979). These areas had been planted to reduce soil erosion caused by visitors.

Concern was expressed for the actual survival of the Vancouver Island marmot (*Marmota vancouverensis*) in a similar mountain habitat which was increasingly used for recreation (Dearden and Hall, 1983). Here the construction of logging roads allowed easy access for recreationists who were skiing and snowmobiling in the winter and, in summer, hunting, hiking, using all-terrain vehicles, flower picking and, last but not least, photographing, especially the marmots themselves. Solutions suggested by these authors included relocating the ski areas (expensive), fencing the colonies and the inevitable relocation of the marmots themselves.

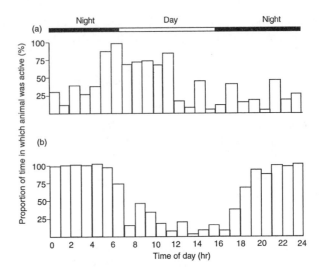

Fig. 24.3 Changes in the activity of two captive elephant shrews (*Elephantulus myurus*) over a period of 24 hours. Periods of dark and light are indicated by the solid and open bars, respectively. (a) A relatively undisturbed animal; (b) an animal maintained under conditions of considerable human disturbance prior to monitoring. (From Woodall, Woodall and Bodero, 1989.)

One factor that is of considerable importance to small mammals that live under the snow in the winter is the compaction of the snow which increases its density, reduces its depth and thus reduces the temperature at ground level (Schmid, 1970) (Fig. 24.4). These observations show a reduction of about 12°C which would drastically increase the energy demand of a hibernating marmot and could easily result in its death. Voles such as the northern bog lemming (*Synaptomys borealis*) which nests on the ground surface and develops runways under the snow (Layser and Burke, 1973) would not only be subject to lower temperatures, but their movement through the compacted snow would also be severely restricted and they could be crushed by the compactive forces exerted by snowmobiles.

Even in urban areas recreation activities can affect wildlife. The number of badgers (*Meles meles*) living in public woods and semi-natural areas in the Copenhagen area declined by one-third from 1973 to 1985 (Aaris-Sørensen, 1987). This was attributed to the effects of disturbance by increasing numbers of people, especially when accompanied by dogs which were not on a leash. The badgers were very sensitive to disturbance near their dens. This has caused the animals to leave the area and, in some cases, they have taken up residence in

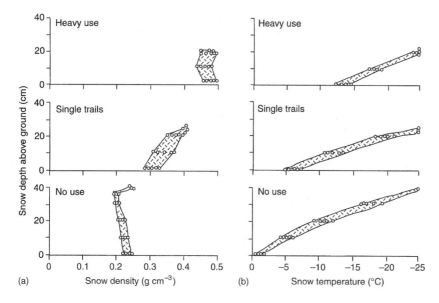

Fig. 24.4 Vertical (a) density and (b) temperature profiles for snow in areas of different levels of use by snowmobiles and an unused control area. The increased density at the top of the unused snow profile was due to a 4 cm thick thaw crust. (From Schmid, 1970.)

the vicinity of railways where there is much more noise but less disturbance by people and dogs.

Bats (Chiroptera) often roost or hibernate in large groups and have critical requirements, especially the cave-dwelling species. Visitors to caves have been responsible for vandalism directed toward the bats, by rock throwing or even burning clustered bats with torches (Humphrey, 1978). More general disturbance during hibernation can be serious. 'Every human visit under a low hibernating Indiana bat (*Myotis sodalis*) cluster caused metabolic heat production. They generally hibernate at air temperatures of between −1.6°C to 1.7°C, arousal, flight and reclustering 30 to 60 minutes later with a concomitant loss of fat reserves' (Humphrey, 1978). Even the mildest stimuli of sound and light from a group of passing cave enthusiasts is sufficient to produce arousal. Well-meaning naturalists should note that 'fat attrition is greater when biologists visit to gather data because observation time is longer' than that of casual visitors and the bats may be handled for measurements. Well-intentioned closure of one entrance to Coach Cave, Kentucky, with a door, caused a decrease in ventilation, allowing the temperature to rise and there was a subsequent fall in the winter population of Indiana bats from 100 000 to a little over

10 000 (Fig. 24.5). The design of appropriate cave gates that will prevent disturbance from visitors, preserve the cave environment and allow free passage for the bats is discussed by White and Seginak (1987).

In general, small mammals' response to habitat change was classified into five groups – pioneers, early immigrants, later immigrants, sporadic visitors and non-immigrants – by Halle (1988). This scheme provides a useful subdivision of species which can be used in conjunction with the Dyksterhuis (1957) classification of increases, decreases and invaders (Chapter 4).

TYPE 3 DISTURBANCE

While many small mammals are undoubtedly the victims of road kills, there appears to be relatively little hunting and killing for recreation in the Western world, although for example, Collins and Lichvar (1986) report that they twice observed drivers stop and shoot at prairie dogs from the roadside during one summer's fieldwork in Wyoming. There are also many negative superstitious ideas about bats, and many are killed for this reason (Racey and Stebbings, 1972). Feral pigs or swine are hunted in Australia and in Florida, where it is the basis of a minor local industry (Degner *et al.*, 1983). In many less-prosperous countries small mammals may be hunted for food or for their skins. In the national parks of western Argentina, the Southern River otter

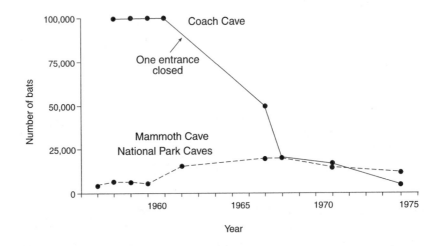

Fig. 24.5 Winter population sizes of Indiana bats (*Myotis sodalis*) at Coach Cave before and after one entrance was closed, and all the occupied caves in nearby Mammoth Cave National Park, Kentucky (from Humphrey, 1978).

(*Lutra provocax*) is absent from some areas and, although human disturbance and tourism are not thought to have a significant effect on their distribution, remote refuges are recommended for this animal (Chehébar *et al.*, 1986). Otter (*Lutra lutra*) hunting was until recently a fashionable sport in England, where otters are now very rare.

Like the rest of animal and plant life on this planet, small mammals, where they have not been able to adapt to anthropogenic changes, are often threatened by the increasing human population, and the remaining natural areas are frequently their last refuge. For this reason, recreation impacts on this group need to be monitored carefully and, if necessary, access to some areas either restricted or stopped altogether.

24.2 Summary

25 Large mammals

The larger mammals are one of the most directly threatened groups of organisms on the planet. The last 50 000 years has seen the extinction of an increasing number of species, including the Irish elk, mammoth (*Mammuthus*), steppe bison, woolly rhinoceros, the mastodon and many African game species (Arms, 1990).

For example, most large mammals have become extinct in North America coincident with the invasion of humans via the northern land bridge (Smith, 1990). The spread of humans over the continent was thought to be the result of population increase and the need to move southward as resources were depleted (Martin, 1973). The question here and in all parts of the globe is, 'What part have humans played in their extinction?' The coincident timing is not doubted, although humans may have spread through North America earlier than originally thought, nor is the fact that some hunting methods, such as driving herds of animals over cliffs (cf. Buffalo Jump in southern Alberta) or surrounding them with fire, have killed large numbers.

In Australia we have seen, for example, the recent extinction of the thylacine (*Thylacinus cynocephalus*), largely due to this marsupial 'wolf' being regarded as a pest and to changes in land use in Tasmania, although it is thought to have been the result of competition with the dingo on the mainland of Australia (Flannery, 1994). Whether or not humans were the direct cause of the loss of all these mammals in the past, or just one of a number of causes, there is little doubt that humans have driven many recent animals to their demise and have endangered many more.

The red data book (IUCN, 1990) lists 140 mammals in the 'endangered' category, and many more are still vulnerable or rare. The sad aspect of too much of the literature is that papers often end with such phrases as 'these animals are unlikely to survive unless immediate measures are taken for their preservation'.

Recreation, especially hunting, has undoubtedly added to the reduction in numbers of the species of large mammals, but it has also been a force for conservation. Ecotourism may, in fact, lead to the survival of some of the species at present in danger of extinction.

In this chapter elephants, dogs and cats are discussed separately, and then a short comment is added on marsupials as there is relatively little information on this group.

The two surviving elephant species – the African elephant (*Loxodonta africana*) and the Asian elephant (*Elaphas maximus*) – tend to be subject to pressures from different kinds of human activity. There are an estimated 500 000 or more African elephants in and out of various reserves. In contrast, there are only about 50 000 Asian elephants, of which 10 000 are working animals and only some of the remainder are in reserves (Tudge, 1994).

TYPE 1 DISTURBANCE

African elephants have been shown to be sensitive to the noise of rifle shots in the Luangwa Valley National Park, Zambia, where residents and non-resident safari clients are allowed to kill game in the adjacent Luangwa Game Management area: 'The frequency and proximity of gunshots may be sufficient to disrupt normal feeding patterns' (Lewis, 1986). There are many accounts (some apocryphal?) of elephants charging hunters on safari, and Asian elephants will attack humans trying to chase them from crops at night (Tudge, 1994). However, there do not appear to be any direct studies of the effects on elephants of recreation-based disturbance.

TYPE 2 DISTURBANCE

The increasing populations of humans have increased the level of competition with elephants for resources (Douglas-Hamilton, 1973). For example Viljoen (1987) commented that 'since 1880 man has caused a decline in elephant distribution and number in South West Africa'. The introduction of national parks and conservation areas which are intended to alleviate the problem, are, however, resented by local people who benefit little from the tourist or safari hunter's dollars (Leader-Williams, Albon and Berry, 1990). Conflict over scarce resources is the essence of the difficulty (Able and Blaikie, 1986) and rising human populations are going to exacerbate these problems. It is also interesting to note that of the 16 000 km² covered by the Luangwa National Parks, only 12% is used for tourism, and the reserve area must be seen as non-profitable by the local population.

TYPE 3 DISTURBANCE

The primary concern for the survival of the African elephant is the large illegal ivory trade. There have been several recent estimates of the effect of this trade on the populations of elephants (Caughley,

Dublin and Parker, 1990; Milner-Gulland and Mace, 1991). The general conclusion is that in spite of the present estimate of a total of 500 000 African elephants (Tudge, 1994), the numbers will fall dramatically, possibly to extinction by the year 2020 as a result of the illegal ivory trade (Caughley, Dublin and Parker, 1990) (Fig. 25.1). The only way to prevent this, if my deductions from Able and Blaikie (1986) are correct, is to increase the distribution of ecotourist revenue amongst the local populations to the point where it is profitable for them to work actively against illegal ivory hunting.

Although many are being electrocuted or shot in defence of crops, Asian elephants are, according to Sukumar (1991), more likely to survive, at least the 10 000 or so working animals and perhaps

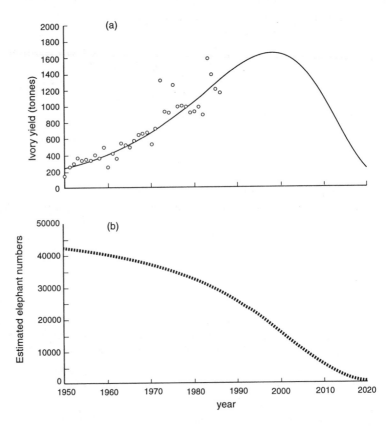

Fig. 25.1 (a) Simulated annual production of ivory (curve) and the reported production (circles) from Africa as a whole. (b) The deduced trend in elephant numbers (indexed by standing crop of ivory) when subjected to that regime of exploitation. One tonne of ivory equals about 100 elephants. (From Caughley, Dublin and Parker, 1990.)

15 000 individuals in Indian reserves and 15 000 elsewhere in Asia (Tudge, 1994). However, Duffey (personal communication) notes that recent press reports say that the Asian elephant is much more vulnerable than the African elephant. Ecotourism may also be one factor in the survival of the Asian and, hopefully, the African elephant.

Dogs are members of the order Carnivora, but I do not discuss *Canis familiaris,* the domestic animal, which undoubtedly provides a great deal of recreation and trade to the pet industry but cannot be considered a part of the natural environment, although feral animals are a problem in some parts of the world. Included here are the wolf (*Canis lupus*), the coyote (*C. latrans*), the dingo (*C. familiaris dingo*) and the fox (*Vulpes vulpes*).

25.2 Dogs

TYPE 1 DISTURBANCE

Dogs have acute senses and can detect humans from considerable distances. They are also curious animals and there are many records of wolves just standing and watching humans or even approaching and sniffing at a passive human (Grace, 1976). Grace recorded 70 encounters with wolves and they either stood, lay down, approached, or circled round him, but mostly they retreated or continued on their original direction and occasionally they howled (Table 25.1). He concluded 'that wolves with experience of man are generally wary of humans even when there are several wolves and a single unarmed man'. Klein (1975, personal communication to Geist, 1975) reported that he had observed a pipeline construction worker feeding a wolf by hand after offering it food over a number of days. However, the hostile attitude of many people to wolves still remains, and stories of wolf attacks on humans from over 100 years ago still circulate or are highlighted by the press, for example in present-day Finland (Pulliainen, 1982). There is also controversy over the concept of reintroducing wolves to the Yellowstone National Park in Wyoming, America.

Foxes also seem to be relatively unafraid of humans. Around Copenhagen they were found to live preferentially in the part of a wood that was within a city boundary, presumably because the supply of food was greater as they frequently fed on garbage and were not hunted in that area (Nielsen, 1989).

In general, dogs have their litters in underground dens, and wolves and coyotes have both been recorded as abandoning the natal dens as a result of human disturbance (Harrison and Gilbert, 1985; Ballard, Whitman and Gardiner, 1987) – the coyotes took their pups up to

Table 25.1   Responses of wolves to a human observer at a remote Ellesmere Island weather station (Grace, 1976)

Observer actions	Responses of wolves							
	Stand	Lie	Approach	Circle	Retreat	Continue[a]	Howl	Total
Stand	2	6	2	3	8	10	3	34
Lie		2	1	2		6	2	13
Approach	1				13			14
Walk		1	4			4		9
Total	3	9	7	5	21	20	5	70

[a] Continue travelling on their original path.

1 km from the original den. However, transfer of pups to other dens also occurred without the provocation of human disturbance (Harrison and Gilbert, 1985). In a military training area in Colorado, coyotes altered their behaviour when manoeuvres were taking place. These equate to some degree with four-wheel-drive use for recreation (Gese, Rongstad and Mytton, 1989). The animals changed their home ranges, moved their centres of activity away from military exercises (presence of vehicles and people, etc.) and increased their diurnal rate of movement (Fig. 25.2). The animals avoided disturbance by moving into cover too rough for vehicular travel (see also Chapters 21 and 22). Of 16 animals, two permanently abandoned their home ranges, fortunately it was not at a time when they had pups, but there may well have been an unnatural increase in territorial conflict with other animals.

TYPE 2 DISTURBANCE

As with other animals, the greatest effect on wild dogs has been the increasing human population and consequent reduction in natural habitat. However, dogs are also able to make use of discarded human foodstuffs. In the Gran Paradiso National Park, Italy, foxes made use of the increase in edible material available during the summer tourist season. They descended to the valley for feeding, and extended their normal territory or home range, so that at least three animals used the same part of the valley (Boitani, Barrasso and Grimod, 1984) (Fig. 25.3). Foxes are also relatively common in towns and cities of northern Europe and have adapted well to the changes in land use that accompanied the expansion of the human population.

In contrast, wolves have been displaced from much of the northern hemisphere which they occupied only a few centuries ago (Piechocki,

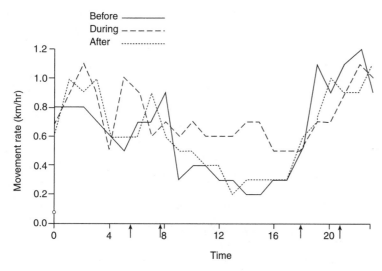

Fig. 25.2 The 24-hour activity pattern of coyotes before (———), during (_____), and after (_ _ _ _ _ _) military manoeuvres were carried out on the Pinon Canyon Manoeuvre Site, Colorado. Movement stopped earlier in the morning and peaked later at night when manoeuvres were on. Arrows indicate twilight periods at sunrise and sunset. (From Gese, Rongstad and Mytton, 1989.)

1994), although there have been some small population increases coincident with changing attitudes to conservation. There are also increasing pressures for reintroduction of wolves in certain areas (Spinney, 1995). They are protected for the full year in 9 out of 16 European countries and part of the year in another two. The Munich Wildlife Society is, among other things, creating a policy for wolves and tourism (Piechocki, 1994). One essential point in predator conservation is that owners of stock must be compensated for losses that result from predation. Fortunately, the rise of ecotourism and its economic benefits are making this a realistic policy.

TYPE 3 DISTURBANCE

Hunting is not uncommon where people have lost animals to wolves, for sport or, until recently, for bounty (in Poland for example; Okarma, 1993). In some countries illegal killing may be a problem. Blanco, Reig and de la Cuesta (1992) estimated that 550–750 wolves are killed every year out of a Spanish population of 1500–2000. The impact of modern technology also adds to this toll. On Ellesmere Island there

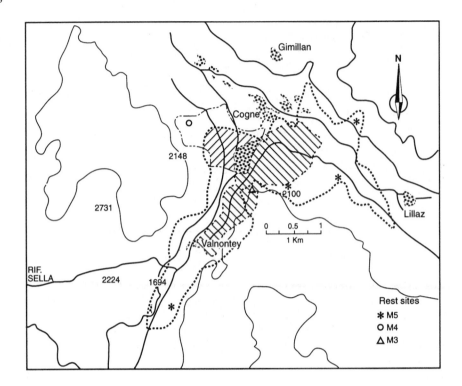

Fig. 25.3 Home ranges of three foxes (M3, M4, M5) showing abnormal overlap (hatched or dotted areas) during the tourist season; Gran Paradiso National Park, Italy (from Boitani, Barrasso and Grimod, 1984).

was a rapid increase in the number of high arctic wolves shot between 1956 and 1971, from 1.7 to 9.6 per hunter per year (Riewe, 1975). At this time there was a bounty of $40 per head and hunters started to use snowmobiles to follow fresh wolf tracks, so the animals were more vulnerable as a result of the hunters increased speed over the terrain. The main threat to Italy's 300 or so wolves was thought by Boitani (1992) to be interbreeding with feral domestic dogs, but worldwide this seems a relatively rare occurrence (Pulliainen, 1982). Dingos are still subject to control programmes including shooting, trapping, poisoning with 1080 and the erection of the 'longest fence in the world' to keep them out of SE Australia (Newsome, 1991).

25.3 Cats (family Felidae) The Felidae are essentially carnivorous, nocturnal hunters with large eyes. These include lions (*Panthera leo*), tigers (*P. tigris*), snow leopard, (*P. uncia*), common leopard (*P. pardus*), jaguar (*Felis onca*), cheetah (*Acinonyx jubata*) and many others.

TYPE 1 DISTURBANCE

The cats have well-developed senses of vision, hearing and smell and, like dogs, generally detect humans from considerable distances. There are records of most of the large cats attacking humans, sometimes for food, particularly when the animal is incapacitated in some way or defending a recent kill (for example, attacks by the cougar or mountain lion (*Felis concolor*) in the USA, but there were only seven verified human deaths from 1900 to 1977; Saile, 1977). Areas of human disturbance were avoided by dispersing animals when they were selecting sites for residence in northern Arizona and south-central Utah (van Dyke *et al.*, 1986). These authors considered that even if human residence, presence or activity is temporary, the area is reduced in quality for the lion population, and that large relatively undisturbed areas may be a requisite for a stable lion population. In southern Utah, mountain lions have also been shown to avoid areas where there were improved dirt roads and hard surface roads, logging activities and few, or no, permanent human disturbance sites (van Dyke, Brocke and Shaw, 1986). Resident mountain lions were most frequently disturbed by distant machine noise (100–1000 m) whereas animals in transit more often encountered close disturbances (less than 100 m) (van Dyke *et al.*, 1986) (Table 25.2), presumably because the transient animals were less familiar with the locations of humans, or they were forced into closer proximity by the resident lions. However, Boyden (1996, personal communication) informed me that mountain lions have been seen in urbanized areas and a jogger was recently killed by one in northern California.

Mountain lions in the immediate vicinity of human disturbance shifted their activity peaks to after sunset and were inactive rather than active at sunrise, in contrast to the increased daytime activity of disturbed coyotes (see Fig. 25.2).

TYPE 2 DISTURBANCE

The Indochinese tiger (*Panthera tigris corbetti*) was found to be threatened by human disturbance, including settlements, especially beside large waterways, dams or reservoirs, as a result of fires, road construction, and loss of forest in Thailand, where an estimated 200–250 animals are surviving (Rabinowitz, 1993). These animals are in danger of further decline due to continued human encroachment within protected areas. A similar story to that of the mountain lion above. The remaining 400 snow leopards in north-western India may be in a better situation because of the 'Project Snow Leopard' of the central Indian government and the 'Snow Leopard Recovery Program'

Table 25.2 Types of human disturbances encountered by resident and 10 transient mountain lions in the Escalante Study Area, Utah, as determined by trailing and on-site surveillance, December 1980–February 1982 (Van Dyke *et al.*, 1986)

	N encounters	
Type of disturbance	Residents	Transients
Distant machine noise		
(more than 100 m but less than 1 km)	14	7
Close disturbance (less than 100 m)	9	17[a]
Idle machines and buildings with no human activity	3	1
Working machines and people less than 100 m	3	12
Human residence or camp	3	4
Total	23	24

[a] Different from resident lions ($\chi^2 = 4.8$, $P < 0.05$).

of the state of Jamma and Kashmir where new parks and reserves are being established (Fox *et al.*, 1991). Large areas of the snow leopard's range lie outside the parks and reserves and, although there is a low human population, little hunting and some uninhabited areas (Mallon, 1991), conservation measures need to be considered for these parts of the animal's range (Fox *et al.*, 1991). Some of these areas are being increasingly used by tourists and this form of development needs to be curtailed unless the income from this industry can be used to create alternative refuges. As Mallon (1991) pointed out, there needs to be an equitable sharing out of natural resources (and, I would add, the profits from ecotourism) between local populations and wildlife.

TYPE 3 DISTURBANCE

A bounty was paid for killing mountain lions in California from 1907 to 1963 and 12 461 were killed in that period (Weaver, 1976). In 1969 the animal was classified as game and a licence was required to kill them. But in 1971 a moratorium was placed on killing them, except for those attacking cattle.

As in so many cases, humans have come to value large and potentially dangerous mammals only when their populations have been drastically reduced. There is little economic gain from hunting them and their conservation value and the indirect income from ecotourism have started to become apparent. They are still under threat from local populations who receive virtually none of the tourist income and hence have no compensation for their cattle and crops destroyed by the animals. For example, a high potential for tourism exists in the Pantanal region of Brazil near the Paraguay River where the jaguar is still present. Historically, direct killing has been the primary cause of jaguar losses in this region, although habitat loss has become increasingly important. Here the local people usually justify jaguar killing on the basis of their cattle-killing habits (Quigley and Crawshaw, 1992). Conservation depends upon the acquisition and protection of large blocks of land (4000–6000 km^2), which Quigley and Crawshaw (1992) estimate would preserve 30–50 adult jaguars (Table 25.3). As local human populations are often responsible for killing large cats because they predate their cattle, adequate compensation must be paid to preserve the animals in these situations.

Hunting alters the age structure of mountain lion populations. In Wyoming when hunting was permitted the oldest adult was only about 7 years old and 27% were killed in their first year. Kittens comprised 50% of the population in the first winter (Logan, Irwin and Skinner, 1986). Such a population is probably more vulnerable to decline than one with a more widely spaced age structure. The young animals appear to be more vulnerable to human disturbance and van Dyke *et al.* (1986) recorded that four animals that died in their study areas were all juveniles.

25.4 A note on marsupials

There are approximately 130 living species of marsupial mammals in Australia. Generally, over the past 200 years marsupial populations have been greatly reduced in both species number and in population sizes, by land use changes, feral placental mammals and, in some cases, by hunting (e.g. the Tasmanian wolf or tiger, *Thylacinus cynocephalus*). It is therefore difficult, and perhaps of low priority, to separate the effects of recreation from all the other pressures.

However, there has been a study of the numbers of swamp wallabies (*Wallabia bicolor*) and grey kangaroos (*Macropus giganteus*) killed on a 20 km section of highway in central Victoria (Coulson, 1982) (Fig. 25.4). The most interesting point to note is the relationship between kill sites and woodland adjacent to the road. This is comparable to situations where mountain goats also crossed a busy road in the Rocky Mountains (Fig. 23.6). Coulson also noted that the warning

Table 25.3 Areas of habitat required for each individual large mammal

Tiger	0.2–4.3 per 100 km^2	(Rabinowitz, 1993)
Snow leopard	Approx 1 per 100 km^2	(from data in Fox *et al.*, 1991)
Jaguar	0.7–1.6 per 100 km^2 with a high degree of overlap	(Quigley and Crawshaw 1992)

signs for traffic appear to have no effect on the number of animals killed. The writer has observed the rapid turnover of dead animals on the roadside – a corpse only lasting about 3 weeks in summer. The numbers killed are therefore likely to be much greater than reported by Coulson, and he noted that the great majority of road kills go unrecorded, many animals probably dying away from the road.

The impact of road kills on the kangaroo population may also be affected by the fact that 10 of the 25 kills were adult males, which may have a long-term effect on the social organization of kangaroo mobs. The similarity of location for road kills (Fig. 25.4) suggests that while information is lacking specifically on marsupials in Australia, it may be realistic to apply lessons learned from management of wildlife in North America, in particular the bovids and the deer, sheep and goats. However, the marsupials do have different behaviour patterns and field observations are also required.

25.5 Summary   There are many reports of declines of mammal populations in less developed countries (Kayanja, 1984; Leader-Williams, Albon and Berry, 1990; Fox, Nurbu and Chundawat, 1991; Loggers, Thévenot and Aulagnier, 1992; Tulgat and Schaller, 1992; Ross, Srivastava and Pirta, 1993) and some part of this decline may be directly associated with recreation, especially hunting. There is also a growing awareness that these animals may all represent tourist dollars. However, until the income derived from tourism reaches the indigenous people, who have traditionally utilized the animals for food and now as a cash crop, there will be continuous poaching. The adoption of firearms, outboard motors, dogs and headlamps by many indigenous hunters

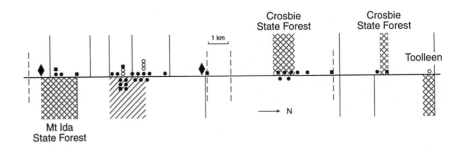

Fig. 25.4 Schematic diagram of the 20 km section of highway surveyed for road kills. ◆, Kangaroo warning signs; ●, grey kangaroo road-kills; ○, grey kangaroo kills reported; ■, swamp wallaby bicolor road kills. Hatching indicates woodland; cross-hatching, open forest.———Minor roads; --------creek crossing. (From Coulson, 1982.)

(Redford and Robinson, 1985) is also a great concern, especially where 'cash cropping' is the motive. Park guards and the like can only protect small areas and small populations. Most of the undeveloped countries cannot afford full protection for their national parks, let alone the marginal areas that are often essential to mammal populations (e.g. elephants and rhinoceros in Luangwa Valley, Zambia (Leader-Williams, Albon and Berry, 1990) and African mammals in general (Kayanja, 1984)).

Foxes appear to be able to live in close proximity to humans and the impact of recreation is most likely positive in terms of population sizes, although in some of the more backward countries they are still hunted for sport! Wolves occupy a reduced area worldwide but in Canada and North America there is still a large (50 000?) population living segregated from humans in the extensive northern wilderness. In Europe and more heavily populated countries wolves come more often into conflict with humans and damage compensation payments are required for their conservation (Piechocki, 1994). There is evidence that human attitudes are changing, especially with the increasing economic benefit of ecotourism, and most species of wild dog are being increasingly valued.

Survival of many species is no longer just a question of good management but a political problem requiring political will for its solution. A few species may be saved by the intervention of the developed world, but in the long run the survival of many large mammal species depends on a fundamental change in world economy and, to some extent, the tourist dollars being channelled to the right place.

26 The impacts of water-based recreation

Water plants and animals are likely to be affected by many human activities, including sewage disposal, land drainage and various landuse practices, which, in turn, may be influenced by an influx of visitors to an area for recreation (Liddle and Scorgie, 1980). Those activities taking place on or near the water are considered in the next three chapters.

The effects of recreational activities on aquatic animals are less well understood than the effects on plants, partly because animals react to the presence of humans, and to the results of their activities, in very different ways. They may be disturbed by sight and sound, as well as by any associated pollution or other change in the environment. Furthermore, animals are often strongly dependent on plants for food, shelter, breeding sites, or simply for somewhere to sit, so that they may suffer indirectly if plants themselves are affected. This applies equally to zooplankton in the open water and to birds and mammals at the margins of a water body.

Sometimes the effects of recreational activities are clear, for example when groups of birds feeding or roosting on the water take flight at the approach of a boat. However, unless an animal or plant is particularly conspicuous, or the subject of special interest (e.g. angling), the effects may not be noticed. When more than one factor is involved, as in the case of an enclosed water body used for multi-recreational activities, it may be virtually impossible to isolate the cause of any observable effect, except in the clear-cut instance of wildfowl being disturbed by boats or fishermen (Chapter 20).

The many possible ways of classifying the impacts of recreation will be influenced by the amount and quality of information available, as well as by the nature of the subject. A useful distinction can be made between shore- and water-based activities (between fishing from the bank and boating, for example) and this type of user-orientated system, which has been widely used (see Satchell and Marren, 1976), has been adopted for the first part of this chapter.

Recreation affects water quality through chemical, physical and biotic changes or introductions into the environment. The fact that species-rich, oligotrophic, aquatic systems are often the 'targets' of recreation activity leads to the most vulnerable systems receiving high levels of use, and hence they show the greatest changes in qualitative, if not quantitative, terms.

The physical impacts of recreation on fresh waters were also divided into those derived from water-based activities and those caused by shore-based pursuits by Liddle and Scorgie (1980), and the discussion that follows is largely based on their paper.

26.1 Physical impacts of freshwater recreation

WATER-BASED ACTIVITIES

Physical impacts

The physical forces associated with water-based activities originate mainly from boats, and include wash, turbulence, propeller action (cutting effects), direct contact and also disturbance by sight and sound. The effects of other activities, such as swimming, are insignificant in comparison, except when they are particularly concentrated in space and/or time.

The forces originating from boats all interact (Fig. 26.1) and the final effect of any one component is difficult to quantify. Their relative importance will also vary according to the dimensions and type of habitat involved. For example, the height of the wash from a boat will decline as it spreads out over a large body of water, so the size of the water body will have a strong influence on the effect of wash on marginal macrophytes. The consequences of the impacts will also depend upon the time of year in which they occur in relation to the phenology of the plants and the activities of animals. In temperate zones many water plants spend the winter in a resting stage, for example the turions of *Potamogeton* spp., and are not likely to be damaged at that time. However, boating, other recreation activities and management operations may help to disperse reproductive structures or vegetative fragments of plants and thus aid their survival. For example, the spread of the introduced pondweed (*Elodea nuttallii*), in the UK since 1974 may have occurred in this way (T.C.E. Wells, personal communication).

Wash

Motorboats are often high-powered and the wash they create can cause considerable erosion of plant roots. The susceptibility of a number of plants to erosion was investigated experimentally by Haslam (1978),

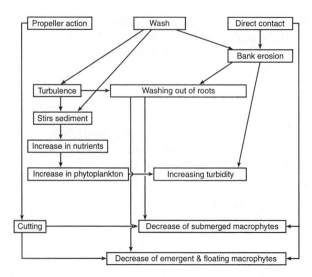

Fig. 26.1 The impacts of boats on plants (from Liddle and Scorgie, 1980).

who placed them in four groups (Table 26.1). It is notable that in those heavily used waterways which still have marginal vegetation, the species present are often those in Haslam's group 4, which are difficult to erode.

The effect on plants of water movement caused by boats is in some respects similar to that of current flow in faster-moving waters. It is a common observation that the leaves of yellow waterlily *(Nuphar lutea)* are smaller in faster-flowing waters. Wash was thought by Sukopp (1971) to cause damage to reedbeds when boats sailed into gaps and turned close to the plants. The down wash from waves striking boats moored over soft sediments may also clear vegetation from a circular area as the boat swings around the anchor.

Turbulence and turbidity

Propeller action may create turbulence in the water and the extent of this will depend upon its size, design, position in relation to the hull and the horsepower of the motor driving it. Boats propelled by oars or paddles impart relatively little energy to the system, although Liddle (1973a) noted that 0.5 m² of sward of shore-weed (*Littorella uniflora*) in a shallow oligotrophic lake could be uprooted by the turbulence from one oar stroke.

Increases in the turbidity of waters used by boats for recreation have often been reported, but there seems to be little quantitative evidence

Table 26.1 Susceptibility of aquatic plants to erosion (after Haslam, 1978) (Liddle and Scorgie, 1980)

1. Very easily eroded
   Agrostis stolonifera (submerged)   *Epilobium hirsutum* (rooting fragments)
   Ceratophyllum demersum   *Rorippa amphibia*
   Elodea canadensis   *Rorippa nasturtium-aquaticum* agg.
2. Easily eroded
   Callitriche spp.   *Myriophyllum spicatum*
   Epilobium hirsutum   *Sparganium erectum*
   Myosotis scorpioides   *Zannichellia palustris*
3. Rather difficult to erode
   Apium nodiflorum   *Potamogeton perfoliatus*
   Berula erecta   *Schoenoplectus lacustris*
   Potamogeton crispus   *Sparganium emersum*
4. Difficult to erode
   Glyceria maxima   *Potamogeton pectinatus*
   Nuphar lutea   *Ranunculus calcareus*
   Oenanthe fluviatilis   *Ranunculus fluitans*
   Phalaris arundinacea   *Ranunculus penicillatus*
   Phragmites communis   *Ranunculus trichophyllus*

Susceptibility to erosion was determined by directing a horizontal jet of water from upstream on to the soil at the base of the plants, and the time taken for the plant to be eroded was noted. The habitats used were typical for each species.

for this. Lagler *et al.* (1950) found that there was no recordable increase in turbidity due to the effects of outboard motors in their experimental ponds, although there was considerable movement of the bottom sediments. They did record some redistribution of benthic invertebrates, but not damage, as a result of turbulence caused by outboards operating in shallow water. Moss (1977) reported that in the Norfolk Broads, UK, the turbidity of the waters was not strongly correlated with the amount of use by boats, but that the opacity was due to the phytoplankton. Many factors must determine the amount of turbidity directly caused by boats – the amount of clay in the sediment, the depth of the water and the size and horsepower of the craft being among the most important.

The physical effect of four-wheel-drive vehicles fording shallow rivers on sediment suspension (turbidity) and deposition was investigated by Brown (1989). There was a clear relationship between the number of vehicles crossing the rivers and the amount of sediment deposited downstream of the ford (Fig. 26.2). The mean deposit was 970 g m^{-2} 30days^{-1}, which amounts to 1.26 tonnes every 30 days in a channel 15 m wide and 75 m long (Brown, 1989). This amount of deposited material will clearly have a major impact on the fauna in the substrate as well as any water plants and algae growing on the stream bed.

Propeller action (cutting)

The edges of propellers can act as a set of rotating knives, as demonstrated by the effect of the occasional collisions with swimmers. Propellers have been observed by Crossland (1976) to remove about 10 cm from the top of the submergent spiked water milfoil (*Myriophyllum spicatum*), and Liddle and Scorgie (1980) found that an outboard motor attached to a boat driven through a patch of yellow waterlily (*Nuphar lutea*) will cut through the petioles, leaving a very jagged end. On a run of 50 m, 15 leaves were detached and many more were overturned. Lagler *et al.* (1950) found that prolonged use of an outboard motorboat, operating in water 75 cm deep, with the propeller 35 cm from the bottom, removed all plants from a strip 1.5 m wide; and that the silt had been washed to the sides of the strip, leaving sand and gravel in the centre.

Direct contact of boats

Boats may also dissipate their kinetic energy by direct collision with the marginal vegetation or bank. A survey of the interactions of the various factors which may cause bank erosion and affect the substrate was given in Bruschin and Dysli (1973) (Fig. 26.3).

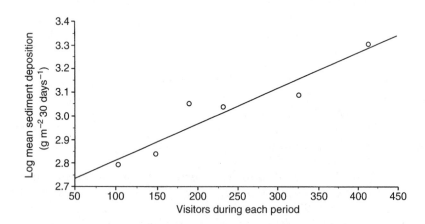

Fig. 26.2 The mean sediment deposition downstream from the point where ORVs crossed the Crooked and Wongungarra Rivers, Victoria, Australia (from Brown, 1989).

Damage to emergent macrophytes by boats running into them at right angles to the shore line, and by boats turning, leaving isolated patches of plants, was recorded by Sukopp (1971). He also noted that gaps caused in this way were then enlarged by moored boats being moved to and fro by wash from other craft. Boat berthing, launching and beaching are reported by Rees and Tivy (1977) to have an abrasive action on the beds and shores of Scottish lochs. This activity can eliminate extensive areas of emergent vegetation where heavy use occurs. A similar observation was made by Marnell, Foster and Chilman (1978) at the shallow mouths of creeks and on gravel bars in the Missouri Ozarks, where shallow-rooted species such as *Hamemelis vernalis* and *Amsonia illustris* may be disrupted or locally eliminated. However, Rees and Tivy (1977) considered that floating-leaved plants are relatively immune to damage because boat users tend to avoid these communities where oars, fishing lines and even propellers can become entangled. Sukopp (1971) also observed that arrowhead (*Sagittaria latifolia*) was not damaged as people 'take care of this conspicuous plant'.

In some areas the intensity of recreational boating has become so great that vessels have to be treated like road traffic in order to minimize impacts. Jaakson (1988) developed a technique for scoring potential impact based on vessel speed (slow = 1 or fast = 2), number of visualized traffic lanes used (1–3), wake (small = 1 or large = 3) and operation (stop = 1.5 and U-turn = 2). The scores ranged from 1 to 24, but the interesting result was that water-skier speedboats had over two times the impact score of the runabout cruiser category and nearly four times the impact of canoes and kyaks (Table 26.2). The same author developed an analysis of 'sight line shadows' for waterways which were also used in the management plan.

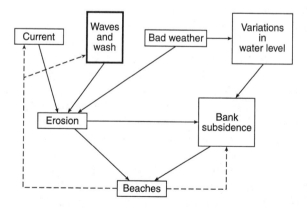

Fig. 26.3 Factors affecting erosion, and the consequences, in flowing water (after Bruschin and Dysli, 1973, from Liddle and Scorgie, 1980).

Table 26.2   Rideau river boating impact scores: 9-day data (Jaakson, 1988)

Boat type	Impact score			
	Total 9 days	Average per hour	Total percentage	Mean per boat pass
Runabout/cruiser	3278	44.9	54%	2.4
Water-skier	2371	32.5	39%	5.1
Canoe/kayak	359	4.9	6%	1.4
Houseboat	43	0.6	1%	1.1
Total	6052	82.9	100%	2.9

IMPACTS ON SHORES AND BANKS

Physical impacts

Recreational activities that take place on the shores of water bodies include angling, bird watching, swimming, camping, picnicking and walking. Since these activities produce broadly similar physical effects on aquatic plants and animals, they are considered together. The effects of management for recreation and the effects of disturbance on animals are treated separately.

Trampling

Walking in and out of the water is an activity associated with many forms of aquatic recreation. The forces exerted by walking have been described in detail by Harper, Warlow and Clarke (1961), who resolved them into vertical, horizontal and tangential components and showed that the force of the impact is partly determined by the hardness of the substratum (Chapter 2). Some forms of recreation produce additional effects as people deliberately clear marginal vegetation to gain access to the water. Anglers (coarse fishing), for example, will normally cut bank and marginal vegetation at the beginning of the fishing season, and may also remove submerged aquatic plants with the aid of a drag line. Furthermore, they may dig out sloping banks to make provision for keep nets and other equipment. The tall bank vegetation may be changed to a short sward, containing such species as rye grass (*Lolium perenne*), meadow grass (*Poa pratensis*), great plantain (*Plantago major*) and common knot grass (*Polygonum aviculare*). At one site on the river Ouse near Huntingdon, Liddle and Scorgie (1980) found that 30% of the area of the bank vegetation had been changed in this way near an access track, and that 20% was changed 300 m further away. This may increase the diversity of the river-bank vegetation but it breaks up a continuous habitat into a series of small units. In those places where algae had been dredged from the river and deposited on the banks, about 0.5 m² of the vegetation had been killed.

Marginal vegetation may also be damaged by people walking parallel to the water's edge or seeking access to the water for activities such as swimming, scuba diving or fly fishing. The damage may be extensive, changing whole communities, as Sukopp (1971) observed at the margins of the Havel River in West Berlin, or less widespread, as reported from some Scottish lochs by Rees (1978). The vegetation fringing the Havel River was subjected to wear by as many as 350 000 people on 1 day on 95 km of shore, which, because of restricted access, resulted in 9 people m^{-1} of usable shoreline. Slight disturbance at first allowed room for annual and short-lived species, especially where the margins were a managed meadow (Fig. 26.4), but the reed stands vanished with intensive use, especially for bathing, and this was followed by erosion of the bank. Sukopp (1971) recorded that a total 31% of the reed swamps disappeared from the shores of part of the Havel River in the 5 years between 1962 and 1967.

The same author analysed the responses of 22 macrophytes as reported in a number of studies and found that, while sewage pollution had mainly negative or neutral effects, there were many studies showing negative (as well as positive) responses to impacts that he placed in a category of 'general influences' (Table 26.3).

At the other extreme, Rees (1978) noted that paths made by fishermen and wildfowlers were usually between 30 and 45 cm wide, and they were typically parallel to the shore at the junction of two different plant communities. The substratum on which these pathways develop beside Scottish lochs (lakes) was usually silty with a high organic content and often with stands of reeds (*Phragmites* spp.), reed grasses (*Phalaris arundinacea* and *Glyceria maxima*) and sedges (*Carex* spp. especially *C. rostrata*) (Rees, 1978). This author observed that on little-used pathways the dominant emergent species were still present. They were replaced on pathways of intermediate use by harder-wearing species, including bent grasses (*Agrostis* spp.) and meadow grasses *(Poa* spp.), with amphibious periscaria (*Polygonum amphibium*), common knot grass (*P. aviculare*) or forget-me-not (*Myosotis* spp.) in the margins; the heavily used pathways largely consisted of bare mud with occasional invading species. The introduction of members of the common path flora was restricted to common knot grass in this case, but the often-observed increase in species number under conditions of light trampling was recorded by Rees (1978) (Fig. 26.5) (see also Figs 4.9a, 4.10).

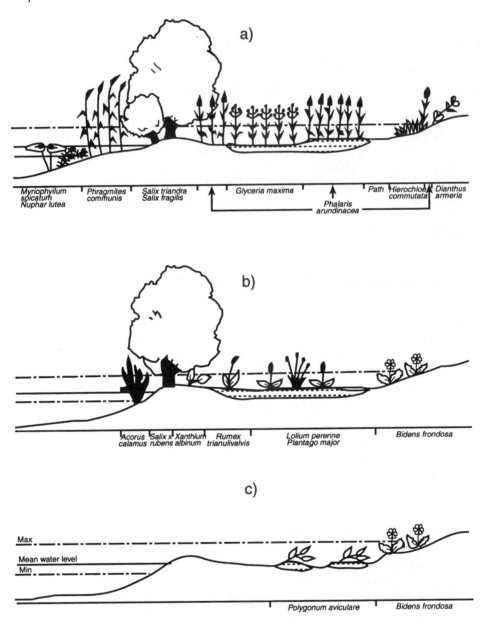

Fig. 26.4 Influence of different intensities of use for recreation on the marginal vegetation of a water meadow: (a) little use, (b) intermittent light use, and (c) prolonged heavy use (after Sukopp, 1971, from Liddle and Scorgie, 1980).

Table 26.3 Influence of humans on littoral vegetation as indicated in the literature (after Sukopp, 1971); figures show numbers of studies (Liddle and Scorgie, 1980)

Species	General influences			Boating			Grazing			Eutrophi- cation			Sewage		
	+	−	ø	+	−	ø	+	−	ø	+	−	ø	+	−	ø
Equisetum fluviatile	1	3		1			1							1	1
Typha latifolia	6	1	2							1			1		
Typha angustifolia	6	1	1	1			1								
Sparganium erectum			2											1	1
Sparganium emersum	1	2		2			2			1					
Alisma platago-aquatica	2						2							1	1
Sagittaria sagittifolia	1	2	1				2	1		1					
Sagittaria latifolia	3	1													
Butomus umbellatus	3	1					1			1					
Typhoides arundinacea	1														
Phragmites communis		4	2		5			2		5			2		
Glyceria maxima	5									1					
Eleocharis palustris	2				1		1							1	1
Scirpus lacustris	1	1	1					2						1	1
Cladium mariscus		2	1						1						
Carex pseudocyperus		4													
Acorus calamus	2	1								1					
Ranunculus lingua		6					1								
Rorippa amphibia	2	2								1					
Cicuta virosa		3													
Berula erecta		1										4			
Sium latifolium	1														

+ = Increase, − = decrease, ø = no change.

An experiment in which various loch shore dominants were subjected to different levels of trampling during the summer of 1976 was carried out by Rees and Tivy (1977). They summarized their results by giving the vegetation types a susceptibility rating of low, medium or high. They found that at low intensities of wear, common bent (*Agrostis tenuis*) and bottle sedge (*Carex rostrata*) on an organic substratum were stimulated to increase growth, but in all other cases increasing intensities of wear depressed the amount of plant cover. The taller reed grasses on very wet sites were more susceptible to damage than the shorter sedges on drier, firmer substrata. They attributed this, in part, to the dense growth of tough leaves of the sedges, which were flattened, but not as easily broken up by trampling as the brittle leaves of common reed (*Phragmites communis*), or the soft leaves of reed sweet grass (*Glyceria maxima*). It is notable that those species which are susceptible to trampling damage are not the same

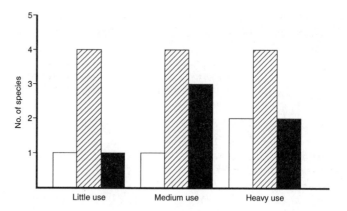

Fig. 26.5 The effect of trampling on the number of species in waterside vegetation. The unshaded, hatched and shaded histograms refer to the general vegetation, path edges and paths, respectively (data from Rees, 1978) (Liddle and Scorgie, 1980).

as those species which are easily eroded. For example, Sukopp (1971) commented that common reed is able to stand wave action caused by boats but not mechanical damage caused by treading (Table 26.1).

26.2 Recreational release of sewage

Sewage resulting from water-based recreational activities may be discharged directly into the water, particularly from boats, or, in the case of visitors based ashore, it may undergo some form of treatment before being discharged. Both the quantity and quality of effluent discharged are dependent on several factors, including the type, extent and location of activity, and whether refuse from visitors can be processed by existing sewage works (Liddle and Scorgie, 1980).

The amount of sewage and other pollutants released into waters by land- and water-based recreation activities varies from the low levels of leaching from faeces shallowly buried in individual 'cat holes' to the urban levels of discharge from large tourist developments. The major constituents of whole and settled sewage (Table 26.4) indicate that a potentially large amount of nutrients, especially nitrogen and phosphorus, may be released into waters where overnight stays by visitors occur. In camping, and to some extent boating, the amount released will be reduced, but the large amounts may be released from any form of accommodation development. The release of sewage into freshwater areas can clearly cause serious problems, but the nature and extent of the damage depends to some extent on the 'natural' status of the water body receiving the effluent, as well as the quality and quantity of the effluent itself (Liddle and Scorgie, 1980).

Table 26.4 Major constituents of domestic whole and settled sewage (Liddle and Scorgie, 1980)

Whole sewage (domestic)[a]	gC person^{-1} day^{-1}	Settled sewage (domestic)[b]	g person^{-1} day^{-1}
Carbohydrates	6.0	Chloride	10.4
Amino acids	4.2	Sodium	13.7
Higher fatty acids	9.9	Potassium	2.7
Soluble acids	3.7	Calcium	14.9
Esters	4.5	Magnesium	0.7
Anionic surfactants	2.1	Iron	0.2
Amino sugars	0.1	Copper	0.04
Amide	0.18	Nickel	0.01
Creatinine	0.42	Zinc	0.09
Faeces	17.0	Chromium	0.03
Urine	5.0	Manganese	0.08
Dishwashing and		Lead	0.01
food preparation	8.0	Nitrogen[b] (Ammonia)	10.0
Personal and clothes	7.0	Nitrate[c] (NO_3)	0.4
washing		Phosphorus[c]	1.5

gC = grams of carbon.
[a] Painter and Biney (1959).
[b] Painter (1958).
[c] Bond and Staub (1974).

As well as the bacterial input of raw sewage, when there is body contact with the water, swimming or falling in, there can be a considerable input of bacteria. Hanes and Fossa (1970) found from experiments that the mean number of all types of bacteria released by 30 minutes of swimming was 553×10^{-9} per person. In detail, 11×10^{-5} coliform A, 23×10^{-7} coliform B and 3.09×10^{-6} enterococci per person were counted.

The authors of a study in the Boundary Waters Canoe Area (Minnesota, USA and Canada) estimated that visitors left behind 360 000 lb (163 000 kg) of solid wastes, equal to 1 ton of phosphates and 13 tons of nitrogen (Barton, 1969). In addition, heavy campsite and trail use accelerated soil erosion, adding to the solid and chemical loads of the waters (Table 26.5). At some sites during the heavy-use period the pollution spread through the waters affecting 'control' sample sites 200 ft (60 m) from the main point of activity, but during low-use periods many of the control sites had zero coliform bacteria while the main use points had over 18 MPN 100ml^{-1} coleform bacteria (Table 26.5) (King and Mace, 1974). A further example dating from as early as 1969 showed that faecal coliform bacteria and faecal streptococci were significantly higher in areas of the Ross Barnett reservoir

Table 26.5 Pollution of heavily used and less used waters (based on the data of King and Mace, 1974)

			Turbidity (JTU)	Phosphates (mg l⁻¹)	Coleform bacteria (MPN 100 ml⁻¹)
Moose Lake chain	Controls	A	4.42	0.052	0.73
mean of six sites,		JJ	10.89	0.03	1.22
heavily used	Campsites	A	4.17	0.046	6.8
		JJ	5.97	0.047	18.73
Lake Isabella	Controls	A	16.33	0.069	3.16
mean of three		JJ	16.72	0.034	3.8
sites, less used	Campsites	A	19.67	0.75	4.13
		JJ	31.33	0.034	6.07

JTU, Jackson turbidity units; MPN, Mean platable number; A, 18–28 August 1970; JJ, 30 June to 9 July 1970.

used for recreation and that 'the marinas are the prime areas of pollution in the study area' (Barbaro *et al.*, 1969). In contrast, waters such as the heavily used Norfolk Broads are already quite eutrophic and in 1977 Moss estimated that visitors' sewage contributed only 2% of the total phosphorus entering the Broads. The amount of water flow and overland flow will often affect the bacterial concentrations. Measurements at Mississippi Lake, Ontario, showed a tenfold increase in the levels of faecal coliforms and faecal streptococci after rainfall (Hendry and Toth, 1982).

The changes in water quality resulting from point source discharge of high levels of organic effluent are well documented (Fig. 26.6). Broadly, the responses of the freshwater biota to organic pollution are as follows:

MICRO-ORGANISMS

Bacteria that use waste as a substrate increase. Viruses derived directly from sewage effluent also increase. Sewage fungus develops if the waters are heavily polluted. Sewage fungus itself is a community of micro-organisms including bacteria and fungi.

ALGAE

Heavy pollution eliminates algae. Some filamentous algae (e.g. *Stigeoclonium tenue*) become common in less polluted zones in some English rivers (Butcher, 1947). *Cladophora*, another filamentous alga, also

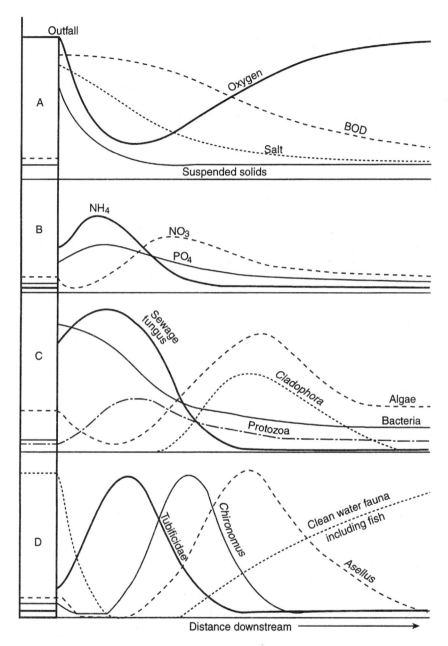

Fig. 26.6 Schematic representation of the changes in water quality and the populations of organisms in a river below a discharge of organic effluent (from Hynes, 1960). A, Physical changes; B, chemical changes; C, changes in micro-organisms; D, changes in macroinvertebrates. (From Mason, 1991.)

became dominant in the 'recovery zone', forming dense blankets over the substratum which provided cover and food for invertebrates (Mason, 1991). However, it may deoxygenate the water at night, suffocating fish that may be present. Some diatoms reappear next (e.g. *Nitzschia palea* and *Gomphonema parvulum*) but Liddle, Happy-Wood and Buse (1979) found the number of desmids almost inversely related to pollution effects in oligotrophic Welsh lakes. The blue-green alga *Chamaesiphon* spp., the green alga *Ulvella frequens* and the diatom *Cocconeis placentula* appeared when the pollution had dispersed.

HIGHER PLANTS

Macrophytes, higher plants that grow at the edge or floating in fresh waters, are also adversely affected by organic pollution. The suspended solids and subsequent planktonic algal blooms limit light penetration, this may then shade the macrophytes (Liddle and Scorgie, 1980) but floating-leaved macrophytes will themselves also limit light penetration (Liddle, Happy-Wood and Buse, 1979). Species vary in their response to pollution, some sensitive ones are eliminated, others may increase (Haslam, 1987). Haslam classed five species as being tolerant of pollution in European rivers: monkey flower (*Mimulus guttatus*), curled pondweed (*Potamogeton crispus*), clubrush (*Schoenoplectus lacustris*) and the bur-reeds (*Sparganium emersum* and *S. erectum*). The only species that increased its range was pondweed (*Potamogeton pectinatus*). Untreated sewage in quantity may, as would be expected, totally eliminate all macrophytes (Hawks, 1978).

INVERTEBRATES

In general, those organisms associated with the silted regions of rivers are the most tolerant of organic pollution, while species associated with eroding substrata and swiftly flowing water are the most sensitive (Mason, 1991). The gills of sensitive species may become clogged (e.g. mayflies (Ephemeroptera) and stoneflies (Plecoptera), Figs 26.6 and 26.7) and the higher metabolic rates of invertebrates from swiftly flowing waters may be inhibited by reduced oxygen levels (e.g. *Dinocras*, Fig. 26.7). In highly contaminated waters only tubificid worms may survive, and in extreme conditions only *Tubifex tubifex* flourishes, forming dense monocultures of up to 10^6 m^{-2} (Mason, 1991). In less-polluted conditions the midge larvae (blood worms) of *Chironomus riparius* and other chironomid species may become dominant, followed by the isopod *Asellus aquaticus* (Figs 26.6, 26.7).

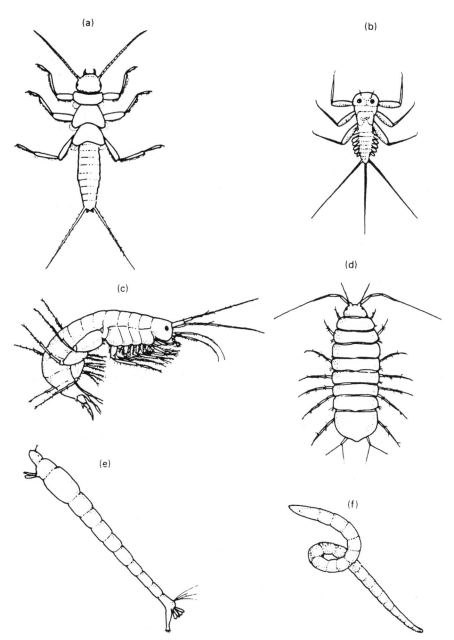

Fig. 26.7 Some species of freshwater invertebrates, in order of increasing tolerance to organic pollution (adapted from Macan, 1959). (a) *Dinocras cephalotes* (Plecoptera); (b) *Ecdyonurus venosus* (Ephemeroptera); (c) *Gammarus pulex* (Amphipoda); (d) *Asellus aquaticus* (Isopoda); (e) *Chironomus riparius* (Diptera); (f) *Tubifex tubifex* (Oligochaeta). (From Mason, 1991.)

WATER-BORNE HUMAN DISEASES

In one sense these are the impact of the environment on recreators but because the causative organisms are often released into the waters by such people, I have included a brief survey here. The main diseases associated with freshwater organisms (Table 26.6) were divided into four groups by Cairncross and Feachem (1983):

1. True water-borne diseases contracted by drinking (e.g. cholera, typhoid, hepatitis).
2. Those contracted by a lack of personal hygiene (e.g. class 1 plus other diarrhoeal diseases, trachoma, ringworm and those carried by lice and mites).
3. Those caused by helminths (flukes and flatworms) (e.g. tapeworm, bilharziasis, trematode infection, etc.).
4. Those that require a water-related insect vector (yellow fever, malaria, river blindness, filariasis, etc.) but not polluted water.

The impact of these human-vectored organisms is varied. Bacteria such as *Salmonella* may infect many animals and have long survival times in the environment. They have been recorded 117 km downstream from the nearest source of domestic animal pollution in the Red River, North Dakota (Geldreich, 1972) and have survived for 51 weeks in experimentally inoculated faeces buried at both 5 cm and 20 cm below the soil surface in a mountain environment in North Montana (Temple, Camper and Lucas, 1982). Buried faeces are potential sources of water pollution in the wilderness camping areas. Many human pathogens may be carried by rodents, and their urine may carry the organisms into the water (e.g. *Leptospira icterohaemorrhagiae*, the bacteria that causes Weil's disease) (Mason, 1991). The effects of these pathogens on wildlife do not seem to have been reported. Another example is *Campylobacter*, which causes diarrhoea, and is carried by birds. The effect of varying concentrations of enterococcus in the water and the proportion of swimmers who develop gastrointestinal symptoms was investigated by Cabelli *et al.* (1982) (Fig. 26.8).

The European, Canadian and United States' standards for recreational waters are given in Table 26.7, but Kay and Wyer (1994) stated: 'Good quality scientific evidence on which to base advice is almost non-existent and the extent to which pathogen presence equates to health risk is not well understood'. It should be noted that only faecal coliform bacteria are now considered an appropriate measure and total coleform is not precise enough. Some authors even suggest that only measures of specific disease-causing bacteria give a true indication of significant pollution (McNeill, 1991).

Table 26.6 Some water-related diseases and their causative organisms (Mason, 1991)

Causative organisms	Disease or symptoms
Bacteria	
Salmonella typhi	Typhoid fever
S. paratyphi	Paratyphoid fever
Salmonella spp.	Gastroenteritis
Shigella spp.	Bacterial dysentery
Vibrio cholerae	Cholera
Escherichia coli	Gastroenteritis
Leptospira icterohaemorrhagiae	Weil's disease
Campylobacter spp.	Intestinal infections
Francisella tularensis	Chills, fever, weakness
Mycobacterium	Tuberculosis
Viruses	
Enteroviruses	Many diseases, including poliomyelitis, respiratory diseases, meningitis and infectious hepatitis
Rotavirus	Diarrhoea and enteritis
Protozoa	
Entamoeba histolytica	Amoebic dysentery
Giardia lamblia	Diarrhoea, malabsorption
Naegleria fowleri	Amoebic meningoencephalitis
Cryptosporidium sp.	Diarrhoea
Helminths	
Diphyllobothrium latum	Tapeworm infection
Taenia saginata	Tapeworm infection
Schistosoma spp.	Bilharzia
Clonorchis sinensis	Trematode infection
Dracunculus medinensis	Guinea worm

See McNeill (1991) (Table 1) for a more detailed list of tropical water-borne pathogens.

26.3 Release of petrol and oil

Outboard motors are probably the most common means of propulsion for motorboats used for recreation, the majority are now of the four-stroke cycle type. These are much 'cleaner' than the two-stroke engines in common use until the early 1980s. The following discussion is adapted from Liddle and Scorgie (1980).

The substances emitted by outboard motors are derived from petrol and lubricating oil. Both petrol and oil consist mainly of hydrocarbon compounds with small amounts of additives. Oils contain elements such as zinc, sulphur, phosphorus and other unspecified additives (Jackivicz and Kuzminski, 1973a).

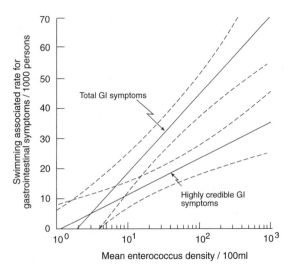

Fig. 26.8 Regression of swimming-associated gastrointestinal (GI) symptom rates on the mean enterococcus densities in the water. Data are from all US studies. The 95% confidence limits (CL) for the lines are as shown. (From Cabelli *et al.*, 1982.)

There is little quantitative information on what substances actually appear in the aquatic environment during the operation of outboard motors. Jackivicz and Kuzminski (1973a) working with two-stroke motors suggested that water vapour, carbon oxides, nitrogen and sulphur emitted from the combustion chamber, in the unburned fuel mixture and partial oxidation products are discharged below the water surface. Various investigators have reported values for the volatile and non-volatile fractions of oil, phenols, lead, chemical oxygen demand (COD) and biological oxygen demand (BOD) in two-stroke outboard motor-exhausted water (Jackivicz and Kuzminski, 1973a). They have estimated that the total discharge of hydrocarbons from one two-stroke outboard engine, running for 1 day, would be equivalent to the waste material (sewage) produced by a population of 400 people, assuming that the products contain 85% degradable carbon.

Lagler *et al.* (1950) reported no effects on populations of fish in experimental ponds, that could be attributable to outboard motor exhausts. Jackivicz and Kuzminski (1973b) reviewed the available information and concluded that while pollution from outboard motors can exhibit a toxic effect in sufficiently high concentrations, and may affect reproduction of fish, under the conditions of normal use there is nothing to suggest that there is a problem. They called for more

Table 26.7 European, Canadian and United States' standards for recreational waters (sources: EEC, 1976; Canadian Government, 1983) (from Kay and Wyer, 1994)

(a) North American standards

Agency	Regime	Faecal coliform standard
Toronto Health Department	Daily	GM[a] < 100 100 ml^{-1} No sample to exceed 400 100 ml^{-1}
Canadian Federal	5/30 days	GM < 200 100 ml^{-1} Resample if any sample exceeds 400 100 ml^{-1}
USEPA[b]	5/30 days	GM < 200 100 ml^{-1} < 10% only to exceed 400 100 ml^{-1}

(b) European standards

	Total coliform 100 ml^{-1}	Faecal coliform 100 ml^{-1}
	(fortnightly sampling)	
Guide (recommended) 80% of samples should not exceed this figure	500	100
Imperative (mandatory) 95% of samples should not exceed this figure	10 000	2000

[a] GM, Geometric mean.
[b] USEPA, United States Environmental Protection Agency (1986).

research to relate laboratory and field observations. As Tanner (1973) pointed out, pollution can pose a serious threat to wildlife, but the pollution from recreational activities is generally small compared to other sources.

According to the Dartington Amenity Research Trust (1974), the effects of pollution from outboard motors and fuel spillage are likely to be lost in a river like the Yorkshire Ouse, which carries large volumes of water to dilute the 'relatively small' quantities of discharged material. They consider that pollution caused by the discharge of crude sewage and litter directly into the river is more serious. The available evidence suggests that, as far as effects on plants are concerned, this appears to be the case. Lagler *et al.* (1950) concluded that if no oil pollution could be discerned in their experimental ponds, it is probable that this is 'almost never an item of concern from ordinary outboard use in natural waters'. Oil from outboard motors may affect plants indirectly, particularly phytoplankton, by lowering the oxygen content of the water, particularly of the first few centimetres of a lake.

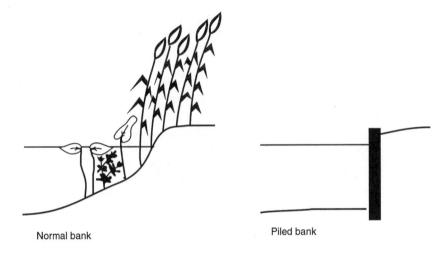

Fig. 26.9 Destruction of marginal reed beds or mangrove habitat by piling (from Liddle and Scorgie, 1980).

Stewart and Howard (1968) have shown that this can decrease phytoplankton production.

26.4 Impacts of management for recreation

The increased use of freshwater areas for recreation has led to a demand both for more space and for 'better' management of existing facilities. Aquatic plants and animals are thus threatened by the development of marinas and the restoration of canals, as well as by routine dredging and weed control operations. While the creation of large areas of solid substrata, such as wooden or steel piles, may benefit certain organisms, piling a previously sloping bank removes the normal habitat of marginal plants (Fig. 26.9). The creation of 'landing holes' and the laying out of 'wild animal bays' to overcome this problem was suggested by von Schneider and Wölfel (1978). Mechanical weed control and dredging appear to have only minimal long-term effects (Pearson and Jones, 1975, 1978; Scorgie, 1978), although animals living at the margins and on the water surface may be severely affected by management programmes that cause disturbance and wholesale changes in the habitat.

26.5 Impacts of recreation in marine waters

ORGANIC POLLUTION, SEDIMENTATION AND PHYSICAL IMPACTS

The marine situation differs from fresh waters in four major characteristics:

- The volume of water is much greater, although where embayment occurs and in places of restricted flow such as marinas, canal estates and some estuaries the effective dilution factor may be limited.
- Tidal effects are present or much greater than in even the largest lakes.
- The wave action (kinetic energy) is generally much greater.
- The water is saline and carries more dissolved minerals than fresh waters, salt lakes excepted.

The natural levels of nitrogen and phosphorus are often very low in unpolluted estuarine or coastal systems such as saltmarshes, mangroves or coral reefs. Indeed Boto (1991) remarked that: 'Pristine tropical mangrove systems can be considered to be in a finely balanced state for the major nutrient elements of nitrogen and phosphorus. This has significant implications for situations in which substantial extraneous inputs of nutrients may occur. As for many unpolluted tropical systems most of the nitrogen and the phosphorus is "bound" in various particulate and dissolved organic forms in the soils and plant tissue.'

As for freshwater systems the primary pollution from recreation is sewage and some detergent or chemical toilet discharges. There are many records of high levels of faecal coliform bacteria at bathing beaches but these are usually the consequences of the volume of urban discharge and the nature of the sewage treatments. For example in 1973 over half the coastal sewage outfalls in the UK did not reach low watermark (Department of the Environment, 1973).

The discharge of sewage associated with recreation activities is generally a very small fraction of the total discharge and the resulting concentrations are low except in situations of restricted flow and/or specific recreation developments such as boat marinas or localized tourist development. However, Kocasoy (1989) recorded total coliform bacteria at a number of beaches in the south-west of the Anatolean Peninsula, Turkey, one of which was free from development. The sample data sheet from an unnamed site shows generally low concentrations except for one occasion in June 1986 when concentrations reached a maximum of 3700 total coliform 100ml^{-1}. This author gave a fairly complex predictive equation derived from multiple regressions of his data.

A study of the impact of sewage on a coral reef community at Kaneohe Bay, Hawaii, showed that there were very marked changes and loss of the biota (Smith et al., 1981). Hawker and Connell (1989) have summarized the nature of the impact of sewage on coral reefs, although the degree of change will depend on the level of pollution (Fig. 26.10; Table 26.8).

One factor that is important in these situations is the position of the discharge point in relation to water flow or currents. Where the

Table 26.8 Summary of physicochemical and biological effects of sewage discharges on coral systems (Hawker and Connell, 1989)

Characteristic	Change with increasing enrichment
Biomass	Increase
Primary production	Increase
Coral numbers	Decrease with *Porites compressa* among the most sensitive: numbers are effectively zero in extreme situations
Chlorophyll *a*	Large increase
Filter and detritus feeders	Large increase
Benthic algae	Large increase
Sediments	Medium size, low in organic matter to fine and high in organic matter
Sediment redox potential	High, with high dissolved oxygen in the interstitial water, to patches with low oxygen and some anaerobic areas
Water characteristics	pH, DO[a], BOD[b] little affected
Turbidity	Increase
Occurrence of blooms	Large increase

[a] DO, dissolved oxygen.
[b] BOD, biological oxygen demand.

residence time for the pollutants is low (hours) there is not time for a build-up of nutrients, plankton biomass and eutrophic waters (Hawker and Connell, 1991). Sewage discharge also involves particulate material and Pastorok and Bilyard (1985) have shown that species richness, percentage cover and mean colony size of corals are each inversely related to sedimentation (Fig. 26.11) (Chapter 18).

The use of a herbicide, Irgarol 1051, in antifowling paint on recreational boats has also given rise to some concern around the coasts of northern Europe and the south of France (Pearce, 1995). Concentrations of up to 0.6 mg l[-1] have been detected in the Medway and Hamble estuaries, UK, and 1.7 mg l[-1] near Monaco and Antibes, France. The possible induced ecological changes that may occur are at present unknown.

26.6 Summary

The physical impacts of water-based recreation can be very intense, although not as widespread as those of pollution. Macrophytes and shore vegetation may be destroyed by trampling from shore-based fishermen, campers and people gaining access to the water, either for swimming or launching boats.

The physical damage may be the result of trampling, sliding boats (on or off trailers) down banks or the collision of boat hulls and keels

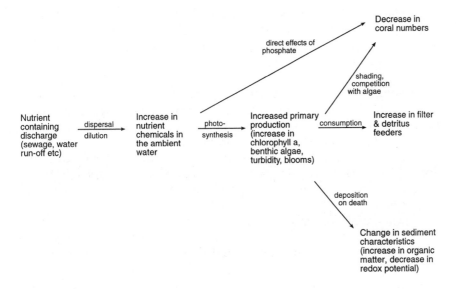

Fig. 26.10 The general effects of nutrient-containing discharge on a coral reef system (from Hawker and Connell, 1989).

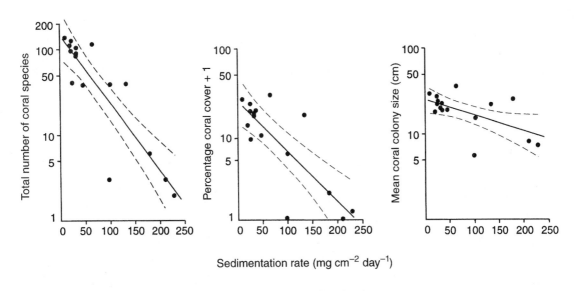

Sedimentation rate (mg cm^{-2} day^{-1})

Fig. 26.11 Coral species richness, percentage cover and colony size as a function of sedimentation rate, Guam; dotted lines are 95% confidence limits (data from Randall and Birkeland, 1978) (Pastorok and Bilyard, 1985).

with the plants. High-speed propellers may act as rotating knives, cutting floating macrophytes, and the turbulence created by motor-driven craft may increase the suspended material and hence the turbidity of the water column. Wash from powerboats will also erode unprotected banks in lakes and rivers.

Although the amounts of pollution released by recreators are generally low, it does become a significant factor in oligotrophic situations such as mountain lakes, mangroves and some coral reefs. This is particularly evident where there is any form of development, which includes running water leading to 'normal' quantities of sewage being produced. In remote areas even the location of individual 'cat holes' may be significant and they should be as far from water bodies or streams as possible.

Hawker and Connell (1991) suggested that an increase to three times the ambient level of nutrients is the maximum that should occur in coral reef communities, and that sedimentation rates above 10–30 mg cm^{-2} day^{-1} should not be exceeded in most areas.

However, the impact of small quantities of sewage on wildlife and plant growth in many habitats does not seem to have been much investigated (Holmes and Dobson, 1976) and this is a problem of increasing concern.

Fish 27

In this chapter both bony fishes (Osteichthyes) and cartilaginous fishes (Chondrichthyes) are discussed. As with some of the terrestrial vertebrates there is an extensive literature on the effects of commercial harvesting of these animals and there is often no clear distinction between this and recreation impacts on the populations. There is, for example, evidence that recreational fishing has been involved in the reduction of coral trout (*Plectropomus leopardus*), tiger trout (*P. melanoleucus*) and oceanic trout (*Plectropomus* spp.) on the Great Barrier Reef (Goeden, 1982). However, this author was unable to define the proportions caught by recreational and professional fishermen, except to show that the numbers of these 'keystone' species decline the nearer one gets to the town of Cairns in northern Queensland, Australia (population 55 000).

Fishing is one of, and possibly the most widespread of, outdoor recreation activities. It is not unusual for recreational fishing organizations to number their members in thousands. There is no doubt that fishing has a considerable effect on the environment, even when no fish are actually caught. These effects include driving off-road to reach river or coastal fishing sites, all the normal camping impacts, plus the direct effects on the fish populations and, in the case of coarse fishing, the fresh waters have, in the past, been polluted by the lead weights lost from their lines, although these are now being phased out (Duffey, personal communication). Additionally, in Australia and New Zealand many streams are stocked with trout (for fly fishing) which have been bred in captivity. These, when introduced into the streams and lakes, often impact on the native species already present. Some large fish have also attracted a great deal of attention from 'sports' fishermen and sadly there are many records of fish not being found in areas where they used to be plentiful, and of fish reaching sizes in the past that are not found today.

27.1 Introduced fish

Possibly the greatest impact of recreational fishing is the introduction of exotic fish which are imported to many countries and released in natural waters. Fish are also translocated away from the catchment in which they evolved to locations in which other slightly different ecotypes are living. Translocation of fish is not a new phenomena and there are reports from the fourteenth century of monks from Slapton Village introducing pike (*Esox lucius*) to Slapton Ley, the largest body of fresh water in the south-west of England (Bregazzi, Burrough and Kennedy, 1982).

In Australia 21 species of exotic freshwater fish have become established and the same number translocated (McKay, 1989). Of these, nine species were introduced for 'sporting' purposes (Arthington, 1989) (Table 27.1). In contrast, 84 exotic species have been collected from open waters in the USA and 39 have established breeding populations (Radonski *et al.*, 1984). In some areas the number of species introduced exceeds the numbers of endemic species. For example, in the San Francisco Bay drainage basin there are 20 introduced species and only 11 endemic fish (Leidy and Fiedler, 1985).

The common or European carp (*Cyprinus carpio*) and brown trout (*Salmo trutta*), both from Europe, are the most widespread introductions for 'sport' fishing in the USA. Trout are eagerly sought after by anglers although they are often competitive with native trout populations. Brown trout support considerable recreational fisheries and stocking of rivers and lakes is practised in the USA and in many other countries (Radonski *et al.*, 1984). However, as Arthington (1989) commented, 'The impact of trout has been of lesser concern mainly because they are prized for "sport" fishing; yet [in Australia] there is growing evidence of their adverse effects'. In contrast, carp are not popular and their introduction is considered in the USA as a 'monumental mistake' by Radonski *et al.* (1984).

Table 27.1 Introduced freshwater sports fish that have established self-maintaining populations in Australian inland waters. Compiled from various sources (see Arthington, 1986) (modified from Arthington, 1989)

Family/species	Common name
Salmonidae	
Salmo trutta	Brown trout
Salmo gairdneri	Rainbow trout
Salvelinus fontinalis	Brook trout
Oncorhynchus tshawytscha	Chinook salmon
Salmo salar	Atlantic salmon
Percidae	
Perca fluviatilis	European perch or redfin
Cyprinidae	
Cyprinus carpio	European carp
Tinca tinca	Tench
Rutilus rutilus	Roach

Introduced species impact on the local populations through competition for space, competition for food and direct predation. Trout are highly aggressive and territorial. Competition for space may be through the introduced species utilizing the spawning sites, competing for positions that provide the best cover and the best access to food (Crowl, Townsend and McIntosh, 1992). This type of competition may be difficult to observe. McIntosh, Townsend and Crowl (1992) found that competition between brown trout in the Shag River (South Island of New Zealand) and *Galaxias vulgaris* did not occur by day, but at night the native species was kept out of its favoured feeding positions in stream channels, and as a consequence the numbers of the native species were reduced.

Competition for food may also occur, for example, where there is direct overlap between trout and native eels (*Anguilla* spp.) (Hopkins, 1965). More recently both brown and rainbow trout have been implicated in the decline of the Australian native Macquarie perch (*Macquaria australasica*) through competition for food (Cadwallader and Rogan, 1977).

According to Crowl, Townsend and McIntosh (1992) a number of early accounts verified the consumption of native species by the predatory introduced trout, 'However it is noteworthy that many of the records are relatively old, probably because trout no longer co-exist with native fishes in many areas', the prey fish having altered their habitat occupancy to avoid their predators.

Hybridization has apparently occurred between two introduced varieties of European carp (*Cyprinus carpio*) in Australia, to produce the vigorous Boolarra strain which spread rapidly in the mid 1960s and 1970s (Fig. 27.1).

The impact of this alien on native species will not be known until the populations have stabilized; however, if the same pattern occurs in these fish as has been observed in the spread of hybrid plants, then considerable changes will result from their invasion.

Imported fishes may also bring with them parasites and diseases. Experiments have shown that Macquarie perch and other native species in Australia are extremely susceptible to fatal diseases after exposure to the epizootic haematopoietic necrosis virus (Langdon, 1989), which may be carried by redfin (*Perca fluviatilis*) and trout (Lintermans, 1990). Very careful quarantine is essential when stocks are imported from other countries.

It is fairly evident that importing and release of exotic species of both animals and plants for recreation has only rarely fulfilled expectations and almost inevitably has severe or catastrophic effects on the native flora and fauna.

Eradication of introduced fishes is almost always unsuccessful. All introduced fish have the potential to impact the Australian (or any other country's) environment. Once ecosystems are changed they usually remain changed.

(Kailola, 1990)

27.2 Fish trampling   The effect of anglers on fish populations may operate through any form of contact between fisher and fish. Experimental trampling of eggs in the manner of a fly fisherman wading through redds in which eggs had been laid was conducted in Montana (Roberts and White, 1992). They found that twice-daily trampling killed up to 96% of

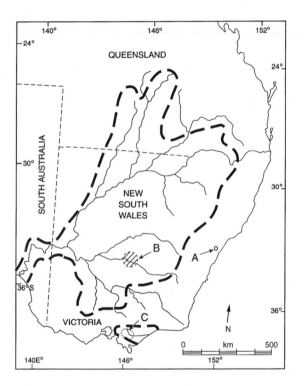

Fig. 27.1   South-eastern Australia, showing state boundaries and the distribution of three strains of European carp, *Cyprinus carpio*. A, Prospect Reservoir strain; B, Murrumbidgee Irrigation Area (Riverina) strain; and C, initial distribution of the Boolara strain, first introduced in 1960. Dashed lines indicate the distribution of the Boolara strain in 1977 (redrawn from Pollard *et al.*, 1980, in Arthington, 1989).

eggs, after the point of chorion softening, and pre-emergent fry. A single wading, perhaps a more realistic simulation, killed up to 43% (Fig. 27.2). The implications for native fish with similar egg-laying habits are clear and could be severe where suitable habitat is limited.

One benefit that can accrue from recreational fishing in fresh water is the result of the most preferred 'sports' fish being trout (Table 27.2), because these can only survive in high-quality waters (Table 27.3). There is strong pressure on the authorities to raise the quality of sewage discharge and other water inflows to the fishes' habitat (Vaughan *et al.*, 1982). Large-mouth and small-mouth bass are creating the same situation in warmer waters (Tables 27.2, 27.3).

27.3 Fishing and water quality

Apart from direct killing for sport, a large number of sharks are killed where beaches used for swimming are protected by nets set permanently offshore. Resorts which attract tourists because of their beaches, such as the Gold Coast, Australia, and the beaches on the Natal Coast of South Africa, cannot afford the possibility of shark attacks on swimmers and their possible reputation as dangerous places. Protective nets are one solution, in addition to lifeguards and aerial spotting, to the

27.4 Protective shark nets

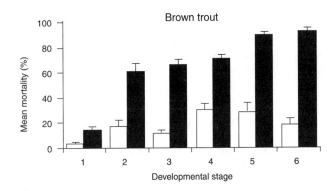

Fig. 27.2 Adjusted mean mortality (%) and standard errors for brown trout, eggs and pre-emergent fry in the control (clear bars) (*N* = 8) and test (solid bars) (*N* = 8) egg baskets exposed to wading treatments 1–6 in the laboratory. Test egg baskets were waded on twice daily. Treatments were as follows: 1, fertilization to start of eyed-egg stage; 2, start of eyed-egg stage to start of hatching; 3, fertilization to start of hatching; 4, start of hatching to start of emergence; 5, start of eyed-egg stage to start of emergence; 6, fertilization to start of emergence. (From Roberts and White, 1992.)

problem of keeping dangerous sharks offshore. Unfortunately, apart from the legitimacy of killing animals for pleasure, a large number of other marine animals and harmless sharks are also entangled in the shark nets and die.

When shark nets were first placed off the Sydney beaches in 1937, 1500 sharks were caught in the first 17 months (Coppleson, 1962). Catches from a decade later were only 95 sharks per year (Reid and

Table 27.2 Angler preference classes for fish species and associated water quality requirements (Vaughan *et al.*, 1982)

Broad species group	Assigned preference	Principal species	Water quality requirements[a]
I. Cold-water gamefish	High (cold water)	Rainbow trout	DO greater than 5.0 mg l^{-1}; temperature less than 18°C; TSS less than 100 mg l^{-1}
		Brook trout Brown trout Lake trout	
II. Warm water gamefish–panfish	Medium	Yellow perch White bass Bluegill Crappie Rock bass Pumpkinseed Lake herring Northern pike	DO greater than 3.0 mg l^{-1}; temperature between 18°C and 32°C; TSS below 100 mg l^{-1}
	High (warm water/ cool water)	Large-mouth bass Small-mouth bass Walleye Striped bass	
III. Rough fish	Low	Carp Freshwater drum Buffalo Bullhead Catfish	DO between 2.0 mg l^{-1} and 3.0 mg l^{-1} or DO above 3.0 mg l^{-1} and temperature above 32°C or DO above 3.0 mg l^{-1}; TSS greater than 100 mg l^{-1}
IV. No fish	Not applicable	Not applicable	pH above 10.0 or below 5.0, or DO below 2.0 mg l^{-1}; or temperature above 34°C or summer streamflow zero.

[a] DO, 90-day average summer dissolved oxygen concentration; Temperature, 90-day average summer temperature; TSS, annual average total suspended solids concentration.

Krogh, 1992). There has also been a steady reduction in the size of sharks caught, as shown in Reid and Krogh's analysis of data from the periods 1950–70 and 1972–90. It is possible that the role of older individuals in shark populations have been overlooked in our youth-centred society, and this change in age structure may be very significant. Also particularly noticeable is the drop in numbers of whaler (*Carcharhinus* spp.), white pointer (*Charcharodon carcharias*) and grey nurse (*Carcharis taurus*) sharks (Table 27.4). Sharks are generally long lived and slow growing and give birth to only a few offspring, and therefore have a low capacity to recover from overfishing (Walker, 1993).

Given the worldwide decline in shark numbers, reflected by the lower catches per unit effort and even the demise of some shark fisheries (Walker, 1993), and the large numbers of other animals that die in the nets, it is clearly time to review and find a less damaging way to protect the beaches (Wilde, 1994) and tourist dollars.

The catch per unit effort (number per km of net per year) in the Natal nets was at first very high, but this declined and between 1978 and 1990 an average of 1326 sharks were caught per year (Dudley and Cliff, 1993) (Fig. 27.3). It appears that when nets are installed the local shark population is killed or reduced drastically in numbers and after that mainly migrants are caught in the nets (Dudley and Cliff, 1993).

Table 27.3 Fishery types as determined by water quality using Nielsen's rules (Vaughan *et al.*, 1982)

	Exogenously determined water quality conditions			
	Cold water temperature, $\leq 18°C$		Warm water temperature, between 18°C and 32°C	
Water quality improvements traced to policy	Low solids water TSS ≤ 100 mg l⁻¹	High solids water TSS >100 mg l⁻¹	Low solids water TSS ≤ 100 mg l⁻¹	High solids water TSS > 100 mg l⁻¹
Dissolved oxygen > 5.0 mg l⁻¹	↑ Cold water gamefish	↑	↑	↑
> 3.0 mg l⁻¹	↑		Warm water gamefish/panfish	
> 2.0 mg l⁻¹	Rough fish	Rough fish	Rough fish	Rough fish
< 2.0 mg l⁻¹	Unfishable[a]	Unfishable[a]	Unfishable[a]	Unfishable[a]

[a] Other unfishable conditions include pH greater than 10 or less than 5; temperature greater than 32°C; summer flow zero. TSS, Total suspended solids.

Table 27.4 The numbers of various species of shark caught in nets protecting Sydney's beaches between 1950–70 and 1972–90 (data from Reid and Krogh, 1992)

Species	1950–70	1972–90
Whalers (*Carcharhinus* spp.)	1477	418
Hammerhead (*Sphyrna* spp.)	848	1332
White pointer (*Carcharodon carcharias*)	249	85
Tiger (*Galeocerdo cuvier*)	112	110
Angel (*Squatina australis*)	705	684
Grey nurse (*Carcharis taurus*)	300	30

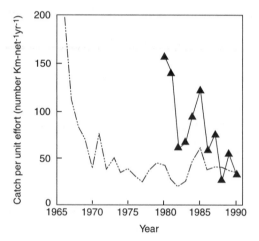

Fig. 27.3 Catch per unit effort, all shark species combined, at all installations on the Natal coast and at the remote Richards Bay (triangles) (from Dudley and Cliff, 1993).

There is also a real concern that shark netting is having a considerable impact on other groups of animals besides its effect on sharks themselves. Wilde (1994) reported that between 1962 and 1988 the southern Queensland nets additionally killed 500 dolphins, 500 dugongs, 3500 marine turtles and 10 000 rays besides 1000 sharks. The Natal shark nets appear to be even more damaging, killing in 10 years over 600 turtles, nearly 800 dolphins and 3500 rays and their relatives (Table 27.5) as well as other birds and fish (Dudley and Cliff, 1993). Fortunately some of these animals were also released alive.

Table 27.5 Catches in the Natal shark nets between 1981 and 1990, of animals other than potentially dangerous sharks (modified from Dudley and Cliff, 1993)

Species	Common name	Caught (number)	Released (%)
Birds			
Sula capensis	Cape gannet	13	0
Phalacrocorax sp.	Cormorant	1	0
Spheniscus demersus	Jackass penguin	<u>1</u>	0
		15	
Turtles			
Eretmochelys imbricata	Hawksbill turtle	18	28
Lepidochelys olivacea	Olive ridley turtle	11	27
Caretta caretta	Loggerhead turtle	426	35
Chelonia mydas	Green turtle	14	34
Cheloniidae	Turtle	11	73
Dermochelys coriacea	Leatherback turtle	<u>68</u>	35
		674	
Cetaceans			
Sousa plumbea	Indo-Pacific humpbacked dolphin	61	2
Delphinus delphis	Common dolphin	363	4
Tursiops truncatus	Bottlenosed dolphin	349	1
Stenella coeruloealba	Striped dolphin	3	33
Stenella longigrostris	Spinner dolphin	1	0
Lagenodelphis hosei	Fraser's dolphin	1	0
Pseudorca crassidens	False killer whale	1	0
Delphinidae	Dolphin	9	0
Balaenoptera acutorostrata	Minke whale	<u>4</u>	25
		792	
Sharks			
Rhizoprionodon acutus	Milkshark	46	6
Mustelus mosis	Hardnosed smooth-hound	2	50
Halaelurus lineatus	Banded catshark	1	0
Rhinocodon typus	Whale shark	7	57
Squatina africana	African angelshark	<u>329</u>	45
		385	
Batoids			
Aetobatus narinari	Spotted eagleray	140	80
Myliobatis aquila	Eagleray	37	54
Pteromylaeus bovinus	Bullray	378	61
Rhinoptera javanica	Flapnose ray	411	58
Manta birostris	Manta	525	66
Mobula spp.	Devilray	142	60
Dasyatidae	Stingray	65	74
Dasyatis marmorata	Blue stingray	8	88
Gymnura natalensis	Backwater butterflyray	496	78
Himantura gerradi	Sharpnose stingray	14	93
Himantura uarnak	Honeycomb stingray	19	84
Torpedo sinuspersici	Marbled electric ray	4	25

Table 27.5 *continued*

Species	Common name	Caught (number)	Released (%)
Torpediniformes	Electric ray	6	100
Rhina ancylostoma	Bowmouth guitarfish	1	100
Rhynchobatus djiddensis	Giant guitarfish	1220	75
Pristis microdon	Smalltooth sawfish	2	100
Pristis pectinata	Largetooth sawfish	9	67
Pristis spp.	Sawfish	<u>10</u>	70
		3487	
Teleosts			
Sphyraena spp.	Barracuda	3	
Trachinotus blochii	Snubnose pompano	12	
Lichia amia	Garrick	115	
Scomberoides spp.	Queenfish	18	
Caranx ignobilis	Giant kingfish	2	
Carangidae	Kingfish	2	
Thunnus albacares	Yellowfin tuna	46	
Euthynnus affinis	Eastern little tuna	20	
Katsuwonus pelamis	Skipjack tuna	43	
Scomberomonis commerson	King mackerel	6	
Scomberomonis plurilineatus	Queen mackerel	5	
Scombridae	Tuna bonito	12	
Rachycentron canadum	Prodigal son	9	
Argstrosomus hololepidotus	Kob	32	
Atractoscion aequidens	Geelbek	15	
Makaira indica	Black marlin	9	
Istiophorus platypterus	Sailfish	1	
Elops machnata	Ladyfish (springer)	10	
Epinephelus lanceolatus	Brindlebass	4	
Sparodon durbanensis	White musselcracker	3	
Cymatoceps nasutus	Black musselcracker	7	
Oplegnathus spp.	Knifejaw	2	
Tripterodon orbis	Spadefish	1	
Epinephelus tukula	Potato bass	1	
Pomadasys kaakan	Javelin grunter	3	
	Unidentified fish	10	
		<u>373</u>	
Crustaceans			
Panulirus homarus	Crayfish	1	

The effect of recreational fishing on fish populations is hard to separate from that of commercial harvesting. However, given the evidence of reductions in both size and number of popular 'sport' fish caught, it seems likely that recreational fishing is having some direct impact on populations.

There is no doubt that introduced freshwater 'sports' fish are having a direct effect on native species and have probably caused irrevocable changes in riparian ecology in many parts of the world.

Protective shark netting is also having a direct effect on the target species as well as many other marine animals. Alternative methods are available and, although more costly, should perhaps be adopted in the same way that manufacturers are required to spend money to prevent their wastes from polluting our environment.

27.5 Summary

28 Aquatic mammals

This chapter is mainly concerned with the Cetacea and manatees, as these are the groups about which there is most information on visitor or visitor-like impact. Other species, such as the sea otters (*Lutra felina, Enhydra lutris*) and the platypus (*Ornithorhynchus anatinus*), should also be considered under this heading although their reactions are likely to be very different to those of the animals discussed here.

28.1 The whales (order Cetacea)

The whales are truly aquatic, with no hind limbs, paddle-like fore limbs, expanded flattened tail 'flukes' and practically hairless skin under which is a thick layer of oily blubber (Young, 1962). They have very large lungs and some species may remain submerged for over 1 hour (Milligan and Williams, 1959).

The living Cetacea are divided into two suborders. The Odontoceti are the toothed whales and include dolphins (Delphinoidea), porpoises (Phocaena), sperm whales (*Physeter macrocephalus*), killer whales (*Orcinus orca*) and some lesser known species such as the 'horned' narwhale (*Monodon monoceros*), white whales (*Delphinapterus leucas*) and pilot whales (*Globicephala macrorhynchus*). The Mysticeti are the whalebone whales, which have no teeth but hundreds of plates of whalebone or 'baleen' which hang from the upper jaw. These plates sieve the krill, *Euphausia* spp. (small, shrimp-like animals), from the water and they are then licked off the plates by the tongue and swallowed. The well-known blue whale (*Balaenoptera musculus*), which is almost extinct because of hunting, is a member of this group, as are the Greenland or bowhead whale (*Balaena mysticetus*), the humpback whale (*Megaptera novaeangliae*) and the grey whale (*Eschrichtius robustus*).

28.2 Toothed whales: dolphins

Probably the most well-known cetacean is the dolphin (*Delphinus delphis*). In many parts of the world there has been a decline in the number of sightings of these animals. Tragenza (1992) estimated that over the 50 years from 1935 to 1985 there had been a 90% decline in the number of 'smaller Cetaceans' (dolphins and porpoises) seen from the Cornish coast. The decline has been greater for the larger groups and has been occurring throughout the study period. There has also been an increase in the proportion of single animals. Several causes for the dolphin decline have been put forward, including increasing disturbance by boats, destruction by fishing gear, reduction of food supply from overfishing, changes in distribution of prey, habitat destruction and pollution (Evans, 1987).

TYPE 1 DISTURBANCE

There are many accounts of interactions between dolphins and humans. Often the exchange is actively sought by the animals and this interaction is only a disturbance in the sense that their behaviour is changed by the humans or their artefacts. One individual bottlenosed dolphin (*Tursiops truncatus*) interacted so uniquely and extensively around the coasts of the Isle of Man, Wales and Cornwall that two papers were written on its behaviour (Lockyer, 1978; Webb, 1978). Variously known as 'Donald', 'Bubbles' or 'Beaky' the animal had a range of activities which included putting its head on the gunwale of small boats, biting the paddles and capsizing canoes, frequently playing with and apparently making sexual advances to a small tender (dinghy) (including erection of his penis and ejaculation) (Lockyer, 1978). He appeared to be excited by noise, sometimes from boat or helicopter engines or noise from the occupants of small boats when he played with them. On these occasions he would sometimes leap clear of the water several times in succession. He could be summoned from a distance of over half a mile by the rattling of chains in the water. According to Lockyer (1978) he always approached divers or swimmers with a friendly or playful attitude, but he did not enjoy being chased or harassed. He would allow himself to be petted and stroked, or his head or jaw to be rubbed. If swimmers prevented him reaching his favourite boat or buoy, he would rush at them or surface beneath them but at no time did the dolphin actually harm anyone by biting or hitting them. At one time he developed the 'game' of picking up moorings of boats and towing them around, sometimes for hundreds of yards (Webb, 1978) (Fig. 28.1).

The natural sociability of bottlenosed dolphins is well known and there are many other records of wild and solitary animals seeking human company, and they appear to be able to recognize individual people. Lockyer (1978) discussed examples from New Zealand, Florida, Scotland, South Carolina and Monkey Mia, West Australia. However, not all individuals or species react to humans in this manner. The Ganges River dolphin (*Platanista gangetica*) in the Karnali River,

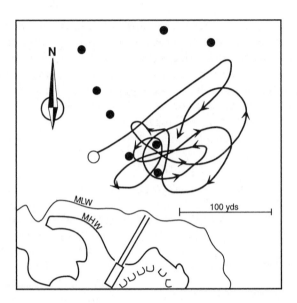

Fig. 28.1   Path taken by the yacht 'Slim 3' (estimated weight 3 tons) during a tow by dolphin at Coverack on 12 August 1977 between 19.00 and 19.40 hours. Reconstructed from 26 photographs. ●, Mooring buoys. (From Webb, 1978.)

Nepal, is relatively unaffected by row boats or canoes, but has not been seen in former habitat areas where motorized ferries have been introduced (Smith, 1993). A very different reaction to motors to that shown by 'Beaky' who appeared to be fascinated by pulsed sounds and vibrations even to the point of exhaling into turning propellers, causing them to cavitate and the engine to race, so altering the pitch of the noise produced (Lockyer, 1978). The Ganges dolphins also avoided areas of noisy shoreline activity and did not reappear until the activity ceased (Smith, 1993). Hua, Zhao and Zhang (1989) also reported noise avoidance by the Chinese river dolphins (*Lipotes vexillifer*).

TYPE 3 DISTURBANCE

One of the recreation-based causes of cetacean deaths are the shark nets permanently suspended offshore in Natal, South Africa and the Gold Coast, Australia. As many as 100 dolphins may be killed in these nets each year (Chapter 27).

   The other toothed whale on which I have been able to find information is the killer whale (*Orcinus orca*). This large carnivore, 5–6 m in length, is well known for feeding on dolphins and seals, but apparently it does not attack humans. However, there are several accounts of these animals attacking boats. The Italian yacht 'Guia III' was holed beneath the waterline by killer whales off the coast of Brazil and sank in about 15 minutes (Di Sciara, 1977). There were four or five killer whales near the boat as it sank, but they made no attempt to attack the crew as they swam to and from the life raft. Other reports of similar attacks included chasing and lifting a small fishing boat out of the water after the whale had been struck by a harpoon (perhaps this is understandable) (Di Scaria, 1977). In 1962 off Bellingham on the west coast of the USA, a female killer whale was captured with a rope. Twenty minutes later a male whale arrived and together the two whales attacked the craft of the captors (Caldwell and Caldwell, 1966). Di Scaria (1977) also reported how a 13 m wooden schooner was attacked 100 miles (161 km) east of the Galapagos Islands by three killer whales and sank in about 60 seconds. The crew, a family on a round-the-world voyage were unharmed. Whether these incidents are human recreators having an impact on the whales, or the whale's recreation having an impact on the humans seems to be an open question!

These are often very large animals and many species have been hunted nearly to extinction. It is incredible that some humans are so callous that whales are still being hunted. As far as I can tell, there is no recreational whale hunting, so I shall only be considering type 1 disturbance to these animals.

28.3 Whalebone whales

TYPE 1 DISTURBANCE

There is a notable increase in awareness of the almost mystical presence of whales and this is leading to increased pressure from tourist visits, especially to inshore areas in eastern Australia, New Zealand, California and Hawaii.

The humpback whale (*Megaptera novaeangliae*) has two populations: one, estimated to be 500–740 animals (10 000 according to Salden (1988) or 5000 according to Lockyer (1993)), in the Pacific, overwintering and calving around Hawaii. The other is in the north Atlantic, estimated to be 800–3000 animals (most estimates are less than 1500 but Lockyer (1993) puts it at 5500) (Hay, 1985). The original numbers were estimated at 10 000–15 000 in the Pacific (Salden puts the number at 120 000 in the nineteenth century) but now hunting is prohibited. The Atlantic humpbacks breed and calve on the offshore Silver and Navidad, Caribbean banks which comprise only about 3400 km². About 85% of the population spends the winter in these areas and they are highly susceptible to human disturbance and pollution (Hay, 1985).

There is little doubt that humpback whales are sensitive to boat sounds. Watkins and Goebel (1984) found that in Glacier Bay, Alaska, during the summer months 'they seemed generally unconcerned with boats beyond 150 m but at about that distance they often reacted sharply as if the boats were suddenly more noticeable. In contrast the whales in Icy Strait [further towards the open sea] appeared to be aware of the boats at greater distances but they could be approached quite closely'. A tidal current boundary line apparently acted as an acoustic barrier in some situations. However, Salden (1988) reported that humpbacks avoided the inshore areas of the four-island area of Hawaii, where parasail and jet skis are common.

As Hay (1985) quoted, 'humpback whales are a valuable tourist resource since they are quite tame and concentrate in certain inshore areas during summer', or in the case of eastern Australia, the early spring. The numbers of the Pacific population appear to be increasing, but the curtailment of whale watching in the waters of the Great Barrier Reef was recommended by Corkeron (1992), and Stevens and Chaloupka (1992) stated that harassment from some commercial and private vessels remains a problem in Harvey Bay, Queensland. Other baleen whales show considerable tolerance of boats, but often avoid rapidly or erratically moving vessels (Ray et al., 1978).

Sound can normally travel considerable distances through water, especially the low frequencies to which the baleen whales are particularly sensitive (Richardson et al., 1985). These whales, including the bowheads, produce low-frequency calls (Thompson, Winn and Perkins, 1979) which, when they are recorded and the pitch raised to the human level, make such wonderful and mysterious music. Although functions have rarely been documented, the calls seem important for communication between whales (Herman and Tavolga, 1980). The effects of interference with these signals may include short-term behavioural reactions, masking of communication, masking of other important environmental sounds, physiological effects including stress and short- or long-term displacement (Myrberg, 1978).

The affects of aircraft overflights and boats on the bowhead whales (*Balaena mysticetus*) summering in the Beaufort Sea, were studied by Richardson et al. (1985). The experimental aircraft was a Norman Britten Islander and this was flown at varying altitudes over groups of whales. When the aircraft approached or circled at or below an altitude of 305 m above sea-level, there were conspicuous reactions from the bowheads. These occurred infrequently when it was at an altitude of 457 m and were not detected at all when it was at or above 610 m. Most reactions were hasty dives when the aircraft first approached or descended. Hasty dives involved an unusually quick or abbreviated sequence of pre-dive events. The intervals between blowing were recorded in the absence and presence of aircraft and no difference could be detected, but when an aircraft circled specific animals and gradually descended, there was a sequential shortening of the blow interval (Fig. 28.2). In the authors' opinion, this was the clearest evidence that aircraft do, in fact, have a prolonged effect on these animals. In general, the effect of the aircraft was most noticeable when it first arrived and the bowheads were 'presumably undisturbed' when the aircraft was above 457 m.

Sometimes the whales appeared to be less sensitive to aircraft. Payne *et al.* (1983) found that southern right whales (*Eubalaena australis*) rarely reacted strongly to small aircraft circling at 65–150 m above sea-level but this may have been because they were feeding (Richardson *et al.*, 1985). Grey whales (*Eschrichtius robustus*) were reported to avoid a location where recorded helicopter noise was played back into the water at intervals of 10 seconds to 2 minutes (Malme *et al.*, 1983). The best recommendation would appear to be to fly at heights over about 500 m to avoid the danger of disturbing baleen whales.

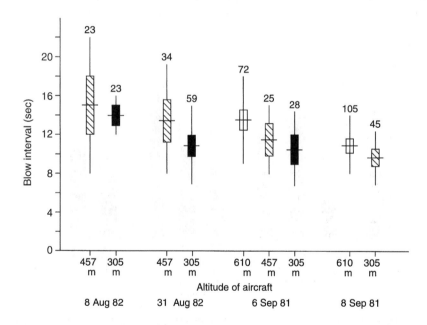

Fig. 28.2 Blow intervals of bowheads observed from the Islander aircraft at different altitudes on four dates. Calves are excluded. The mean ± 1 SD, ± 95% confidence interval, and sample sizes are shown. (From Richardson *et al.*, 1985.)

Richardson *et al.* (1985) also found that the bowhead whales had a strong and generally consistent reaction to boats. The orientation of the whales was observed from an aircraft flying at 457–762 m above sea-level as the boats approached. Two boats were used, both were diesel powered, one a 16 m crew boat with a top speed of 41 km h^{-1} and the other a 13 m fishing boat with a top speed of 15 km h^{-1}. When the boats were between 0.8 and 3.4 km distant from the whales, those animals directly in its path began swimming rapidly away, generally attempting to outrun it (Fig. 28.3). When the boats were within a few hundred metres, whales usually turned and swam away from the boat's track. Whales ceased swimming away when the boats were 0.8–5 km distant. Whales that had originally been closely grouped, sometimes were scattered widely after the boat moved away and, in one case, they remained apart for at least 1 hour.

The distances over which the whales were disturbed were considerable. Unless the population or species is known to accept the presence of boats either from habituation, natural tolerance or even curiosity, it is clearly desirable that whale watching is done from some distance. Richardson *et al.* (1985) also commented that 'boat-based observers sometimes do not detect whales that move away from the boats when they are still a few kilometres distant'. Thus, other species may also react more strongly than the literature suggests.

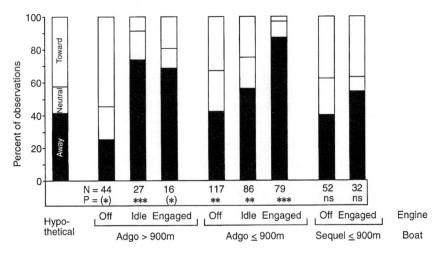

Fig. 28.3 Orientations of bowheads with respect to boats less than 900 m and more than 900 m away. Adgo, fast crew boat; Sequel, slow fishing boat. *P* values are for chi-square comparisons with hypothetical orientations (those expected if whales were randomly oriented): ns means *P* > 0.01, (*) 0.1 > *P* > 0.05; ** 0.01 > *PRW* > 0.001; *** *P* < 0.01. (From Richardson *et al.*, 1985.)

The interactions between cetaceans and humans are more complex than most human/animal contacts. In spite of that fact, many species have been hunted nearly or completely to extinction. We clearly must treat those remaining with the greatest respect and the *Guide to Watching Whales in Canada* (Breton, 1986), suggests some approaches that are needed for all species and areas. Perhaps one day we will realize that the dream of communicating with animals is possible, at least at some level and with some cetacean species.

These are fully aquatic mammals which have very large lungs, a skin with scarce hair, a layer of blubber, absence of hind limbs, paddle-like fore limbs and a horizontally flattened tail (Young, 1962) (Fig. 28.4) – all features similar to those of the whales, but this is thought to be the result of convergent evolution rather than a close relationship (Milligan and Williams, 1959). There are three species of manatees *(Trichechus)*, one each from the Caribbean, Brazil and West Africa. There is only one species of the dugong, and this occurs from the Red Sea, across the Indian Ocean to the Philippines and Australia. They are all herbivorous.

28.4 Manatees and dugongs (order Sirenia)

Fig. 28.4   The West Indian manatee, order Sirenia (*Trichechus manatus*) (from Parker and Haswell, 1962).

Shackley (1992) described the West Indian manatee (*Trichechus manatus*) as a huge, slow-moving, gentle mammal, most of whose time is spent grazing seagrass beds in the shallow, slow-moving coastal waters and estuaries of southern Florida. She added that adults may reach 13 ft (4 m) in length and weigh over 3000 lb (1360 kg). Their reproductive rate is low, with one calf being born every 2–3 years after 13 months' gestation. They are harmless with no natural fear of humans and they often approach divers in a spirit of curiosity (Hartman, 1979). There are only about 1200–1500 surviving, all concentrated in Florida during the winter months (Shackley, 1992). These animals have come to be a major tourist attraction. However, recreational boating and diving, using the same waters as the manatees, are two of Florida's main industries. In 1992 there were 720 000 registered boats and some 100 000 additional boats not requiring registration. By the year 2000 it is estimated that there will be 1.75 million boats on Florida's waterways (Shackley, 1992).

TYPE 1 DISTURBANCE

Because manatees have now become tourist attractions in their own right, they 'cause problems for themselves' because they will often voluntarily interact with humans, making it difficult to avoid physical encounters (Shackley, 1992). Harassment by divers and photographers is often noted. They are also attracted to marinas and popular boating areas because they receive food from visitors in defiance of published codes of conduct (Shackley, 1992). Manatees are often harassed by tourists circling over them in boats with or without the motor running. Divers also follow manatees and touch them, although it is prohibited. Divers are advised to use snorkel and not scuba gear but this recommendation is often ignored 'because of the differential profit margins of the dive shops'. Finally, a new threat has arisen of manatee watching from helicopters (Shackley, 1992).

During the winter months the manatees gather around a series of warm springs, as during cold periods the general water temperature may fall below 16°C, which is the lower lethal limit for these slow-moving animals with their low metabolic rates. This is the same time of year as the maximum Florida tourist visitation. Kochman, Rathbun and Powell (1985) suggested that manatee movement from the Main Spring to Magnolia Spring during January and February may be the result of human disturbance at Main Spring. If the animals are dispersed by harassment at this time, the adults, and especially perinatal individuals, may perish through starvation followed by hypothermia. This was especially noticed in the survey carried out between 1976 and 1981 (O'Shea *et al.*, 1985) (Table 28.1).

Table 28.1 Distribution of winter (Jan–Mar) manatee deaths among major cause-of-death categories in Florida, 1977–81 (modified from O'Shea *et al.,* 1985)

Cause-of-death	Totals
Collision with boats	26
Sluice gate entrapment	3
Perinatal/early juvenile	11
Undetermined	126
Total	166

TYPE 3 DISTURBANCE

Direct contact, especially with boats, is an increasing cause of manatee deaths and the numbers rose from 26 killed in the 1976–81 period, to 46 in 1988 (Shackley, 1992) (Fig. 28.5). Most deaths seem to result from collisions with boats or with vessels driven by large horsepow-ered engines and nearly half of all boat-killed manatees die from the impact trauma without propeller cuts (Beck, Bonde and Rathbun, 1982). When the oncoming boats are travelling slowly, manatees can detect them and avoid collision if not taken by surprise (Hartman, 1979). In 1994 16 manatees died after lock gates closed on them but experiments are being carried out with acoustic detectors to stop gates closing when manatees are present (Pain, 1996). In 1989 a total of 166 animals were known to have died from all causes. This is over 10% of the population and greater than the replacement rate. The manatee could become extinct by the year 2000 (Shackley, 1992) and this, to a considerable degree, will be the result of 'manatee tourism' developed in the first place because of the well-intentioned publicity by people trying to preserve this animal. There is a strong preserva-tion movement which has succeeded in introducing boat speed limits and laws prohibiting behaviour that might harm or harass manatees (Shackley, 1992), but with such a low population number *Trichechus manatus* is a very vulnerable animal.

Dugong (*Dugong dugong*) are similar in behavioural characteristics to manatees but they occur in greater numbers than the West Indian manatee. The main human impact, apart from destruction or pollu-tion of the seagrass beds on which they feed, appears to be hunting.

As tourism continues to increase, more and more 'wild' habitats become open to human visitors. Antarctic visitation is one of the more recent of these and the number of visitors has increased year by year (Enzenbacher, 1992) (Fig. 28.6). Fortunately the tourist industry has

28.5 A note on other aquatic mammals

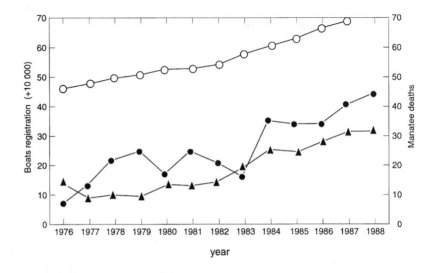

Fig. 28.5 Boat registration and manatee mortality in Florida since 1976: ○, boat registrations (× 10 000); ●, manatees killed by collision with boats or barges; ▲, perinatal deaths (from Shackley, 1992).

formed the International Association of Antarctic Tour Operators (IAATO) and they are providing a regulatory framework for Antarctic tourism. One aspect of their guidelines is that visitors must maintain a distance of at least 15 m from fur seals and 4.5 m from crawling (or true) seals, penguins and nesting birds. They have also produced a general guide to management practices with respect to this special environment. Perhaps this can be used as an example for those habitats and animals where the biota does not appear so vulnerable but are nevertheless heavily impacted.

28.6 Summary Whales are increasingly the basis of a segment of the ecotourist industry, especially in Hawaii and eastern Australia. Many of the whale species have been hunted nearly to extinction and the numbers of many other species have declined. Further harassment must be avoided if their populations are to recover.

Dolphins interact closely with humans and some tourist resorts are based on this attraction. However, some species, such as the Ganges River dolphin, are quite shy of humans. Killer whales are also an attraction, especially in aquaria, but they are also on record as having attacked small boats, but not the people in them.

The large whalebone whales, notably the humpbacks, are often the

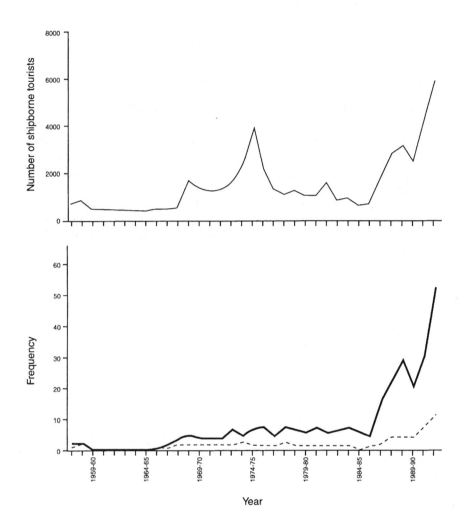

Fig. 28.6 (a) Numbers of shipborne tourists, and (b) total numbers of tourist cruises offered (———) and ships used to carry tourists (- - - -), each summer season in the Antarctic from 1958 to 1992. Note the lack of data available on shipborne tourism from 1959 to 1965. The sudden rise in the number of shipborne tourists during the 1974–75 season is attributable to cruises offered by ships with capacities of 474 and 800, respectively. (From Enzenbacher, 1992.)

'quarry' of whale-watching trips, and there is some evidence that they are being driven from their nursery refuges and mating areas around Hawaii. Bowhead whales are sensitive to the noise of low-flying aircraft; they are likely to dive as the noise increases, and their blow intervals are shortened. Boats have also been shown to disturb and scatter groups of bowhead whales. The distances at which disturbance occurs may be several kilometres.

The West Indian manatee has become a major attraction in Florida waterways, and harassment, together with often fatal collisions with fast-moving boats, has become a threat to the small remaining population of this species.

Seals are also an attraction to tourists and in the Antarctic the tourist association has drawn up a code of conduct suggesting that tourists must maintain a distance of at least 15 m from fur seals and 4.5 m from crawling (true) seals.

Overall, aquatic mammals are an important element in the ecotourist spectrum but the tour operators are perhaps more conscious than most of their conservation obligations and perhaps the sustainability of their industry.

Conclusion 2

In this chapter I summarize some of the salient areas of recreation ecology and briefly suggest some problems for future research. It is not possible to refer to all the work that has a bearing on recreation impacts, but I have attempted to point out some commonalities that exist between plant, soil and animal responses, in the hope that these comparisons may suggest new syntheses of knowledge leading to new and inventive solutions to problems in recreation ecology.

BIOMASS AND PLANT COVER

29.1 General consequences of recreation impacts

Associated with recreation activities there is usually a reduction in biomass as a result of physical impacts. This is especially evident in the case of trampled plants (Fig. 3.4) and I suspect would also occur with total animal biomass if the removal of the top carnivores (lions, bears, wolves, etc.) and large herbivores (elephants, buffaloes, megatheres, etc.) is taken into account. However, as Duffey has pointed out, 'There is recent evidence that growing tourist interest in seeing large carnivores and herbivores should ensure their survival in some areas because it provides income for local people' (Duffey, personal communication). Consideration of the cultural context of recreation of course extends the nature of the impacts into activities such as hunting and deliberate killing for non-recreational purposes. Corals on reef flats are also severely reduced in biomass as a result of trampling impacts, but this is much reduced on the reef crest. Freshwater macrophyte biomass is also vulnerable to the effects of physical impacts (Fig. 26.4).

In contrast, where eutrophication is a significant factor biomass may increase as a result of nutrient enrichment, this is evident in phytoplankton and some other algae (e.g. *Cladophora*, Fig. 26.6) in fresh waters and in nettle beds (*Urtica dioica*) in terrestrial situations.

There has been considerable discussion on the rate of change but the general consensus is that the decline in biomass is curvilinear, with some exceptions showing a regular or straight line relationship between trampling or impact intensity and biomass.

Conclusion

The certain consequence of recreation impacts is a reduction in the number of species of living organisms present in the impacted areas. Again this can be seen with respect to trampled plants (Figs 4.7–4.9, 4.11), birds (Fig. 20.21) and aquatic situations (Fig. 26.6). In many situations the result of very high intensity use can be the elimination of practically all of the original species in the impact zone. However, in some situations the initial phase of impacts can cause a slight increase in species number (Fig. 4.9). This is due to the introduction of ruderal or panglobal species such as annual meadow grass (*Poa annua*), green couch grass (*Cynodon dactylon*) and members of the genus *Plantago* in trampled areas (Table 4.13) or the common cockroach (*Periplaneta*) and the European sparrow (*Passer domesticus*) or the house mouse (*Mus musculus*). These are associated with human disturbance and can either survive in the altered environmental conditions where the natives have been reduced in numbers or, in the case of the animals, they can utilize human food wastes.

Following this initial increase there is a steady reduction in species as the intensity of use increases. This decline may also be curvilinear with a steep fall in numbers followed by a more gradual decline, the impact is apparently reduced by the introduction of other species (e.g. Fig. 3.23).

It should be noted that a claim to have maintained or even increased species diversity in an area through some form of development or rehabilitation needs to be examined in terms of the balance between natural and introduced species in that particular habitat.

MORPHOLOGY

There is some analysis of the special morphological features of plants that survive trampling. A survey of the life forms (Table 6.3) of dune plants growing on paths and tracks showed that rosette and semi-rosette hemicryptophytes occurred more commonly on the paths and tracks than any other life form. In trailside vegetation early blooming, graminoid forms and therophytes were also found to survive trampling. The importance of protected buds was emphasized and plants with protected meristems also survived better than those with meristems exposed. There is also little doubt that smaller plants survive better in trampled areas, as plants that are not able to grow in rosette, creeping or other low-growing form do not survive long on new pathways and rarely appear in the trampled flora. Another morphological feature is reflected in the differential survival of monocotyledonous and dicotyledonous species at some sites (Fig. 6.1).

The larger soil animals appear to be at a disadvantage in the surface layers, presumably getting crushed because they are larger than the interstitial spaces. Work on earthworms has shown that at greater depths larger individuals are more common in compacted soils, possibly because they have the strength to push through the matrix. However, 'Habitat destruction is the main influence causing a decline in ground living invertebrates. In any case, the recreational effect is minimal compared with the destruction of this form of life by agriculture, forestry and industrialisation' (Eric Duffey, personal communication).

Smaller terrestrial animals generally seem to survive better in well-used recreation areas, possibly because humans see larger animals as threats and kill them or drive them off as they will with 'dangerous' forms such as snakes, usually without knowing which ones are venomous.

EVOLUTION

That 'trampling' has existed as a selective force ever since the first animals crawled up the beach is not in doubt. It is, therefore, not surprising to find some plant species well adapted to survive in trampled areas (Chapter 6). There is also evidence that evolution to survive trampling is still occurring, as shown by the work of Warwick and Briggs on lawn species which generally survive because of phenotypic plasticity, and the study of *Poa annua* which has developed genotypes that survive trampling.

The evidence for plant adaptation is usually taken to be the morphological or anatomical characters (see above) which appear to aid survival. However, the combination of these features with physiological features such as growth rates also appears to be involved (Sun and Liddle, 1993a,b), although only slightly understood and very little researched.

The work by Emetz (1985c) on the carabid beetle also suggests that trampling has a selective effect on different genotypes of invertebrates, as demonstrated by pits on the beetles' elytra. It is not clear how these different forms are adapted (or not) to survive trampling. It could be some morphological, physiological or behavioural characteristics which provide an advantage in this case, but Eric Duffey feels that 'some other factor may be involved'. With larger animals it seems likely that behavioural changes, including the ability to habituate positively to the presence and artefacts of humans, are a primary consequence of human recreation. The use of garbage dumps by grizzly bears, until the tips were closed, is a good example. The less dramatic adaptations of rats (*Rattus rattus*), house mice (*Mus musculus*), European sparrows

(*Passer domesticus*) and, in a curiously unexpected way, the fighting ability of trout when they struggle for their lives at the end of a line, have all led to an increase in these species around the world, although only the trout are solely spread by recreation activities. Now that there is a worldwide conservation movement the characteristics of being a large and/or furry mammal that appeals visually to humans (e.g. giant pandas (*Ailuropoda*); koala bear (*Phascolarctos cinereus*); lion (*Felis leo*)) can operate, in a broadly interpreted way, as a survival mechanism in the recreation context. Many national parks and breeding stations, such as the San Diego Zoo (USA), Dubbo Zoo (Australia) and the Slimbridge Wildfowl Trust (UK) only exist because the preservation of species fulfils a human desire to 'enjoy' the animals, either by seeing them or their pictures. In these cases the evolution of the animal's particular form was not induced by recreation but has turned into an advantage *post hoc* as our society has developed.

PHYSICAL SYSTEMS

It has become increasingly evident that the physical systems of our planet are dependent upon living organisms to maintain their stability (Lovelock, 1979) and this is as true at the microscale as it is for global phenomena. Recreation impacts removing plants have been shown to lead to higher soil temperatures (Figs 12.14, 12.15) and in some situations to drier soils. This in turn will impact on the nutrient and water cycling systems (Figs 9.5, 9.6). The reduction in organic content of soil (Figs 10.7, 10.8b) as a direct result of trampling, or indirectly through the removal of fallen wood for fires, will also change the porosity and nutrient cycling processes as well as characteristics such as thermal conductivity, moisture retention and availability.

The compaction of reef flat corals, loss of freshwater macrophytes, reduction in vegetative cover in terrestrial situations and surface litter all reduce the extent and physical structure of animals' habitats. The natural physical shelter from wave action, high temperatures, desiccation or predators is thus reduced or removed by recreation impacts with a concomitant loss of faunal diversity.

QUANTIFICATION OF IMPACTS

Recreation ecology started with qualitative observations such as those of Meinecke (1928), Klecka (1937) and Bates (1935, 1938). However from the 1960s onwards there has been a strong quantitative element in the observational and experimental approaches of workers such as Bayfield (1971b), Cole (1978), Blom (1977) and many others.

The most obvious quantification has been that of trampling, the number of passages, in experimental work on both plants and soils (Figs 3.3, 3.4) and the relative distribution of walkers across paths (Figs 3.2, 3.8, 4.5, 4.6) or some other relative measure in observational research. The interaction of trampling intensity with both season and frequency have also received attention, but no clear generalization is evident. For glasshouse experiments in which plants have been grown in pots, trampling has often been simulated by dropped weights, giving a similar impact to the human footfall.

The impacts on animals are less readily quantified, but the distance at which the animal shows a particular response has frequently been used. Flyaway distances for birds (Figs 20.1, 20.6; Table 20.1, 20.8, 20.10), distances that stimulate a raised heart beat in sheep (Figs 23.2, 23.3), distances at which boats caused whales to take evasive action and distances and/or altitudes at which aircraft or helicopters cause mountain sheep or whales to react by moving away from the source of sound or visual irregularity. Sound, or the size of the disturbance hemisphere or volume (Fig. 17.3) has not often been measured but Brown's (1980) work on crested terns is an exception (Fig. 17.6).

QUANTIFICATION OF RESPONSES

Responses have been the most regularly quantified aspect of recreation ecology. For plants most aspects of growth, including height, cover, biomass, various parameters of morphology such as leaf or tiller number, size and strength, as well as numbers and ages of individuals, species present and other diversity measures, have all been used (Chapters 3–8).

The most common measures of soil changes have been bulk density and penetration resistance but particle size distribution, nutrient status, pH, water content and organic content have also been recorded (Chapters 9–16).

The responses of larger animals have often been recorded as 0 or 1 measures, such as presence or absence, flight reaction or no reaction. More graded observations include the degree of disturbance to birds or deer (Fig. 22.2), length of time between whales blowing (Fig. 28.2) and, in the case of invertebrates, the numbers surviving (Figs 13.5, 18.6). Disturbance to sessile animals such as coral has also been recorded as percentage cover or weight detached from the substrate (Figs 18.12, 18.16).

THEORETICAL DEVELOPMENTS

Recreation ecologists have utilized 'mainstream' ecological theory in much of their work. Probably the most discussed theories are those of resistance or vulnerability, survival, recovery and tolerance (Chapter 8) (see for example Cole, 1995a,b). Conceptual models have also been used frequently. The author, together with Dr John Barkham, developed an early model which formed the basis of the presentation of Wall and Wright (1977) (Fig. 1.2) and the physiological reaction model (Fig. 7.1) is another example. A functional model of the trampling process was developed as a research tool (Fig. 8.3) and a more practically oriented model of soil changes has been published (Fig. 15.15).

Theoretical constructs of animal disturbance as a result of recreation activities include the categorization of impacts into three levels or types (Fig. 17.1) and the hemispheric volume or disturbance volume concept (Fig. 17.3) presented as a potential basis for the quantification of both impacts and responses.

29.2 Future research   Three kinds of research are still clearly required. One is an extension of our basic knowledge of the effects of recreation impacts, the second is research into management problems and the third, perhaps not strictly research but very important, is the communication of results in a form and context that can be immediately used by management.

The first involves a development of our basic knowledge of the reactions of organisms and communities to recreation impacts. This is particularly needed in relation to the physiological reactions of plants and animals to disturbance. There is almost no knowledge of how the metabolism of plants is changed by disturbance and very little with respect to animals. We also need more detailed understanding of how the interactions between species are affected by this kind of impact; for example, are the size-limiting effects on plants of soil compaction due to changed mycorrhizal associations?

There is also a need for a quick, low-cost methodology that can lead to vulnerability maps, perhaps combining Cole and Bayfield's (1993) protocol with Kuss and Morgan's (1986) mapping technique (Fig. 15.17). The result would be extremely useful for management planning. Detailed tabulation of the different community vulnerabilites for large-scale planning processes and impact statements, probably developed from the work proposed above, would also find immediate application.

There has also been very little research synthesizing the changes that occur to vegetation, soils and animals from the same impact. This synthesis is necessary in order to provide a complete picture of what happens, for example, when a party of ecotourists drives across a sand island in, say, half a dozen four-wheel-drive vehicles. I doubt if even the impacts of such an event in any one habitat have all been quantified in terms of ground pressure, tangential forces, decibels, strength of scent and exhaust gases, etc., let alone with respect to the responses of the plants and animals.

There is also a need to link the ecological research into the social conditions of the region. If the local people living adjacent to a national park do not receive any income from its use by ecotourists problems will persist as they continue to use its natural resources in their traditional manner. This was exemplified for the Royal Chitwan National Park in Nepal (Nepal and Weber, 1993) (Fig. 29.1).

It is also essential that the ecological knowledge of recreation impacts feeds into the local and national political and planning systems so that decisions driven by considerations other than the well-being of the environment do not lead to its destruction. There is a continuing grave

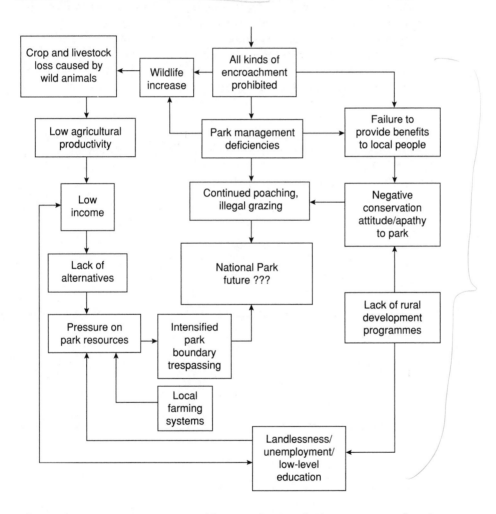

Fig. 29.1 Some management problems in the Royal Chitwan National Park, Nepal. The crucial role of management and adequate financial support is apparent in this presentation (from Nepal and Weber, 1993).

~~There is~~ risk that the ecotourist facilities will be developed beyond the carrying capacity of ~~the~~ its environment; this can only result in an infrastructure with no function in a degraded habitat, about as useful as a dry well in the desert!

Finally it must be emphasized that it is essential to convey the present and future knowledge of recreation ecology to management and decision makers in a readily useable form (see, for example, Kuss *et al.*, 1990). Without this stage the research becomes an esoteric occupation with little purpose outside the scientific culture. However, this cannot be done without a willingness by grant authorities to finance the preparation of the necessary documents, books, CD-ROMs and multimedia disks, or whatever is required for this purpose.

It is my intention that this book should contribute to this process. One difficulty is that with the very incomplete level of knowledge that we have at present the advice offered must be piecemeal and often lacking for critical parts of the environment.

29.3 Finis Recreation ecology may then be regarded *has been regarded as* as the science of a destructive process, but if people's reverence for life is increased by visiting the natural environment, then it is possible that the overall effect will be beneficial for the survival of the world's biota.

but also as a beannificial. process. with the increase of Human visiting the natural Environment, that shows It shows possible. with Reverence for nature. & The environment.

which in turn could save the biota.

References

Aaris-Sørensen, J. (1987) Post and present distribution of badgers (*Meles meles*) in the Copenhagen area. *Biol. Conserv.*, **41**, 159–65.

Abeles, F.B. (1973) *Ethylene in Plant Biology*, Academic Press, London.

Able, N. and Blaikie, P. (1986) Elephants, people, parks and development: the case of the Luangwa Valley, Zambia. *Environ. Manage.*, **10**, 735–51.

Ainsworth-Davis, J.R. (1903) *The Natural History of Animals*, Gresham, London.

Aitchison, S.W. (1977) Some effects of a campground on breeding birds in Arizona, in *Importance, Preservation and Management of Riparian Habitat* (eds R.R. Johnson and D.A. Jones), USDA Forest Service, Fort Collins, Colorado, pp. 175–82.

Akers, S.W. and Mitchell, C.A. (1984) Seismic stress effects on vegetative and reproductive development of 'Alaska' pea. *Can. J. Bot.*, **62**, 2011–15.

Albertson, F.W. (1937) Ecology at a mixed prairie in West Central Kansas. *Ecol. Monogr.*, **7**, 481–550.

Alstrup, V. (1978) Lichen genera *Stereocaulon* and *Leprocaulon* in Denmark. *Botanisk Tidsskrift*, **73**, 185–90.

Altman, M. (1956) Patterns of herd behaviour in free-ranging elk of Wyoming *Cervus canadensis nelsoni*. *Zeologica*, **41**, 65–71.

Altman, M. (1958) The flight distance in free-ranging big game. *J. Wildl. Manage.*, **22**, 207–9.

Anderson, D.E., Rongstad, O.J. and Mytton, W.R. (1989) Response of nesting red-tailed hawks to helicopter overflights. *Condor*, **91**, 296–9.

Anderson, D.J. (1961) The structure of some upland plant communities in Caernarvonshire. II. The pattern shown by *Vaccinium myrtillus* and *Calluna vulgaris*. *J. Ecol.*, **49**, 731–8.

Anderson, D.W. and Keith, S.O. (1980) The human influence on seabird nesting success: conservation implications. *Biol. Conserv.*, **18**, 65–80.

Anthony, G.R. and Isaacs, F.B. (1989) Characteristics of bald eagle nest sites in Oregon. *J. Wildl. Manage.*, **53**, 148–59.

Archer, T. (1976) The organisation of aggression and fear in vertebrates, in *Perspectives in Ethology*, Vol. 2 (eds P.P.G. Baleson and P. Klopter), Plenum Press, New York, pp. 231–98.

Arianoutsou, M. (1988) Assessing the impacts of human activities on nesting of Loggerhead Sea-turtles (*Caretta caretta* L.) on Zákynthos Island, Western Greece. *Environ. Conserv.*, **16**, 327–34.

Arms, K. (1990) *Environmental Science*, Saunders College Publishing, Chicago.

Arndt, W. and Rose, C.W. (1966) Traffic compaction of soil and tillage requirements. IV. The effect of traffic compaction on a number of soil properties. *J. Agric. Engin. Res.*, **11**, 170–87.

Arnold, G.W. and Weeldenburg, J.R. (1990) Factors determining the number and species of birds in road verges in the wheatbelt of Western Australia. *Biol. Conserv.*, **53**, 295–315.

Arthington, A.H. (1986) Introduced cichlid fishes in Australian inland waters, in *Limnology in Australia* (eds O. De Deckker and W.D. Williams), CSIRO, Melbourne and Dr W. Junk, Dordrecht, pp. 239–48.

Arthington, A.H. (1989) Impacts of introduced and translocated freshwater fishes in Australia, in *Exotic Aquatic Organisms in Asia* (ed. S.S. De-Silva), Asian Fisheries Society, Manila, Philippines, pp. 7–20.

Aspinall, R.J. and Pye, A.M. (1987) The effect of trampling on limestone grassland in the Malham area of North Yorkshire. *J. Biogeog.*, **14**, 105–15.

Aune, K. and Stivens, T. (1983) Rocky Mountain Front grizzly bear. *Monitoring and Investigation Bureau of Land Management*, US Forest Service. (Monographed report quoted in Kolstad *et al.*, 1986.)

Baiderin, V.V. (1978) Effect of winter recreation on the soil and vegetation of slopes in the vicinity of Kazan. *Ekologiya*, **1**, 93–7.

Balát, F. (1969) Influence of repeated disturbance on the breeding success in the mallard *Anas platyrhynchos* Linn. *Zoologické Listy.*, **18**, 247–52.

Ballard, W.B., Whitman, J.S. and Gardiner, C.L. (1987) Ecology of an exploited wolf population in south central Alaska. *Wildl. Monogr.*, **98**, 1–54.

Bally, R. and Griffiths C.L. (1989) Effects of human trampling on an exposed rocky shore. *Intern. J. Environ. Studies*, **34**, 115–25.

Balph, D.F. and Malecheck, J.C. (1985) Cattle trampling of crested wheatgrass under short duration grazing. *J. Range Manage.*, **38**, 226–7.

Bamford, A.R., Davies, S.J.J.F. and van Delft, R. (1990) The effects of model power boats on waterbirds at Herdsman Lake, Perth, Western Australia. *Emu*, **90**, 260–5.

Bangs, E.E., Spraker, T.H., Berns, V.D. and Baily, T.N. (1982) Effects of increased human populations on wildlife resources of the Kenai Peninsula, Alaska, USA, in *Transactions of the North American Wildlife and Natural Resources Conference* (ed. V. Sabal), Wildlife Management Institute, Washington, DC, Vol. 47, pp. 605–16.

Bannister, P. (1976) *Introduction to Physiological Plant Ecology*, Blackwell Scientific Publications, Oxford.

Barbaro, R.D., Carroll, B.J., Tebo, L.B. and Walters, L.C. (1969) Bacteriological water quality of several recreational areas in the Ross Barnett reservoir. *J. Wat. Pollut. Control. Fed.*, **41**, 1330–9.

Barley, K.P. (1959) Earthworms and soil fertility. *Anst. J. Ag. Res.* **10**, 171–85.

Barnes, K.K., Carleton, W.M., Taylor, H.M. *et al.* (eds) (1971) *Compaction of Agricultural Soils*, American Society of Agricultural Engineers, Michigan.

Barton, M.A. (1969) Water pollution in remote recreational areas. *J. Soil Water Conserv.*, **24**, 132.

Bashore, T.L., Tzilkowski, W.M. and Bellis, E.D. (1985) Analysis of deer–vehicle collision sites in Pennsylvania. *J. Wildl. Manage.*, **49**, 769–74.

Bates, G.H. (1935) The vegetation of footpaths, sidewalks, cart-tracks and gateways. *J. Ecol.*, **23**, 470–87.

Bates, G.H. (1938) Life forms of pasture plants in relation to treading. *J. Ecol.*, **26**, 452–5.

Batten, L.A. (1977) Sailing on reservoirs and its effects on water birds. *Biol. Conserv.*, **11**, 49–58.

Bayfield, N.G. (1971a) Thin wire tramplometers – a simple method for detecting variations in walker pressure across paths. *J. Appl. Ecol.*, **8**, 533–6.

Bayfield, N.G. (1971b) Some effects of walking and skiing on vegetation at Cairngorm, in *The Scientific Management of Animal and Plant Communities for Conservation* (eds. E. Duffey, S.A. Watt), Blackwell Scientific Publications, Oxford, pp. 469–85.

Bayfield, N.G. (1973) Use and deterioration of some Scottish hill paths. *J. Appl. Ecol.*, **10**, 635–44.

Bayfield, N.G. (1974) Burial of vegetation by erosion debris near ski lifts on Cairngorm, Scotland. *Biol. Conserv.*, **6**, 246–51.

Bayfield, N.G. (1976) Effect of substrate type and microtopography on establishment of a mixture of bryophytes from vegetative fragments. *Bryol.*, **79**, 199–207.

Bayfield, N.G. (1979a) Some effects of trampling on *Molophilus ater* (Meigen) (*Diptera: Tipulidae*). *Biol. Conserv.*, **14**, 219–32.

Bayfield, N.G. (1979b) Recovery of four mountain heath communities on Cairngorm, Scotland, from disturbance by trampling. *Biol. Conserv.*, **15**, 165–79.

Bayfield, N.G. and Brookes, B.S. (1979) Effects of repeated use of an area of Heather, *Calluna vulgaris* (L.) Hull Moor at Kindrogan, Scotland, for teaching purposes. *Biol. Conserv.*, **14**, 31–41.

Bayfield, N.G. and Lloyd, R.J. (1971) The condition of the footpath, in *Pennine Way Survey* (eds N.G. Bayfield and I. Brotheton), Countryside Commission, Cheltenham, pp. 34–57.

Bayfield, N.G., Urquhart, U.H. and Cooper, S.M. (1981) Susceptibility of four species of *Cladonia* to distrubance by trampling in the Cairngorm Mountains, Scotland. *J. Appl. Ecol.*, **18**, 303–10.

Beard, J.B. (1982) *Turfgrass Management for Golf Courses*, Bargess Publishing Co., Minneapolis.

Beauchamp, K.A. and Gowing, M.M. (1982) A quantitative assessment of human trampling effects on a rocky intertidal community. *Mar. Env. Res.*, **7**, 279–93.

Beck, C.A., Bonde, R.K. and Rathbun, G.B. (1982) Analyses of propeller wounds on manatees in Florida. *J. Wildl. Manage.*, **46**, 531–5.

Beeching, J.M. (1975) Recreation induced changes in moorland ecology, with special reference to Kinder Scout, Bleaklow Plateau, Derbyshire. M.Sc. Thesis, University of Manchester.

Begon, M., Harper, J.L. and Townsend, C.R. (1990) *Ecology: Individuals, Populations and Communities*, Blackwell Scientific Publications, London.

Behrend, D.F. and Lubeck, R.A. (1968) Summer flight behaviour of white-tailed deer in two Adirondack forests. *J. Wildl. Manage.*, **32**, 615–18.

Bélanger, L. and Bédard, J. (1989) Responses of staging greater snow geese to human disturbance. *J. Wildl. Manage.*, **53**, 713–19.

Bell, D.V. and Austin, L.W. (1985) The game fishing season and its effects on overwintering wildfowl. *Biol. Conserv.*, **33**, 65–80.

Bell, K.L. and Bliss, L.C. (1973) Alpine disturbance studies: Olympic National Park, USA. *Biol. Conserv.*, **5**, 25–32.

Bellamy, D., Radforth, J. and Radforth, N.W. (1971) Terrain, traffic and tundra. *Nature*, **231**, 429–32.

Bellis, E.D. and Graves, H.B. (1971) Deer mortality on a Pennsylvanian interstate highway. *J. Wildl. Manage.*, **35**, 232–7.

Belsky, A.J., Carson, W.P., Jensen, C.L. and Fox, G.A. (1993) Overcompensation by plants: herbivore optimisation or red herring? *Evolutionary Ecology*, **7**, 109–21.

Bernardino, F.S. Jr and Dalrymple, G.H. (1992) Seasonal activity and road mortality of the snakes of the Pa-hay-okee Wetlands of Everglades National Park, USA. *Biol. Conserv.*, **62**, 71–5.

Biddington, N.L. and Dearman, A.S. (1985) The effect of mechanically induced stress on the growth of cauliflower, lettuce and celery seedlings. *Ann. Bot.*, **55**, 109–19.

Biro, R.L., Hunt, E.R. Jr, Erner, Y. and Jaffe, M.J. (1980) Thigmomorphogenisis: changes in cell division and elongation in the internodes of mechanically-perturbed or ethrel-treated bean plants. *Ann. Bot.*, **45**, 655–64.

Bjorkhem, U., Fries, J., Hyppel, A., *et al.* (1974) Damage by heavy vehicles in thinnings. *Skogs-och Lantbruk-Akademien Tidskrift*, **113**, 304–23.

Black, C.A. (1968) *Soil–Plant Relationships*, 2nd edn, Wiley, New York.

Black, R.A. and Mack, R.N. (1984) A seasonal leaf abscission in *Populus* induced by volcanic ash. *Oecol.* (Berlin), **64**, 2950–99.

Blacklock, L. (1977) Encounters in the wild. *National Wildl.*, **15**, 25–9.

Blackman, G.E. and Rutter, A.J. (1950) Physiological and ecological studies in the analysis of plant environment. V. An assessment of the factors controlling the distribution of the Bluebell (*Scilla non-scripta*) in different communities. *Ann. Bot.*, **14**, 487–520.

Blakesley, J.A. and Reese, K.P. (1988) Avian use of campground and non-campground sites in riparian zones. *J. Wildl. Manage.*, **52**, 399–402.

Blanchard, B.M. and Knight, R.R. (1991) Movements of Yellowstone grizzly bears. *Biol. Conserv.*, **58**, 41–67.

Blanco, J.C., Reig, S. and de la Cuesta, L. (1992) Distribution, status and conservation problems of the wolf *Canis lupus* in Spain. *Biol. Conserv.*, **60**, 73–80.

Blodget, B.G. (1978) The effect of off-road vehicles on least terns and other shore birds. *National Park Service Cooperative Research Unit, Report No 26*, Institute for Man and Environment, University of Massachusetts, Amherst.

Blom, C.W.P.M. (1977) Effects of trampling and soil compaction on the occurrence of some *Plantago* species in coastal sand dunes. II. Trampling and seedling establishment. *Oecol. Plant.*, **12**, 363–81.

Blom, C.W.P.M. (1979) Effects of trampling and soil compaction on the occurrence of some *Plantago* species in coastal sand dunes. Ph.D. Thesis, University of Nijmegen, The Netherlands.

Blom, C.W.P.M. (1983) Plasticity of life characteristics in two different populations of *Plantago maritima* L. *Oecol. Plant.*, **4**, 377–94.

Blong, B. (1967) *Desert Bighorn and People in the Santa Rosa Mountains*, Cal-Neva Wildlife: Transactions, CA Wildlife Society, Smartsville.

Bluhdorn, D.R. (1985) Human trampling: its effects on the ground cover vegetation of Rose Gum Flat Picnic Ground, Brisbane Forest Park. Honours Thesis, Griffith University, Brisbane.

Bocker, E.L. and Ray, T.D. (1971) Golden eagle population studies in the south west. *Condor*, **73**, 463–7.

Boitani, L. (1992) Wolf research and conservation in Italy. *Biol. Conserv.*, **61**, 125–32.

Boitani, L., Barrasso, P. and Grimod, I. (1984) Ranging behaviour of the red fox in Gran Paradiso National Park (Italy). *Bollettino di Zoologia*, **51**, 275–84.

Bond, R. and Staub, C. (eds) (1974) *Handbook of Environmental Control. IV. Waste Water Treatment and Disposal*, CRC Press, Clevland, Ohio.

Boorman, L.A. and Fuller, R.M. (1977) Studies on the impact of paths on the dune vegetation at Winterton, Norfolk, England. *Biol Conserv.*, **12**, 203–16.

Boradale, L.A., Potts, F.A., Eastham, L.E.S. and Saunders, J.T. (1963) *The Invertebrata*, 4th edn, Cambridge University Press, Cambridge.

Boto, K.G. (1991) Nutrients and mangroves, in *Pollution in Tropical Aquatic Systems* (eds. D.W. Connell and D.W. Hawker), CRC Press, Boca Raton, London, pp. 129–46.

Boucher, D.H., Aviles, J., Chepote, R., *et al.* (1991) Recovery of trailside vegetation from trampling in a tropical rainforest. *Environ. Manage.*, **15**, 257–62.

Bowles, J.M. and Maun, M.A. (1982) A study of the effects of trampling on the vegetation of Lake Huron sand dunes at Pinery Provincial Park. *Biol. Conserv.*, **24**, 273–83.

Bradshaw, A.D. (1987) Restoration: an acid test for ecology, in *Restoration Ecology* (eds W.R. Jorden, M.E. Gilpin and J.D. Aber), Cambridge University Press, Cambridge, pp. 23–9.

Bradshaw, H.D. and Chadwick, M.J. (1980) *The Restoration of Land: The Ecology and Reclamation of Derelict and Degraded Land*, Blackwell, London.

Brady, N.C. (1974) *The Nature and Properties of Soils*, Macmillan, New York.

Branson, F.A. (1953) Two new factors affecting resistance of grasses to grazing. *J. Range Manage.*, **6**, 165–71.

Bratton, S.P. (1985) Effects of disturbance by visitors on two woodland orchid species in Great Smoky Mountains National Park, USA. *Biol. Conserv.*, **31**, 211–27.

Bratton, S.P., Hickler, M.G. and Graves, J.H. (1979) Trial erosion patterns in Great Smoky Mountain National Park. *Environ. Manage.*, **3**, 431–45.

Brattstrom, B.H. and Bondello, M.C. (1983) Effects of off-road vehicle noise on desert vertebrates, in *Environmental Effects of Off-Road Vehicles* (eds R.H. Webb and H.G.Wilshire), pp. 167–206, Springer-Verlag, New York.

Braun-Blanquet, J. (1928) *Pflanzensoziologie. Grundzuge Vegetationskunde*, Springer, Berlin.

Braun-Blanquet, J. (1932) *Plant sociology: The study of plant communities* (translated and edited by C.D. Fuller and H.S. Conard), Hafner, London.

Bregazzi, P.R., Burrough, R.J. and Kennedy, C.R. (1982) The natural history of Slapton Ley Nature Reserve XIV: the history and management of the fishery. *Field Studies* 5, 581–9.

Breton, M. (1986) *Guide to Watching Whales in Canada*. Department of Fisheries and Oceans, Ottawa, Ontario.

British Transport Docks Board (1972) *Creation of Wash by Pleasure Craft*, Southall British Transport Docks Board Report No. R236, London.

Brodhead, J.M. and Godfrey, P.J. (1977) Off road vehicle impact in Cape Cod National Seashore: disruption and recovery of dune vegetation. *Intern. J. Biometrol.*, **21**, 299–306.

Brody, S. (1945) *Bioenergetics and Growth*. Hafner Publishing, New York.

Broekhuysen, G.J. (1973) Behavioural responses of dabchicks *Podiceps ruficollis* to disturbances while incubating. *Ostrich*, **44**, 111–17.

Brooks, R.P. and Croonquist, M.J. (1990) Wetland, habitat, and trophic response guilds for wildlife species in Pennsylvania. *J. Penn. Acad. Sci.*, **64**, 93–102.

Brown, A.L. (1990) Measuring the effect of aircraft noise on sea birds. *Environ. Intern.*, **16**, 587–92.

Brown, J.K. (1989) Sediment deposition out of a plume resulting from 4WD activity at low level fords in the Victorian Central Highlands. Student report to Department of Geography and Environmental Science, Monash University, Clayton, Victoria 3168.

Brown, J.M. Jr, Kalisz, S.P. and Wright, W.R. (1977) Effects of recreational use on forested sites. *Environ. Geol.*, **1**, 425–31.

Brown, L. (ed.) (1993) *The New Shorter Oxford English Dictionary*, Clarendon Press, Oxford.

Bruschin, J. and Dysli, M. (1973). *Erosion des rives due aux oscillations du plan d'eau d'une retenue – Le Rhône à l'aval de Genève*, Communication des Laboratoires d'hydraulique et de Geotechnique de l'Ecole Polytechnique Federal de Lausanne, Lausanne.

Buechner, H.D. (1960) The bighorn sheep in the United States, its past, present and future. *Wildl. Monogr.*, **4**.

Buehler, D.A., Mersmann, T.J., Fraser, J.D. and Seegar, J.K.D. (1991) Effects of human activity on bald eagle distribution on the Northern Chesapeake Bay. *J. Wildl. Manage.*, **55**, 282–90.

Burden, R.F. and Randerson, P.F. (1972) Quantitative studies of the effects of human trampling on vegetation as an aid to the management of semi-natural areas. *J. Appl. Ecol.*, **9**, 439–57.

Burger, H. (1932) Physikalische Eigenschaften von Wald und Freilandboden. IV. Mitteilung, Ferienlager und Waldboden. *Mitteilung der Schweizerischen Central Anstralt fur das Forstliche Versuchswesen*, **17**, 299–322.

Burger, T. (1981a) The effect of human activity on birds at a coastal bay. *Biol. Conserv.*, **21**, 231–41.

Burger, T. (1981b) Behavioural responses of herring gulls *Larus argentatus* to aircraft noise. *Environ. Poll.* (Series A), **24**, 177–84.

Burger, T. and Galli, T. (1980) Factors affecting distribution of gulls (*Larus* spp.) on two New Jersey coastal bays. *Environ. Conserv.*, **20**, 59–65.

Burt, G.R. (1970) *Travel on Thawed Tundra*, Institute of Arctic Environment and Engineering, University of Alaska.

Bury, R.B. and Marlow, R.W. (1973) The desert tortoise: will it survive? *Environ. J.*, **June**, 9–12.

Busack, S.D. and Bury, R.B. (1974) Some effects of off-road vehicles and sheep grazing on lizard populations in the Mojave Desert. *Biol. Conserv.*, **6**, 179–83.

Butcher, R.W. (1947) Studies on the ecology of rivers VII. The algae of organically enriched waters. *J. Ecol.*, **35**, 186–91.

Bykov, A.V. (1985) Peculiarities of small mammal populations in recreational forests of southern Moscow district. *Lesovedenie*, **4**, 47–52.

Cabelli, V.J., Dafour, A.P., McCabe, L.J. and Levin, M.A.A. (1982) Swimming associated gastroenteritus and water quality. *Am. J. Epidemiol.*, **115**, 606–16.

Cadwallader, P.L. and Rogan, P.L. (1977) The Macquarie perch *Macquaria australasica* (Pisces: Percichthyidae), of Lake Eildon, Victoria. *Aust. J. Ecol.*, **2**, 409–18.

Cairncross, S. and Feachem, R.G. (1983) *Environmental health engineering in the tropics*, Wiley, Chichester.

Cairns, D. (1980) Nesting density, habitat structure and human disturbance as factors in black guillemot reproduction. *Wilson Bulletin*, **92**, 352–61.

Cairns, J. Jr (1990) Lack of theoretical basis for predicting rate and pathways of recovery. *Environ. Manage.*, **14**, 517–26.

Caldwell, M.C. and Caldwell, D.K. (1966) Epimeletic behaviour in Cetacia, in *Whales, Dolphins and Porpoises* (ed. K.S. Norris), University of California Press, Berkeley, pp. 755–89.

Calef, G.W. (1976) Numbers beyond counting, miles beyond measure. *Audobon*, **78**, 42–61.

Camberlein, G. (1976) Impact du camping sauvage sur un milieu naturel: frequentation et evolution du systeme dunaire de Lampaul-Poloudalmezeau Nord-Finistere. *OPIE*, **23**, 4–17.

Cameron, R.D. (1983) Issue: caribou and petroleum development in Arctic Alaska. *Arctic.* **36**, 227–31.

Canadian Government (1983) *Guidelines for Canadian Water Quality*, Ministry of Health and Welfare, Ottawa.

Canaway, P.M. (1976) Fundamental techniques in the study of turf grass wear: an advance report on research. *J. Sports Turf Res. Inst.*, **51**, 104–15.

Cannell, R.Q. (1977) Soil aeration and compaction in relation to growth and soil management, in *Applied Biology*, Vol. II (ed. T.H. Coaker), pp. 1–83, Academic Press, London.

Carbaugh, B., Vaughan, T.P., Bellis, E.D. and Graves, H.B. (1975) Distribution and activity of white-tailed deer along an interstate highway. *J. Wildl. Manage.*, **39**, 570–81.

Carlson, G.E. (1966) Growth of clover leaves after complete or partial leaf removal. *Crop Sci.*, **6**, 419–22.

Carlson, L.H. and Godfrey, P.J. (1989) Human impact management in a coastal recreation and natural area. *Biol. Conserv.*, **49**, 141–56.

Caughley, G., Dublin, H. and Parker, I. (1990) Projected decline of the African elephant. *Biol. Conserv.*, **54**, 157–64.

Cederlund, G. (1981) Daily and seasonal activity pattern of roe deer in a boreal habitat. *Swedish Wildlife Research Viltrevy*, **11**, 315–53.

Chamberlin, W.D. (1982) Avian population density in the maritime forest of two South Carolina Barrier Islands. *Am. Birds*, **36**, 142–5.

Chancellor, W.J. (1971) Effects of compaction on soil strength, in *Compaction in Agricultural Soils* (eds K.K. Barns *et al.*), American Society of Agricultural Engineers, Michigan, pp. 190–2l2.

Chang, H. and Xiao, Q. (1988) Selection of winter habitat of red deer in Dailing region. *Acta Theriol. Sinica*, **8**, 81–8.

Chappell, H.G., Ainsworth, J.F., Cameron, R.A.D. and Redfern, M. (1971) The effect of trampling on a chalk grassland ecosystem. *J. Appl. Ecol.*, **8**, 869–82.

Charman, D.J. and Pollard, A.J. (1993) Long term vegetation recovery after vehicle track abandonment on Dartmoor, south-west England, UK. *The Bulletin*, **25**, 22–8, British Ecological Society.

Chase, A. (1983) The last bears of Yellowstone. *Atlantic Monthly*, **Feb.**, 63–73.

Chehébar, C.E., Gallur, A., Giannico, G. *et al.* (1986) A survey of the Southern River Otter (*Lutra provocax*) in Lanin, Puelo and Los Alerces National Parks, Argentina, and evaluation of its conservation status. *Biol. Conserv.*, **38**, 293–304.

Cieslinski, T.J. and Wagar, J.A. (1970) Predicting the Durability of Forest Recreation Sites in Northern Utah – Preliminary Results, Research Note INT-117, USDA Forest Service, Intermountain Forest and Range Experiment Station, Ogden, Utah.

Clapham, A.R., Tutin, T.G. and Warburg, E.F. (1962) *Flora of the British Isles*, 2nd edn, Cambridge University Press, London.

Clark, F.E. (1967) Bacteria in the soil, in *Soil Biology* (eds A. Burges and F. Raw), Academic Press, London, pp. 15–19.

Clarkson, J.R. and Clifford, H.T. (1987) Germination of *Jedda multicaulis* J.R. Clarkson (*Thymelaeaceae*). An example of cryptogeal germination in the Australian flora. *Aust. J. Bot.*, **35**, 715–20.

Clement, C.R., Hopper, M.J., Jones, L.H.P. and Leafe, E.L. (1978) Uptake of nitrate by *Lolium perenne* from flowing nutrient solution. II. Effect of light, defoliation and relationship to CO_2 flux. *J. Exp. Bot.*, **29**, 1173–83.

Clevenger, G.A. and Workman, G.W. (1975) The effects of campgrounds on small mammals in the Canyon Lands and Arches National Parks, Utah, in *The 42nd North American Wildlife and Natural Resources Conferences* (ed. K. Sabol), Wildlife Management Institute, Washington, DC, pp. 473–84.

Clifford, H.T. (1956) Seed dispersal on footwear. *Proc. Bot. Soc. British Isles*, **2**, 129–31.

Clifford, H.T. (1959) Seed dispersal by motor vehicles. *J. Ecol.*, **47**, 322–5.

Cluzeau, D., Binet, F., Vertes, F. *et al.* (1992) Effects of intensive cattle trampling on soil–plant-earthworms system in two grassland types. *Soil Biol. Biochem.*, **24**, 1661–5.

Cohron, G.T. (1971) Forces causing soil compaction, in *Compaction of Agricultural Soils* (eds K.K. Barns *et al.*), American Society of Agricultural Engineers, Michigan, pp. 106–22.

Cole, D.N. (1978) Estimating the susceptibility of wildland vegetation to trailside alteration. *J. Appl. Ecol.*, **15**, 281–6.

Cole, D.N. (1981) Vegetational changes associated with recreational use and fire suppression in the Eagle Cap Wilderness, Oregon: some management implications. *Biol. Conserv.*, **20**, 247–70.

Cole, D.N. (1982) *Wilderness campsite impacts: effect of amount of use*, USDA Forest Service. Intermountain Forest and Range Experiment Station, Research Paper INT-284, Missoula, Montana.

Cole, D.N. (1983a) *Assessing and monitoring back country trail conditions*, USDA Forest Service. Intermountain Forest and Range Experiment Station, Research Paper INT-303, Ogden, Utah.

Cole, D.N. (1983b) *Campsite conditions in the Bob Marshall Wilderness, Montana*, USDA Forest Service, Intermountain Forest and Range Experimental Station, Research Paper INT-312, Ogden, Utah.

Cole, D.N. (1987a) Effects of three seasons of experimental trampling on five montane forest communities and a grassland in Western Montana, USA. *Biol. Conserv.*, **40**, 219–44.

Cole, D.N. (1987b) *Research on soil and vegetation in wilderness: a state-of-knowledge review*. Proceedings – National Wilderness Research Conference: Issues, State-of-Knowledge, Future Directions (compiler R.C. Lucas), Intermountain Research Station, Ogden, Utah, pp. 135–77.

Cole, D.N. (1992) Modelling wilderness campsites: factors that influence amount of impact. *Environ. Manage.*, **16**, 255–64.

Cole, D.N. (1993) *Trampling effects on mountain vegetation in Washington, Colorado, New Hampshire and North Carolina*, US Dept of Agriculture, Forest Services, Intermountain Research Station, Research Paper 1NT-464, Ogden, Utah.

Cole, D.N. (1995a) Experimental trampling of vegetation. I. Relationship between trampling intensity and vegetation response. *J. Appl. Ecol.*, **32**, 203–14.

Cole, D.N. (1995b) Experimental trampling of vegetation. II. Predictors of resistance and resilience. *J. Appl. Ecol.*, **32**, 215–24.

Cole, D.N. and Bayfield, N.G. (1993) Recreational trampling of vegetation: standard experimental procedures. *Biol. Conserv.*, **63**, 209–15.

Cole, D.N. and Dalle-Molle, J. (1982) *Managing campfire impacts in the back country*, USDA Forest Service, Intermountain Forest and Range Experiment Station, General Technical Report INT-135, Ogden, Utah.

Cole, D.N. and Fichtler, R.K. (1983) Campsite impact on three western wilderness areas. *Environ. Manage.*, **7**, 275–88.

Cole, G.F. (1974) Management involving grizzly bears and humans in Yellowstone National Park, 1970–1973. *Bioscience*, **24**, 1–11.

Coles, J.M. and Hibbert, F.A. (1968) Prehistoric roads and tracks in Somerset, England: I Neolithic. *Proc. Prehist. Soc.*, **34**, 238–58.

Collins, E.I. and Lichvar, R.W. (1986) Vegetation inventory of current and historic black-footed ferret habitat in Wyoming. *Great Basin Naturalist Memoirs*, **8**, 85–93.

Connell, J.H. (1978) Diversity in tropical rain forests and coral reefs. *Science*, **199**, 1302–10.

Connell, J.H. (1979) Tropical rain forests and coral reefs as open non-equilibrium systems, in *Population Dynamics* (eds R.M. Anderson, B.D. Turner and L.R. Taylor), Blackwell, Oxford, pp. 141–63.

Connell, J.H. and Slatyer, R.O. (1977) Mechanisms of succession in natural communities and their role in community stability and organisation. *Am. Natural.*, **111**, 1119–44.

Constantine, T. (1961) The behaviour of ships moving in restricted waterways. *Proc. Inst. Civil Eng.*, **19**, 549–61.

Cooke, A.S. (1974) The effects of fishing on waterfowl at Grafham Water. *Camb. Bird Club Rep.* **48**, 40–6.

Cooke, A.S. (1980) Observations on how close certain passerine species will tolerate an approaching human in rural and suburban areas. *Biol. Conserv.*, **18**, 85–8.

Coombs, E.A.K. (1976) *The impact of camping on vegetation in the Bighorn Crags, Idaho Primitive Area.* MS Thesis, University of Idaho, Moscow, Idaho.

Cooper, W.C. (1972) Trauma-induced ethylene production by citrus flowers, fruit and wood, in *Plant Growth Substances 1970* (ed. D.J. Carr), Springer-Verlag, Berlin, pp. 543–8.

Cooper, W.C., Rasmussen, G.K. and Waldon, E.S. (1969) Ethylene production in freeze-injured citrus trees. *J. Rio Grande Valley Hortic. Soc.*, **23**, 29–37.

Coppleson, V.M. (1962) *Shark Attack*, 2nd edn, Angus and Robertson, Sydney.

Corbett, J.R. (1969) *The Living Soil: The Processes of Soil Formation*, Martindale Press, West Como, New South Wales.

Corkeron, P.J. (1992) *Managing whale watching for hump-back whales.* Abstracts 5th Annual Conference, Australasian Wildlife Management Society, Queensland University of Technology, Brisbane, Queensland.

Cornish, P.S., So, H.B. and McWilliam, J.R. (1984) Effects of soil bulk density and water regimen on root growth and uptake of phosphorus by ryegrass. *Aust. J. Agric. Res.*, **35**, 631–44.

Cott, M.B. (1969) Tourists and crocodiles in Uganda. *Oryx*, **10**, 153–60.

Coulson, G.M. (1982) Road-kills of Macropods on a section of highway in Central Victoria. *Aust. Wildl. Res.*, **9**, 21–6.

Craig, R.J., Mitchell, E.S. and Mitchell, J.E. (1988) Time and energy budgets of bald eagles wintering along the Connecticut River. *J. Field Ornithol.*, **59**, 22–32.

Craighead, F.C. Jr and Mindell, D.P. (1981) Nesting raptors in Western Wyoming. *J. Wildl. Manage.*, **45**, 865–72.

Craighead, J.J. and Craighead, F.C. (1971) Grizzly bear–man relationships in Yellowstone National Park. *Biosci.*, **21**, 845–57.

Crawford, A.K. and Liddle, M.J. (1977) The effect of trampling on neutral grassland. *Biol. Conserv.*, **12**, 135–42.

Crawford, R.J.M. and Fairall, N. (1984) Male rock hyraxes (*Procavia capensis*) return to former home ranges after translocation. *Koedoe*, **27**, 151–3.

Crawshaw, P.G.J. and Schaller, G.B. (1980) Nesting of Paraguayan Caimen *Caimen yacare* in Brazil. *Papers Auulsos Zool.* (São Paulo), **33**, 283–92.

Croonquist, M.J. and Brooks, R.P. (1991) Use of avian and mammalian guilds as indicators of cumulative impacts in riparian-wetland areas. *Environ. Manage.*, **15**, 701–14.

Crossland, R. (1976) Part of discussion in Proceedings of the British Crop Protection Council Symposium on Aquatic Herbicides, 1976, Oxford, p. 49.

Crowl, T.A., Townsend, C.R. and McIntosh, A.R. (1992) The impact of introduced brown and rainbow trout on native fish: the case of Australasia. *Rev. Fish Biol. Fisheries*, **2**, 217–41.

Cryer, M., Whittle, G.N. and Williams, R. (1987) The impact of bait collection by anglers on marine intertidal invertebrates. *Biol. Conserv.*, **42**, 83–93.

Curtis, J.T. (1959) *The Vegetation of Wisconsin*, University of Wisconsin Press, Madison.

Curtis, S. (1974) How to track wildlife on skiis. *Backpacker*, **2**, 40–3.

Cypher, E.A., Yahner, R.H., Storm, G.L. and Cypher, B.C. (1986) Flora and fauna survey in a proposed recreational area of Valley Forge National Historical Park, Pennsylvania, USA. *Proc. Penn. Acad. Sci.*, **60**, 47–50.

Daele, L.J. van and Daele, H.A. van (1982) Factors affecting the productivity of ospreys nesting in West-Central Idaho. *Condor*, **84**, 292–9.

Dale, D. and Weaver, T. (1974) Trampling effects on vegetation of the trail corridors of North Rocky Mountain forests. *J. Appl. Ecol.*, **11**, 767–72.

Darwin, C. (1881) *The Formation of Vegetable Mould Through the Action of Worms, with Observations of Their Habits*, Murray, London.

Davids, R.C. (1978) Polar Bears aren't pets, but this town is learning how to live with them. *Smithsonian*, **8**, 70–9.

Davies, W. (1938) Vegetation of grass verges and other excessively trodden habitats. *J. Ecol.*, **26**, 28–49.

Davis, G.E. (1977) Anchor damage to a coral reef on the coast of Florida. *Biol. Conserv.*, **11**, 29–34.

Dawson, J.O., Hinz, P.N. and Gordon, J.C. (1974) Hiking-Trail impact on Iowa streamvalley Forest Preserves. *Iowa State J. Res.*, **48**, 329–37.

Dearden, P. and Hall, C. (1983) Non-consumptive recreation pressures and the case of the Vancouver Island marmot (*Marmota vancouverensis*). *Environ. Conserv.*, **10**, 63–6.

De Forge, J.R. (1972) Man's invasion into the Bighorn's habitat. *Desert Bighorn Council Trans.*, **16**, 112–16.

De Forge, J.R. (1976) Stress: is it limiting Bighorn? *Desert Bighorn Council Trans.*, **20**, 30–1.

Degner, R.L., Rodan, L.W., Mathis, W.K. and Gibbs, E.P.J. (1983) The recreational and commercial importance of feral swine in Florida: relevance to the possible introduction of African swine feaver into the USA. *Preventive Veterinary Medicine*, **1**, 371–81.

De Graaf, R.M. and Thomas, J.W. (1976) Wildlife habitat in or near human settlements, in *Trees and Forests for Human Settlement*, Center for Urban Forestry Studies, University of Toronto, pp. 54–62.

Deitz, D.C. and Hines, T.C. (1980) Alligator nesting in North-Central Florida. *Copeia*, 1980, (2), 249–58.

Denniston, R.H. (1956) Ecology, behaviour and population dynamics of the Wyoming or Rocky Mountain Moose *Alces alces shirasi*. *Zoologica* NY, **41**, 105–18.

Department of the Environment (1973) *Report of a survey of the discharges of foul sewage to coastal waters of England and Wales*, HMSO, London.

de Vos, A. (1958) Summer observations on moose behaviour in Ontario. *J. Mammal.*, **39**, 128–39.

Dewey, R.A. and Nellis, D.W. (1980) Seabird Research in the US Virgin Islands, in *The 45th North American Wildlife Conference*, Wildlife Management Institute, Washington, DC, pp. 445–52.

Di Sciara, G.N. (1977) A killer whale (*Orcinus orca* L.) attacks and sinks a sailing boat. *Soc. Ital. Sci. Nat. Museo. civ. Stor. nat. e Acquario civ.*, **68**, 218–20.

Dodd, C.K. Jr, Enge, K.M. and Stuart, J.N. (1989) Reptiles on highways in North-Central Alabama, USA. *J. Herpetology*, **23**, 197–200.

Dorrance, M.J., Savage, P.J. and Huff, D.E. (1975) Effects of snowmobiles on white-tailed deer. *J. Wildl. Manage.*, **39**, 563–9.

Dotzenko, A.D., Papamichas, N.T. and Romine, D.S. (1967) Effect of recreational use on soil and moisture conditions in Rocky Mountain National Park. *J. Soil Water Conserv.*, **22**, 196–7.

Douglas-Hamilton, I. (1973) On the ecology and behaviour of the Lake Manyara elephants. *E. Afr. Wildl. J.*, **11**, 401–3.

Dregne, H.E. (1983) Soil and soil formation in arid regions, in *Environmental Effects of Off-road Vehicles: Impacts and Management in Arid Regions* (eds R.H. Webb and H.G. Wilshire), Springer-Verlag, New York, pp. 15–50.

Drolet, C.A. (1976) Distribution and movements of white-tailed deer in Southern New Brunswick in relation to environmental factors. *Can. Field Nat.*, **90**, 123–36.

Drozdz, A. and Osiecki, A. (1973) Intake and digestibility of natural feeds by roe-deer. *Bialowieza*, **18**, 81–91.

Dudley, S.F.J. and Cliff, G. (1993) Some effects of shark nets in the Natal nearshore environment. *Environ. Biol. Fishes*, **36**, 243–55.

Duffey, E. (1975) The effects of human trampling on the fauna of grassland litter. *Biol. Conserv.*, **7**, 255–74.

Duffey, E. (1982) *National Parks and Reserves of Western Europe*, MacDonald, London.

Duffus, D.A. and Dearden, P. (1990) Non-consumptive wildlife – oriented recreation: a conceptual framework. *Biol. Conserv.*, **53**, 213–31.

Duggeli, M. (1937) Wie wirkt das oftere Betreten des Walbodens auf einzelene physikalische und biologishe Eigenschaften. *Schweizerische Zeitschrift fur Furstwesen*, **88**, 151–65.

Dunnet, G.M. (1977) Observations on the effects of low-lying aircraft at seabird colonies on the coast of Aberdeenshire, Scotland. *Biol. Conserv.*, **12**, 55–63.

Dunstan, T.C. (1968) Breeding success of osprey in Minnesota from 1963 to 1968. *Loon*, **40**, 109–12.

Dunstan, T.C. (1973) The biology of ospreys in Minnesota. *Loon*, **45**, 108–13.

Dyke, F.G. van, Brocke, R.H., Shaw, H.G. *et al.* (1986) Reactions of mountain lions to logging and human activity. *J. Wildl. Manage.*, **50**, 95–102.

Dyke, F.G. van, Brocke, R.H. and Shaw, H.G. (1986) Use of road track counts as indices of mountain lion presence. *J. Wildl. Manage.*, **50**, 102–9.

Dyksterhuis, E.J. (1957) The savannah concept and its use. *Ecology*, **38**, 435–42.

Earner, Y. and Jaffe, M.J. (1982) Thigmomorphogenesis: the involvement of auxin and abscisic acid in growth retardation due to mechanical perturbation. *Plant Cell Physiol.*, **23**, 935–41.

Eavis, B.W. (1972) Soil physical conditions affecting seedling root growth. I Mechanical impedance, aeration and moisture availability as influenced by bulk density and moisture levels in a sandy loam soil. *Plant Soil*, **36**, 613–22.

Eberhardt, L.L., Blanchard, B.M. and Knight, R.R. (1994) Population trend of the Yellowstone grizzly bear as estimated from reproductive and survival rates. *Can. J. Zool.*, **72**, 360–3.

Eckert, R.E. Jr, Wood, M.K., Blackburn, W.H. and Petersen, F.F. (1979) Impacts of off-road vehicles on infiltration and sediment production of two desert soils. *J. Range Manage.*, **32**, 394–7.

Eckstein, R.G., O'Brien, T.F., Rongstad, O.J. and Bollinger, J.G. (1979) Snowmobile effects on movements of white-tailed deer: a case study. *Environ. Conserv.*, **6**, 45–51.

Edgar, B. (1990) Suffering symbol of the Mojave. *Pacific Discovery*, **18**, 17–21.

Edington J.M. and Edington, M.A. (1977) *Ecology and Environmental Planning*, Chapman & Hall, London.

Edington J.M. and Edington, M.A. (1986) *Ecology and Environmental Planning*, Chapman & Hall, London.

Edmond, D.H. (1958) Some effects of soil physical condition on ryegrass growth. *N.Z. J. Agric. Res.*, **1**, 652–9.

Edmond, D.B. (1962) Effects of treading pasture in summer under different soil moisture levels. *N.Z. J. Agric. Res.*, **5**, 389–95.

Edmond, D.B. (1964) Some effects of sheep treading on the growth of 10 pasture species. *N.Z. J. Agric. Res.*, **7**, 1–16.

Edmond, D.B. (1974) Effects of sheep treading on measured pasture yield and physical condition of four soils. *N.Z. J. Exp. Agric.*, **2**, 38–43.

Edwards, C.A. and Lofty, J.R. (1977) *Biology of Earthworms*, 2nd edn, Chapman & Hall, London.

EEC (1976) Council directive of 8 Dec. 1975 concerning the quality of boiling water (76/160/EEC). *Official Journal of the European Communities*, **L/31**, 1–7.

Elgmork, K. (1978) Human impact on the brown bear population (*Ursus arctos* L.). *Biol. Conserv.*, **13**, 81–103.

Elgmork, K. (1983) Influence of holiday cabin concentrations on the occurrence of brown bears (*Ursus arctos* L.) in south-central Norway. *Acta Zool. Fennica*, **174**, 161–2.

Elgmork, K. (1987) *The cryptic brown bear populations of Norway.* Proceedings of the International Conference on Bear Research Management, Vol. 7, pp. 13–16.

Elgmork, K. (1988) Reappraisal of the brown bear status in Norway. *Biol. Conserv.* **46**, 163–8.

Elwell, H.A. and Stocking, M.A. (1976) Vegetal cover to estimate soil erosion hazard in Rhodesia. *Geoderma*, **15**, 61–70.

Emetz, V.M. (1983) Changes in the density and structure of a population of *Pterostichus melanarius* (Coleoptera Carabidae) under the influence of recreation. *Zoologicheskii Zhurnal*, **62**, 1505–9 [in Russian].

Emetz, V.M. (1984a) Dynamics of phenotypical composition and level of asymmetry of the number of fossae on elytra of imagos in a population of *Pterostichus oblongopunctatus* (Coleoptera Carabidae) on recreation territory. *Zoologicheskii Zhurnal*, **63**, 218–21 [in Russian].

Emetz, V.M. (1984b) Changes in indices of migration of *Pterostichus oblongopunctatus* under the influence of recreation. *Zoologicheskii Zhurnal*, **63**, 1808–13 [in Russian].

Emetz, V.M. (1984c) Changes in the phenotypic structure of a population of the pitted ground beetle in places changed by recreation. *Ekologia (Sverdlobsk: nauka)*, **5**, 85–8 [in Russian].

Emetz, V.M. (1985a) Certain changes in a population of the ground beetle *Pterostichus oblongopunctatus* F. (Coleoptera Carabidae) under the influence of recreation. *Moskovskoe obshchestvo ispytatelei prirody Biulleten*, **90**, 61–8 [in Russian].

Emetz, V.M. (1985b) Long-term changes in variability characteristics of the polymorphic trait (number of pits on elytra) in imago groupings of ground-beetle *Pterostichus oblongopunctatus* F. (Coleoptera Carabidae) on recreational and little visited plots of oak grove. *Entomologicheskoye Obozreniye*, **64**, 85–8 [in Russian].

Emetz, V.M. (1985c) Long-term dynamics of variability parameters of groups of *Pterostichus oblongopunctatus* F. (Coleoptera Carabidae) adults on the basis of a polymorphic character (number of pits on elytra) in the recreational and less visited areas of an oak grove. *Entomol. Rev.*, **64**, 113–16 [in English].

Emetz, V.M. (1986) Changes in spatial distribution of adults in a population of the ground beetle *Pterostichus oblongopunctatus* F. (Coleoptera Carabidae) under the influence of recreation. *Zhurnal Obshchei Biologii*, **47**, 125–7 [in Russian].

Emlen, J.T. (1974) An urban bird community in Tucson Arizona: derivation, structure and regulation. *Condor*, **76**, 184–97.

Emmons, R. (1984) *Turfgrass Science and Management*, Delmar Press, Albany, New York.

Engelaar, W. (1994) Roots, nitrification and nitrate acquisition in waterlogged and compacted soils. Ph.D. Thesis, Katholieke Universiteit, Nijmegen.

Engelaar, W.M.H.G, Jacobs, M.H.H.E. and Blom, C.W.P.M. (1993) Root growth of *Rumex* and *Plantago* species in compacted and waterlogged soils. *Acta Bot. Neerl.*, **42**, 25–35.

Enzenbacher, D.J. (1992) Antarctic tourism and environmental concerns. *Mar. Pollut. Bull.*, **25**, 258–65.

Erwin, R.M. (1980) Breeding habitat use by colonially nesting waterbirds in two Mid-Atlantic US regions under different regimes of human disturbance. *Biol. Conserv.*, **18**, 39–51.

Espmark, Y. (1974) Social behaviour of roe deer at winter feeding stations. *Appl. Anim. Ethol.*, **1**, 35–47.

Etchberger, R.C., Krausman, P.R. and Mazaika, R. (1989) Mountain sheep habitat characteristics in the Pusch Ridge Wilderness, Arizona. *J. Wildl. Manage.*, **53**, 902–7.

Evans, P.G.H. (1987) *The Natural History of Whales and Dolphins*, Christopher Helm, London.

Evans, P.S. (1967a) Leaf strength studies of pasture grasses I. Apparatus, techniques and some factors affecting leaf strength. *J. Agric. Sci.* (Cambridge), **69**, 171–4.

Evans, P.S. (1967b) Leaf strength studies of pasture grasses II. Strength, cellulose content and schlerenchyma tissue proportions of eight grasses grown as single plants. *J. Agric. Sci.* (Cambridge), **69**, 175–81.

Evans, R. (1980) Mechanics of water erosion and their spatial and temporal controls: an empirical viewpoint, in *Soil Erosion* (eds M.S. Kirkby and R.P.C. Morgan), Wiley, Chichester.

Fairbanks, W.S., Bailey, J.A. and Cook, R.S. (1987) Habitat use by a low-elevation, semi-captive Bighorn sheep population. *J. Wildl. Manage.*, **51**, 912–15.

Fedorova, V.G. (1985) Midges of the genus *Culicoides* (Ceratopogonidae) and evaluation of the critical level of their abundance in the Novgorod region. *Parazitologiya*, **19**, 123–27.

Feldhamer, G.A., Gales, J.E., Harmon, D.M. *et al.* (1986) Effects of interstate highway fencing on white-tailed deer activity. *J. Wildl. Manage.*, **50**, 497–503.

Fellin, D.G. (1980) Effect of silvicultural practices, residue utilisation and prescribed fire on some forest floor arthropods, in *Environmental Consequences of Timber Harvesting in Rocky Mountain Coniferous Forests*, USDA Forest Service, General Technical Report INT-90. Intermountain Forest and Range Experiment Station, Ogden, Utah, pp. 287–316.

Fenn, D.B., Gogue, C.J. and Burge, R.E. (1976) *Effects of Campfires on Soil Properties*, US Dept Interior, Natl. Park Serv., Ecol. Serv. Bull. 5, Washington, DC.

Ferguson, M.A.D. and Keith, L.B. (1982) Influence of nordic skiing on distribution of moose and elk in Elk Island National Park, Alberta. *Can. Field Nat.*, **96**, 69–78.

Fetcher, N. and Shaver, G.R. (1983) Life histories of tillers of *Eriophorem vaginatum* in relation to tundra disturbance. *J. Ecol.*, **71**, 131–47.

Fichtler, R.K. (1980) The relationship of recreational impacts on back country campsites to selected Montana habitat types. M.S. Thesis, University of Montana, Missoula.

Fish, E.B., Brothers, G.L. and Lewis, R.B. (1981) Erosional impacts of trails in Guadalupe Mountains National Park, Texas. *Landscape Planning*, **8**, 387–98.

Fisher, J.B. (1985) Induction of reaction wood in Terminalia (*Combretaceae*): roles of gravity and stress. *Ann. Bot.*, **55**, 237–48.

Fisher, J.B. and Stevenson, J.W. (1981) Occurrence of reaction wood in branches of dicotyledons and its role in tree architecture. *Bot. Gaz.*, **142**, 82–95.

FitzPatrick, E.A. (1971) *Pedology: A Systematic Approach to Soil Science*, Oliver and Boyd, Edinburgh.

Flannery, T. (1994) *The Future Eaters*, Reed Books, Chatswood, New South Wales.

Fournier, F. (1972) *Soil Conservation*, Nature and Environment Series, Council of Europe, Strasburg.

Fox, J.L., Nurbu, C. and Chundawat, R.S. (1991) The Mountain Ungulate of Ladakh, India. *Biol. Conserv.*, **58**, 167–90.

Fox, J.L., Nurbi, C. and Chundawat, R.S. (1991). The mountain ungulates of Ladakh, India. *Biol. Conserv.*, **58**, 167–90.

Fox, J.L., Sinha, S.P., Chundawat, R.S. and Das, P.K. (1991). Status of the snow leopard *Panthera unica* in Northwest India. *Biol. Conserv.*, **55**, 283–98.

Fraser, D. and Thomas, E.R. (1982) Moose–vehicle accidents in Ontario, relation to highway salt. *Wildl. Soc. Bull.*, **10**, 261–5.

Fraser, J.D., Frenzel, L.D. and Mathison, J.E. (1985) The impact of human activities on breeding bald eagles in north-central Minnesota. *J. Wildl. Manage.*, **49**, 585–92.

Freddy, D.J., Bronaugh, W.M. and Fowler, M.C. (1986) Responses of mule deer to disturbance by persons afoot and snowmobiles. *Wildl. Soc. Bull.*, **14**, 63–8.

Frederick, F.H. and Henderson, J.M. (1970) Impact force measurement using preloaded transducers. *Am. J. Vet. Res.*, **31**, 2279–83.

Frederiksen, P. (1976) Turistslitge I et Klitlandskab Skallingen 1976. *Geogr. Tidsskr.*, **76**, 68–77.

Freitag, D.R. (1971) Methods of measuring soil compaction, in *Compaction in Agricultural Soils* (eds K.K. Barnes *et al.*), American Society of Agricultural Engineers, Michigan, pp. 45–103.

Frenkel, R.E. (1972) Trampled vegetation and floristic convergence in the tropics. *Association of Pacific Coast Geographers Yearbook*, **34**, 87–98.

Gaddy, L.L. and Kohlsaat, T.L. (1987) Recreational impact on the natural vegetation avitauna and herpetotauna of four south Carolina Barrier Islands, USA. *Nat. Areas J.*, **7**, 55–64.

Gadgil, M. and Solbrig, O.T. (1972) The concept of r- and K- selection: evidence from wild flowers and some theoretical considerations. *Am. Natural.*, **106**, 14–31.

Garland, G.G. (1990) Technique for assessing erosion risk from mountain footpaths. *Environ. Manage.*, **14**, 793–8.

Garton, E.O., Bowen, C.W. and Foin, T.C. (1977) The impact of visitors on small mammal communities of Yosemite National Park, in *Visitor Impacts on National Parks: The Yosemite Ecological Impact Study* (ed. T.C. Foin Jr), Vol. 10, Institute of Ecology, University of California, Davis, pp. 44–50.

Geist, V. (1959) Diurnal activity of moose. *Mem. Soc. pro Fauna et Flora Fennica*, **35**, 95–100.

Geist, V. (1963) On the behaviour of the North American moose (*Alces Alces Andersoni* Peterson 1950) in British Columbia. *Behaviour*, **20**, 377–416.

Geist, V. (1971a) Is big game harrassment harmful? *Oil Week.*, **22**, 12–13.

Geist, V. (1971b) Bighorn Sheep Biology. *Wildl. Soc. News*, **136**, 61.

Geist, V. (1971c) A behavioural approach to the management of wild ungulates, in *The Scientific Management of Animal and Plant Communities* (eds E. Duffey and A.S. Watt), Blackwell, Oxford, pp. 413–24.

Geist, V. (1972) On the management of large mammals in national parks. *Park News*, **8**, 16–24.

Geist, V. (1975) *Harrassment of Large Mammals and Birds*, Report to the Berger Commission. Faculty of Environmental Design, University of Calgary, p. 62.

Geist, V. (1978) Behaviour, in *Big Game of North America: Ecology and Management* (eds J.L. Schmidt and D.L. Gilbert), Stackpole Books, Harrisburg, PA, pp. 283–96.

Geldreich, E.E. (1972) Waterborne pathogens, in *Water Pollution Microbiology* (ed. R. Mitchel), Wiley Interscience, New York, pp. 207–41.

Gendebian, J.F. and Mövzer Bruijns, M.F. Recreatiegeroeligheid van vogles *De Veverde Natuur.*, **73**, 85–8.

Georgii, B. and Schröder, W. (1978) Radiolele metrisch geniessene. Aktioitât weiblichen Rotwildes) (*Cervus eliphas* L). *Jagdwiss.*, **24**, 9–23.

Gese, E.M., Rongstad, O.J. and Mytton, W.R. (1989) Changes in coyote movements due to military activity. *J. Wildl. Manage.*, **53**, 334–9.

Ghazanshahi, J., Huchel, T.D. and Devinny, J.S. (1983) Alteration of southern California rocky shore ecosystems by public recreational use. *J. Environ. Manage.*, **16**, 379–94.

Gilbertson, D. (1983) The impacts of off-road vehicles in the Coorong Dune and Lake Complex of South Australia, in *Environmental Effects of Off-road Vehicles Impacts and Management in Arid Regions* (eds R.H. Webb and H.G. Wilshire), Springer-Verlag, New York, pp. 355–74.

Gill, J.A., Sutherland, W.J. and Watkinson, A.R. (1996) A method to quantify the effects of human disturbance on animal populations. *J. Appl. Ecol.*, **33**, 786–92.

Gillett, W.H., Hayward, J.L. Jr and Stout, J.F. (1975) Effects of human activity on egg and chick mortality in a glaucus-winged gull colony. *Condor*, **77**, 492–5.

Gillham, M.E. (1956) Ecology of the Pembrokeshire Islands. IV. Effects of treading and burrowing by birds and mammals. *J. Ecol.*, **44**, 51–82.

Gimingham, C.H. (1972) *Ecology of Heathlands*, Chapman & Hall, London.

Gittings, R. (1969) The application of ordination techniques, in *Ecological Aspects of Mineral Nutrition of Plants* (ed. I. Rorison), Blackwell, Oxford, pp. 37–66.

Gleson, H.A. (1926) The individualistic concept of plant association. *Bull. Torrey Bot. Club*, **53**, 7–26.

Godfrey, P.J. (1975) *The ecological effects of off-road vehicles in Cape Cod National Seashore, Massachusetts (Phase II)*, University of Massachusetts–National Park Service Cooperative Research Unit, Report No. 18.

Godfrey, P.J., Leatherman, S.P. and Buckley, P.A. (1978) *Impact of off-road vehicles on coastal ecosystems*. Proceedings of a Symposium on Coastal Zones 1978, pp. 581–600.

Goeden, G. (1982) Intensive fishing and a 'keystone' predator species: ingredients for community instability. *Biol. Conserv.*, **22**, 273–81.

Goeschl, J.D., Rappaport, L. and Pratt, H.K. (1966) Ethylene as a factor regulating the growth of pea epicotyls subject to physical stress. *Plant Physiol.*, **41**, 877–84.

Goldsmith, F.B. and O'Connor, F.B. (1975) *Ivinghoe Beacon: Experimental Restoration Project*. Report to the Countryside Commission, Cheltenham.

Goldsmith, F.B., Munton, R.J.C. and Warren, A. (1970) The impact of recreation on the ecology and amenity of semi-natural areas: methods of investigation used in the Isles of Scilly. *Biol. J. Linn. Soc.*, **2**, 287–306.

Gompel, J. van (1979) The occurrence of the short-eared owl *Asio flammeus* on the Belgian coast. *Gerfaut.*, **69**, 83–110.

González, L.M., Bustamante, J. and Hiraldo, F. (1992) Nesting habitat selection by the Spanish imperial eagle *Aquila adalberti*. *Biol. Conserv.*, **59**, 45–50.

Goodrich, J.M. and Berger, J. (1994) Winter recreation and hibernating black bears *Ursus americanus*. *Biol. Conserv.*, **67**, 105–10.

Goryshina, T.K. (1983) The effect of trampling caused by recreation on the internal structure of the leaf and thallus of some plants. *Ekologiya*, **2**, 11–18.

Goss, M.J. and Drew, M.C. (1972) *Effect of mechanical impedance on growth of seedlings*, Agricultural Research Council Letcombe Laboratory Annual Report 1971, 35–42.

Graber, D. and White, M. (1978) Management of black bears and humans in Yosemite National Park. *California and Nevada Wildlife*, 1978, 42–51.

Grabherr, G. (1985) Damage to vegetation by recreation in the Austrian and German Alps, in *The Ecological Impact of Outdoor Recreation on Mountain Areas in Europe and North America* (eds N. Bayfield and G.C. Barrow), Recreation Ecology Research Group Report 9, Wye College, Ashford, England, pp. 74–91.

Grabherr, G., Mair, A. and Stimpfl, H. (1987) Wachstums – und reproduktionsstrategien von hochgebirgspflanzen and ihre bedeutung für die begrünung von schipisten and anderen hochalpinen erosionsflächen. *Verhandlungen der Gesellschaft für Ökologie (Graz 1985)*, **XV**, 1987.

Grabherr, G., Mair, A. and Stimpfl, H. (1988) Vegetation dynamics in alpine meadows and chances of revegetation of skiruns and other eroded surfaces in high alpine regions, in *Ingenicurbiologie - Erosionsbekämpfung im Hochgebirge*, Jarbuch 3 der Gesellschaft für Ingenieurbiologie 1988, 94–113.

Grable, A.R. (1971) Effects of compaction on content and transmissions of air in soils, in *Compaction of Agricultural Soils* (ed. K.K. Barnes *et al.*), American Society of Agricultural Engineers, Michigan, pp. 154–64.

Grable, A.R. and Siemer, E.G. (1968) Effects of bulk density, aggregate size, and soil water suction on oxygen diffusion, redox potentials and elongation of corn roots. *Proc. Soil Sci. Soc. Am.*, **32**, 180–6.

Grace, E.S. (1976) Interactions between men and wolves at an Arctic outpost on Ellesmere Island. *Can. Field Nat.*, **90**, 149–56.

Grace, J. (1977) *Plant Response to Wind*, Academic Press, London.

Grace, J. and Russell, G. (1982) The effect of wind and a reduced supply of water on the growth and water relations to *Festuca arundinacea* Schrub. *Ann. Bot.*, **49**, 217–25.

Grace, J., Pitcairn, C.E.R., Russell, G. and Dixon, M. (1982) The effects of shaking on the growth and water relations of *Festuca arundinacea* Schrub. *Ann. Bot.*, **49**, 207–15.

Grant, S.A. and Hunter, R.F. (1966) The effects of frequency and season of clipping on the morphology, productivity and chemical composition of *Calluna vulgaris* (L) Hull. *New Phytol.*, **65**, 125–33.

Grau, G.A. and Grau, B.L. (1980) Effects of hunting on hunter effort and white-tailed deer (*Odocoileus virginiatus*) behaviour. *Ohio J. Sci.*, **80**, 150–5.

Gref, R. and Ericsson, A. (1984) Wound-induced changes of resin acid concentrations in living bark of Scots Pine seedlings. *Can. J. For. Res.*, **15**, 92–6.

Greller, A.M., Goldstein, M. and Marcus, L. (1974) Snowmobile impact on three alpine tundra plant communities. *Environ. Conserv.*, **1**, 101–10.

Grier, J.W. (1969) Bald eagle behaviour and productivity responses to climbing to nests. *J. Wildl. Manage.*, **33**, 961–6.

Grime, J.P. (1973) Control of species density in herbaceous vegetation. *J. Environ. Manage.*, **1**, 151–67.

Grime, J.P. (1974) Vegetation classification by reference to strategies. *Nature*, **250**, 26–31.

Grime, J.P. (1977) Evidence for the existence of three primary strategies in plants and its relevance to ecological and evolutionary theory. *Am. Natural.*, **111**, 1169–94.

Grime, P.J. (1979) *Plant Strategies and Vegetation Processes*, Wiley, Chichester.

Grime, J.P., Hodgson, J.G. and Hunt, R. (1988) *Comparative Plant Ecology*, Unwin Hyman, London.

Grubb, M.M. (1978) Effects of increased noise levels on nesting herons and egrets. *Proc. Colonial Waterbird Group*, 1978, 49–54.

Grubb, T.G. (1976) Nesting bald eagle attacks researcher. *Auk*, **93**, 842–3.

Grubb, T.G. and King, R.M. (1991) Assessing human disturbance of breeding bald eagles with classification tree models. *J. Wildl. Manage.*, **55**, 500–11.

Haesler, V. (1989) The situation of the invertebrate fauna of coastal dunes and sandy coasts in the western Mediterranean (France, Spain), in *Perspectives in Coastal Dune Management* (ed. F. van der Meulen, P.D. Jungerus and J.H. Visser), S.P.B. Academic Publishing, The Hague, 125–31.

Hale, M.E. Jr (1983) *The Biology of Lichens*, Edward Arnold, London.

Hall, C.N. and Kuss, F.R. (1989) Vegetation alteration along trails in Shenandoah National Park, Virginia. *Biol. Conserv.*, **48**, 211–27.

Hall, T.D., Weinman, H., Murray, S.M. and Weinbrenn, C. (1948) *Experiments with* Cynodon dactylon *and other species at the South African Turf Research Station*, African Explosives and Chemical Industries Ltd and South African Turf Research Fund, Johannesburg.

Halle, S. (1988) Die Säugetierfauna in einem jungen rekultivierungsgebiet des rheinischen Braunkohlenreniers. *Zeitschrift für Angewandte Zoologie*, **75**, 421–8.

Halme, E. and Niemelä, J. (1993) Carabid beetles in fragments of coniferous forest. *Ann. Zool. Fennici*, **30**, 17–30.

Hammitt, W.E. and Cole, D.N. (1987) *Wildland Recreation: Ecology and Management*, Wiley, New York.

Hand, J.C. (1980) Human disturbance in western gull, *Larus occidentalis* Ivens, colonies and possible amplification by intraspecific predation. *Biol. Conserv.*, **18**, 59–63.

Hanes, N.B. and Fossa, A.J. (1970) *A quantitative analysis of the effects of bathers on recreational water quality.* Proceedings of the 5th International Conference on Water Pollution Research, HA-9/1 to HA-9/9.

Hansen, C.G. (1971) Overpopulation as a factor in reducing desert bighorn populations. *Desert Bighorn Council Trans.*, **15**, 46–52.

Harder, R., Schumacher, W., Firbas, F. and Denffer, D. (1965) *Strasburger's Text Book of Botany*, New English Edition (eds P. Bell and D. Coombe), Longmans, London.

Harper, F.C., Warlow, W.J. and Clarke, B.L. (1961) *The forces applied to the floor by the foot in walking. I Walking on a level surface*, Research Paper 32, National Building Studies, HMSO, London.

Harris, M.P. and Norman, F.I. (1981) The distribution and status of coastal colonies of seabirds in Victoria, Australia. *Natural History Museum, Victoria,* **42**, (part 2), 89–106.

Harris, W.L. (1971) The soil compaction process, in *Compaction of Agricultural Soils* (eds K.K. Barnes *et al.*), American Society of Agricultural Engineers, Michigan, pp. 9–44.

Harris (1973) Unsourced photocopy in the author's collection.

Harrison, D.J. and Gilbert, J.R. (1985) Denning ecology and movements of coyotes in Maine during pup rearing, *J. Mammal.,* **66**, 712–19.

Hartman, D.S. (1979) *Ecology and Behaviour of the Manatee (*Trichechus manatus*) in Florida,* The American Society of Mammalogists, Special Publication No. 5.

Haslam, S.M. (1978) *River Plants: The Macrophyte Vegetation of Watercourses,* Cambridge University Press, Cambridge.

Haslam, S.M. (1987). *River Plants of Western Europe,* Cambridge University Press, Cambridge.

Hawker, D.W. and Connell, D.W. (1989) An evaluation of the tolerance of corals to nutrients and related water quality characteristics. *Intern. J. Environ. Studies,* **34**, 179.

Hawker, D.W. and Connell, D.W. (1991). Standards and criteria for pollution control in coral reef areas, in *Pollution in Tropical Aquatic Systems* (eds D.W. Connell and D.W. Hawker), CRC Press, Ann Arbor, pp. 169–91.

Hawkins, J.P. and Roberts, C.M. (1992) Effects of recreational scuba diving on fore-reef slope communities of coral reefs. *Biol. Conserv.,* **62**, 171–8.

Hawkins, J.P. and Roberts, C.M. (1993) Effects of recreational scuba diving on coral reefs: trampling on reef flat communities. *J. Appl. Ecol.,* **30**, 25–30.

Hawks, H.A. (1978) River bed animals tell tales of pollution, in *Biosurveillance of River Water Quality* (collators J.G. Hughes and H.A. Hawkes), Proceedings of section K of the British Association for the Advancement of Science, Aston, 1997, pp. 55–77.

Hay, K.A. (1985) Status of the humpback whale, *Megaptera novaeangliae,* in Canada. *Can. Field Nat.,* **99**, 425–32.

Heiligenberg, van den T. (1987) Effects of mechanical and manual harvesting of Lugworms *Arenicola marina* L. on the benthic fauna of tidal flats in the Dutch Wadden Sea. *Biol. Conserv.,* **39**, 165–77.

Helgath, S.F. (1975) *Trail deterioration in the Selway-Bitterroot Wilderness,* USDA Forest Service Research Note INT-193, Ogden, Utah.

Hell, von P. and Bevilaqua, F. (1988) Coexistence of man with brown bear (*Ursus arctos*) in Western Carpathians. *Z. Jagdwiss,* **34**, 153–63.

Hellgren, E.C. and Vaughan, M.R. (1989) Demographic analysis of a black bear population in the Great Dismal Swamp. *J. Wildl. Manage.,* **53**, 969–77.

Hendreck, K.A. (1938) *Earthworms for Gardeners and Fishermen,* CSIRO, Division of Soils, Australia.

Hendry, G.S. and Toth, A. (1982) Some effects of land use on bacteriological water quality in a recreational lake. *Water Res.,* **16**, 105–12.

Herman, L.M. and Tavolga, W.N. (1980) The communication systems of cetacians, in *Cetacean Behaviour: Mechanisms and Functions* (ed. L.M. Herman), J. Wiley, New York, pp. 149–209.

Herrero, S. (1976) Conflicts between man and grizzly bears in the national parks of North America. *Int. Conf. Bear Res. and Manage.*, 3, 121–45.

Herrero, S. (1980) Social behaviour of black bears at a garbage dump in Jasper National Park. *Int. Conf. Bear Res. and Manage.*, 5, 54–70.

Heuchert, J.C., Marks, J.S. and Mitchell, C.A. (1983) Strengthening of tomato shoots by gyratory shaking. *J. Am. Soc. Hort. Sci.*, 108, 801–5.

Hicks, L.L. and Elder, J.M. (1979) Human disturbance of Sierra Nevada Bighorn Sheep. *J. Wildl. Manage.*, 43, 909–15.

Hill, J.N.S. and Sumner, M.E. (1967) Effect of bulk density on moisture characteristics of soils. *Soil Sci.*, 103, 234–8.

Hill, M.O. (1973) Diversity and evenness: a unifying notation and its consequences. *Ecology*, 54, 427–32.

Hill, M.O. (1979) *TWINSPAN – a FORTRAN Programme for Arranging Multivariate Data in an Ordered Two-way Table by Classification of the Individuals and Attributes*, Cornell University, Ithaca, New York.

Hill, M.O., Bunce, R.G.M. and Shaw, M.W. (1975) Indicator Species Analysis, a divisive polythetic method of classification, and its application to a survey of native pine woods in Scotland. *J. Ecol.*, 63, 597–613.

Hinkley, B.S., Iverson, R.M. and Hallet, B. (1983) Accelerated water erosion in ORV-use areas, in *Environmental Effects of Off-road Vehicles* (eds R.M. Webb and H.G. Wilshire), Springer-Verlag, New York, pp. 81–96.

Hiraki, Y. and Ota, Y. (1975) The relationship between growth inhibition and ethylene production by mechanical stimulation in *Lilium longiflorum*. *Plant and Cell Physiol.*, 16, 185–9.

Hjulstrom, F. (1935) Studies of the morphological activity of rivers as illustrated by the River Fyries. *Bull. Geological Institute, Uppsala University*, 25, 221–527.

Holling, C.S. (1973) Resilience and stability of ecological systems. *Ann. Rev. Ecol. Syst.*, 4, 1–24.

Holmes, D.O. and Dobson, H.E.M. (1976) *Ecological Carrying Capacity Research: Yosemite National Park, Part I. The Effects of Human Trampling and Urine on Subalpine Vegetation, and Survey of Past and Present Backcountry Use*, California University, Berkeley.

Holroyd, J.C. (1967) Observations of Rocky Mountain goats on Mount Wardle, Kooteney National Park. *Can. Field Nat.*, 81, 1–78.

Hood, R.E. and Inglis, J.M. (1974) Behavioural responses of white-tailed deer to intensive ranching operations. *J. Wildl. Manage.*, 38, 488–98.

Hopkins, C.L. (1965) Feeding relationships in a mixed population of freshwater fish. *N.Z. J. Sci.*, 8, 149–57.

Horikawa, Y. and Miyawaki, A. (1954) *Habitat Segregation of the Weeds as an Indicator of the Soil Hardness*, Science Report of Yokohama, National University, Section 2, No. 3, Yokohama.

Hosier, P.E. and Eaton, T.E. (1980) The impact of vehicles on dune and grassland vegetation on a south-eastern North Carolina barrier beach, *J. Appl. Ecol.*, 17, 173–82.

Hosier, P.E., Kockhar, M. and Thayer, V. (1981) Off-road vehicle and pedestrian effects on the sea approach of hatchling Loggerhead Turtles. *Environ. Conserv.*, 8, 158–61.

Howe, M.A., Geissler, P.H. and Harrington, B.A. (1989) Population trends of North American shorebirds based on the International Shorebird Survey. *Biol. Conserv.*, **49**, 185–99.

Howell, R. (1985) The effect of bait digging on the bioavailability of heavy metals from superficial intertidal marine sediments. *Mar. Pollut. Bull.*, **16**, 292–5.

Hua, Y., Zhao, Q. and Zhang, G. (1989) The habitat and behaviour of *Lipotes vexillifer*, in *Biology and Conservation of River Dolphins* (eds W.F. Perrin, R.L. Brownell Jr, Zhou Kaiya and Liu Jiankang), Occ. Pap. IUCN Species Survival Comm. No. 3, pp. 92–8.

Hubbell, D.S. and Gardner, J.L. (1948) Effects of aeration, compaction, and water-logging on soil structure and microflora. *J. Am. Soc. Agron.*, **40**, 832–40.

Hudson, N.W. (1971) *Soil Conservation*, Batsford, London.

Hulbert, I.A.R. (1990) The response of ruddy shelduck *Tadorna ferruginea* to tourist activity in the Royal Chitwan National Park of Nepal. *Biol. Conserv.* **52**, 113–23.

Hulsman, K. (1984) *Seabirds of the Capricornia Section, Great Barrier Reef Marine Park.* Royal Society of Queensland Symposium, pp. 53–60.

Hume, R.A. (1972) Reactions of goldeneyes to boating. *Brit. Birds.*, **69**, 178–9.

Humphrey, S.R. (1978) Status, winter habitat and management of the endangered Indiana bat, *Myotis sodalis. Florida Scientist*, **41**, 65–76.

Hunt, G.L. (1972) Influence of food distribution and human disturbance on the reproductive success of herring gulls. *Ecology*, **53**, 1051–61.

Huxley, T. (1970) *Footpaths in the Countryside*, Countryside Commission for Scotland, Perth.

Hylgaard, T. (1982) Recovery of plant communities on coastal sand dunes disturbed by human trampling. *Biol. Conserv.*, **19**, 15–25.

Hylgaard, T. and Liddle, M.J. (1981) The effect of human trampling on a sand dune ecosystem dominated by *Empetrum nigrum. J. Appl. Ecol.*, **18**, 559–69.

Hynes, H.B.N. (1960) *The Biology of Polluted Waters*, Liverpool University Press, Liverpool.

Ingelog, T., Olsson, M.T. and Bodvarsson, H. (1977) *Effects of long-term trampling and vehicle driving on soil, vegetation and certain soil animals of an old Scots pine stand*, Research Notes. No. 27, Royal College of Forestry, Uppsala.

Irby, L.R., Swenson, S.E. and Stewart, S.T. (1989) Two views of the impacts of poaching on bighorn sheep in the Upper Yellowstone Valley, Montana, USA. *Biol. Conserv.*, **47**, 259–72.

IUCN (1990) *1990 IUCN Red List of Threatened Animals*, International Union for Conservation of Nature and Natural Resources, Gland, Switzerland.

Iversen, F.M. (1986) The impact of disturbance on the lapwing's *Vanellus vanellus* incubation. *Danish Ornithol. Forum Tidsskr.* **80**, 97–102.

Jaakson, R. (1988) River recreation boating impacts. *J. Waterway, Port, Coastal, and Ocean Engin.*, **114**, 363–7.

Jackivicz, T.P. and Kuzminski, L.N. (1973a) The effects of the interaction of

outboard motors with the aquatic environment – a review. *Environ. Res.*,
6, 436–54.

Jackivicz, T.P. and Kuzminski, L.N. (1973b) A review of outboard motor
effects on the aquatic environment. *J. Wat. Pollut. Control Fed.*, **45**,
1759–70.

Jackson, M.J. and James, R. (1979) The influence of bait digging on cockle
Cerastodesma edule populations in North Norfolk. *J. Appl. Ecol.*, **16**,
667–79.

Jacobsen, T. and Kushlan, J.A. (1986) Alligators in natural areas, choosing
conservation policies consistent with local objectives. *Biol. Conserv.*, **36**,
181–96.

Jaffe, M.J. (1973) Thigmomorphogenesis: the response of plant growth and
development to mechanical stimulation. *Planta* (Berlin), **114**, 143–57.

Jaffe, M.J. and Biro, R. (1979) Thigmomorphogenesis: the effect of mechanical perturbation on the growth of plants with special reference to anatomical changes, the role of ethylene, and interaction with other environmental
stresses, in *Stress Physiology of Crop Plants* (eds H. Mussell and R. Staples),
Wiley, New York, pp. 25–59.

Jeppesen, J.L. (1984) Human disturbance of roe deer and red deer:
preliminary results, in *Multiple-use Forestry in the Scandinavian countries*
(eds O. Saastamoiren, S.G. Hultman, N. Elerstioch and L. Mattisson),
Communication Institute Foresta, pp. 113–18.

Jeppesen, J.L. (1987) Impact of human disturbance on home range movements and activity of red deer *Cervus elaphus* in a Danish environment.
Danish Rev. Game Biol., **13**, 1–38.

Jim, C.Y. (1987) Trampling impacts of recreationists on picnic sites in a Hong
Kong country park. *Environ. Conserv.*, **14**, 117–27.

Johnson C.D., Beley, J.R., Ditsworth, T.M. and Butt, S.M. (1983) Secondary
succession of arthropods and plants in the Arizona Sonoran Desert,
in response to transmission line construction. *J. Environ. Manage.*, **16**,
125–37.

Johnson, I.R. (1983) Nitrate uptake and respiration in roots and shoots: a
model. *Physiol. Plant.*, **58**, 145–7.

Jones, F.G.W., Dunning, R.A. and Humphries, K.P. (1955) The effects of
defoliation and loss of stand upon yield of sugar beet. *Ann. Appl. Biol.*,
43, 63–70.

Jonkel, C. and Servheen, C. (1977) *Bears and People. Western Wildlands*,
Vol. 4, Montana Forest and Conservation Experiment Station, Missoula,
pp. 22–5.

Jónsdóttir, I.S. (1991) Effects of grazing on tiller size and population dynamics
in a clonal sedge (*Carex bigelowii*). *Oikos*, **62**, 177–88.

Jurko, A. (1983) Trampling effects on species diversity and leaf characteristics of vegetation in the High Tatras Mountains. *Ekologia* (CSSR), **2**,
281–93.

Kailola, P.J. (1990) Translocated and exotic fishes: towards a cooperative role
for industry and government, in *Introduced and Translocated Fishes and
their Ecological Effects* (ed. D.A. Pollard), Bureau of Rural Resources
Proceedings No. 8, Australian Government Printing Service, Canberra, pp.
31–7.

Kammermeyer, K.E. and Marchington, R.L. (1977) Seasonal change in circadian activity of radio monitored deer. *J. Wildl. Manage.*, **41**, 315–17.

Kar, S. and Ghildyal, B.P. (1975) Rice root growth in relation to size, quantity and rigidity of pores. *Plant Soil*, **43**, 627–37.

Kardill, L. (1974) *Damage to the vegetation caused by orienteering*, Research Notes 4, Stockholm Royal College of Forestry, Stockholm.

Kay, A.M. and Liddle, M.J. (1984) *Tourist Impact on Reef Corals*, 1984 report to the Great Barrier Reef Marine Park Authority, Townsville, Queensland.

Kay, A.M. and Liddle, M.J. (1987) Resistance, survival and recovery of trampled corals on the Great Barrier Reef. *Biol. Conserv.*, **42**, 1–18.

Kay, A.M. and Liddle, M.J. (1989) Impact of human trampling in different zones of a coral reef flat. *Environ. Manage.*, **4**, 509–20.

Kay, D. and Wyer, M. (1994) Making waves: recreational water quality. *Biologist*, **41**, 17–20.

Kayanja, F.I.B. (1984) Conservation of African mammals in the aftermath of commercial poaching. *Acta Zool. Fennica.*, **172**, 195–6.

Keating, K.A. (1986) Historical grizzly bear *Ursus arctos*. Trends in Glacier National Park, Montana, USA. *Wildl. Soc. Bull.*, **14**, 83–7.

Keller, V. (1989) Variations in the response of great crested grebes *Podiceps cristatus* to human disturbance – a sign of adaptation? *Biol. Conserv.*, **49**, 31–45.

Keller, V.E. (1991) Effects of human disturbance on eider ducklings *Somateria mollissima* in an estuarine habitat in Scotland. *Biol. Conserv.*, **58**, 213–28.

Kellomaki, S. (1973) Ground cover response to trampling in a spruce stand of *Myrtillus* type. *Silva Fennica*, **7**, 96–113.

Kellomaki, S. and Saastamoinen, V.L. (1975) Trampling tolerance of forest vegetation. *Acta Forest. Fennica*, **147**, 5–19.

Kemper, W.D., Stewart, B.A. and Porter, L.K. (1971) Effects of compaction on soil nutrient status, in *Compaction of Agricultural Soils* (eds K.K. Barnes *et al.*), American Society of Agricultural Engineers, Michigan, pp. 178–89.

Kendal, P.C. McM. (1982) The effect of season and/or duration of trampling on vegetation in an open forest of South East Queensland. Master of Science Thesis, Griffith University, Brisbane.

Ketchledge, E.H. and Leonard, R.E. (1970) The impact of man on the Adirondack High Country. *The Conservationist*, **25**, 14–18.

Ketchledge, E.H., Leonard, R.E., Richards, N.A. *et al.* (1985) *Rehabilitation of Alpine Vegetation in the Adirondack Mountains of New York State*, USDA Forest Service, Northeastern Forest Experiment Station Research Paper, NE-553, Broomall, Pennsylvania.

Kevan, P.G. (1971) *Vehicle tracks on high arctic tundra: an all year case history around Hazen Camp, Ellesmere Island, NWT*, Defence Research Board of Canada, Ottawa.

Keymer, R. and Ellis, W.M. (1978) Experimental studies on plants of *Lotus Corniculatus* L. from Anglesey polymorphic for cyanogensis. *Heredity*, **40**, 189–206.

King, J.G. and Mace, A.C. Jr (1974) Effects of recreation on water quality. *J. Wat. Pollut. Control Fed.*, **46**, 2453–9.

Kirkby, M.J. (1980) The problem, in *Soil Erosion* (eds M.J. Kirkby and R.C.P. Morgan), Wiley, Chichester, pp. 1–16.

Klecka, A. (1937) Vliv seslapavani na asociaci travnatych porostu. *Sbornik Coskoslovenske Akademie Zemedelske*, **12**, 715–24.

Knapp, P.A. (1992) Secondary plant succession and vegetation recovery in two western Great Basin Desert ghost towns. *Biol. Conserv.*, **60**, 81–9.

Knight, R.R. and Blanchard, B.M. (1995) *Yellowstone Grizzly Bear Investigations*, Report of the Interagency Study Team 1994, USDI, National Biological Service, Wyoming.

Kocasoy, G. (1989) A method for prediction of extent of microbial pollution of seawater and carrying capacity of beaches. *Environ. Manage.*, **13**, 469–75.

Kochman, H.I., Rathbun, G.B. and Powell, J.A. (1985) Temporal and spatial distribution of manatees in Kings Bay, Crystal River, Florida. *J. Wildl. Manage.*, **49**, 921–4.

Kolattukudy, P.E. (1978) Chemistry and biochemistry of the aliphatic components of suberin, in *Biochemistry of Wounded Plant Tissues* (ed. G. Kahl), Walter de Gruyter, Berlin, pp. 43–84.

Kolstad, M., Mysterud, I., Kvam, T. *et al.* (1986) Status of the brown bear in Norway: Distribution and population 1978–82. *Biol. Conserv.*, **38**, 79–99.

Kopischke, E.D. (1972) Effects of snowmobile activity on the distribution of white-tailed deer in south-central Minnesota. *Minnesota Department of Natural Resources, Division of Game and Fish, Game Research, Quarterly Report*, **32**, (3), 139–46.

Kostrowicki, A.S. (1970) Zastasowanie metod geobotanicznych Wocenie phzyda tnosci Ferenu dla potrzeb rekreacji i Wypoczynku. *Prezeglecl Geograticzny Warszawa*, **43**, 631–45.

Krausman, P.R., Leopold, B.D. and Scarbrough, D.L. (1986) Desert mule deer response to aircraft. *Wildl. Soc. Bull.*, **14**, 68–70.

Krebs, C.J. (1994) *Ecology: The Experimental Analysis of Distribution and Abundance*, 4th edn, Harper and Row, New York.

Kuhnelt, W. and Walker, N. (1976) *Soil Biology with Special Reference to the Animal Kingdom*, Faber & Faber, London.

Kury, C.R. and Gochfeld, M. (1975) Human interference and gull predation in cormorant colonies. *Biol. Conserv.*, **8**, 23–34.

Kuss, F.R. (1983) Hiking boot impact on woodland trails. *J. Soil Water Conserv.*, **38**, 119–21.

Kuss, F.R. (1986) *The effect of two hiking intensities on woodland trail wear*. Proceedings of National Wilderness Research Conference; Current Research, Intermountain Research Station, General Technical Report INT-2l2, Ogden, Utah.

Kuss, F.R., Graefe, A.R. and Vaske, J.J. (1990) *Visitor Impact Management*: A Review of Research. Report to the National Parks and Conservation Association, Washington, DC.

Kuss, F.R. and Hall, C.N. (1989) Vegetation alteration along trails in Shenandoah National Park, Virginia. *Biol. Conserv.*, **48**, 211–27.

Kuss, F.R. and Hall, C.N. (1991) Ground flora trampling studies: five years after closure. *Environ. Manage.* **15**, 715–27.

Kuss, F.R. and Jenkins, W.A. (1985) *Effects of footgear design on trail wear: a summary of five years of research*. Proceedings of the southeastern Recreation Research Conference, 1984 (ed. L.M. Anderson), Asheville, North Carolina, pp. 39-49.

Kuss, F.R. and Morgan, J.M. (1980) Estimating the physical carrying capacity of recreation areas: a rationale for application of the universal soil loss equation. *J. Soil Water Conserv.*, **35**, 87–9.

Kuss, F.R. and Morgan, J.M. (1984) Using the USLE to estimate the physical carrying capacity of natural areas for outdoor recreation planning. *J. Soil Water Conserv.*, **39**, 383–7.

Kuss, F.R. and Morgan, J.M. (1986) A first alternative for estimating the physical carrying capacities of natural areas for recreation. *Environ. Manage.*, **10**, 255–62.

Lagler, K.F., Hazzard, A.S., Hazen, W.E. and Tomkins, W.A. (1950) Outboard motors in relation to fish behaviour, fish production and angling success. *Trans. N. Am. Wildl. Conf.*, **15**, 280–303.

Lagocki, H.F.R. (1978) Surface soil stability as defined and controlled by a drainage criteria, in *Modification of Soil Structure* (eds W.W. Emerson, R.D. Bond and A.R. Dexter), Wiley, New York, pp. 233–7.

Lamb, T.W. (1960) *Compacted Clay*, American Society of Civil Engineers Transactions 125, Parts I and II.

Lambert, J.D.H. (1972) *Botanical Changes Resulting from Seismic and Drilling Operations, Mackenzie Delta Area*, Arctic Land Use Research Program 1971/1972, Carleton University, Ottawa.

Lamont, B.B., Rees, R.G., Witkowski, E.T.F. and Whitten, V.A. (1994) Comparative size, fecundity and ecophysiology of roadside plants of *Banksia hookeriana*. *J. Appl. Ecol.*, **31**, 137–44.

Lance, A.N., Baugh, I.D. and Love, J.A. (1989) Continued footpath widening in the Cairngorm Mountains, Scotland. *Biol. Conserv.*, **49**, 201–14.

Langdon, J.S. (1989) *Prevention and Control of Fish Diseases in the Murray–Darling Basin*. Proceedings of the Workshop on Native Fish Management, Murray–Darling Commission, Canberra.

La Page, W.F. (1962) Recreation and the forest site. *J. Forest.*, **60**, 319–21.

La Page, W.F. (1967) *Some Observations on Campground Trampling and Ground Cover Response*, USDA, Forest Service, Research Paper NE68, Syracuse, New York.

Larcher, X.X. (1973) *Okologie der Pflanzen*, Ulmer, Stuttgart.

Larson, P.R. (1965) Stem form of young *Larix* as influenced by wind and pruning. *Forest Sci.*, **II**, 412–24.

Law, R. (1975) Colonisation and the evolution of life histories in *Poa annua*. Ph.D. Thesis, University of Liverpool.

Law, R., Bradshaw, A.D. and Putwain, P.D. (1977) Life history variation in *Poa annua*. *Evolution*, **31**, 233–46.

Layser, E.F. and Burke, T.E. (1973) Northern bog lemming and its unique habitat in north eastern Washington. *Murrelet*, **54**, 7–8.

Leader-Williams, N., Albon, S.D. and Berry, P.S.M. (1990) Illegal exploitation of black rhinoceros and elephant populations: patterns of decline, law enforcement and patrol effort in Luangwa Valley, Zambia. *J. Appl. Ecol.*, **27**, 1055–87.

Lee, T.H., McGlasson, W.B. and Edwards, R.A. (1970) Physiology of disks of irradiated tomato fruit. I. Influence of cutting and infiltration on respiration, ethylene production and ripening. *Radiation Botany*, **10**, 521–9.

Legg, M.H. and Schneider, G. (1977) Soil deterioration on campsites: northern forest types. *Soil Sci. Soc. Am. J.*, **41**, 437–41.

Leidy, R.A. and Fiedler, P.L. (1985) Human disturbance and patterns of fish species diversity in the San Francisco Bay drainage, California. *Biol. Conserv.*, **33**, 247–67.

Leney, F.M. (1974) The ecological effects of public pressure on picnic sites. Ph.D. Thesis, University of Aberdeen.

Leslie, D.M. Jr and Douglas, C.L. (1980). Human disturbance at water sources of desert bighorn sheep. *Wildl. Soc. Bull.*, **8**, 284–90.

Letey, J. (1961) Aeration compaction and drainage. *California Turfgrass Culture*, **11**, 17–21.

Lewis, D.M. (1986) Disturbance effects on elephant feeding: evidence for compression in Luangwa Valley, Zambia. *Afr. J. Ecol.*, **24**, 227–41.

Liddle, M.J. (1973a). A survey of twelve lakes in the Gwydyr Forest region of Snowdonia, Internal Report to the Nature Conservancy, Caernarvonshire.

Liddle, M.J. (1973b) The effects of trampling and vehicles on natural vegetation. Ph.D. Thesis, University College of North Wales.

Liddle, M.J. (1975a) A selective review of the ecological effects of human trampling on natural ecosystems. *Biol. Conserv.*, **7**, 17–36.

Liddle, M.J. (1975b) A theoretical relationship between the primary productivity of vegetation and its ability to tolerate trampling. *Biol. Conserv.*, **8**, 251–5.

Liddle, M.J. (1984) Classification of the interactions between animals and plants with particular reference to human trampling, in *The Ecological Basis of Interactions between Orgamons* (eds M.J. Liddle and J.C. Tothill), AES Monograph 1/84, School of Australian Environmental Studies, Griffith University, Brisbane, pp. 113–28.

Liddle, M.J. (1987) Impact of recreational use, in *An Island in Suburbia: The National and Social History of Toohey Forest* (eds C.P. Catterall and C.J. Wallace), Institute of Applied Environmental Research, Griffith University, Brisbane, pp. 23–9.

Liddle, M.J. and Chitty, L.D. (1981) The nutrient budget of horse tracks on an English lowland heath. *J. Appl. Ecol.*, **18**, 841–8.

Liddle, M.J. and Elgar, M.A. (1984) Multiple pathways in diaspore dispersal, exemplified by studies of Noogoora Burr (*Xanthium occidentali* Bertol, (*Compositae*)). *Bot. J. Linn. Soc.*, **88**, 303–15.

Liddle, M.J. and Greig-Smith, P.J. (1975a) A survey of tracks and paths in a sand dune ecosystem. I Soils. *J. Appl. Ecol.*, **12**, 893–908.

Liddle, M.J. and Greig-Smith, P.J. (1975b) A survey of tracks and paths in a sand dune ecosystem. II Vegetation. *J. Appl. Ecol.*, **12**, 909–930.

Liddle, M.J. and Kay, A.M. (1987) Resistance, survival and recovery of trampled corals on the Great Barrier Reef. *Biol. Conserv.*, **42**, 1–18.

Liddle, M.J. and Moore, K.G. (1974) The microclimate of sand dune tracks: the relative contribution of vegetation removal and soil compression. *J. Appl. Ecol.*, **11**, 1057–68.

Liddle, M.J. and Scorgie, H.R.A. (1980) The effects of recreation on freshwater plants and animals: a review. *Biol. Conserv.*, **17**, 183–206.

Liddle, M.J. and Thyer, N. (1986) Trampling and fire in a subtropical dry sclerophyll forest. *Environ. Conserv.*, **13**, 33–9.

Liddle, M.J., Budd, C.S.J. and Hutchings, M.J. (1982) Population dynamics and neighbourhood effects in establishing swards of *Festuca rubra. Oikos*, **38**, 52–9.

Liddle, M.J., Happy-Wood, C.M. and Buse, A. (1979) A survey of the biota environment and use for recreation of twelve lakes in Snowdonia. *Biol. J. Linn. Soc.* **11**, 77–101.

Lieb, T.W. and Mossman, A.S. (1974) Elk drowning. *Pacific Northwest Bird and Mammal Soc.*, **55**, 39–40.

Lincoln, R.J., Boxshall, G.A. and Clark, P.F. (1982) *A Dictionary of Ecology, Evolution and Systematics*, Cambridge University Press, Cambridge.

Lindburg, S. and Petterson, S. (1985) Effects of mechanical stress on uptake and distribution of nutrients in barley. *Plant Soil*, **83**, 295–309.

Lintermans, M. (1990) Introduced fish of the Canberra region – recent range expansions, in *Introduced and Translocated Fishes and their Ecological Effects* (ed. D.A. Pollard), Bureau of Rural Resources Proceedings No. 8, Australian Government Publishing Service, Canberra, pp. 50–60.

Little, A.D. (1974) *Betridings-experimenten in een Duinvallei: Effecten op Mesofauna en Vegetatie*, Series B: Biologische aspecten No. 5, Vrij Universiteit, Amsterdam.

Lockyer, C. (1978) The history and behaviour of a solitary wild but sociable bottlenose dolphin (*Tursiops truncatus*) on the west coast of England and Wales. *J. Nat. Hist.*, **12**, 513–28.

Lockyer, C.H. (1993) *Status of the Great Whales*. Abstracts, Sixth International Theriological Congress, University of New South Wales.

Logan, K.A., Irwin, L.L. and Skinner, R. (1986) Characteristics of a hunted mountain lion population in Wyoming. *J. Wildl. Manage.*, **50**, 648–54.

Loggers, C.O., Thévenot, M. and Aulagnier, S. (1992) Status and distribution of Moroccan wild ungulates. *Biol. Conserv.*, **59**, 9–18.

Lonsdale, W.M. and Lane, A.M. (1992) Vehicles as vectors of weed seeds in Kakadu National Park, in plant invasions: The media of environmental weeds in Australia, *Kowari*, **2**, 167–9.

Lovelock, J.E. (1979) *Gaia: A New Look at Life on Earth*, Oxford University Press, Oxford.

Lowry, W.P. (1967) *Weather and Life: An Introduction to Biometeorology*, Academic Press, London.

Lubchenco, J. and Gains, S.D. (1981) A unified approach to marine plant herbivore interactions. I. Populations and Communities. *Ann. Rev. Ecol. Syst.*, **12**, 405–38.

Luckenbach, R.A. and Bury, R.B. (1983) Effects of off-road vehicles on the biota of the Algodones Dunes, Imperial County, California, *J. Appl. Ecol.*, **20**, 265–86.

Lull, H.S. (1959) *Soil Compaction on Forest and Range Lands*, Miscellaneous Publication No. 768, United States Department of Agriculture, Washington, DC.

Lutz, H.J. (1945) Soil conditions of picnic grounds in public forest parks. *J. Forest.*, **43**, 121–7.

Lynch, J.F. and Johnson, N.K. (1974) Turnover and equilibria in insular ovifaunas, with special reference to the California Channel Islands. *Condor*, **76**, 370–84.

Macan, T.T. (1959) *A Guide to Freshwater Invertebrate Animals*, Longman, London.

McArthur, R.H. and McArthur, J.W. (1961) On bird species diversity. *Ecology*, **42**, 594–8.

MacArthur, R.A., Geist, V. and Johnston, R.H. (1982a) Cardiac and behavioural responses of mountain sheep to human disturbance. *J. Wildl. Manage.*, **46**, 351–8.

MacArthur, R.A., Geist, V. and Johnston, R.H. (1982b) Physicological correlates of social behaviour in Bighorn sheep: a field study using electrocardiogram telemetry. *J. Zool. Soc. Lond.*, **196**, 401–15.

McCann, L. (1956) Ecology of the mountain sheep. *Am. Midland Natural.*, **56**, 297–324.

McCullough, D.R. (1982) Behaviour, bears and humans. *Wildl. Soc. Bull.*, **10**, 27–33.

McGlasson, W.B. and Pratt, H.K. (1964) Effects of wounding on respiration and ethylene production by Cantaloupe fruit tissue. *Plant Physiol.*, **39**, 128–32.

MacInnes, C.D. and Misra, R.K. (1972) Predation on Canada goose nests at McConnell River, north-west territories. *J. Wildl. Manage.*, **36**, 414–22.

McIntosh, A.R., Townsend, C.R. and Crowl, T.A. (1992) Competition for space between introduced brown trout (*Salmo trutta L.*) and a native galaxid (*Galaxias vulgaris Stokell*) in a New Zealand stream. *J. Fish. Biol.*, **41**, 63–81.

McIntyre, J. (1986) A louder voice in the wilderness. The eerie call of the loon is returning to many of the nations lakes and rivers. *National Wildl.*, **24**, 46–51.

McKay, R.J. (1989) Exotic and translocated freshwater fishes in Australia, in *Exotic Aquatic Organisms in Asia*. Proceedings of the Workshop on Introduction of Exotic Aquatic Organisms in Asia, Asian Fisheries Society, Manila, Special Publication 3, pp. 21–34.

Mackerron, D.K.L. (1976) Wind damage to the surface of strawberry leaves. *Ann. Bot.*, **40**, 351–4.

McLaren, G. and Cameron, K.C. (1996) *Soil Science: Sustainable Production and Environmental Protection*. Oxford University Press, Auckland.

McLellan, B.N. (1989a) Dynamics of a grizzly bear population during a period of industrial resource extraction. I. Density and age–sex composition. *Can. J. Zool.*, **67**, 1856–60.

McLellan, B.N. (1989b) Dynamics of a grizzly bear population during a period of industrial resource extraction. II. Mortality rates and causes of death. *Can. J. Zool.*, **67**, 1861–4.

McLellan, B.N. and Shackleton, D.M. (1988) Grizzly bears and resource extraction industries: effects of roads on behaviour, habitat use and climography. *J. Appl. Ecol.*, **25**, 451–60.

McNaughton, S.J. and Wolf, L.L. (1973) *General Ecology*, Holt, Rinehart and Winston, New York.

McNeill, A.R. (1991) Recreational water quality, in *Pollution in Tropical Aquatic Systems* (eds D.W. Connell and D.W. Hawker), CRC Press Inc., Ann Arbor, pp. 193–216.

Mader, H.J., Schell, C. and Kornacker, P. (1990) Linier barriers to arthropod movements in the landscape. *Biol. Conserv.*, **54**, 209–22.

Madsen, J. (1985) Impact of disturbance on field utilisation of pink-footed geese in West Jutland, Denmark. *Biol. Conserv.*, **33**, 53–63.

Magdefrau, K. (1982) Lifeforms of bryophytes, in *Bryophyte Ecology* (ed. A.J.E. Smith), Chapman & Hall, London.

Mainini, B., Neuhaus, P. and Ingold, P. (1993) Behaviour of marmots (*Marmota marmota*) under the influence of different hiking activities. *Biol. Conserv.*, **64**, 161–4.

Mallon, D.P. (1991) Status and conservation of large mammals in Ladakh. *Biol. Conserv.*, **56**, 101–19.

Malme, C.I., Miles, P.R., Clark, C.W., *et al.* (1983) *Investigations of the Potential Effects of Underwater Noise from Petrolium Industry Activities on Migrating Grey Whale Behaviour*, BBN Rep. No. 5366, MMS Alaska, OCS Region, Anchorage, Alaska.

Marion, J.L. and Merriam, L.C. (1985) Recreational impacts on well-established campsites, in *The Boundary Waters Canoe Area Wilderness. Sta. Bull.*, AD-SB-2502, University of Minnesota, St. Paul.

Marnell, L., Foster, D. and Chilman, K. (1978) *River recreation research conducted at Ozark National Scenic Riverways* 1970–1978: *a summary of research projects and findings*, Van Buren, Missouri, National Park Service.

Marshall, C. and Sagar, G.R. (1968) The distribution of assimilates in *Lolium multiflorum Lam.* Following differential defoliation. *Ann. Bot.*, **32**, 715–19.

Martin, P.S. (1973) The discovery of America. *Science*, **179**, 969–74.

Mason, C.F. (1991) *Biology of Freshwater Pollution*, 2nd edn, Longman Scientific and Technical, Essex.

Mathisen, J.E. (1968) Effects of human disturbance on nesting of bald eagles. *J. Wildl. Manage.*, **32**, 1–6.

Meerts, P. and Vekemans, X. (1991) Phenotypic plasticity as related to trampling within a natural population of *Polygonum aviculare* subsp. *Aequale. Acta Oecol.* **12**, 203–12.

Meinecke, P. (1928) *The Effect of Excessive Tourist Travel on the California Redwood Parks*, California Department of Natural Resources, Division of Parks, Sacramento.

Merritt R.W. and Newson, H.D. (1978) Ecology and management of arthropod populations in recreational lands, in *Perspectives in Urban Entomology* (eds G.W. Frankie and C.S. Koehler), Academic Press, New York pp. 125–62.

Mikola, J., Miettinen, M., Lehikoinen, E. and Lehtilä, K. (1994) The effects of disturbance caused by boating on survival and behaviour of velvet scoter *Melanitta fusca* ducklings. *Biol. Conserv.*, **67**, 119–24.

Milligan, H.N. and Williams, G.E. (1959) *A Survey of the Animal Kingdom*, Horniman Museum and Library, London.

Milner-Gulland, E.J. and Mace, R. (1991) The impact of the ivory trade on the African Elephant *Loxodonta africana* population as assessed by data from the trade. *Biol. Conserv.*, **55**, 215–29.

Mitchell, J.K. and Baubenzer, G.D. (1980) Soil loss estimation, in *Soil Erosion* (eds M.S. Kirkby and R.P.C. Morgan), Wiley, Chichester, pp. 17–62.

Moen, A.N. (1976) Energy conservation by white-tailed deer in the winter. *Ecology*, **57**, 192–8.

Morgan, J.M. and Kuss, F.R. (1986) Soil loss as a measure of carrying capacity in recreation environments. *Environ. Manage.*, **10**, 263–70.

Morgan, R.P.C. (1979) *Soil Erosion*, Longmans, London.

Morgan, R.P.C. (1985) The impact of recreation on mountain soils: towards a predictive model for soil erosion, in *The ecological impacts of outdoor recreation on mountain areas in Europe and North America* (eds N.G. Bayfield and G.C. Barrow), Recreation Ecology Research Group Report No. 9, Ashford, Kent, pp. 112–21.

Morgantini, L.E. and Hudson, R.J. (1980) Human disturbance and habitat selection in elk, in *North American Elk and Behaviour, Ecology and Management* (eds M.S. Bayee and L.D. Haydon-Wing), The University of Wyoming, Lanomie, pp. 132–9.

Mortensen, C.O. (1989) Visitor use impacts within the Knobstone Trail Corridor. *J. Soil. Water Conserv.*, **4**, 156–9.

Moss, B. (1977) Conservation problems in the Norfolk Broads and rivers of East Anglia, England – phytoplankton, boats and the causes of turbidity. *Biol. Conserv.*, **12**, 95–114.

Muller, W.H. (1979) *Botany: A Functional Approach*, Macmillan, New York.

Muller-Dombois, D. and Ellenberg, H. (1974) *Aims and Methods of Vegetation Ecology*, Wiley, New York.

Munguira, M.L. and Thomas, J.A. (1992) Use of road verges by a butterfly and burnet populations, and the effects of roads on adult dispersal and mortality. *J. Appl. Ecol.*, **29**, 316–29.

Muratori, A. (1968) How outboards contribute to water pollution. *Conservationist*, **22**, 6–8.

Mutchler, C.K. and Young, R.A. (1975) Soil detachment by raindrops, in *Present and Prospective Technology for Predicting Sediment Yields and Sources*, US Department of Agriculture, Agricultural Research Service Publication, ARS-S-40, Washington, DC, pp. 113–17,

Myrberg, A.A. (1978) Ocean noise and the behaviour of marine animals: relationships and implications, in *Effects of Noise on Wildlife* (ed. J.L. Fletcher and R.G. Busnel), Academic Press, New York, pp. 169–208.

Neel, P.L. and Harris, R.W. (1972) Tree seedling growth: effects of shaking. *Science*, **175**, 918–19.

Nelson, M. (1966) Problems of recreational use of game ranges. *Desert Bighorn Council Trans.*, **10**, 13–20.

Nepal, S.K. and Weber, K.E. (1993) *Struggle for Existence: Park–People Conflict in the Royal Chitwan National Park, Nepal*, Division of Human Settlements Development, Asian Institute of Technology, Bankok, Thailand.

Neumann, P.W. and Merriam, H.G. (1972) Ecological effects of snowmobiles. *Can. Field Nat.*, **86**, 207–12.

Newsome, A.E. (1991) Dingo. *Canis familiaris dingo*, in *The Australian Museum, Complete Book of Australian Mammals* (ed. R. Strahan), Collins, Angus and Robertson, Sydney.

Newson, H.D. (1977) Arthropod problems in recreation areas. *Ann. Rev. Entomol.*, **22**, 333–53.

Newton, I., Davis, P.E. and Moss, D. (1981) Distribution and breeding of red kites *Milvus milvus* in relation to land use in Wales, UK *J. Appl. Ecol.*, **18**, 173–86.

Newton, J. and Pugh-Thomas, M. (1979) The effects of trampling on the soil Acari and Collembola of a heathland. *Intern. J. Environ. Studies*, **13**, 219–23.

Nickerson, N.H. and Thibodeau, F.R. (1983) Destruction of *Ammophila breviligulata* by pedestrian traffic: quantification and control. *Biol. Conserv.*, **27**, 277–87.

Niel, D. (1990) Potential for coral stress due to sediment resuspension and deposition by reef walkers. *Biol. Conserv.*, **52**, 221–7.

Nielsen, S.M. (1989) Occurrence of fox (*Vulpes vulpes*) in an urban area. *Flora og Fauna* (Århus), **95**, 35–42.

Nisbet, I.C.T. and Drury, W.H. (1972) Measuring breeding success in common and roseate terns. *Bird Bonding*, **43**, 97–106.

Niven, B.S. (1987) The logical synthesis of an animal's environment. I Primitive terms and definitions. *Aust. J. Zool.*, **35**, 597–606.

Nixon, C.M. (1989). Winter refuges – the key to survival for farmland deer. *The Illinois Nat. Hist. Surv. Rep.*, **290**, 3–4.

Oberdorfer, E. (1960) *Pflenzen sozialogische studen in Chile: ein vergleich mit Europa*, Cramer, Weinheim.

O'Connor, K.F. (1956) Influences of treading on grasslands. Ph.D. Thesis, Cornell University.

O'Gara, B.W. and Harris, R.B. (1988) Age and condition of deer killed by predators and automobiles. *J. Wildl. Manage.*, **52**, 316–20.

Ogden, J.C. (1978) Status and nesting biology of the American crocodile, *Crocodylus acutus*, (Reptilia Crocodilidae) in Florida. *J. Herpetol.*, **12**, 183–96.

Okarma, H. (1993) Status and management of the wolf in Poland. *Biol. Conserv.*, **66**, 153–8.

Onadeko, S.A. (1992) The American alligator and concern for visitors at Brazos Bend State Park. *J. Environ. Manage.*, **35**, 261–9.

Oniki, Y. (1976) Effects of humans on nests and birds in and near to tropical reserves. *Acta Amazonica*, 7, 555–7.

Opdam, P. and Helmrich, V.R. (1984) Vogelmeenschappen van heide en hougveen: een lypologische beschrijuing. *Limosa*, **57**, 47–62.

O'Shea, T.J., Beck, C.A., Bonde, R.K. *et al.* (1985) An analysis of manatee mortality patterns in Florida 1976–81. *J. Wildl. Manage.*, **49**, 1–11.

Owens, N.W. (1977) Responses of wintering brent geese to human disturbance. *Wildfowl*, **28**, 5–14.

Ozoga, J.J., Verme, L.J. and Bienz, C.S. (1982) Parturition behaviour and territoriality in white-tailed deer: impact on neonatal mortality. *J. Wildl. Manage.*, **46**, 1–11.

Pain, S. (1996) Where Manatees may safely swim. *New Scientist*, 20 July, p. 22.

Paine, R.T. (1971) A short-term experimental investigation of resource partitioning in a New Zealand rocky intertidal habitat. *Ecology*, **52**, 1096–106.

Painter, H.A. (1958) Some characteristics of a domestic sewage. *Water Waste Treat. J.*, **6**, 496–8.

Painter, H.A. and Biney, M. (1959). Composition of domestic sewage. *J. Biochem. Microbial. Technol. Engng.*, **1**, 143–62.

Parish, D.H. (1971) Effects of compaction on nutrient supply to plants, in *Compaction in Agricultural Soils* (eds K.K. Barnes *et al.*), American Society of Agricultural Engineers Monograph, Michigan, pp. 277–91.

Parker, K.L., Robbins, C.T. and Hanley, T.A. (1984) Energy expenditure for locomotion by mule deer and elk. *J. Wildl. Manage.*, **48**, 474–88.

Parker, T.J. and Haswell, W.A. (1963) *A Text-Book of Zoology*, Macmillan, London [Revised by A.J. Marshall].

Paruk, J.D. (1987) Habitat utilization by bald eagles wintering along the Mississippi River. *Trans. Illinois Acad. Sci.*, **80**, 333–42.

Pastorok, R.A. and Bilyard, G.R. (1985) Effects of sewage on coral-reef communities. *Mar. Ecol. Prog. Ser.*, **21**, 175–89.

Patric, J.H. and Brink, L.K. (1977) Soil erosion and its control in the Eastern Forest, in *Soil Erosion: Prediction and Control* (ed. G. Foster), Soil Conservation Society of America, Ankeny, Iowa.

Patterson, I.J. (1988) Responses of Apennine chamois to human disturbance. *Z. Säugeitierkunde*, **53**, 245–52.

Payne, R., Brazier, O., Dorsey, E.M. *et al.* (1983) External features in Southern Right Whales (*Eubalanea australis*) and their use in identifying individuals, in *Communication and Behaviour of Whales* (ed. R. Payne), Westview Press, Boulder, CO, pp. 371–445.

Peace, T.R. and Gilmour, J.S.L. (1949) The effect of picking on the flowering of bluebell, *Scilla non-scripta*. *New Phytologist*, **48**, 115–17.

Pearce, F. (1995) Alternative antifouling widespread in Europe. *New Scientist*, 14 Jan. 1995, 7.

Pearson, R.G. and Jones, N.V. (1975). The effects of dredging operations on the benthic community of a chalk stream. *Biol. Conserv.*, **18**, 273–8.

Pearson, R.G. and Jones, N.V. (1978). The effects of weed-cutting on the macro-invertebrate fauna of a canalised section of the river Hull, a northern English chalk stream. *J. Environ. Manage.*, **7**, 91–7.

Pentecost, A. and Rose, F. (1985) Changes in the cryptogam flora of the Wealden sandrocks, 1688–1984. *Biol. J. Linn. Soc.*, **90**, 217–30.

Perring, F.H. (1967) Changes in chalk grassland caused by galloping, in *The Biotic Effects of Public Pressures on the Environment* (ed. E. Duffey), Monks Wood Experimental Station, Symposium No. 3, Abbots Ripton.

Petrak, von M. (1988) Cross country skiing and red deer (*Cervus elephus* Linné 1758) in the Eifel. *Z. Jagdwiss.*, **34**, 105–14.

Pfister, C., Harrington, B.A. and Lavine, M. (1992) The impact of human disturbance on shorebirds at a migration staging area. *Biol. Conserv.*, **60**, 115–26.

Pickard, B.G. (1973) Action potentials in higher plants. *Botanical Review*, **39**, 172–201.

Piearce, T.G. (1984) Earthworm populations in soils disturbed by trampling. *Biol. Conserv.*, **29**, 241–52.

Piechocki, R. (1994) Who's afraid of the wandering wolf? *New Scientist*, April, 19–21.

Pierce, G., Spray, C.J. and Stuart, E. (1993) The effect of fishing on the distribution and behaviour of waterbirds in the Kakut area of Lake Songkla, Southern Thailand. *Biol. Conserv.*, **66**, 23–34.

Pijl, L. van der (1972) *Principles of Dispersal in Higher Plants*, 2nd edn, Springer-Verlag, Berlin.

Ploeg , S.W.F. van der and Wingerden, W.K.R.E. (1974) *The influence of trampling on spiders*. Proceedings of the 6th Arachn. Congress, pp. 167–72.

Povey, A. and Keough, M.J. (1991). Effects of trampling on plant and animal populations on rocky shores. *Oikos*, **61**, 355–68.

Pradhan, P. and Trepathi, R.S. (1980) Competition between *Trifolium repens* and *Paspalum dilatatum* as related to trampling. *Acta Oecologia – Oecol. Plant.*, **4**, 345–53.

Prager, R.L. (1911–1915) *A Biological Survey of Clare Island in the County of Mayo, Ireland, and of the Adjoining District. 10 Phanerogamia and Pteridophyta*, Williams and Norgate, London, Vols 5 and 6, pp. 31–106.

Preston, R.D. (1955) The mechanical properties of the cell wall, in *Handbuch der Pflanzen-Physiologie* (ed. W. Ruckland), Springer-Verlag, Berlin, Vol. I, 745–51.

Price, R. and Lent, P.C. (1972) Effects of human disturbance on Dall sheep. *Alaska Co-operative Wildl. Res. Unit Quart. Rep.*, **23**, (3), 23–8.

Prihar, S.S. and Aggarwal, G.C. (1975) A new technique for measuring emergence force of seedlings and some laboratory and field studies with corn (*Zea mays L.*). *Soil Sci.*, **120**, 200–4.

Prindiville-Gains, E. and Ryan, M.R. (1988) Piping plover habitat use and reproductive success in North Dakota, USA. *J. Wildl. Manage.*, **52**, 266–73.

Probine, M.L. (1963) The plant cell wall. *Tuatara*, 115–41.

Progulske, D.R. and Baskett, T.S. (1958) Mobility of Missouri deer and their harassment of dogs. *J. Wildlife Manage.*, **22**, 184–92.

Prose, D.V. (1985) Persisting effects of armoured military manoeuvres on some soils of the Mojave Desert. *Environ. Geol. Wat. Sci.*, **7**, 163–70.

Prose, D.V., Metzger, S.K. and Wilshire, H.G. (1987) Effects of substrate disturbance on secondary plant succession; Mojave Desert, California. *J. Appl. Ecol.* **24**, 305–13.

Pulliainen, E. (1972) Distribution and population structure of the bear (*Ursus aratos* L) in Finland. *Ann. Zool. Feunici*, **9**, 199–207.

Pulliainen, E. (1982) Experiences in the protection of the large predators in Finland. *Intern. J. Study Anim. Probl.*, **3**, 33–41.

Puntieri, J.G. (1991) Vegetation response on a forest slope cleared for a ski-run with special reference to the herb *Alstroemeria aurea* Graham (*Alstroemeriaceae*), Argentina. *Biol. Conserv.*, **56**, 207–21.

Quigley, H.B. and Crawshaw, P.G. Jr (1992) A conservation plan for the jaguar *Panthera onca* in the Pantanal region of Brazil. *Biol. Conserv.*, **61**, 149–57.

Quinn, N.W., Morgan, R.P.C. and Smith, A.J. (1980) Simulation of soil erosion induced by human trampling. *J. Environ. Manage.*, **110**, 155–65.

Rabinowitz, A. (1993) Estimating the Indochinese tiger *Panthera tigris corbetti* population in Thailand. *Biol. Conserv.*, **65**, 213–17.

Racey, P.A. and Stebbings, R.E. (1972) Bats in Britain – a status report. *Oryx*, **11**, 319–27.

Radonski, G.G., Prosser, N.S., Martin, R.G. and Stroud, R.M. (1984) Exotic fishes and sport fishing, in *Distribution, Biology and Management of Exotic*

Fishes (eds W.R. Courtney Jr and J.R. Stauffer Jr), Johns Hopkins University Press, Baltimore, pp. 313–21.

Raghavan, G.S.U., McKyes, E., Gendron, G. *et al.* (1978) Effects of soil compaction on development and yield of corn (maize). *Can. J. Plant Sci.,* **58**, 435–43.

Randall, R.H. and Birkeland, C. (1978) Guam's Reefs and Beaches. Part II Sedimentation Studies at Fouka Bay and Ylig Bay, University of Guam Marine Laboratory, Tech. Rep. No. 47.

Randall, R.M. and Randall, B.M. (1981) Roseata tern breeding biology and factors responsible for low chick production in Algoa Bay, South Africa. *Ostrich,* **52**, 17–24.

Ranwell, D.S. (1959) Newborough Warren Anglesey. I The dune system and dune slack habitat. *J. Ecol.,* **47**, 571–602.

Raunkier, C. (1934) *The Life Forms of Plants and Statistical Plant Geography,* Oxford University Press, Oxford.

Ray, G.C., Mitchell, E.D., Wartzok, D. *et al.* (1978) Radiotracking of a fin whale (*Balaenoptera physalus*). *Science,* **202**, 521–4.

Ream, C.H. (1979) Human–wildlife conflicts in backcountry: possible solutions, in *Recreation Impacts on Wildlands* (eds R. Ittner, D.R. Potter and J.K. Agee), USDA Forest Service, Pacific North West Region, Seattle, pp. 153–63.

Reaves, C.A. and Cooper, A.W. (1960) Stress distribution in soils under tractor loads. *Agric. Engin.,* **41**, 20, 21, 31.

Redford, K.H. and Robinson, J.G. (1985) Hunting by indigenous people and conservation of game species. *Cultural Survival Quart.,* **9**, 41–4.

Reed, D.F. (1981) Mule deer behaviour at a highway underpass exit. *J. Wildl. Manage.,* **45**, 542–3.

Rees, J.R. (1978) A 'people-counter' for unsurfaced wetland footpaths. *Environ. Conserv.,* **5**, 66–8.

Rees, J. and Tivy, J. (1977) Recreational impact on lochshore vegetation. *J. Scott. Assoc. Geogr. Teach.,* **6**, 8–24.

Reese, J.G. (1972) Osprey nesting success along the Choptank River, Maryland. *Chesapeake Sci.,* **13**, 233–5.

Reid, D.D. and Krogh, M. (1992) Assessment of catches from protective shark meshing off New South Wales beaches between 1950 and 1990. *Aust. J. Mar. Freshwater Res.,* **43**, 283–96.

Reijnen, R., Foppen, R., Braak, C.T. and Thissen, J. (1995) The effects of car traffic on breeding bird populations in woodland. III. Reduction of density in relation to the proximity of main roads. *J. Appl. Ecol.,* **32**, 187–202.

Richards, P.W. (1952) *The Tropical Rainforest,* Cambridge University Press, London.

Richardson, W.J., Fraker, M.A., Würsig, B. and Wells, R.S. (1985) Behaviour of Bowhead Whales (*Balaena mysticetus*) summering in the Beaufort Sea: reactions to industrial activities. *Biol. Conserv.,* **32**, 195–230.

Richens, V.B. and Lavigne, G.R. (1978) Response of white-tailed deer to snowmobiles and snowmobile trails in Maine. *Can. Field Nat.,* **92**, 334–44.

Richter, G. and Negendank, J.F.W. (1977) Soil erosion processes and their measurement in the German area of the Moselle River. *Earth Surface Processes,* **2**, 26l–78.

Rickard, W. (1972) *Preliminary ecological evaluation of the effects of aircushion vehicles tests on the Arctic Tundra of Northern Alaska*, USA CRREL Special Report 182, Cold Regions Research and Engineering Laboratory, Hanover, New Hampshire.

Rickard, W.E. and Brown, J. (1974) Effects of vehicles on Arctic Tundra. *Environ. Conserv.*, **1**, 55–62.

Rickard, W.E. and Slaughter, C.W. (1973) Thaw and erosion on vehicular trails in perma frost landscapes. *J. Soil. Water Conserv.*, **28**, 263–6.

Ridley, H.N. (1930) *The Dispersal of Plants Throughout the World*, Reeve, Ashford.

Riewe, R.R. (1975) The high arctic wolf in the Jones Sound region of the Canadian High Arctic. *Arctic*, **28**, 209–12.

Robert, H.C. and Ralph, C.J. (1975) Effects of human disturbance on the breeding success of gulls. *Condor*, **77**, 495–9.

Roberts, B.C. and White, R.G. (1992) Effects of angler wading on survival of trout eggs and pre-emergent fry. *North Am. J. Fish. Manage.*, **12**, 450–9.

Robinson, S.K. (1990) Effects of forest fragmentation on nesting songbirds. *Illinois Nat. Hist. Surv. Rep.*, **296**, 1–2.

Roblin, G. (1979) *Mimosa pudica*: a model for the study of excitability in plants. *Biol. Rev.*, **54**, 135–53.

Roblin, G. (1985) Analysis of the variation potential induced by wounding in plants. *Plant Cell Physiol.*, **26**, 455–61.

Rogers, L.L., Wilker, G.A. and Scott, S.S. (1991) Reactions of black bears to human menstrual odors. *J. Wildl. Manage.* **55**, 632–4.

Rogers, R.W. (1977) Lichens of hot arid and semi-arid lands, in *Lichen Ecology* (ed. M.R.D. Seaward), Academic Press, London, pp. 211–52.

Rogers, R.W. and Lange, R.T. (1971) Lichen populations on arid soil crusts around sheep watering places in South Australia. *Oikos*, **22**, 93–100.

Roos, G. Th. de (1981) *The Impact of Tourism upon some Breeding Wader Species on the Isle of Vlieland in The Netherlands Wadden Sea*, Mededelingen Landbouwhogeschool, Wageningen, pp. 81–114.

Root, J.D. and Knapik, L.J. (1972) *Trail conditions along a portion of the Great Divide trail route, Alberta and British Colombia Mountains*, Research Council of Alberta, Edmonton, Alberta.

Ross, C., Srivastava, A. and Pirta, R.S. (1993) Human influences on the population density of Hanuman langurs *Presbytis entellus* and rhesus macaques *Macaca mulatta* in Shimla, India. *Biol. Conserv.*, **65**, 159–63.

Roxburgh, S.H., Watkins, A.J. and Bastow, W.J. (1993) Lawns have vertical stratification. *J. Vegetat. Sci.*, **4**, 699–704.

Russell, A. (1967) *Grizzly Country*, Jarrolds, London.

Russell, A. (1968) The people vs. the grizzlies. *Field and Stream*, **62**, 60–1, 113–19 and 151.

Russell, E.W. (1973) *Soil Conditions and Plant Growth*, 10th edn, Longman, London.

Russell, G. and Grace, J. (1978) The effect of wind on grasses V. Leaf extension, diffusive conductance and photosynthesis in the wind tunnel. *J. Exp. Bot.*, **29**, 1249–58.

Russell, R.S. (1977) *Plant root systems: their function and interaction with the soil*, McGraw-Hill, London.

Rutherford, G.K. and Scott, D.C. (1979) The impact of recreational land use on soil chemistry in a Provincial Park. *Park News*, 5, 22–5.

Sage, R.W. Jr, Tierson, W.C., Mattfield, G.F. and Behrend, D.F. (1983) White-tailed deer visibility and behaviour along forest roads. *J. Wildl. Manage.*, 47, 940–53.

Saile, B. (1977) Cougar attacks: new crisis for the big cats. *Outdoor Life*, 160, 66–8 and 126–8.

Salden, D.R. (1988) Humpback whale encounter rates offshore of Maui, Hawaii. *J. Wildl. Manage.*, 52, 301–4.

Salt, G., Hollick, F.S.J., Raw, F. and Brian, M.V. (1948) Arthropod populations of pasture soil. *J. Anim. Ecol.*, 17, 139–50.

Salter, P.J. and Williams, J.B. (1965) The influence of texture on the moisture characteristics of soils. II Available water capacity and moisture release characteristics. *J. Social Sci.*, 16, 310–l7.

Sambek, J.W. van and Pickard, B.G. (1976) Mediation of rapid electrical, metabolic, transpirational and photosynthetic changes by factors released from wounds. III. Measurements of CO_2 and H_2O flux. *Can. J. Bot.*, 54, 2662–71.

Samways, M.J. and Moore, S.D. (1991) Influence of exotic conifer patches on grasshopper (Oorthoptera) assemblages in a grassland matrix at a recreational resort, Natal, South Africa. *Biol. Conserv.*, 57, 117–37.

Saris, F.J.A. (1976) *Breeding populations of birds and the recreation in the Duivelsbert*, Institute for Environmental Studies, Free University Working Paper No. 66.

Satchell, J.E. and Marren, P.R. (1976) *The effects of recreation on the ecology of natural landscapes*, Nature and environment series No. 11, Council of Europe, Strasbourg.

Saunders, E. (1937) *A Beast Book for the Pocket. The Vertebrates of Britain, Wild and Domestic other than Birds or Fishes*, Oxford University Press, London.

Saunders, P.R., Howerd, G.E. and Stanley-Saunders, G.H. (1980) *Effect of different boot sole configurations on forest soils*, Department of Recreation and Park Administration. Extension/Research Paper RPA 1980–3, Clemson University, Clemson, South Carolina.

Schaerffenberg, B. (1941) Zur Biologie des Maulwurfs (*Talpa europaea* L.). *Z. Saugetierk.*, 22, 272–7.

Schemnitz, S.D. and Schortemeyer, J.D. (1974) *The influence of vehicles on Florida Everglades vegetation*, Ecological Report, Florida State Game and Fresh Water Fish Commission, Tallahassee, Florida.

Schick, C.S. (1890) Birds found breeding on Seven Mile Beach, New Jersey. *Auk*, 7, 326–9.

Schlesinger, C.A. and Shine, R. (1994) Choosing a rock: perspectives of a bush-rock collector and saxicolous lizard. *Biol. Conserv.*, 67, 49–56.

Schmid, W.D. (1970) *Modification of the Subnivian Microclimate by Snow Mobiles*. Proceedings of Snow and Ice in Relation to Wildlife Symposium, (ed. A.O. Haugen), Iowa Cooperative Wildlife Research Unit Iowa State University, pp. 251–7.

Schneider, E. Von and Wölfel, H. (1978) Suggested modes of protection for wild animals in connection with the construction of shipping canals and canalised inland waterways. *Zeitschrift für Jagdwissenschaft*, **24**, 72–88.

Schneider, W. (1977) *Where the Grizzly Walks*, Mountain Press, Missoula.

Scorgie, H.R.A. (1978) Effects of aquatic herbicides on freshwater ecosystems, Internal Report to the Nature Conservancy Council.

Seely, M.K. and Hamilton, W.J. III (1978) Durability of vehicle tracks on three Namib Desert substrates, South Africa. *J. Wildl. Res.*, **8**, 107–11.

Sennstam, B. and Stålfelt, F. (1976) Rapport angående 1975 års femdager-sorienterings inverkan på klöuviltet, Rapport 12:35, p. 35.

Settergren, C.D. and Cole, D.M. (1970) Recreation effects on soil and vegetation in the Missouri Ozarks. *J. Forest.*, **68**, 231–3.

Shackley, M. (1992) Manatees and tourism in southern Florida: opportunity or threat? *J. Environ. Manage.*, **34**, 257–65.

Shearman, R.C. and Beard, J.B. (1975a) Turfgrass wear tolerance mechanisms. I Wear tolerance of seven turfgrass species and quantitative methods for determining turfgrass wear injury. *Agron. J.*, **67**, 208–11.

Shearman, R.C. and Beard, J.B. (1975b) Turfgrass wear tolerance mechanisms. II Effects of cell wall constituents on turfgrass wear tolerance. *Agron. J.*, **67**, 211–15.

Shearman, R.C. and Beard, J.B. (1975c) Turfgrass wear tolerance mechanisms. III Physiological, morphological and anatomical characteristics associated with turfgrass wear tolerance. *Agron. J.*, **67**, 215–18.

Sherrod, S.K., White, C.M. and Williamson, F.S.L. (1976) The biology of the bald eagle on Amchitka Island, Alaska, USA. *Living Bird*, **15**, 143–82.

Shields, L.M. and Dean, H.C. (1949) Microtome compression in plant tissue. *Am. J. Bot.*, **36**, 408–16.

Shimwell, D.W. (1971) *Description and Classification of Vegetation*, Sidgwick and Jackson, London.

Shoop, C.R., Ruckdeschel, C.A. and Thompson, N.B. (1985) Sea turtles in the south east USA. Nesting activity as derived from aerial and ground surveys. *Herpetologica*, **41**, 252–9.

Sibaoka, T.(1950) Action potential and conduction of excitation in the leaf of *Mimosa pudica*. *Sci. Rep. Tohoku University, 4th Series (Biology)*, **18**, 362–9.

Silcock, R.G. (1980) Seeding characteristics of tropical pasture species and their implications for ease of establishment. *Tropical Grasslands*, **14**, 174–80.

Singer, F.J. (1975) *Behaviour of mountain goats, elk and other wildlife in relation to US Highway 2, Glacier National Park*, Prepared for Federal Highway Administration and Glacier National Park, West-Glacier, Montana.

Singer, F.J. (1978) Behaviour of mountain goats in relation to US Highway 2, Glacier National Park, Montana. *J. Wildl. Manage.*, **42**, 591–7.

Skidmore, E.L. (1966) Wind and sandblast injury to seedling green beans. *Agron. J.*, **58**, 311–15.

Slaughter, C.W., Racine, C.H., Walker, D.A. *et al.* (1990) Use of off-road vehicles and mitigation of effects in Alaska permafrost environments: a review. *Environ. Manage.*, **14**, 63–72.

Smith, B.D. (1993) 1990 status and conservation of the Ganges River dolphin (*Platanista gangetica*) in the Karnali River, Nepal. *Biol. Conserv.*, **66**, 159–69.

Smith, N.S. and Krausman, P.R. (1988) Desert bighorn sheep: a guide to selected management practices. A literature review and synthesis including appendices on assessing condition, collecting blood, determining age, constructing water catchments and evaluating bighorn range. *US Fish Wildl. Serv. Biol. Rep.*, **88**, (35), 1–27.

Smith, R.L. (1990) *Ecology and Field Biology*, Harper and Rowe, New York.

Smith, S.V., Kimmerer, W.S., Laws, E.A. *et al.* (1981) Kaneohe Bay sewage diversion experiment: perspectives on ecosystem responses to nutritional perturbation. *Pac. Sci.*, **35**, 279.

Snyder, H.A. and Snyder, N.F.R. (1974) Increased mortality of Cooper's Hawks accustomed to man. *Condor*, **76**, 315–16.

Šomšák, L., Kubiček, F., Háberová, I. and Majzlánová, E. (1979) The influence of tourism upon the vegetation of the High Tatras. *Biologia (Bratislava)*, **34**, 571–82.

Sparrow, S.D., Wooding, F.J. and Whiting, E.H. (1978) Effects of off-road vehicle traffic on soils and vegetation in the Denali Highway region of Alaska. *J. Soil. Water Conserv.*, **33**, 20–7.

Spinney, L. (1995) Return to the wild. *New Scientist*, 14 Jan., 35–8.

Ssemakula, J. (1983) A comparative study of hoof pressures of wild and domestic ungulates. *Afr. J. Ecol.*, **21**, 325–8.

Stalmaster, M.V. and Newman, J.R. (1978) Behavioural responses of wintering bald eagles to human activity. *J. Wildl. Manage.*, **42**, 506–13.

Stancyk, S.E. and Ross, J.P. (1978) An analysis of sand from green turtle nesting beaches on Ascension Island. *Copeia*, 1978, 93–9.

Stebbins, R.C. (1990) A desert at the crossroads. *Pacific Discovery California Academy of Sciences*, Winter 1990, 3–16, 32.

Steiner, A.J. and Leatherman, S.P. (1981) Recreational impacts on the distribution of ghost crabs *Ocypode quadrata* Fab. *Biol. Conserv.*, **20**, 111–22.

Stelmock, J.J. and Dean, F.C. (1979) Vegetation trampling effects analysis – 1975 plots, Mount McKinley National Park, Alaska. Unpublished Report, US Department of the Interior, National Park Service, Mount McKinley National Park, Alaska.

Stephens, H.V. (1978) *Guide for Predicting Rainfall–Erosion Losses from Agricultural Land in Maryland and Delaware*, Technical Note Conservation Planning. Soil Conservation Service, US Department of Agriculture, College Park, Maryland.

Stevens, T. and Chaloupka, M. (1992) *Whale watching in Hervey Bay Marine Park and the Whitsundays: Implications for Management*. Abstracts, 5th Annual Conference, Australasian Wildlife Management Society, Queensland University of Technology, Brisbane, Queensland.

Stewart, R. and Howard, H.H. (1968) Water pollution by outboard motors. *Conservationist*, **22**, 6–8.

Stillingfleet, B. (1759) *Observations on grasses in miscellaneous tracts relating to natural history, husbandry and physics*, London.

Stockwell, C.A., Bateman, G.C. and Berger, J. (1991) Conflicts in national parks: a case study of helicopters and Bighorn sheep time budgets at the Grand Canyon. *Biol. Conserv.*, **56**, 317–28.

Stoddart, D.R., Cowx, D., Peet, C. and Wilson, J.R. (1982) Tortoises and tourists in the Western India Ocean, the Carieuse experiment. *Biol. Conserv.*, **24**, 67–80.

Streeter, D. (1975) *Preliminary Observations on Rates of Erosion on Chalk Downland Footpaths*, Institute of British Geographers, Annual Meeting, Oxford.

Studlar, S.M. (1980) Trampling effects on bryophytes: trail surveys and experiments. *Bryol.*, **83**, 301–13.

Studlar, S.M. (1983) Recovery of trampled bryophyte communities near Mountain Lake, Virginia. *Bull. Torrey Bot. Club*, **110**, 1–11.

Sukopp, H. (1971) Effects of man, especially recreational activities, on littoral macrophytes. *Hydrobiologia*, **12**, 331–40.

Sukumar, R. (1991) The management of large mammals in relation to male strategies and conflict with people. *Biol. Conserv.*, **55**, 93–102.

Sun, D. (1991) Plant resistance to and recovery from trampling in relation to morphology, age and soil fertility. Ph.D. Thesis, Griffith University.

Sun, D. and Liddle, M.J. (1991) Field occurrence, recovery, and simulated trampling resistance and recovery of two grasses. *Biol. Conserv.*, **57**, 187–203.

Sun, D. and Liddle, M.J. (1993a) Trampling resistance, stem flexibility and leaf strength in nine Australian grasses and herbs. *Biol. Conserv.*, **65**, 35–41.

Sun, D. and Liddle, M.J. (1993b) The morphological responses of some Australian tussock grasses and the importance of tiller number in their resistance to trampling. *Biol. Conserv.*, **65**, 43–9.

Sun, D. and Liddle, M.J. (1993c) A survey of trampling effects on vegetation and soil in eight tropical and subtropical sites. *Environ. Manage.*, **17**, 497–510.

Sun, D. and Liddle, M.J. (1993d) Plant morphological characteristics and resistance to simulated trampling. *Environ. Manage.* **17**, 511–21.

Tackett, J.L. and Pearson, R.W. (1964) Oxygen requirements for cotton seedling root penetration of compacted soil cores. *Proc. Soil. Sci. Soc. Am.*, **28**, 600–4.

Tanaka, Y. and Uritani, I. (1979) Effect of auxin and other hormones on the metabolic response to wounding in sweet potato roots. *Plant Cell Physiol.*, **20**, 1557–64.

Tanner, M.F. (1973) *Water resources and recreation*, Study 3, Sports Council, London.

Tayler, N.J.C. (1991) Short-term behavioural responses of Svalbard reindeer *Rangifer tarandus platyrhynchus* to direct provocation by a snowmobile. *Biol. Conserv.*, **56**, 179–94.

Taylor, H.M., Roberson, G.M. and Parker, J.J. Jr (1966) Soil strength – root penetration relations for medium-to-coarse-textured soil materials. *Soil Sci.*, **102**, 18–22.

Taylor, K.C., Reader, R.J. and Larson, D.W. (1993) Scale-dependant inconsistencies in the effects of trampling on a forest understory community. *Environ. Manage.*, **17**, 239–48.

Temple, K.L., Camper, A.K. and Lucas, R.C. (1982) Potential health hazard from human wastes in wilderness. *J. Soil. Water Conserv.*, **37**, 357–9.

Tevis, L. Jr (1959) Man's effect on Bighorn in the San Jacmito – Santa Resa Mountains. *Desert Bighorn Council Trans.*, **3**, 69–74.

Thompson, T.J., Winn, H.E. and Perkins, P.J. (1979) Mysticete sounds, in *Behaviour of Marine Animals*, Vol. 3, *Cetaceans* (eds H.E. Winn and B.L. Olla), Plenum Press, New York, pp. 403–31.

Thyer, N. (1982) Trampling effects and management strategies in a dry schlerophyll forest of South East Queensland. Masters Thesis, Griffith University, Queensland.

Tietje, W.D. and Ruff, R.L. (1983) Responses of black bears to oil development in Alberta. *Wildl. Soc. Bull.*, **11**, 99–112.

Tilghman, N.G. (1987) Characteristics of urban woodlands affecting winter bird diversity and abundance. *Forest Ecol. Manage.*, **21**, 163–75.

Titus, J.R. and van Druff, L.W. (1981) Response of the common loon to recreational pressure in the Boundary Waters canoe area, northeastern Minnesota. *Wildl. Monogr.*, **79**, 5–59.

Tothill, J.C. and Hacker, J.B. (1973) *The Grasses of South East Queensland*, University of Queensland Press, St. Lucia, Queensland.

Tragenza, N.J.C. (1992) Fifty years of cetacean sightings from the Cornish coast, S.W. England. *Biol. Conserv.*, **59**, 65–70.

Trefethan, J.B. (ed.) (1975) *The Wild Sheep in Modern North America*. Proceedings Workshop on the Management Biology of North American Wild Sheep, University of Montana.

Tremblay, J. and Ellison, L.N. (1979) Effects of human disturbance on breeding of black-crowned night herons. *Auk*, **96**, 364–9.

Trew, M.J. (1973) The effects and management of trampling on coastal sand dunes. *J. Environ. Plan. Pollut.*, **1**, 38–49.

Trouse, A.C. Jr (1971) Soil conditions as they affect plant establishment, root development and yield. C, Effects of soil aeration on plant activities, in *Compaction in Agricultural Soils*, (eds K.K. Barns *et al.*), American Society of Agricultural Engineers, Michigan, pp. 241–52.

Tsuyuzaki, S. (1990) Species composition and soil erosion on a ski area in Hokkaido, Northern Japan. *Environ. Manage.*, **14**, 203–7.

Tsuyuzaki, S. (1993) Recent vegetation and prediction of the successional sere on ski grounds in the highlands of Hokkaido, northern Japan. *Biol. Conserv.*, **63**, 255–60.

Tudge, C. (1994) Asia's elephants, no place to hide. *New Scientist*, 15 Jan. 1994, 34–7.

Tulgat, R. and Schaller, G.B. (1992) Status and distribution of wild Bacterian camels *Camelus bacterianus ferrus*. *Biol. Conserv.*, **62**, 11–19.

Turgeon, A.J. (1991) *Turfgrass Management*, Prentice Hall, Englewood Cliffs, New Jersey.

Tuxen, R. (1950) *Mitteilungen der Floristisch-soziologischen arbeits-gemein-schaft*, Glenewinkel, Stolzenau, Vol. 2.

Underwood, A.J. (1992) Beyond BACI: The detection of environmental impacts on populations in the real, but variable, world. *J. Exp. Mar. Biol. Ecol.*, **161**, 145–78.

United States Environmental Protection Agency (USEPA) (1986) *Ambient*

Water Quality Criteria for Bacteria, 1986 EPA 440/5-84-002, Office of Water Regulations and Standards Division, Washington, DC, 20460.

US Department of Agriculture (undated), Washington, DC.

Valentine, S. and Dolan, R. (1979) Footstep-induced sediment displacement in the Grand Canyon. *Environ. Manage.*, **3**, 531–3.

Vareschi, V. (1980) *Vegetations Okologie der Tropen*, Verl E. Ulmer, Stuttgart.

Vaughan, W.J., Russell, C.S., Gianessi, L.P. and Nielsen, L.A. (1982) Measuring and predicting water quality in recreation related terms. *J. Environ. Manage.*, **15**, 363–80.

Veen, B.W. (1981) Relation between root respiration and root activity. *Plant Soil*, **63**, 73–6.

Venning, F.D. (1948) Stimulation by wind motion of collenchyma formation in celery petioles. *Bot. Gazette*, **110**, 511–14.

Viljoen, P.J. (1987) Status and present distribution of elephants in the Kaokoveld, South West Africa/Namibia. *S. Afr. J. Zool.*, **22**, 247–57.

Vines, A.E. and Rees, N. (1964) *Plant and Animal Biology*, Vol. I, Sir Isaac Pitman and Sons, London.

Vollmer, A.T., Maza, B.G., Medica, P.A. *et al.* (1976) The impact of off-road vehicles on a desert ecosystem. *Environ. Manage.*, **1**, 15–129.

Vosin, J.F. (1986) Evolution des puplements d'orthoptères dans le canton D'Aime (Savoie). *Trav. Sci. Parc Nation. Vanoise*, 229–54.

Wagar, J.A. (1964) *The Carrying Capacity of Wild Lands for Recreation*, Forest Science Monograph 7, The Society of American Foresters, Washington, DC.

Wakelyn, L.A. (1987) Changing habitat conditions on Bighorn sheep ranges in Colorado. *J. Wildl. Manage.*, **51**, 904–12.

Walker, T.I. (1993) *Conserving the Shark Stocks of Southern Australia*. Shark Conservation Workshop, Conservation and Fisheries Session, Zoological Parks Board of New South Wales, Mosman.

Walker, W.S. (1957) The effect of mechanical stimulation on the collenchyma of *Apium graveolens* L. *Iowa Acad. Sci.*, **64**, 177–86.

Walker, W.S. (1960) The effects of mechanical stimulation and etiolation on the collenchyma of *Datura stramonium*. *Am. J. Bot.*, **47**, 717–24.

Walker-Simmons, M. and Ryan, C.A. (1977) Wound induced accumulation of trypsin inhibitor activities in plant leaves. *Plant Physiol.*, **59**, 437–9.

Wall, G. and Wright, C. (1977) *The Environmental Impact of Outdoor Recreation*, Department of Geography Publication Series No. 11, University of Waterloo, Ontario.

Walther, F.R. (1969) Flight behaviour and avoidance of predators in Thomson's Gazelle (*Gazella Thomsoni* Guenther 1884). *Behaviour*, **34**, 184–221.

Wanek, W.J. (1974) *The ecological impact of snowmobiling in Northern Minnesota*. Proceedings of the 1973 Symposium, Snowmobile and off the road vehicle research (ed. D.F. Holecek), Technical Report No. 9, College of Agriculture and Natural Resources, Michigan State University, Michigan, pp. 57–76.

Ward, A.L. (1973) *Elk Behaviour in Relation to Multiple Uses on the Medicine Bow National Forest*. Proceedings of the Annual Conference of the Western Association of the State Game and Fish Commissioners.

Ward, A.L., Cupal, J.J., Lea, A.L. *et al.* (1973) Elk behaviour in relation to cattle grazing, forest recreation and traffic. *North Am. Wildl. Nation. Res. Conf. Trans.*, **38**, 327–37.

Waring, P.F., Khalifa, M.M. and Treharne, K.J. (1968) Rate limiting processes in photosynthesis at saturating light intensities. *Nature*, **220**, 435–57.

Warkentin, B.P. (1971) Effects of compaction on content and transmission of water in soils, in *Compaction of Agricultural Soils* (eds K.K. Barnes *et al.*), American Society of Agricultural Engineers, Michigan, pp. 126–53.

Warner, R.E. (1992) Nest ecology of grassland passerines on road rights-of-way in Central Illinois. *Biol. Conserv.*, **59**, 1–7.

Warwick, S.I. and Briggs, D. (1978a) The genecology of lawn weeds. I. Population differentiation in *Poa annua* L. in a mosaic environment of bowling greens, lawns and flower beds. *New Phytol.*, **81**, 711–23.

Warwick, S.I. and Briggs, D. (1978b) The genecology of lawn weeds. Evidence for disruptive selection in *Poa annua* L. in a mosaic environment of bowling greens, lawns and flower beds. *New Phytol.*, **81**, 725–37.

Warwick, S.I. and Briggs, D. (1979) The genecology of lawn weeds. III. Cultivation experiments with *Achillea millifolium* L., *Bellis perennis* L., *Plantago lanceolata* L., *Plantago major* L. and *Prunella vulgaris* L. collected from lawns and contrasting grassland habitats. *New Phytol.*, **83**, 509–36.

Warwick, S.I. and Briggs, D. (1980a) The genecology of lawn weeds. IV. Adaptive significance of variation in *Bellis Perennis* L. as revealed in a transplant experiment. *New Phytol.*, **85**, 275–88.

Warwick, S.I. and Briggs, D. (1980b) The genecology of lawn weeds. V. The adaptive significance of different growth habit in lawn and roadside populations of *Plantago major* L. *New Phytol.*, **85**, 289–300.

Warwick, S.I. and Briggs, D. (1980c) The genecology of lawn weeds. VI. The adaptive significance of variation in *Achillea millifolium* L. as investigated by transplant experiments. *New Phytol.*, **85**, 451–60.

Warwick, S.I. and Briggs, D. (1980d) The genecology of lawn weeds. VII. The response of different growth forms of *Plantago major* L. and *Poa annua* L. to simulated trampling. *New Phytol.*, **85**, 461–9.

Watkins, W.A. and Goebel, C.A. (1984) Sonar observations explain behaviours noted during boat manoeuvres for radio tagging of humpback whales (*Megaptera novaea ngliae*) in the Glacier Bay area. *Cetology*, **48**, 1–8.

Watson, A. (1976) Human impact on animal populations in the Cairngorms. *Landscape Research News*, **1**, 14–15.

Watson, A. (1979) Bird and mammal numbers in relation to human impacts at ski lifts on Scottish hills. *J. Appl. Ecol.*, **16**, 753–64.

Watson, A. (1985) Soil erosion and vegetation damage near ski lifts at Cairngorm, Scotland. *Biol. Conserv.*, **33**, 363–81.

Watson, A. (1988) Dotterel *Charadrius morinellus* numbers in relation to human impact in Scotland. *Biol. Conserv.*, **43**, 245–56.

Watson, A., Payne, S. and Rae, R. (1989) Golden eagles *Aquila chrysetos*: land use and food in northeast Scotland. *Ibis*, **131**, 336–48.

Watson, E.V. (1967) *The structure and life of Bryophytes*, 2nd edn, Hutchinson University Library, London.

Weaver, R. (1976) *Synopsis of California Mountain Lion Study*. Transactions of the Mountain Lion Workshop, Nugget, Sparks, Nevada, US Fish and Wildlife Service, Region 1.

Weaver, T. and Dale, D. (1978) Trampling effects of hikers, motorcycles and horses in meadows and forests. *J. Appl. Ecol.*, **15**, 451–7.

Weaver, T., Dale, D. and Hartley, E. (1979) The relationship of trail condition to use, vegetation, user, slope, season and time, in *Recreational Impact on Wildlands* (eds R. Ittner, D. Potter, J. Agee and S. Anschell), US Forest Service PNW-R6-001-1979.

Webb, N.G. (1978) Boat towing by a bottlenose dolphin. *Carnivor Seattle*, **1**, 122–30.

Webb, R.H. (1982) Off-road motorcycle effects on a desert soil. *Environ. Conserv.*, **9**, 197–208.

Webb, R.H. (1983) Compaction of desert soils by off road vehicles, in *Environmental Effects of Off-road Vehicles: Impact and Management in Arid Regions* (eds R.H. Webb and H.G. Wilshire), Springer Verlag, New York, pp. 5l–80.

Webb, R.H. and Wilshire, H.G. (1980) Recovery of soils and vegetation in a Mojave Desert ghost town, Nevada, USA. *J. Arid Environ.*, **3**, 291–303.

Webb, R.H., Steiger, J.W. and Wilshire, H.G. (1986) Recovery of compacted soils in Mojave Desert ghost towns. *Soil. Sci. Soc. Am. J.*, **50**, 1341–4.

Webb, R.H., Ragland, H.C., Godwin, W.H. and Jenkins, D. (1978) Environmental effects of soil property changes with off-road vehicle use. *Environ. Manage.*, **2**, 219–33.

Weiner, J. (1975) Model of the energy budget of an adult roe deer (*Capreolus capreolus*) in an agricultural area. *Behaviour*, **71**, 246–90.

Weinmann, H. (1952) *Carbohydrate Reserves in Grasses*. Proceedings of the 6th International Grassland Congress, Pennsylvania, pp. 655–60.

Wells, R.E. and Wells, F.B. (1961) *The Bighorn of Death Valley*, United States Department of the Interior National Parks Service, Washington.

Werf, S. van der (1972) Effected van recreatie op de vegetatie in natuurterreinen. *Natuur en Landschap*, **2**, 210–20.

Werschkul, D.F., McMahon, E. and Leitschuk, M. (1976) Some effects of human activities on the great blue heron in Oregon. *Wilson Bull.*, **88**, 660–2.

Westman, W.E. (1978) Measuring the inertia and resilience of ecosystems. *BioScience*, **28**, 705–10.

Westoby, M. (1989) Selective forces exerted by vertebrate herbivores on plants. *Tree*, **4**, 115–17.

Whisler, F.D., Engle, C.F. and Baughman, N.M. (1965) *The effects of soil compaction on nitrogen transformations in the soil*. Bulletin, West Virginia University Agricultural Experimental Station, No. 5l6T.

White, D.H. and Seginak, J.T. (1987) Cave gate designs for use in protecting endangered bats. *Wildl. Soc. Bull.*, **15**, 445–9.

White, R.E. (1979) *Introduction to the Principals and Practice of Soil Science*, Blackwell, Oxford.

Whitehead, F.H. and Luti, R. (1962) Experimental studies of the effect of wind on plant growth and anatomy. I. Zea mays. *New Phytol.*, **61**, 56–8.

Whittaker, R.H. (1975) *Communities and Ecosystems*, 2nd edn, Macmillan, New York.

Whittaker, R.H. and Feeney, P.P. (1971) Allelochemicals: chemical interactions between species. *Science*, **171**, 757–70.

Wier, T.E., Stocking, M.G., Barbour, M.G. and Rost, T.L. (1982) *Botany, An Introduction to Plant Biology*, 6th edn, John Wiley, New York.

Wilde, C. (1994) Shark control an environmental problem. *Griffith Gazette*, **9**, (3), 3 [Anonymous report].

Willard, B.E. (1971) How many is too many? Detecting the evidence of overuse in State parks. *Landscape Architect*, **61**, 118–23.

Willard, B.E. and Marr, J.W. (1970) Effects of human activities on alpine tundra ecosystems in Rocky Mountain National Park, Colorado. *Biol. Conserv.*, **2**, 257–65.

Williams, A.T. and Randerson, P.F. (1989) Nexus: ecology, recreation and management of a dune system in South Wales, in *Perspectives in Coastal Dune Management* (eds F. van der Meulen, P.D. Jungerius and J.H. Visser), SPB Academic Publishing bv, The Hague, pp. 217–27.

Williams, S.L. (1988) *Thalassia testudinum* productivity and grazing by green turtles in a highly disturbed seagrass bed. *Marine Biology* (Berlin), **98**, 447–56.

Williamson, C.E. (1950) Ethylene, A metabolic product of diseased or injured plants. *Phytopathol.*, **40**, 205–8.

Willis, A.J., Folks, B.F., Hope-Simpson, J.F. and Yemm, E.W. (1959) Braunton Burrows: The dune system and its vegetation II. *J. Ecol.*, **47**, 249–88.

Wilshire, H.G. (1983) The impact of vehicles on desert soil stabilizers, in *Environmental Effects of Off-road Vehicles* (eds R.H. Webb and H.G. Wilshire), Springer Verlag, New York, pp. 31–50.

Wilshire, H.G. (1989) *Human Causes of Wind Erosion in California's Desert*, The Desert Protective Council Inc., Box 4294, Palm Springs, CA 92263.

Wilshire, H.G. and Nakata, J.K. (1976) Off-road vehicle effects on California's Mojave Desert. *California Geol.*, **29**, 123–32.

Wilson, L.D. (1969) The forgotten desert Bighorn habitat requirement. *Desert Bighorn Council Trans.*, **13**, 108–13.

Wischmeier, W.H. and Smith, D.D. (1978) *Predicting rainfall erosion losses – a guide to conservation planning*, Agricultural Hand Book 537, US Department of Agriculture, Washington, DC.

Witherington, B.E. and Bjorndal, K.A. (1991) Influences of artificial lighting on the seaward orientation of hatchling Loggerhead turtles *Caretta caretta*. *Biol. Conserv.*, **55**, 139–49.

Wolcott, T.G. and Wolcott, D.L. (1984) Impact of off-road vehicles on microinvertebrates of a mid-Atlantic beach. *Biol. Conserv.*, **29**, 217–40.

Woodall, P.F., Woodall, L.B. and Bodero, D.A.V. (1989) Daily activity patterns in captive elephant shrews (Macroscelididae). *Afr. J. Ecol.*, **27**, 63–76.

Woodard, T.N., Gulierrez, R.J. and Rutterford, W.H. (1974) Bighorn lamb production, survival and mortality in south-central Colorado. *J. Wildl. Manage.*, **38**, 771–4.

Woodland, D.J. and Hooper, N.A. (1977) The effect of human trampling on coral reefs. *Biol. Conserv.*, **11**, 1–4.

Wu, L., Till-Bottraud, I. and Torres, A. (1987) Genetic differentiation in temperature-enforced seed dormancy among golf course populations of *Poa annua* L. *New Phytol.*, **107**, 623–31.

Wynn, S. and Loucks, O.L. (1975) A social and environmental history of human impact on Parfrey's Glen. *Trans. Wisconsin Acad. Sci. Arts and Lett.*, **63**, 26–53.

Yalden, D.W. (1992) The influence of recreational disturbance on common sandpipers *Actitis hypoleucos* breeding by an upland reservoir, in England. *Biol. Conserv.*, **61**, 41–9.

Yalden, P.E. and Yalden, D.W. (1990) Recreational disturbance of breeding golden plovers. *Pluvialis apricarius. Biol. Conserv.*, **51**, 243–62.

Yang, S.F. and Pratt, H.K. (1978) The physiology of ethylene in wounded plant tissues, in *Biochemistry of Wounded Plant Tissues* (ed. G. Kahl), Walter de Gruyter, Berlin, pp. 595–622.

Yang, S.J. and De Jong, E. (1971) Effect of soil water potential and bulk density on water uptake patterns and resistance to flow of water in which plants. *Can. J. Soil. Sci.*, **51**, 211–20.

Young, J.Z. (1962) *The Life of Vertebrates*, Oxford University Press, Oxford.

Young, M.R. and Pendlebury, J.B. (1969) Survey of the effects of public pressure on parts of Dovedale and the Goyt Valley, Derbyshire, Internal Report Nature Conservancy (Midland Region), MS, p. 54.

Younger, V.B. and Nudge, F.J. (1976) Soil temperature, air temperature and defoliation effects on growth and nonstructural carbohydrates of Kentucky bluegrass. *Agron. J.*, **68**, 257–60.

Yur'eva, N.D., Matveva, V.G. and Trapido, I.L. (1976) Influence of recreation on soil invertebrate groups in birch woods round Moscow. *Lesovedenie*, **2**, 27–34.

Zande, A.N. van der (1984) *Outdoor recreation and birds: conflict or symbiosis?* Ph.D. Thesis, Rijksuniversiteit, Leiden.

Zande, A.N. van der and Vos, P. van der (1984) Impact of a semi-experimental increase in recreation intensity on the densities of birds in groves and hedges on a lake shore in The Netherlands. *Biol. Conserv.*, **30**, 237–59.

Zande, A.N. van der, Ter Keurs, W.J. van der and Weijden, W.J. van der (1980) The impact of roads on the densities of four bird species in an open fields habitat – evidence of a long-distance effect. *Biol. Conserv.*, **18**, 299–321.

Zande, A.N. van der, Berkhuizen, J.C., van Latesteijn, H.C. *et al.* (1984) Impact of outdoor recreation on the density of a number of breeding bird species in woods adjacent to urban residential areas. *Biol. Conserv.*, **30**, 1–39.

Zedler, J.B. (1978) Public use effects in the Cabrillo National Monument intertidal zone. Project Report, San Diego State University Biology Department, San Diego, CA [cited in Ghazanshahi *et al.*, 1983].

Zielski, A. (1978) Forest associations of Brodinca Lake district and the effect of forest exploitation and tourism. *Studia Scoetatis Scientiarum Torunensis, Sectio D, (Botanica)*, **10**, 191–273.

Zimmerman, P.W. and Wilcoxen, F. (1935) Several chemical growth substances which cause initiation of roots and other responses in plants. *Boyce Thompson Institute, Contributions*, **7**, 209–29.

Zunino, F. and Herrero, S. (1972) The status of the brown bear (*Ursus arctos*) in Abruzzo National Park, Italy, 1971. *Biol. Conserv.*, **4**, 263–73.

General index

Page numbers appearing in **bold** refer to figures and page numbers appearing in *italic* refer to tables.

Locations index

Species index